The One-Dimensional
Heat Equation

ENCYCLOPEDIA
OF MATHEMATICS
and Its Applications

GIAN-CARLO ROTA, Editor
Department of Mathematics
Massachusetts Institute of Technology
Cambridge, Massachusetts

Editorial Board

GIAN-CARLO ROTA, *Editor*
ENCYCLOPEDIA OF MATHEMATICS AND ITS APPLICATIONS

GIAN-CARLO ROTA, *Editor*
ENCYCLOPEDIA OF MATHEMATICS AND ITS APPLICATIONS

GIAN-CARLO ROTA, *Editor*
ENCYCLOPEDIA OF MATHEMATICS AND ITS APPLICATIONS

Other volumes in preparation

GIAN-CARLO ROTA, *Editor*

ENCYCLOPEDIA OF MATHEMATICS AND ITS APPLICATIONS

Volume 23

Section: Analysis
Felix E. Browder, *Section Editor*

The One-Dimensional Heat Equation

John Rozier Cannon
Washington State University
Pullman, Washington

Foreword by

Felix E. Browder
University of Chicago

1984

Addison-Wesley Publishing Company
Advanced Book Program
Menlo Park, California

Reading, Massachusetts · London · Amsterdam · Don Mills, Ontario · Sydney · Tokyo

CAMBRIDGE UNIVERSITY PRESS
Cambridge, New York, Melbourne, Madrid, Cape Town, Singapore, São Paulo, Delhi

Cambridge University Press
The Edinburgh Building, Cambridge CB2 8RU, UK

Published in the United States of America by Cambridge University Press, New York

www.cambridge.org
Information on this title: www.cambridge.org/9780521302432

First published 1984 by Addison Wesley
First published by Cambridge University Press 1984
This digitally printed version 2008

A catalogue record for this publication is available from the British Library

Library of Congress Cataloguing in Publication data

Cannon, John Rozier, 1938–
   The one-dimensional heat equation.

   (Encyclopedia of mathematics and its applications; v. 23)
   Bibliography: p.
   Includes indexes.
   1. Heat equation.   I. Title.   II. Series.
QA377.C28 1984   515.3'53   83-25819
ISBN 0-201-13522-1

ISBN 978-0-521-30243-2 hardback
ISBN 978-0-521-08944-9 paperback

To my family,
Joyce,
Carolyn, Sue, and Paul

Contents

Editor's Statement

A large body of mathematics consists of facts that can be presented and described much like any other natural phenomenon. These facts, at times explicitly brought out as theorems, at other times concealed within a proof, make up most of the applications of mathematics, and are the most likely to survive change of style and of interest.

This ENCYCLOPEDIA will attempt to present the factual body of all mathematics. Clarity of exposition, accessibility to the nonspecialist, and a thorough bibliography are required of each author. Volumes will appear in no particular order, but will be organized into sections, each one comprising a recognizable branch of present-day mathematics. Numbers of volumes and sections will be reconsidered as times and needs change.

It is hoped that this enterprise will make mathematics more widely used where it is needed, and more accessible in fields in which it can be applied but where it has not yet penetrated because of insufficient information.

GIAN-CARLO ROTA

Foreword

The one-dimensional heat equation, first studied by Fourier at the beginning of the 19th century in his celebrated volume on the analytical theory of heat, has become during the intervening century and a half the paradigm for the very extensive study of parabolic partial differential equations, linear and nonlinear. The present volume is a systematic development of a variety of aspects of this paradigm, of which many have not yet received an extension to the multidimensional space-variable case. Of particular interest are the discussions of free-boundary-value problems such as the one-phase Stefan problem, inverse problems, and some classes of not-well-posed problems.

This type of treatment using concrete analytic machinery for the detailed study of this very familiar and widely applicable partial differential equation should prove valuable as a textbook for courses that try to present basic aspects of partial differential equations in simple but useful cases (like the heat equation in one dimension), where the basic concepts are relatively unobscured by the technical problems and complications encountered in the more general classes of equations. The treatment is reasonably complete and can be followed by scientists who do not necessarily have the mathematical experience necessary for some of the more elaborate treatises on general parabolic equations. In addition, the relative completeness of the presentation for this case makes the volume suitable as a reference book for specific results in this area, for which a reference by specialization of more general results is inappropriate. As the author remarks, a variety of analytical techniques are brought to bear under reasonably simple hypotheses, thus illuminating the relative power and effect of the different classes of methods (and, in particular, the contrast between constructive methods and the use of *a priori* bounds).

In summary, the volume is a useful contribution to the effort to bring the material of the research literature in analysis into a form useful to applied mathematicians and mathematically oriented specialists in the sciences.

FELIX E. BROWDER

Preface

For more than two decades, part of my research has been directed at various questions involving the heat equation. In this volume I have interwoven much of my research and the research of others with the classical material, at a presentation level suitable for upper-division and beginning graduate mathematics, engineering, and science students. However, I have also intentionally written the material as a monograph and/or information source book. After the first six chapters of standard classical material, each chapter is written as a self-contained unit, except for an occasional reference to elementary definitions, theorems, and lemmas in previous chapters. Consequently, I believe that material can be drawn from the book as needed for a variety of courses, such as a standard course in partial differential equations, a course in initial-boundary-value problems for the heat equation, a course in not-well-posed problems and their numerical solution, a course in free-boundary-value problems, a course in parameter identification, and several others.

The treatment begins with a chapter of preliminary material in order to reduce the need for other reference material. This is followed by six chapters containing the standard basic material for the heat equation, such as the weak maximum principle, elementary solutions, and fundamental solution, and the usual initial- and/or boundary-value problems. One exception to the usual material is the treatment of the noncharacteristic Cauchy problem in Chapter 2. Utilizing the solution representations in Chapters 3 through 6, a fair number of initial-boundary-value problems are reduced to an equivalent system of integral equations in Chapter 7. A relevant treatment of integral equations is presented in Chapter 8. Chapter 9 deals with time-periodic problems. The property of analyticity of solutions is dealt with in Chapter 10, and estimation machinery is developed for utilization in the estimation of continuous dependence of solutions of some not-well-posed problems presented in Chapter 11. Chapter 12 discusses some numerical methods for not-well-posed problems. Chapter 13 contains a treatment of the inverse problem of the determination of an unknown coefficient from overspecified boundary data. Chapter 14 begins the discussion of moving-boundary problems and introduces a stiffening of the level of presentation through Chapter 16. Chapter 15 contains some general properties of solutions that are useful in the discussion of the general initial-boundary-value problem via the Perron–Poincaré method in Chapter 16. The results in Chapter 15 also find application in the discussion of the one-phase Stefan problem in Chapters 17 and 18. Chapter 19 contains a discussion of the inhomogeneous heat equation. Chapter 20 is an application of Chapter 19 to some special quasilinear parabolic equations. Finally, the volume concludes with a bibliography to the literature on the heat equation. The bibliography is divided by time periods, and each of the more recent time periods is subdivided by topics.

The study of partial differential equations has been in progress for more than 200 years and the object of this study, $u_t = u_{xx}$, for more than 150 years. Partial differential equations is a branch of analysis which, including the calculus, has been under development for more than 300 years. Almost every branch of analysis can be and has been brought to bear upon the study of solutions of partial differential equations.

I have assumed that the reader has had a typical undergraduate science and engineering mathematics sequence, which usually attempts to survey calculus, vector analysis, Fourier series and transforms, complex analysis, ordinary and partial differential equations. I have made no attempt to redo a reader's mathematical education. Also, I make no apology for the appearance of "rigor" or "lack of rigor" in any discussion throughout the text.

I am happy to acknowledge the assistance and/or support of the following:

1. My family, to whom this work is dedicated.

2. Mrs. Suzy Crumley, for typing the entire manuscript.

3. Mrs. Sylvia Becker, Mr. Dana Lohrey, and Mrs. Pamela Wample, for retyping various portions of the manuscript.

4. Professor Antonio Fasano, Professor Hector Fattorini, Professor Ruben Hersh, Professor Karl-Heinz Hoffman, Professor B. Frank Jones, Professor Joseph Kohn, Professor Mario Primicerio, and Professor John van der Hoek, for reading and commenting on the entire manuscript.

5. Professor Murray Cantor, Professor Paul DuChateau, Professor Emmanuele DiBenedetto, Professor Craig Evans, Professor Tinsley Oden, Professor Charles Radin, Professor Ralph Showalter, and Professor Margaret Waid, for reading and commenting on various portions of the manuscript.

6. Professor Gian-Carlo Rota, Mr. Carl W. Harris, Mr. Charles Hibbard, Mr. B. E. Vance, and Ms. Lore Heinlein, for advice and helpful criticism.

7. Addison-Wesley Publishing Company; The National Science Foundation, U.S.A.; Washington State University, Pullman, Washington; The University of Texas at Austin, Texas; Colorado State University, Fort Collins, Colorado; Hahn–Meitner–Institut für Kernforschung, Berlin; Consiglio Nazionale delle Ricerche, Italia; and Istituto Matematico, U. Dini, Firenze, for various forms of support.

To all of the above, my sincerest thanks.

JOHN ROZIER CANNON

Chapter 0

Preliminaries

0.0. SOME INEQUALITIES

We begin with the fact that $(a-b)^2 = a^2 - 2ab + b^2 \geq 0$, from whence follows

$$2ab \leq a^2 + b^2 \qquad (0.0.1)$$

for any real a and b. Dividing by 2 and replacing a by $\sqrt{2\varepsilon}\, a$ and b by $b/\sqrt{2\varepsilon}$, we obtain

$$ab \leq \varepsilon a^2 + (4\varepsilon)^{-1} b^2 \qquad (0.0.2)$$

for any real a, b and $\varepsilon > 0$.

From the inequality

$$0 \leq \sum (a_i \lambda + b_i)^2 = \left(\sum a_i^2\right)\lambda^2 + 2\left(\sum a_i b_i\right)\lambda + \left(\sum b_i^2\right), \qquad (0.0.3)$$

it follows that the discriminant of the quadratic in λ is nonnegative. This yields Schwarz's inequality:

$$\left|\sum a_i b_i\right| \leq \left(\sum a_i^2\right)^{1/2}\left(\sum b_i^2\right)^{1/2}. \qquad (0.0.4)$$

The proof generalizes easily to definite integrals:

$$\left|\int fg\, dx\right| \leq \left(\int f^2\, dx\right)^{1/2}\left(\int g^2\, dx\right)^{1/2}. \qquad (0.0.5)$$

The calculus can be employed to show that for nonnegative a, b, and α, $0 < \alpha < 1$,

$$(a+b)^{\alpha} \le a^{\alpha} + b^{\alpha}. \tag{0.0.6}$$

Clearly, inequality (0.0.6) is valid for a or b or both equal to zero. So we can assume that $0 < a < b$. Using the Mean-Value Theorem,

$$(a+b)^{\alpha} = b^{\alpha}\left(1 + \frac{a}{b}\right)^{\alpha} = b^{\alpha}\left\{1 + \alpha(1+\xi)^{\alpha-1}\left(\frac{a}{b}\right)\right\}$$

$$\le b^{\alpha}\left\{1 + \alpha\left(\frac{a}{b}\right)^{\alpha-1}\left(\frac{a}{b}\right)\right\} = \alpha a^{\alpha} + b^{\alpha} < a^{\alpha} + b^{\alpha}.$$

By induction it follows for $a_i \ge 0$ and $0 < \alpha < 1$, that

$$\left(\sum a_i\right)^{\alpha} \le \sum a_i^{\alpha}. \tag{0.0.7}$$

We will need the inequality

$$(a+b)^{-2\beta} \le a^{-\beta}b^{-\beta}, \tag{0.0.8}$$

where $0 < \beta < 1$ and $0 < a, b$. This follows from $(a+b)^2 > ab$ since x^{β} is monotone increasing.

We also recall the elementary estimates

$$\exp\{-x\} \le p! x^{-p} \tag{0.0.9}$$

for any nonnegative integer p, and for $x > 0$,

$$x^{\alpha}\exp\{-\beta x\} \le \left(\frac{\alpha}{\beta}\right)^{\alpha}\exp\{-\alpha\} \le \left(\frac{\alpha}{\beta}\right)^{\alpha} \tag{0.0.10}$$

for any positive α and β. From (0.0.10) it follows that

$$x^{\alpha}\exp\{-\gamma x\} \le C\exp\{-\beta x\} \tag{0.0.11}$$

for some positive C depending on the positive α, γ, and β, provided that $0 < \beta < \gamma$.

0.1. SEQUENCES AND SERIES OF CONTINUOUS FUNCTIONS, THE ASCOLI–ARZELA THEOREM, AND THE WEIERSTRASS APPROXIMATION THEOREM

The set of functions $f_n = f_n(x)$, $n = 1, 2, \ldots$, which are defined over a common domain, is called a *sequence* of *functions*. If for each $x \in D$, the common domain of definition,

$$\lim_{n \to \infty} f_n(x) = f(x), \tag{0.1.1}$$

then we say that the sequence *converges pointwise* to the function f. Note that buried in the symbolic statement is the fact that to each ε there corresponds $N = N(\varepsilon, x)$ such that $|f(x) - f_n(x)| < \varepsilon$ for all $n > N$. The notion of *uniform convergence* is simply the removal of the x dependence from N.

DEFINITION 0.1.1. We say that the sequence $\{f_n\}$ *converges uniformly to f* on the domain D if for each and every $\varepsilon > 0$ there exists an $N = N(\varepsilon)$ such that

$$|f(x) - f_n(x)| < \varepsilon$$

holds for all $n > N$ and all x in D.

Remark. Note that the phrase "all x in D" would be superfluous if we replaced $|f(x) - f_n(x)|$ by $\sup_{x \in D}|f(x) - f_n(x)|$. Note also that only trivial modifications are necessary to extend the definition to functions $f = f(x, t)$.

DEFINITION 0.1.2. We say that a sequence $\{f_n\}$ is *uniformly Cauchy* on the domain D if for each and every $\varepsilon > 0$ there exists an $N = N(\varepsilon)$ such that

$$|f_m(x) - f_n(x)| < \varepsilon$$

for all $m, n > N$ and all x in D.

Theorem 0.1.1. *There exists a continuous function f which is the unique uniform limit of the uniformly Cauchy sequence $\{f_n\}$ on the domain D.*

Proof. We give the proof for $D \subset \mathbb{R}$. The unicity of the limit function f follows from the unicity of the $\lim_{n \to \infty} f_n(x)$ of the Cauchy sequence of real numbers $f_n(x)$. The uniformity of the convergence follows from allowing m to tend to infinity in the expression $|f_m(x) - f_n(x)| < \varepsilon$ for all $m, n > N(\varepsilon)$ and all x in D. It remains only to show that the limit function is continuous. Select $n_1 > N(\varepsilon/3)$ so that

$$|f(x) - f_{n_1}(x)| < \frac{\varepsilon}{3} \qquad \text{for all } x \in D.$$

As f_{n_1} is continuous, it follows that to $\varepsilon/3$ there corresponds a $\delta = \delta(f_{n_1}, x_0, \varepsilon/3)$ such that

$$|f_{n_1}(x) - f_{n_1}(x_0)| < \frac{\varepsilon}{3}$$

for all $x \in D \cap I(x_0, \delta(f_{n_1}, x_0, \varepsilon/3))$, where $I = \{x \mid x_0 - \delta < x < x_0 + \delta\}$. Thus,

$$|f(x) - f(x_0)| \le |f(x) - f_{n_1}(x)| + |f_{n_1}(x) - f_{n_1}(x_0)|$$
$$+ |f_{n_1}(x_0) - f(x_0)|$$
$$< \frac{\varepsilon}{3} + \frac{\varepsilon}{3} + \frac{\varepsilon}{3} = \varepsilon$$

holds for all $x \in D \cap I(x_0, \delta(f_{n_1}, x_0, \varepsilon/3))$. As x_0 and ε are arbitrary, f is continuous. $\qquad \square$

By the limit (or sum) of a series of functions f_k defined on a common domain D, we mean

$$f(x) = \lim_{n \to \infty} s_n(x),$$

where

$$s_n(x) = \sum_{k=1}^{n} f_k(x).$$

Symbolically, we write

$$f(x) = \sum_{k=1}^{\infty} f_k(x),$$

and say that the series is convergent.

By absolute convergence, we mean $\lim_{n \to \infty} s_n(x)$ exists where

$$s_n(x) = \sum_{k=1}^{n} |f_k(x)|.$$

Uniform and *absolute uniform* convergence are notions that follow directly from the definitions for sequences.

We turn now to the notions of *uniformly bounded* and *equicontinuity*. Sometimes uniformly bounded is replaced by the phrase *equibounded*.

DEFINITION 0.1.3. A sequence of functions f_n is *uniformly bounded* if there exists a positive constant C such that $|f_n(x)| < C$ for all $x \in D$ and all n.

DEFINITION 0.1.4. A sequence of functions f_n is equicontinuous if at each and every $x_0 \in D$ for each and every $\varepsilon > 0$ there exists a $\delta = \delta(x_0, \varepsilon) > 0$ such that for all n

$$|f_n(x) - f_n(x_0)| < \varepsilon$$

for all $x \in D \cap I(x_0, \delta(x_0, \varepsilon))$.

We can state now the Ascoli–Arzela Convergence Theorem.

Theorem 0.1.2 (Ascoli–Arzela). *From every uniformly bounded equicontinuous sequence of functions defined on a compact domain, a uniformly Cauchy subsequence of functions can be selected.*

Proof. We give the proof for the domain $[a, b]$. The crux of the proof is the fact that $\mathbb{R} = \overline{\mathbb{Q}}$ where $\mathbb{Q}$ is the set of rational numbers. Likewise, $\mathbb{R}^2 = \overline{\mathbb{Q}}^2$. Both $\mathbb{Q}$ and $\mathbb{Q}^2$ are countable. We return now to $[a, b]$ and let $\{r_n\}$ denote a list of the rationals in $[a, b]$. We note that, given any x in $[a, b]$, there exists a subsequence $\{r_{nk}\}$ such that $\lim_{k \to \infty} r_{nk} = x$.

Now, we consider $\{f_n(r_1)\}$. This sequence of real numbers is bounded via the uniform boundedness assumption. Hence, the Bolzano–Weierstrass theorem implies that we can select a Cauchy subsequence $\{f_{nk}(r_1)\}$ from $\{f_n(r_1)\}$. Rename this subsequence $\{f_{1k}\}$ and consider $\{f_{1k}(r_2)\}$. By the same reasoning we can select a Cauchy subsequence $\{f_{2k}(r_2)\} \subset \{f_{1k}(r_2)\}$. Continuing in this fashion, we obtain a sequence of sequences $\{f_{mk}\}$, $m = 1, 2, 3, \ldots$, such that $\{f_{mk}\} \subset \{f_{m-1k}\}$ and $\{f_{mk}(r_m)\}$ is a Cauchy sequence of real numbers. We define now the subsequence of our theorem by the choice $\{f_{kk}\}$ of the diagonal of the sequences written as a matrix array. Since for fixed m, $m = 1, 2, \ldots$,

$$\{f_{kk}(r_m)\}_{k=m}^\infty \quad \text{is a subsequence of} \quad \{f_{mk}(r_m)\}_{k=1}^\infty,$$

we know that $\{f_{kk}(r_m)\}_{k=1}^\infty$ is Cauchy and converges to the same limit; that is,

$$\lim_{k \to \infty} f_{kk}(r_m) = \lim_{k \to \infty} f_{mk}(r_m).$$

Hence, $\{f_{kk}(r_m)\}$ is a Cauchy sequence for each r_m.

We shall use the equicontinuity now to show that $\{f_{kk}(x)\}$ is uniformly Cauchy for all $x \in [a, b]$. Let $\delta = \delta(x, \varepsilon/3)$ denote the δ for $\varepsilon/3$ and x in the definition of equicontinuity. Consider the open intervals

$$I(r_m, \delta(r_m, \varepsilon/3)), \quad \text{where} \quad m = 1, 2, 3, \ldots$$

Clearly, $[a, b] \subset \cup_{m=1}^\infty I(r_m, \delta(r_m, \varepsilon/3))$. Since $[a, b]$ is compact, the Theorem of Heine–Borel implies that there exist p rationals r_i, $i = 1, \ldots, p$ such that $[a, b] \subset \cup_{i=1}^p I(r_i, \delta(r_i, \varepsilon/3))$. Hence, for any $x \in [a, b]$, there exists an i_0 such that $x \in I(r_{i_0}, \delta(r_{i_0}, \varepsilon/3))$. Thus,

$$|f_{mm}(x) - f_{nn}(x)| \le |f_{mm}(x) - f_{mm}(r_{i_0})| + |f_{mm}(r_{i_0}) - f_{nn}(r_{i_0})|$$
$$+ |f_{nn}(r_{i_0}) - f_{nn}(x)|$$
$$\le \tfrac{2}{3}\varepsilon + |f_{mm}(r_{i_0}) - f_{nn}(r_{i_0})|$$

by the equicontinuity of the original sequence of functions. Since each sequence $\{f_{kk}(r_i)\}$, $i = 1, \ldots, p$, is a Cauchy sequence, for $\varepsilon > 0$ there exists $N_i = N_i(\varepsilon/3)$ such that

$$|f_{mm}(r_i) - f_{nn}(r_i)| < \frac{\varepsilon}{3}$$

for all m and $n > N_i$. Selecting $N = N(\varepsilon) = \max_{1 \le i \le p} N_i(\varepsilon/3)$, it follows that

$$|f_{mm}(r_i) - f_{nn}(r_i)| < \frac{\varepsilon}{3}$$

for all $i = 1, \ldots, p$, when $m, n > N$. Thus,

$$|f_{mm}(x) - f_{nn}(x)| < \varepsilon$$

holds for all $x \in [a, b]$ when $m, n > N = N(\varepsilon)$. $\qquad\square$

We conclude this section with the statement of the Weierstrass Approximation Theorem.

Theorem 0.1.3 (Weierstrass Approximation Theorem). *Let u be a continuous function over the compact domain D. For each ε > 0, there exists a polynomial p_ε such that*

$$|u(s) - p_\varepsilon(s)| < \varepsilon$$

for all $s \in D$.

Proof. We discuss the proof for u defined on $[0, 2\pi]$. First we extend u outside of $[0, 2\pi]$ by setting $u(x) \equiv u(0)$, $x < 0$, and $u(x) \equiv u(2\pi)$, $x > 2\pi$. Next, we can let

$$v(x, h) = h^{-1} \int_x^{x+h} u(\xi) \, d\xi$$

and note that u is the uniform limit of v as h tends to zero since u is uniformly continuous. Likewise, v is the uniform limit of

$$w(x, k) = k^{-1} \int_x^{x+k} v(\xi, h) \, d\xi$$

for each fixed h. Thus, we can assume that u is twice continuously differentiable in $[0, 2\pi]$. For u this smooth, its Fourier series converges absolutely and uniformly. Thus,

$$u(x) = \frac{a_0}{2} + \sum_{n=1}^{\infty} (a_n \cos nx + b_n \sin nx).$$

Hence, u can be approximated uniformly by

$$\frac{a_0}{2} + \sum_{n=1}^{N} (a_n \cos nx + b_n \sin nx)$$

for some large N. Since

$$\sin x = \sum_{k=0}^{\infty} (-1)^k \frac{x^{2k+1}}{(2k+1)!} \quad \text{and} \quad \cos x = \sum_{k=0}^{\infty} (-1)^k \frac{x^{2k}}{(2K)!},$$

we truncate the series for $\sin nx$ and $\cos nx$ for $n = 1, \ldots, N$, and obtain a polynomial that approximates u uniformly over $[0, 2\pi]$. □

0.2. THE LEBESGUE DOMINATED-CONVERGENCE THEOREM, LEIBNIZ'S RULE, AND FUBINI'S THEOREM

The Lebesgue Dominated-Convergence Theorem allows us to ignore uniform-convergence hypotheses in exchanging the limit with the integral sign. Here, the integral sign is that of the Lebesgue integral, which is an extension of the Riemann integral. Since the Lebesgue integral coincides with the Riemann integral wherever the latter exists, the reader can view this theorem as a powerful tool for exchanging the limit with the integral sign for Riemann integrals from the calculus.

We state the results for a sequence of functions defined on $[a, b]$.

Theorem 0.2.1 (Lebesgue Dominated-Convergence Theorem). *Let* $\{f_n\}$ *denote a sequence of integrable functions on* $[a, b]$ *such that* $f(x) = \lim_{n \to \infty} f_n(x)$. *Suppose that there exists a positive-valued integrable function g such that* $|f_n(x)| < g(x)$ *for all* $x \in [a, b]$ *and all* $n = 1, 2, \dots$ *Then the limit function* $f(x)$ *is integrable and*

$$\lim_{n \to \infty} \int_a^b f_n(x) \, dx = \int_a^b \lim_{n \to \infty} f_n(x) \, dx = \int_a^b f(x) \, dx.$$

Remark. For the application of this result the reader need only check that the f_n are Riemann-integrable and that $g(x)$ is Riemann-integrable; this implies the Lebesgue integrability of the f_n and g, from whence it follows that f is Lebesgue-integrable and the limits of the Lebesgue integrals are as stated. Quite often, it can be seen directly that the limit function f is Riemann-integrable. Hence, the great utility of the theorem in its application to Riemann integrals of continuous functions is the ability to ignore hypotheses of uniform convergence of the sequence f_n.

We make two applications of the Lebesgue Dominated Convergence Theorem.

Theorem 0.2.2. *Suppose that* $F = F(x, t)$ *is defined on* $[a, b] \times [\alpha, \beta]$ *such that for each* $t \in [\alpha, \beta]$, $F(x, t)$ *is an integrable function of x, and that for each* $x \in [a, b]$, $F(x, t)$ *is continuous in t over* $[\alpha, \beta]$. *Suppose that, for all* $t \in [\alpha, \beta]$, $|F(x, t)| \leq h(x)$ *for some nonnegative integrable function h. Then, the function*

$$H(t) = \int_a^b F(x, t) \, dx$$

is continuous over $[\alpha, \beta]$.

Proof. For $t_0 \in [\alpha, \beta]$, we need only show that $\lim_{t \to t_0} H(t) = H(t_0)$, which is simply

$$\lim_{t \to t_0} \int_a^b F(x, t) \, dx = \int_a^b \lim_{t \to t_0} F(x, t) \, dx.$$

However, this follows immediately from Theorem 0.2.1.   □

The final application is Leibniz's rule.

Theorem 0.2.3 (Leibniz's Rule). *Suppose that* $F = F(x, t)$ *is defined on* $[a, b] \times [\alpha, \beta]$ *such that, for each* $t \in [\alpha, \beta]$, $F(x, t)$ *is an integrable function of x and that for each x,* $(\partial F / \partial t)(x, t)$ *exists and is continuous. Suppose that* *for all $t \in [\alpha, \beta]$,*

$$\left| \frac{\partial F}{\partial t}(x, t) \right| \leq g(x)$$

for some nonnegative integrable function g. Then, $G(t) = \int_a^b F(x, t) \, dx$ is differentiable and

$$G'(t) = \int_a^b \frac{\partial F}{\partial t}(x, t) \, dx$$

for $t \in [\alpha, \beta]$.

   Proof. The difference quotients $h^{-1}\{F(x, t+h) - F(x, t)\}$ are integrable since F is and, by the Mean-Value Theorem,

$$|h^{-1}\{F(x, t+h) - F(x, t)\}| = \left| \frac{\partial F}{\partial t}(x, \xi) \right| \leq g(x).$$

Hence, by Theorem 0.2.1, $(\partial F / \partial t)(x, t)$ is integrable and

$$G'(t) = \lim_{h \to 0} h^{-1}\{G(t + h) - G(t)\}$$

$$= \lim_{h \to 0} \int_a^b h^{-1}\{F(x, t+h) - F(x, t)\} \, dx$$

$$= \int_a^b \frac{\partial F}{\partial t}(x, t) \, dx. \qquad \qquad \square$$

Remark. We shall leave to the reader the formulation of hypotheses which allow us to compute the derivative of

$$G(t) = \int_a^{\varphi(t)} F(x, t) \, dx$$

and conclude that

$$G'(t) = F(\varphi(t), t)\varphi'(t) + \int_a^{\varphi(t)} \frac{\partial F}{\partial t}(x, t) \, dx.$$

   Finally, we wish to recall for the reader Fubini's Theorem on the interchange of the order of integration. We state here a version of Fubini's Theorem that is applicable below.

   Theorem 0.2.4. *Let $f = f(x, y)$ denote an integrable function on the rectangle $D = \{(x, y) : a \leq x \leq b, c \leq y \leq d\}$. If one of the following integrals exists, then the other two exist and*

$$\int_D \int f \, dx \, dy = \int_c^d \left\{ \int_a^b f(x, y) \, dx \right\} dy = \int_a^b \left\{ \int_c^d f(x, y) \, dy \right\} dx.$$

Remark. A useful application of Fubini's Theorem is the result

$$\int_0^t \left(\int_0^\tau g(\tau, \eta) \, d\eta \right) dt = \int_0^t \left(\int_\eta^t g(\tau, \eta) \, d\tau \right) d\eta,$$

where in Theorem 0.2.4 we set $a = c = 0$, $b = d = t$, $x = \tau$, $y = \eta$, and

$$f(\tau, \eta) = \begin{cases} 0, & \eta > \tau, \\ g(\tau, \eta), & \eta \le \tau. \end{cases}$$

0.3. COMPLEX ANALYSIS

We assume that the reader is familiar with the complex numbers $a + ib$, where a and b are real and $i = \sqrt{-1}$ and $|a + ib| = \sqrt{a^2 + b^2}$. We summarize here some of the notions and results for analytic functions of the complex variable $z = x + iy$. We let $f = f(z)$ denote a complex-valued function defined in a domain D of the x, y-plane. Thus,

$$f(z) = u(x, y) + iv(x, y),$$

where u and v are real-valued functions of x and y. A function $f = f(z)$ is *analytic* in D if $f'(z)$ exists at each point of D. Here, we take D to be an open path-connected set in the x, y-plane. Computing the derivative via the difference quotient,

$$f'(z) = \lim_{\Delta z \to 0} \frac{f(z + \Delta z) - f(z)}{\Delta z},$$

leads to the Cauchy–Riemann partial-differential equations. First, with $\Delta z = \Delta x$, we see that

$$f'(z) = \frac{\partial u}{\partial x}(x, y) + i \frac{\partial v}{\partial x}(x, y).$$

Second, with $z = i \Delta y$, we obtain

$$f'(z) = \frac{\partial v}{\partial y}(x, y) - i \frac{\partial u}{\partial y}(x, y).$$

Equating these two results, we obtain the Cauchy–Riemann system

$$\frac{\partial u}{\partial x} = \frac{\partial v}{\partial y}, \qquad \frac{\partial v}{\partial x} = -\frac{\partial u}{\partial y}, \tag{0.3.1}$$

which constitutes a necessary condition on the real and imaginary parts of f for the existence of $f'(z)$. On the other hand if u and v possess continuous first partial derivatives in D and satisfy (0.3.1), then $f'(z)$ exists and is continuous at each point of D.

We shall see that f (and hence) u, and v are infinitely differentiable. Consequently,

$$\Delta u \equiv \frac{\partial^2 u}{\partial x^2} + \frac{\partial^2 u}{\partial y^2} = \frac{\partial^2 v}{\partial x \, \partial y} - \frac{\partial^2 v}{\partial y \, \partial x} = 0$$

in D and likewise for $\Delta v = 0$. The solutions of Laplace's equation $\Delta u = 0$ are called *harmonic* functions. The *Dirichlet* problem is the problem of finding a solution of

$$\begin{cases} \Delta u = 0 & \text{in } D, \\ u = f & \text{on } \partial D, \end{cases}$$

where f is given on ∂D. The reader is referred to the literature for the discussion of this problem. We note, however, that a bounded solution u satisfies a *weak maximum principle*. Namely,

$$\sup_{D} u \leq \sup_{\partial D} f.$$

We shall make use of this fact below.

We recall Green's theorem, which states that

$$\int_{D}\int \left\{ \frac{\partial P}{\partial x} - \frac{\partial Q}{\partial y} \right\} dx \, dy = \oint_{\partial D} Q \, dx + P \, dy,$$

where if ∂D is a closed simple (no loops), rectifiable (finite-length), Jordan curve given by $z = z(t) = x(t) + iy(t)$, $0 \leq t \leq 1$, then the line integral can be computed

$$\oint_{\partial D} Q \, dx + P \, dy = \int_{0}^{1} \left\{ Q(x(t), y(t))\dot{x}(t) + P(x(t), y(t))\dot{y}(t) \right\} dt.$$

Applying Green's theorem to $\oint_{\partial D} f(z) \, dz$ we see that, for f analytic in D,

$$\oint_{\partial D} f(z) \, dz = \oint_{\partial D} (u + iv)(dx + i \, dy)$$

$$= \oint_{\partial D} (u \, dx - v \, dy) + i \oint_{\partial D} (v \, dx + u \, dy)$$

$$= \int_{D}\int \{ -u_y - v_x \} \, dx \, dy + i \int_{D}\int \{ -v_y + u_x \} \, dx \, dy = 0$$

since u and v satisfy (0.3.1). We have obtained Cauchy's Theorem.

Theorem 0.3.1 (Cauchy). *For $f = f(z)$ analytic in the closure of a domain D with rectifiable boundary ∂D,*

$$\oint_{\partial D} f(z) \, dz = 0.$$

An application of this result to the analytic function $H(\zeta) = (\zeta - z)^{-1} f(\zeta)$ in the domain $D - S(z, \varepsilon)$ for each $\varepsilon > 0$ yields the Cauchy–Riemann Representation Theorem.

Theorem 0.3.2 (Cauchy–Riemann). *For $f = f(z)$ analytic in the closure of a simply connected (no holes) domain D with rectifiable boundary ∂D, we have, for each $z \in D$,*

$$f(z) = \frac{1}{2\pi i} \oint_{\partial D} \frac{f(\zeta)\, d\zeta}{(\zeta - z)}. \tag{0.3.2}$$

Moreover, f is infinitely differentiable and

$$f^{(n)}(z) = \frac{n!}{2\pi i} \oint_{\partial D} \frac{f(\zeta)\, d\zeta}{(\zeta - z)^{n+1}}.$$

Remark.   Estimating the integral on the right side, we see that

$$|f^{(n)}(z)| \le (2\pi)^{-1} \Big(\max_{\zeta \in \partial D} |f(\zeta)| \Big) \rho^{-(n+1)} L n!,$$

where ρ is less than the distance from z to ∂D and L is the length of ∂D. Using such estimates we see that

$$f(z) = \sum_{n=0}^{\infty} \frac{f^{(n)}(z_0)}{n!} (z - z_0)^n$$

for all z near z_0. Moreover, the convergence of the series is uniform and absolute for z near z_0.

Another application of the Cauchy–Riemann Representation Theorem is the fact that a uniformly convergent sequence of analytic functions possesses a limit that is an analytic function. This follows from the fact that the limit function satisfies (0.3.2).

We conclude our remarks on complex analysis with the Maximum Modulus Theorem.

Theorem 0.3.3 (Maximum Modulus). *For $f = f(z)$ analytic in the closure of a domain D,*

$$\sup_{z \in D} |f(z)| \le \sup_{z \in \partial D} |f(z)|. \tag{0.3.3}$$

Proof.   We note that $\log|f(z)|$ is the real part of the analytic function $\log f(z)$. Consequently, $\log|f(z)|$ is harmonic in D and (0.3.3) follows from the weak maximum principle for harmonic functions.   □

Remark.   There is an obvious error in the proof above. Namely, f could be zero in D. We ignore that error in favor of reminding the reader about the harmonicity of $\log|f(z)|$, which will be used in subsequent text.

NOTES

Most of the prerequisites for this book can be found in advanced calculus texts. For example, see Bartle [2], Franklin [3], Kaplan [5], and Widder [10]. For material on analytic functions of a complex variable, see Alfors [1], Knopp [7], and Nehari [8]. For convergence of integrals, passing limits through integral signs, and differentiation of integrals, see Jones [4], Klambauer [6], and Titchmarsh [9].

REFERENCES

1. Ahlfors, L. V., *Complex Analysis*. An introduction to the theory of analytic functions of one complex variable. McGraw-Hill, New York-Toronto-London, 1953.
2. Bartle, R. G., *The Elements of Real Analysis*. John Wiley & Sons, New York, London, and Sydney, 1964.
3. Franklin, P., *A Treatise on Advanced Calculus*. John Wiley & Sons, New York, London, and Sydney, 1940.
4. Jones, B. Frank, Jr., *Lebesgue Integration on Euclidean Space*. Prentice-Hall, Inc., Englewood Cliffs, N.J., 1984.
5. Kaplan, W., *Advanced Calculus*. Addison-Wesley Pub. Co., Reading, Mass., 1952.
6. Klambauer, G., *Real Analysis*. American Elsevier Pub. Co., New York, London & Amsterdam, 1973.
7. Knopp, K., *Theory of Functions*, Part I. Dover Publications, 1945.
8. Nehari, Z., *Conformal Mapping*. McGraw-Hill Book Co., New York-Toronto-London, 1952.
9. Titchmarsh, E. C., *The Theory of Functions*. Oxford Univ. Press (Clarendon), London and New York, 1939.
10. Widder, D. V., *Advanced Calculus*. Prentice-Hall, Englewood Cliffs, N.J., 1961.

Chapter 1

Introduction

1.1. DERIVATION OF THE HEAT EQUATION

In *Théorie Analytique de la Chaleur*, Fourier stated his famous law

$$\mathbf{q} = - k \nabla u, \tag{1.1.1}$$

where $\mathbf{q}$ is the *rate of flow of heat energy* per unit time through a unit area (heat flux), the positive constant of proportionality k is called the *conductivity*, and ∇u denotes the gradient of the temperature u. For one space variable x, Eq. (1.1.1) becomes

$$q = - ku_x, \tag{1.1.2}$$

where here and below the subscript x denotes $\partial / \partial x$. For example,

$$u_t \equiv \frac{\partial u}{\partial t} \quad \text{and} \quad u_{xx} \equiv \frac{\partial^2 u}{\partial x^2}.$$

In the absence of work, the change ΔQ of the internal energy of a material can be related to the change Δu of temperature by means of the formula

$$\Delta Q = c\rho \Delta u, \tag{1.1.3}$$

where the positive constant of proportionality c is called the *capacity* and the positive constant ρ is the *mass density* of the material. If a reference temperature—say zero—is chosen, then

$$Q = c\rho u \tag{1.1.4}$$

represents an internal energy per unit volume of material.

The usual derivation of the heat equation employs (1.1.2) and (1.1.4) as follows. Let (x, t) denote the coordinates of a point in space and time. Consider the rectangle

$$R = \{(\xi, \tau): x - \Delta x \leq \xi \leq x + \Delta x \quad \text{and} \quad t - \Delta t \leq \tau \leq t + \Delta t\}. \tag{1.1.5}$$

(See Fig. 1.1.1.) The increase in time $2\Delta t$ of the internal energy of $[x - \Delta x, x + \Delta x] \subset \mathbf{R}$ is

$$c\rho \int_{x-\Delta x}^{x+\Delta x} \{u(\xi, t + \Delta t) - u(\xi, t - \Delta t)\} \, d\xi = c\rho \iint_R \frac{\partial u}{\partial \tau} \, d\xi \, d\tau. \tag{1.1.6}$$

In the absence of work, heat sources, and heat sinks, this change of Q is accounted for by the total flow of heat energy across the boundaries into $[x - \Delta x, x + \Delta x]$, which is given by

$$k \int_{t-\Delta t}^{t+\Delta t} \left\{ \frac{\partial u}{\partial x}(x + \Delta x, \tau) - \frac{\partial u}{\partial x}(x - \Delta x, \tau) \right\} d\tau = k \iint_R \frac{\partial^2 u}{\partial \xi^2} \, d\xi \, d\tau.$$

$$\tag{1.1.7}$$

The conservation of energy yields

$$\iint_R \{c\rho u_\tau - k u_{\xi\xi}\} \, d\xi \, d\tau = 0 \tag{1.1.8}$$

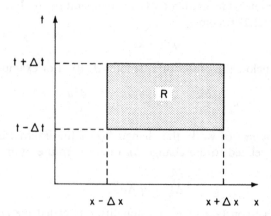

FIGURE 1.1.1

for each Δx and Δt. Consequently, we have

$$c\rho u_t - k u_{xx} = 0 \qquad (1.1.9)$$

at (x, t). The reader should note that this derivation employs a disguised form of the divergence theorem. Dividing by $c\rho$, we obtain

$$u_t - \kappa u_{xx} = 0, \qquad (1.1.10)$$

where

$$\kappa = c^{-1}\rho^{-1}k \qquad (1.1.11)$$

denotes the *thermal diffusivity* of the material. Setting $\tau = \kappa t$ and relabeling τ as t, we can transform Eq. (1.1.10) into the classical equation

$$\mathscr{L}(u) = u_{xx} - u_t = 0. \qquad (1.1.12)$$

1.2. EQUATIONS REDUCIBLE TO THE HEAT EQUATION

As discussed in Sec. 1.1, equations of the form

$$u_t = \kappa u_{xx} \qquad (1.2.1)$$

can be reduced to

$$u_\tau = u_{xx} \qquad (1.2.2)$$

via the transformation $\tau = \kappa t$. Several additional equations that are reducible to the heat equation are listed here, along with the transformation required to effect the reduction.

For equations of the form

$$u_t = a(t)u_{xx}, \qquad a(t) > 0, \qquad (1.2.3)$$

set

$$A(t) = \int_0^t a(\eta)\, d\eta \qquad (1.2.4)$$

and

$$t = \phi(\tau), \qquad (1.2.5)$$

where ϕ is the inverse of the mapping $\tau = A(t)$. Clearly,

$$\phi'(\tau) = [a(\phi(\tau))]^{-1} \equiv [a(t)]^{-1} \qquad (1.2.6)$$

follows from $t = \phi(A(t))$. Setting $U(x, \tau) = u(x, \phi(\tau))$, it follows from the Chain Rule that

$$U_\tau = U_{xx}. \qquad (1.2.7)$$

An equation of the form

$$u_t = u_{xx} - b(t)u_x \qquad (1.2.8)$$

can be handled via the transformation

$$x = \xi + \int_0^t b(\eta)\, d\eta. \tag{1.2.9}$$

Setting $U(\xi, t) = u(\xi + \int_0^t b(\eta)\, d\eta, t)$, it follows that (1.2.8) reduces to

$$U_t = U_{\xi\xi}. \tag{1.2.10}$$

Equations of the form

$$u_t = u_{xx} - c(t)u \tag{1.2.11}$$

can be reduced via the use of the integrating factor $\exp\{\int_0^t c(\eta)\, d\eta\}$ and the change of dependent variable

$$v = u\exp\left\{\int_0^t c(\eta)\, d\eta\right\}. \tag{1.2.12}$$

The reduction of an equation of the form

$$u_t = a(t)u_{xx} - b(t)u_x - c(t)u, \qquad a(t) > 0 \tag{1.2.13}$$

to the heat equation can be achieved by successive application of the three types of transformations displayed above.

The nonlinear equation

$$a(u)u_t = (a(u)u_x)_x, \qquad a(\cdot) > 0 \tag{1.2.14}$$

can be reduced via the change in dependent variable

$$v = \int_0^u a(\xi)\, d\xi. \tag{1.2.15}$$

1.3. ELEMENTARY SOLUTIONS

The classical method of separation of variables involves the search for a solution u in the form

$$u = \varphi(x)\psi(t). \tag{1.3.1}$$

Upon substituting (1.3.1) into $u_t = u_{xx}$ and dividing through by $\varphi\psi$, it follows that

$$\frac{\psi'(t)}{\psi(t)} = \frac{\varphi''(x)}{\varphi(x)} = \lambda^2, \tag{1.3.2}$$

where λ is a complex constant. Solving the resulting ordinary differential equations

$$\begin{cases} \psi'(t) = \lambda^2\psi(t), \\ \varphi''(x) = \lambda^2\varphi(x), \end{cases} \tag{1.3.3}$$

results in the *elementary solutions of exponential type*

$$u = \exp\{\lambda x + \lambda^2 t\}. \tag{1.3.4}$$

When $\lambda = i\beta$,

$$u = \exp\{-\beta^2 t\}\cos\beta x \tag{1.3.5}$$

and

$$u = \exp\{-\beta^2 t\}\sin\beta x \tag{1.3.6}$$

are obtained. From the linearity of $u_t = u_{xx}$, it follows for real λ that

$$u = \tfrac{1}{2}\{\exp\{\lambda x + \lambda^2 t\} \pm \exp\{-\lambda x + \lambda^2 t\}\} \tag{1.3.7}$$

are solutions. These can be rewritten as

$$u = \exp\{\lambda^2 t\}\cosh\lambda x \tag{1.3.8}$$

and

$$u = \exp\{\lambda^2 t\}\sinh\lambda x. \tag{1.3.9}$$

The *heat polynomials* are obtained by expanding Eq. (1.3.4) into a Taylor's series with respect to λ. Thus,

$$\exp\{\lambda x + \lambda^2 t\} = \sum_{n=0}^{\infty} p_n(x,t)\frac{\lambda^n}{n!}. \tag{1.3.10}$$

Since

$$0 = \sum_{n=0}^{\infty} \left\{\frac{\partial p_n}{\partial t} - \frac{\partial^2 p_n}{\partial x^2}\right\}\frac{\lambda^n}{n!} \tag{1.3.11}$$

holds for all λ, it follows that

$$\frac{\partial p_n}{\partial t} = \frac{\partial p_n}{\partial x^2}, \qquad n = 0,1,2,3,\ldots \tag{1.3.12}$$

An explicit formula for $p_n(x,t)$ can be obtained from Cauchy's rule of multiplying two power series together, since

$$\exp\{\lambda x + \lambda^2 t\} = \exp\{\lambda x\}\exp\{\lambda^2 t\}. \tag{1.3.13}$$

Setting

$$\exp\{\lambda x\} = \sum_{n=0}^{\infty} a_n\lambda^n \tag{1.3.14}$$

and

$$\exp\{\lambda^2 t\} = \sum_{n=0}^{\infty} b_n\lambda^n, \tag{1.3.15}$$

where

$$a_n = \frac{x^n}{n!}, \qquad n = 0,1,2,\ldots \tag{1.3.16}$$

and

$$b_n = \begin{cases} \dfrac{t^k}{k!}, & n = 2k, \\ 0, & n = 2k+1, \quad k = 0,1,2,\ldots, \end{cases} \tag{1.3.17}$$

it follows that

$$\exp\{\lambda x + \lambda^2 t\} = \sum_{n=0}^{\infty} c_n \lambda^n, \tag{1.3.18}$$

where

$$c_n = \sum_{j=0}^{n} b_j a_{n-j} = \sum_{k=0}^{[n/2]} \frac{t^k}{k!} \frac{x^{n-2k}}{(n-2k)!}. \tag{1.3.19}$$

Here, $[n/2]$ denotes the largest integer less than or equal to $n/2$. From Eqs. (1.3.10), (1.3.18), and (1.3.19), it is clear that

$$p_n(x,t) = n! \sum_{k=0}^{[n/2]} \frac{t^k}{k!} \frac{x^{n-2k}}{(n-2k)!}. \tag{1.3.20}$$

Remark. Note that these elementary solutions are analytic functions of their arguments.

1.4. METHODS OF GENERATING SOLUTIONS FROM SOLUTIONS

The linearity and homogeneity of $u_t = u_{xx}$ permit an extensive list of methods of generating solutions from other solutions.

Linear Combinations: If u_1 and u_2 are solutions, then $\alpha u_1 + \beta u_2$ is a solution.

Translations: If $u(x,t)$ is a solution, then so is $u(x - \xi, t - \tau)$, where ξ and τ are translation parameters.

Differentiation with Respect to a Parameter: If $u(x, t, \alpha)$ is a solution for each α in $a \leq \alpha \leq b$, then so is $u_\alpha(x, t, \alpha)$, $a \leq \alpha \leq b$.

Integration with Respect to a Parameter: If $u(x, t, \alpha)$, is a solution for each α in $a \leq \alpha \leq b$, then so is $\int_a^b u(x, t, \alpha)\, d\alpha$.

Affine Transformation: If $u(x, t)$ is a solution, then so is $u(\lambda x, \lambda^2 t)$ for any constant λ.

Differentiation with Respect to x and t: If $u(x, t)$ is a solution, then so is $(\partial^{n+m} u / \partial x^n \partial t^m)(x, t)$.

Integration with Respect to x and t: If $u(x, t)$ is a solution, then so is $\int_{x_0}^x u(\xi, t) \, d\xi$, provided that $u_x(x_0, t) = 0$. Also, if $u(x, t)$ is a solution, then so is $\int_a^t u(x, \eta) \, d\eta$, provided that $u(x, a) = 0$.

Convolutions: If $u(x, t)$ is a solution, then so are $\int_a^b u(x - \xi, t) \varphi(\xi) \, d\xi$ and $\int_a^b u(x, t - \xi) \varphi(\xi) \, d\xi$. However, $\int_a^t u(x, t - \xi) \varphi(\xi) \, d\xi$ is a solution only if $u(x, 0) = 0$.

The lack of rigor here should not disturb anyone since the important aspect is the formal application of these methods to derive solutions of problems.

The conditions under which a proposed solution is a solution can always be supplied later and are usually weaker than the conditions necessary to justify the formalism.

1.5. BASIC DEFINITIONS

On a line

$$t \equiv \text{constant}, \tag{1.5.1}$$

u and u_t cannot be specified independently since the equation $u_t = u_{xx}$ forces a compatibility condition on such data. In the general theory of partial differential equations of order n, when u and its normal derivatives up to order $(n - 1)$ cannot be independently specified on a curve, that curve is called a *characteristic*.

The characteristics $t \equiv \text{constant}$ also play a role in the definition of a solution and the notion of an initial-boundary-value problem. In Fig. 1.5.1, we display the type of closed region in $\mathbb{R}^2$ that will concern us throughout this volume. Here and throughout, we assume that s_1 and s_2 are continuous.

DEFINITION 1.5.1.   The *parabolic interior* D_T consists of the interior points depicted in Fig. 1.5.1 as well as the open line segment

$$\{(x, T) : s_1(T) < x < s_2(T)\}$$

on the characteristic $t = T$.

DEFINITION 1.5.2.   The *parabolic boundary* B_T consists of the curve $\{(s_1(t), t) : 0 \le t \le T\}$, the curve $\{(s_2(t), t) : 0 \le t \le T\}$, and the line segment $\{(x, 0) : a \le x \le b\}$ on the characteristic $t = 0$.

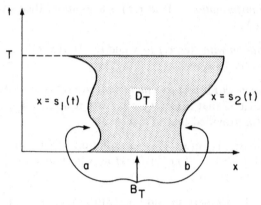

FIGURE 1.5.1

DEFINITION 1.5.3 (Basic).   The function u is said to be a solu-tion of $u_t = u_{xx}$ in D_T if u, u_x, u_t, and u_{xx} are continuous in D_T and $u_t = u_{xx}$ is satisfied as an identity in (x, t) throughout D_T.

An *initial-boundary-value* problem is the determination of a solution for $u_t = u_{xx}$ in D_T that satisfies prescribed conditions upon B_T. For example, u could be specified everywhere on B_T. Other specifications are possible on those portions of B_T that do not coincide with a segment of a characteristic. When B_T coincides with a characteristic, then we are limited to the specifica-tion of u, and we call such a problem an *initial-value problem*. We reserve the name *Cauchy problem* for those problems with the specification of both u and u_x on noncharacteristic curves.

DEFINITION 1.5.4.   Solutions u of $u_t = u_{xx}$ in D_T are called solu-tions of an initial- and/or boundary-value problem, provided that boundary and initial conditions are satisfied in a piecewise-continuous fashion by u.

Remark.   This means that, at a boundary point at which the data has a jump discontinuity, the solution u and whatever derivatives are required must remain bounded, while, at a boundary point at which the data is continuous, the solution and whatever derivatives are required must be continuous at that point and satisfy the given condition.

1.6.   THE WEAK MAXIMUM PRINCIPLE AND SOME APPLICATIONS

A powerful tool for the study of the initial-boundary-value problems below is the Weak Maximum Principle.

Theorem 1.6.1 (The Weak Maximum Principle).   *For a solution u of* $u_t = u_{xx}$ *in a bounded* D_T, *which is continuous in* $D_T \cup B_T$,

$$\max_{D_T \cup B_T} u = \max_{B_T} u. \qquad (1.6.1)$$

Proof.   Set

$$v(x, t) = u(x, t) + \varepsilon x^2, \qquad (1.6.2)$$

where ε is a positive number. Then v assumes its max on B_T. Otherwise, there exists a point $(x_0, t_0) \in D_T$ such that

$$v(x_0, t_0) = \max_{D_T \cup B_T} v. \qquad (1.6.3)$$

Hence, at (x_0, t_0),

$$\mathscr{L}(v) \equiv v_{xx} - v_t \le 0, \qquad (1.6.4)$$

since

$$v_t(x_0, t_0) \ge 0 \quad \text{and} \quad v_{xx}(x_0, t_0) \le 0; \qquad (1.6.5)$$

but

$$\mathscr{L}(v) = \mathscr{L}(u) + \mathscr{L}(\varepsilon x^2) = 2\varepsilon > 0 \quad \text{throughout } D_T \qquad (1.6.6)$$

contradicts the assumption that v assumes its maximum value in D_T.

Since $u \le v$ and $v \le \max_{B_T} v$,

$$u \le \max_{B_T} v \le \max_{B_T} u + \varepsilon \max_{B_T} x^2. \qquad (1.6.7)$$

As ε can be chosen arbitrarily,

$$u \le \max_{B_T} u. \qquad (1.6.8) \quad \square$$

By considering $(-u)$ instead of u, we obtain the following result from the argument above.

Theorem 1.6.2 (The Weak Minimum Principle).   *For a solution u of* $u_t = u_{xx}$ *in a bounded* D_T, *which is continuous in* $D_T \cup B_T$,

$$\min_{D_T \cup B_T} u = \min_{B_T} u. \qquad (1.6.9)$$

Proof.   Set $w = -u$ and use Theorem 1.6.1. $\square$

One of the basic applications of the Weak Maximum (Minimum) Principle is the following Comparison Theorem for solutions of $u_t = u_{xx}$.

Theorem 1.6.3 (The Comparison Theorem).   *If u and v are solutions of* $u_t = u_{xx}$ *in* D_T *that are continuous in* $D_T \cup B_T$ *and if* $u \le v$ *on* B_T, *then* $u \le v$ *in* $D_T \cup B_T$.

Proof.   Set $w = v - u$. Then $\mathscr{L}(w) = 0$ in D_T and $w \ge 0$ on B_T. By the Weak Minimum Principle, $w \ge 0$ in $D_T \cup B_T$. $\square$

Another basic application is the following uniqueness of solutions in D_T.

Theorem 1.6.4 (Uniqueness).   *If u and v are solutions of $u_t = u_{xx}$ in D_T that are continuous in $D_T \cup B_T$ and if $u = v$ on B_T, then $u \equiv v$ in $D_T \cup B_T$.*

Proof.   Set $w = v - u$. Then $\mathcal{L}(w) = 0$ in D_T and $w = 0$ on B_T. By the Comparison Theorem, $w = 0$ in $D_T \cup B_T$.   □

For most physical applications of the theory, the following extensions of the Comparison Theorem and the Uniqueness Theorem are sufficient.

Theorem 1.6.5 (Extended Comparison Theorem).   *If u and v are solutions of $u_t = u_{xx}$ in D_T that are piecewise continuous in $D_T \cup B_T$ with at most a finite number of bounded discontinuities and if $u \leq v$ on B_T except for the points of discontinuity, then $u \leq v$ in D_T.*

Theorem 1.6.6 (Extended Uniqueness Theorem).   *If u and v are solutions of $u_t = u_{xx}$ in D_T that are piecewise continuous in $D_T \cup B_T$ with at most a finite number of bounded discontinuities and if $u = v$ on B_T except for the points of discontinuity, then $u \equiv v$ in D_T.*

Remark.   It is not convenient to prove these theorems here. We must develop some representations of solutions of some simple initial- and/or boundary-value problems. We could first develop these solutions, and prove these theorems prior to their usage. However, this would lead to an inefficient exposition. Hence, we shall outline their proofs as exercises in Chapter 15.

Remark.   The Weak Maximum Principle asserts that, in the absence of heat sources and sinks in a conductor, the temperature at any point in the conductor cannot exceed the maximum of the initial temperatures and the boundary temperatures. This is certainly in line with our experience of the cooling of "hot" objects and the warming of "cold" objects in a "warm" environment. The Comparison Theorem seems to be outside ordinary experience except perhaps for those conductors into which we can poke our fingers.

EXERCISES

1.1.   Let $u = \phi(\lambda)$, where $\lambda = x/2\sqrt{t}$. Show that $u_t = u_{xx}$ if and only if $\phi'' + 2\lambda\phi' = 0$.

1.2. Integrating $\phi'' + 2\lambda\phi' = 0$, show that $u = \text{erf}(x/2\sqrt{t})$ satisfies $u_t = u_{xx}$, where

$$\text{erf}(\lambda) = \frac{2}{\sqrt{\pi}} \int_0^\lambda \exp\{-\rho^2\}\, d\rho.$$

1.3. Show that $u = \text{erfc}(x/2\sqrt{t})$ satisfies $u_t = u_{xx}$, where

$$\text{erfc}(\lambda) = \frac{2}{\sqrt{\pi}} \int_\lambda^\infty \exp\{-\rho^2\}\, d\rho.$$

1.4. By differentiating $\text{erf}(x/2\sqrt{t})$, show that

$$K(x,t) = \frac{1}{\sqrt{4\pi t}} \exp\left\{-\frac{x^2}{4t}\right\}, \qquad t > 0,$$

satisfies $K_t = K_{xx}$.

1.5. *Appell's Transformation*: Show that if $u(x,t)$ is a solution of $u_t = u_{xx}$, then for $t > 0$, so is

$$v(x,t) = K(x,t) u\left(\frac{x}{t}, -\frac{1}{t}\right),$$

where $K(x,t)$ is defined in Exercise 1.4.

1.6. Show that the equation $u_t + uu_x = \mu u_{xx}$, $\mu > 0$, transforms into $\varphi_t = \mu\varphi_{xx} + c(t)\varphi$ via $\varphi = \exp\{-(1/2\mu)\int u\, dx\}$. *Hint*: Use the inverse $u = -2\mu(\varphi_x/\varphi) = -2\mu(\log\varphi)_x$.

1.7. Transform $\varphi_t = \mu\varphi_{xx} + c(t)\varphi$ into the heat equation $v_t = v_{xx}$ via an appropriate transformation.

NOTES

The derivation of the heat equation was given by Fourier [2]. Other derivations can be found in Carslaw and Jaeger [1] and Widder [9]. Gevrey [4] presented examples of equations reducible to the heat equation, and recently, Hopf [5] has studied another such example. Elementary solutions of the heat equation were given by Fourier [2] and can be found in most elementary texts on partial differential equations. For example, see Garabedian [3], Petrovsky [6], Weinberger [8], Widder [9], Young [10], and Zackmanoglou and Thoe [11]. Methods of generating solutions of linear partial differential equations from solutions are standard to most texts on partial differential equations. See those cited above. The weak maximum principle is standard material and the reader may find Protter and Weinberger [7] of interest.

REFERENCES

1. Carslow, H. S., and Jaeger, J. C., *Conduction of Heat in Solids*. Oxford Univ. Press (Clarendon), London and New York, 1948.
2. Fourier, J., *Analytical Theory of Heat* (Great Books of the Western World, Vol. 45). Encyclopedia Britannica, Inc., Chicago-London-Toronto, 1952.
3. Garabedian, P. R., *Partial Differential Equations*. John Wiley & Sons, Inc., New York, London, and Sydney, 1964.
4. Gevrey, M., Sur les équations aux dérivées partielles du type parabolique, *J. Math. Pures Appl.*, **9** (1913), 305–471.
5. Hopf, E., The partial differential equation $u_t + uu_x = \mu u_{xx}$, *Comm. Pure Appl. Math.*, **3** (1950), 201–230.
6. Petrovsky, I. G., *Lectures on Partial Differential Equations*. Interscience Pub., Inc., New York and London, 1954.
7. Protter, M. H., and Weinberger, H. F., *Maximum Principles in Differential Equations*. Prentice-Hall, Inc., Englewood Cliffs, N. J., 1967.
8. Weinberger, H. F., *A First Course in Partial Differential Equations*. Blaisdell Pub. Co. (Ginn & Co.), Waltham, Massachusetts-Toronto-London, 1965.
9. Widder, D. V., *The Heat Equation*. Academic Press, New York 1975.
10. Young, E. C., *Partial Differential Equations: An Introduction*. Allyn and Bacon, Inc., Boston, 1972.
11. Zachmanoglou, E. C., and Thoe, D., *Introduction to Partial Differential Equations with Applications*. The Williams & Wilkins Company, Baltimore, 1976.

Chapter 2

The Cauchy Problem

2.1. INTRODUCTION

We shall begin by considering the simplest Cauchy problem for $u_t = u_{xx}$. This is the problem of determining a function u that satisfies

$$u_t = u_{xx}, \qquad 0 < x < \gamma_1, \qquad |t - t_0| < \gamma_2,$$

$$u(0, t) = f(t), \qquad |t - t_0| < \gamma_2, \qquad (2.1.1)$$

$$u_x(0, t) = g(t), \qquad |t - t_0| < \gamma_2,$$

where γ_i, $i = 1, 2$, are positive constants, and where f and g are known functions. (See Fig. 2.1.1.)

A simple form for a solution is that of a power series. Let

$$u(x, t) = \sum_{j=0}^{\infty} a_j(t) x^j, \qquad (2.1.2)$$

where the coefficients a_j are to be determined. By substituting u into $u_t = u_{xx}$, we see that the condition

$$0 = u_t - u_{xx} = \sum_{j=0}^{\infty} \left\{ a_j'(t) - (j+2)(j+1) a_{j+2}(t) \right\} x^j \qquad (2.1.3)$$

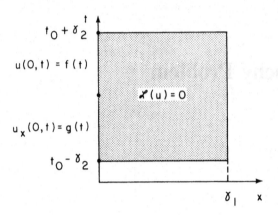

FIGURE 2.1.1

requires that

$$a_{j+2}(t) = \frac{a_j'(t)}{(j+2)(j+1)}, \qquad j = 0,1,2,\ldots \qquad (2.1.4)$$

From the recursion relation (2.1.4), we see that

$$a_{2j}(t) = \frac{a_0^{(j)}(t)}{(2j)!}, \qquad j = 0,1,2,3,\ldots, \qquad (2.1.5)$$

while

$$a_{2j+1}(t) = \frac{a_1^{(j)}(t)}{(2j+1)!}, \qquad j = 0,1,2,3,\ldots \qquad (2.1.6)$$

From the boundary conditions $u(0,t) = f(t)$ and $u_x(0,t) = g(t)$, we note that

$$a_0(t) = f(t) \qquad (2.1.7)$$

and

$$a_1(t) = g(t). \qquad (2.1.8)$$

Thus, we obtain the formal expression

$$u(x,t) = \sum_{j=0}^{\infty} \left\{ f^{(j)}(t) \frac{x^{2j}}{(2j)!} + g^{(j)}(t) \frac{x^{2j+1}}{(2j+1)!} \right\}, \qquad (2.1.9)$$

which we shall analyze in the next section.

2.2. THE HOLMGREN FUNCTION CLASSES

We study now the convergence of the power series (2.1.9). From the ratio test, we see that the series of even powers will converge if

$$\lim_{j \to \infty} \frac{|f^{(j+1)}(t)|\gamma_1^2}{|f^{(j)}(t)|(2j+2)(2j+1)} < 1. \tag{2.2.1}$$

In other words, the convergence is assured if

$$\frac{|f^{(j+1)}(t)|\gamma_1^2}{|f^{(j)}(t)|(2j+2)(2j+1)} \le \rho < 1 \tag{2.2.2}$$

holds for all j sufficiently large. From this we obtain the recursion relation

$$|f^{(j+1)}(t)| < |f^{(j)}(t)|\frac{(2j+2)(2j+1)}{\gamma_1^2}, \tag{2.2.3}$$

which would imply that

$$|f^{(j)}(t)| < |f(t)|\frac{(2j)!}{\gamma_1^{2j}} \tag{2.2.4}$$

if (2.2.2) held for all $j = 0, 1, 2, \ldots$

From this discussion we are motivated to define the following function classes.

DEFINITION 2.2.1 (Holmgren). For the positive constants γ_1, γ_2, and C_1, the Holmgren class $H(\gamma_1, \gamma_2, C_1, t_0)$ is the set of infinitely differentiable functions φ defined on $|t - t_0| < \gamma_2$ that satisfy

$$|\varphi^{(j)}(t)| \le C_1\gamma_1^{-2j}(2j)!, \quad j = 0, 1, 2, 3, \ldots \tag{2.2.5}$$

for all t in $|t - t_0| < \gamma_2$.

2.3. AN EXISTENCE THEOREM

We can now state an existence result.

Theorem 2.3.1. *If f and g belong to $H(\gamma_1, \gamma_2, C_1, t_0)$, then the power series*

$$u(x, t) = \sum_{j=0}^{\infty} \left\{ f^{(j)}(t)\frac{x^{2j}}{(2j)!} + g^{(j)}(t)\frac{x^{2j+1}}{(2j+1)!} \right\} \tag{2.3.1}$$

converges uniformly and absolutely for $|x| \le r < \gamma_1$ and u is a solution of

$$\begin{aligned} u_t &= u_{xx}, & |x| &< \gamma_1, & |t - t_0| &< \gamma_2, \\ u(0, t) &= f(t), & |t - t_0| &< \gamma_2, \\ u_x(0, t) &= g(t), & |t - t_0| &< \gamma_2. \end{aligned} \tag{2.3.2}$$

Proof. Since f and g belong to $H(\gamma_1, \gamma_2, C_1, t_0)$,

$$\left| f^{(j)}(t)\frac{x_j^2}{(2j)!} + g^{(j)}(t)\frac{x^{2j+1}}{(2j+1)!} \right| < C_1\left\{1 + \frac{r}{(2j+1)}\right\}\left(\frac{r}{\gamma_1}\right)^{2j} \quad (2.3.3)$$

for $|x| \le r$ and $|t - t_0| < \gamma_2$. As $r < \gamma_1$, we see that, for $|x| \le r$ and $|t - t_0| < \gamma_2$, the power series $u(x, t)$ is majorized by the convergent power series

$$C_1 \sum_{j=0}^{\infty} \left\{1 + \frac{r}{(2j+1)}\right\}\left(\frac{r}{\gamma_1}\right)^{2j}. \quad (2.3.4)$$

By similar arguments, u_t, u_x, and u_{xx} are shown to exist and equal the result of term-by-term differentiation. Consequently, the formalities of Sec. 2.1 are valid. $\square$

2.4. AN EXAMPLE

We give here an example of a function f that has the property that for each $\gamma_1 > 0$ there exists a $C_1 = C_1(\gamma_1) > 0$ such that $f \in H(\gamma_1, \gamma_2, C_1(\gamma_1), 0)$ for a fixed $\gamma_2 > 0$.
   Set

$$f(t) = \begin{cases} \exp\left\{-\dfrac{1}{t^2}\right\}, & t > 0, \\ 0, & t \le 0. \end{cases} \quad (2.4.1)$$

We see immediately that f is infinitely differentiable and that $\lim_{t \to 0} f^{(j)}(t) = 0$ for $j = 0,1,2,3,\ldots$ Clearly, f is not an analytic function of t for $|t| < \gamma_2$. However, the analyticity of f for $t > 0$ will enable us to estimate the derivatives of f.
   For $t > 0$, consider the circle

$$\Gamma = \left\{z = t + i\tau \,\middle|\, z = t + \frac{t}{2}\exp\{i\theta\}, \quad 0 \le \theta \le 2\pi\right\}. \quad (2.4.2)$$

By the Cauchy–Riemann integral formula, we see that for $t > 0$,

$$f^{(j)}(t) = \frac{j!}{2\pi i}\int_\Gamma \frac{\exp\{-z^{-2}\}\,dz}{(z-t)^{j+1}}. \quad (2.4.3)$$

Since for z on Γ,

$$\operatorname{Re} z^{-2} = \frac{1 + \cos\theta + 4^{-1}\cos 2\theta}{t^2((5/4) + \cos\theta)^2} \ge \frac{1/4}{t^2(9/4)^2} = \frac{\lambda}{t^2}, \quad (2.4.4)$$

where

$$\lambda = \tfrac{1}{4}\left(\tfrac{9}{4}\right)^{-2}, \quad (2.4.5)$$

we see that for $t > 0$,

$$|f^{(j)}(t)| \le j! \left(\frac{2}{t}\right)^j \exp\left\{\frac{-\lambda}{t^2}\right\}. \tag{2.4.6}$$

Since $x^\alpha \exp\{-\beta x\} \le (\alpha/e\beta)^\alpha$ for $\alpha, \beta > 0$, we see that for $\alpha = j/2$ and $\beta = \lambda$,

$$|f^{(j)}(t)| \le j! 2^j \left(\frac{j}{2e\lambda}\right)^{j/2} = j! \left(\frac{2j}{e\lambda}\right)^{j/2}. \tag{2.4.7}$$

We must show now that for $\gamma_1 > 0$, there exists a constant $C_1 > 0$ such that

$$|f^j(t)| \le C_1 \gamma_1^{-2j}(2j)! \tag{2.4.8}$$

holds for all $j = 0,1,2,\ldots$ For this it suffices to consider

$$\lim_{j \to \infty} \frac{j! \gamma_1^{2j}(2j/e\lambda)^{j/2}}{(2j)!} \equiv \lim_{j \to \infty} A_j. \tag{2.4.9}$$

By the ratio test, we see that

$$\lim_{j \to \infty} \frac{A_{j+1}}{A_j} = \lim_{j \to \infty} \frac{(j+1)\gamma_1^2 (2j/e\lambda)^{1/2}(1+1/j)^{(j+1)/2}}{(2j+2)(2j+1)} = 0 \tag{2.4.10}$$

Hence,

$$\sum_{j=1}^{\infty} A_j < \infty. \tag{2.4.11}$$

Consequently,

$$\lim_{j \to \infty} A_j = 0, \tag{2.4.12}$$

which implies the existence of a constant C_1 such that (2.4.8) holds.
    Using $f(t)$ in Theorem 2.3.1, we see that

$$u(x,t) = \sum_{j=0}^{\infty} f^{(j)}(t) \frac{x^{2j}}{(2j)!} \tag{2.4.13}$$

and

$$v(x,t) = \sum_{j=0}^{\infty} f^{(j)}(t) \frac{x^{2j+1}}{(2j+1)!} \tag{2.4.14}$$

converge for all x. These functions will provide us later with examples of non-uniqueness for certain problems.

2.5. CONTINUOUS DEPENDENCE UPON THE DATA

The Cauchy problem for $u_t = u_{xx}$ is not well posed. In the presence of uniqueness of the solution, this means that if the data is made small, then the solution does not become small. In fact, the solution may even increase unboundedly as the data tends to zero.

We can construct an example of this behavior by considering the exponential solutions of the heat equation

$$\exp\{n(1+i)x + n^2(1+i)^2 t\} = \exp\{nx + i(nx + 2n^2 t)\}, \qquad n = 1, 2, \ldots$$
$$(2.5.1)$$

Selecting the real part and multiplying by n^{-2} we obtain the solutions

$$u_n = n^{-2}\exp\{nx\}\cos(nx + 2n^2 t), \qquad n = 1, 2, 3, \ldots \qquad (2.5.2)$$

These functions are, respectively, the solutions of the Cauchy problems

$$\frac{\partial u_n}{\partial t} = \frac{\partial^2 u_n}{\partial x^2}, \qquad 0 < x, \qquad \text{all } t,$$

$$u_n(0, t) = n^{-2}\cos 2n^2 t, \qquad \text{all } t, \qquad (2.5.3)$$

$$\frac{\partial u_n}{\partial x}(0, t) = n^{-1}\{\cos 2n^2 t - \sin 2n^2 t\}, \qquad \text{all } t, \qquad n = 1, 2, 3, \ldots$$

Note that, as n tends to infinity, the data $u_n(0, t)$ and $(\partial u_n/\partial x)(0, t)$ tend uniformly to zero, while for $x > 0$, the function u_n assumes values $n^{-2}\exp\{nx\}$, which tend to infinity as n tends to infinity.

Uniqueness of the solution of the Cauchy problem follows from the analyticity of the solution of the heat equation in the spatial variable x. Consequently, discussion of uniqueness and continuous dependence is postponed until Chapter 11, Sec. 4.

2.6. VARIABLE DATA-BEARING CURVES: EXISTENCE: AN EXERCISE

Exercise 2.1. We consider the problem

$$u_t = u_{xx}, \qquad s(t) < x < s(t) + \gamma_1, \qquad |t - t_0| < \gamma_2,$$
$$u(s(t), t) = f(t), \qquad |t - t_0| < \gamma_2, \qquad (2.6.1)$$
$$u_x(s(t), t) = g(t), \qquad |t - t_0| < \gamma_2,$$

where γ_i, $i = 1, 2$, are positive constants and where s, f, and g are known functions.

Theorem 2.6.1. *For Problem* (2.6.1) *with analytic s, f, and g,*

$$u(x,t) = \sum_{j=0}^{\infty} \frac{\partial^j}{\partial t^j} \left\{ f(t) \frac{(x-s(t))^{2j}}{(2j)!} + (s^{(1)}(t)f(t) \right.$$

$$\left. + g(t)) \frac{(x-s(t))^{2j+1}}{(2j+1)!} \right\}. \tag{2.6.2}$$

Proof. This rearrangement of the Cauchy–Kowalewski series is due to C. Denson Hill [3]. We shall leave its proof as an *exercise* for the reader. □

Exercise 2.2. Use Stirling's approximation on (2.4.7) to derive (2.4.8).

NOTES

In the literature the Holmgren function classes go by the name Gevrey. Gevrey [1] attributes these classes to Holmgren [4,5]. The study of the function $\exp\{-t^{-2}\}$, $t > 0$, in Sec. 2.4 follows the presentation given by Widder [9]. The discussion on continuous dependence in Sec. 2.5 is analogous to that found in Hadamard [2], Payne [6], and Pucci [7,8]. Theorem 2.6.1 is due to Hill [3].

REFERENCES

1. Gevrey, M., Sur les équations aux dérivées partielles du type parabolique, *J. Math. Pures Appl.*, **9** (1913), 305–471.
2. Hadamard, J., *Lectures on Cauchy's Problem in Linear Partial-Differential Equations*. Dover Publications, New York, 1952.
3. Hill, C. Denson, Parabolic equations in one space variable and the noncharacteristic Cauchy problem, *Comm. Pure Appl. Math.*, **20** (1967), 619–633.
4. Holmgren, E., Sur l'équation de la propagation de la chaleur, *Arkiv. Mat. Fysik*, **14** (1908), No. 14, 1–11.
5. Holmgren, E., Sur l'équation de la propagation de la chaleur II, *Arkiv. Mat. Fysik*, **4** (1908), No. 18, 1–27.
6. Payne, L. E., *Improperly Posed Problems in Partial-Differential Equations*. CBMS Regional Conference Series in Applied Math., Soc. Ind. Appl. Math., Philadelphia, PA, 1975.
7. Pucci, C., Alcune limitazioni per le soluzioni di equazioni paraboliche, *Ann. di Mat. Pura ed Applicata*, (IV), **48** (1959), 161–172.
8. Pucci, C., Discussione del problema di Cauchy per le equazioni di tipo ellittico, *Ann. Mat. Pura ed Appl.* (IV), **46** (1958), 131–154.
9. Widder, D. V., *The Heat Equation*, Academic Press, N.Y., N.Y., 1975.

Chapter 3

The Initial-Value Problem

3.1. INTRODUCTION

For the initial-value problem of determining u satisfying

$$u_t = u_{xx}, \qquad -\infty < x < \infty, \qquad 0 < t,$$
$$u(x,0) = f(x), \qquad -\infty < x < \infty, \tag{3.1.1}$$

where f is a known function, we can apply the Fourier Transform method to derive a formal representation of a solution. (See Fig. 3.1.1.)

Let

$$\hat{u}(\alpha, t) = \frac{1}{2\pi} \int_{-\infty}^{\infty} u(\xi, t) \exp\{-i\alpha\xi\} \, d\xi. \tag{3.1.2}$$

Then,

$$-\alpha^2 \hat{u}(\alpha, t) = \frac{1}{2\pi} \int_{-\infty}^{\infty} u_{xx}(x, t) \exp\{-i\alpha x\} \, dx, \tag{3.1.3}$$

and we can convert (3.1.1) into the initial-value problem

$$\hat{u}_t = -\alpha^2 \hat{u}, \qquad 0 < t, \qquad \hat{u}(\alpha,0) = \hat{f}(\alpha). \tag{3.1.4}$$

Integrating (3.1.4), we obtain

$$\hat{u}(\alpha, t) = \hat{f}(\alpha) \exp\{-\alpha^2 t\}. \tag{3.1.5}$$

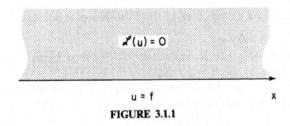

FIGURE 3.1.1

Applying the inverse transform, we get

$$u(x,t) = \frac{1}{2\pi} \int_{-\infty}^{\infty} \exp\{i\alpha x - \alpha^2 t\} \hat{f}(\alpha) \, d\alpha$$

$$= \frac{1}{2\pi} \int_{-\infty}^{\infty} \exp\{i\alpha x - \alpha^2 t\} \int_{-\infty}^{\infty} f(\xi) \exp\{-i\alpha\xi\} \, d\xi \, d\alpha$$

$$= \int_{-\infty}^{\infty} f(\xi) \left\{ \frac{1}{2\pi} \int_{-\infty}^{\infty} \exp\{i\alpha(x - \xi) - \alpha^2 t\} \, d\alpha \right\} d\xi. \quad (3.1.6)$$

It is left as an exercise for the reader to show that

$$\frac{1}{2\pi} \int_{-\infty}^{\infty} \exp\{i\alpha(x - \xi) - \alpha^2 t\} \, d\alpha = K(x - \xi, t), \quad (3.1.7)$$

where

$$K(x,t) = \frac{1}{\sqrt{4\pi t}} \exp\left\{-\frac{x^2}{4t}\right\}, \quad t > 0. \quad (3.1.8)$$

From (3.1.7), (1.3.4), and Sec. 1.4, we see that $K(x, t)$ satisfies $K_t = K_{xx}$.

Hence we are led to a formal representation for a solution to the initial-value problem (3.1.1) in the form

$$u(x,t) = \int_{-\infty}^{\infty} K(x - \xi, t) f(\xi) \, d\xi, \quad t > 0. \quad (3.1.9)$$

We begin our treatment of the initial-value problem by studying the properties of $K(x, t)$.

3.2. THE FUNDAMENTAL SOLUTION $K(x, t)$

For the fundamental solution

$$K(x,t) = \frac{1}{\sqrt{4\pi t}} \exp\left\{-\frac{x^2}{4t}\right\}, \quad t > 0, \quad (3.2.1)$$

we demonstrate the following properties.

Theorem 3.2.1 (Properties of the Fundamental Solution).

A. $K(x,t) > 0$ for $t > 0$.

B. For fixed $t > 0$, K and its derivatives tend to zero exponentially fast as $|x|$ tends to infinity.

C. For any fixed $\delta > 0$, $\lim_{t \downarrow 0} K(x,t) = 0$ uniformly for all $|x| \geq \delta$.

D. For any fixed $\delta > 0$, $\lim_{t \downarrow 0} \int_{|x| \geq \delta} K(x,t)\, dx = 0$.

E. For all $t > 0$, $\int_{-\infty}^{\infty} K(x,t)\, dx = 1$.

F. $\lim_{x \uparrow 0} - \int_0^t (\partial K / \partial x)(x, t - \tau)\, d\tau = -\frac{1}{2}$.

G. $\lim_{x \downarrow 0} - \int_0^t (\partial K / \partial x)(x, t - \tau)\, d\tau = +\frac{1}{2}$.

H. For $t > 0$, K is an analytic function of x and t.

Proof. Property A is evident. Property B follows from the fact that $K_{x^n t^m} = R_{m,n}(x, t^{1/2})K$, where $R_{m,n}$ is a rational function in x and $t^{1/2}$. For Property C, note that, for $|x| \geq \delta$,

$$K(x,t) = \leq \frac{1}{\sqrt{4\pi t}} \exp\left\{ -\frac{\delta^2}{4t} \right\}. \qquad (3.2.2)$$

Note also that, for $|x| \geq \delta$,

$$|K_{x^n t^m}| \leq |R_{m,n}| \frac{1}{\sqrt{4\pi t}} \exp\left\{ -\frac{\delta^2}{4t} \right\}. \qquad (3.2.3)$$

Consequently, K and $K_{x^n t^m}$ tend to zero uniformly for $|x| \geq \delta$ as t tends to zero.

To demonstrate properties D, E, F, and G, we use the result

$$\int_{-\infty}^{\infty} \exp\{-\rho^2\}\, d\rho = \sqrt{\pi}, \qquad (3.2.4)$$

which is left as an exercise for the reader.

Property E follows directly from the change of variable $\rho = x/\sqrt{4t}$ since

$$\frac{1}{\sqrt{4\pi t}} \int_{-\infty}^{\infty} \exp\left\{ -\frac{x^2}{4t} \right\} dx = \frac{1}{\sqrt{\pi}} \int_{-\infty}^{\infty} \exp\{-\rho^2\}\, d\rho = 1. \qquad (3.2.5)$$

For property D,

$$\int_{|x| \geq \delta} K(x,t)\, dx = \frac{1}{\sqrt{\pi}} \int_{|\rho| \geq \delta/\sqrt{4t}} \exp\{-\rho^2\}\, d\rho. \qquad (3.2.6)$$

Since $\delta/\sqrt{4t}$ tends to infinity as t tends to zero, property D follows from the convergence of the integral of (3.2.4).

Under the change of variable $\rho = x/2\sqrt{t - \tau}$,

$$-\int_0^t \frac{\partial K}{\partial x}(x, t - \tau)\, d\tau = \begin{cases} \dfrac{1}{\sqrt{\pi}} \displaystyle\int_{x/(2\sqrt{t})}^{+\infty} \exp\{-\rho^2\}\, d\rho, & x > 0, \\[3mm] -\dfrac{1}{\sqrt{\pi}} \displaystyle\int_{-\infty}^{x/(2\sqrt{t})} \exp\{-\rho^2\}\, d\rho, & x < 0. \end{cases}$$

$$\qquad (3.2.7)$$

Consequently, Property F and Property G follow directly from the convergence of the integral in (3.2.4). Property H is evident since K is the product of one analytic function with the composition of two analytic functions. $\qquad\square$

3.3. CONVERGENCE OF THE CONVOLUTION $u(x,t) = \int_{-\infty}^{\infty} K(x-\xi,t)f(\xi)\,d\xi$

First, we shall investigate the conditions under which the integral

$$u(x,t) = \int_{-\infty}^{\infty} K(x-\xi,t)f(\xi)\,d\xi \qquad (3.3.1)$$

converges and diverges. Using $2ab \le \varepsilon^{-1}a^2 + \varepsilon b^2$, $\varepsilon > 0$, we see that

$$\exp\left\{-\frac{(1+\varepsilon^{-1})x^2}{4t}\right\}\exp\left\{-\frac{(1+\varepsilon)\xi^2}{4t}\right\}$$

$$\le \exp\left\{-\frac{(x-\xi)^2}{4t}\right\} \le \exp\left\{-\frac{(1-\varepsilon^{-1})x^2}{4t}\right\}\exp\left\{-\frac{(1-\varepsilon)\xi^2}{4t}\right\}.$$

$$(3.3.2)$$

Thus, there exist two positive-valued functions $F_1(x,t)$ and $F_2(x,t)$ such that

$$F_1(x,t)\exp\left\{-\frac{(1+\varepsilon)\xi^2}{4t}\right\} \le K(x-\xi,t) \le F_2(x,t)\exp\left\{-\frac{(1-\varepsilon)\xi^2}{4t}\right\}.$$

$$(3.3.3)$$

Now, for

$$f(x) = C_1\exp\left\{C_2|x|^{2+\alpha}\right\}, \qquad \alpha > 0, \qquad (3.3.4)$$

where C_1 and C_2 are positive constants, we see from (3.3.3) that

$$K(x-\xi,t)f(\xi) \ge F_1 \cdot C_1\exp\left\{C_2|\xi|^{2+\alpha} - (4t)^{-1}(1+\varepsilon)\xi^2\right\}. \quad (3.3.5)$$

Since

$$C_2|\xi|^{\alpha} > (4t^{-1})(1+\varepsilon) \qquad (3.3.6)$$

for ξ sufficiently large, it follows that

$$\int_{-\infty}^{\infty} K(x-\xi,t)f(\xi)\,d\xi = \infty \qquad (3.3.7)$$

for all functions f that are asymptotic at $|x| = \infty$ to $C_1\exp\{C_2|x|^{2+\alpha}\}$ for any $\alpha > 0$. Next, we consider

$$f(x) = C_1\exp\left\{C_2x^2\right\}. \qquad (3.3.8)$$

From (3.3.3), it follows that

$$f(\xi)K(x-\xi,t) \le C_1F_2\exp\left\{\left[C_2 - (4t)^{-1}(1-\varepsilon)\right]\xi^2\right\}. \qquad (3.3.9)$$

Since

$$C_2 - (4t)^{-1}(1 - \varepsilon) < 0 \qquad (3.3.10)$$

for $0 < \varepsilon < 1$ and

$$0 < t < \frac{1 - \varepsilon}{4C_2}, \qquad (3.3.11)$$

it follows that for $0 < t < (1/4C_2)$,

$$\int_{-\infty}^{\infty} K(x - \xi, t) f(\xi) \, d\xi < \infty \qquad (3.3.12)$$

for all functions f that are piecewise-continuous and asymptotic at $|x| = \infty$ to $C_1 \exp\{C_2 x^2\}$. Finally, for

$$f(x) = C_1 \exp\{C_2 |x|^{1+\alpha}\}, \qquad 0 \le \alpha < 1, \qquad (3.3.13)$$

we see that

$$K(x - \xi, t) f(\xi) \le C_1 F_2 \exp\{C_2 |\xi|^{1+\alpha} - (4t)^{-1}(1 - \varepsilon)\xi^2\}. \quad (3.3.14)$$

As

$$C_2 - (4t)^{-1}(1 - \varepsilon)|\xi|^{1-\alpha} < 0 \qquad (3.3.15)$$

for $0 < \varepsilon < 1$ and $|\xi|$ sufficiently large, it follows that for all x and $t > 0$,

$$\int_{-\infty}^{\infty} K(x - \xi, t) f(\xi) \, d\xi < \infty \qquad (3.3.16)$$

for all functions f that satisfy

$$|f(x)| \le C_1 \exp\{C_2 |x|^{1+\alpha}\}, \qquad 0 \le \alpha < 1. \qquad (3.3.17)$$

Recalling that

$$\frac{\partial^{m+n} K}{\partial x^n \partial t^m}(x, t) = R_{m,n}(x, t^{1/2}) K(x, t), \qquad (3.3.18)$$

where $R_{m,n}$ is a rational function, it follows from the above discussion and Leibniz's rule that if f satisfies (3.3.17), then u is infinitely differentiable in $\mathbb{R}_+^2 = \{(x, t) | t > 0\}$ and $u_t = u_{xx}$. On the other hand, if f is asymptotic at infinity to $C_1 \exp\{C_2 x^2\}$, then u is infinitely differentiable and satisfies $u_t = u_{xx}$ in every $D_T = \{(x, t) | -\infty < x < \infty$ and $0 < t \le T\}$ such that $T < (4C_2)^{-1}$.

3.4. $\mathrm{Lim}_{t \downarrow 0} u(x, t)$

We shall investigate the continuity of u at $t = 0$. To facilitate this discussion, we demonstrate the following result.

Lemma 3.4.1. *For continuous f satisfying $|f(x)| \le C_1 \exp\{C_2 x^2\}$, where C_1 and C_2 are positive constants,*

$$\lim_{t\downarrow 0} u(x,t) = f(x) \tag{3.4.1}$$

is uniform for all x in a compact subset of $-\infty < x < \infty$.

Proof. From Property E of Theorem 3.2.1, we can write

$$f(x) = \frac{1}{\sqrt{4\pi t}} \int_{-\infty}^{\infty} \exp\left\{-\frac{(x-\xi)^2}{4t}\right\} f(x) \, d\xi. \tag{3.4.2}$$

Hence, for fixed x,

$$u(x,t) - f(x) = J_1(t) + J_2(t) + J_3(t), \tag{3.4.3}$$

where

$$J_1(t) = \frac{1}{\sqrt{4\pi t}} \int_{-\infty}^{x-\delta} \exp\left\{-\frac{(x-\xi)^2}{4t}\right\} [f(\xi) - f(x)] \, d\xi, \tag{3.4.4}$$

we have

$$J_2(t) = \frac{1}{\sqrt{4\pi t}} \int_{x-\delta}^{x+\delta} \exp\left\{-\frac{(x-\xi)^2}{4t}\right\} [f(\xi) - f(x)] \, d\xi, \tag{3.4.5}$$

and

$$J_3(t) = \frac{1}{\sqrt{4\pi t}} \int_{x+\delta}^{\infty} \exp\left\{-\frac{(x-\xi)^2}{4t}\right\} [f(\xi) - f(x)] \, d\xi. \tag{3.4.6}$$

From the continuity of f, it follows that f is uniformly continuous on compact subsets of $-\infty < x < \infty$. For such a subset $\mathscr{X}$ and for each $\varepsilon > 0$ there exists a $\delta_\varepsilon > 0$ such that

$$|f(x) - f(\xi)| < \frac{\varepsilon}{3} \tag{3.4.7}$$

holds for every x and ξ in $\mathscr{X}$ such that $|x - \xi| < \delta_\varepsilon$. It is no loss of generality to assume that $x - \delta_\varepsilon \le \xi \le x + \delta_\varepsilon$ is contained in $\mathscr{X}$. Thus for $\delta = \delta_\varepsilon$, we obtain from Property E of Theorem 3.2.1 that

$$|J_2(t)| \le \int_{x-\delta_\varepsilon}^{x+\delta_\varepsilon} K(x-\xi, t)|f(\xi) - f(x)| \, d\xi \le \frac{\varepsilon}{3} \int_{x-\delta_\varepsilon}^{x+\delta_\varepsilon} K(x-\xi, t) \, d\xi \le \frac{\varepsilon}{3}. \tag{3.4.8}$$

Considering $J_3(t)$, we see that

$$|J_3(t)| \le 2C_1 \frac{1}{\sqrt{4\pi t}} \int_{x+\delta_\varepsilon}^{\infty} \exp\left\{-\frac{(x-\xi)^2}{4t}\right\} \cdot \exp\{C_2 \xi^2\} \, d\xi. \tag{3.4.9}$$

From the transformation $\rho = (x - \xi)/2\sqrt{t}$, we obtain

$$|J_3(t)| \le 2C_1 \int_{-\infty}^{-\delta_\epsilon/(2\sqrt{t})} \exp\left\{ -\rho^2 + C_2 \left(x - 2\sqrt{t}\,\rho\right)^2 \right\} d\rho. \quad (3.4.10)$$

Since we know that

$$C_2\left(x - 2\sqrt{t}\,\rho\right)^2 \le 2C_2 x^2 + 8C_2 t\rho^2 \quad (3.4.11)$$

follows from $2ab \le a^2 + b^2$, we see that

$$|J_3(t)| \le 2C_1 \exp\left\{ 2C_2 x^2 \right\} \int_{-\infty}^{-\delta_\epsilon/(2\sqrt{t})} \exp\left\{ (8C_2 t - 1)\rho^2 \right\} d\rho. \quad (3.4.12)$$

Consequently for $0 < t < (16C_2)^{-1}$,

$$|J_3(t)| \le C \int_{-\infty}^{-\delta_\epsilon/(2\sqrt{t})} \exp\left\{ -\tfrac{1}{2}\rho^2 \right\} d\rho, \quad (3.4.13)$$

where

$$C = 2C_1 \exp\left\{ 2C_2 A^2 \right\} \quad (3.4.14)$$

provided that $|x| < A$. The convergence of the integral

$$\int_{-\infty}^{0} \exp\left\{ -\tfrac{1}{2}\rho^2 \right\} d\rho \quad (3.4.15)$$

implies the existence of a τ_ϵ such that for all $0 < t < \tau_\epsilon$,

$$|J_3(t)| < \frac{\epsilon}{3} \quad (3.4.16)$$

holds uniformly for all $|x| < A$. By a similar argument, there exists a τ_ϵ such that for all $0 < t < \tau_\epsilon$,

$$|J_1(t)| < \frac{\epsilon}{3} \quad (3.4.17)$$

holds uniformly for all $|x| < A$.

By selecting A so that the compact $\mathscr{X}$ is contained in $|x| < A$, we can combine (3.4.8), (3.4.16), and (3.4.17) and see that for each $\epsilon > 0$ there exists a $\tau_\epsilon > 0$ such that for $0 < t < \tau_\epsilon$,

$$|u(x, t) - f(x)| < \epsilon \quad (3.4.18)$$

holds for all x throughout the compact set $\mathscr{X}$.   □

As corollaries of the argument for Lemma 3.4.1, we obtain the following results.

Lemma 3.4.2. *For piecewise continuous f that satisfies $|f(x)| \le C_1 \exp\{C_2 x^2\}$, where C_1 and C_2 are positive constants,*

$$\lim_{t \downarrow 0} u(x, t) = f(x) \quad (3.4.19)$$

is uniform for all x in compact subsets of intervals of continuity of f.

Proof. The proof is left to the reader.   □

Lemma 3.4.3. *For any integrable function f that satisfies* $|f(x)| \leq C_1 \exp\{C_2 x^2\}$, *where* C_1 *and* C_2 *are positive constants,*

$$\lim_{t \downarrow 0} u(x, t) = f(x) \qquad (3.4.20)$$

at the points x of continuity of f.

Proof. The proof is left to the reader. ◻

The uniform convergence of $u(x, t)$ to $f(x)$ on compact subsets of intervals of continuity of f coupled with the continuity of f produces the two-dimensional piecewise continuity of u at $t = 0$. To see this, we consider x_1 in a compact subset of an interval of continuity of f and

$$|u(x_2, t) - f(x_1)| \leq |u(x_2, t) - f(x_2)| + |f(x_1) - f(x_2)|, \qquad (3.4.21)$$

where x_2 is another point in the compact subset. The first term on the left can be made small through a restriction of t sufficiently small while the second term can be made small by restricting x_2 to a neighborhood sufficiently close to x_1. Hence, we see that for each $\varepsilon > 0$ there exists a $\delta_\varepsilon(x_1) > 0$ such that

$$|u(x_2, t) - f(x_1)| < \varepsilon \qquad (3.4.22)$$

for all $t > 0$ and x_2 that satisfy $(x_1 - x_2)^2 + t^2 < (\delta_\varepsilon(x_1))^2$. Hence, u is piecewise two-dimensionally continuous at $t = 0$ and has the specified data f as its limit.

3.5. AN EXISTENCE THEOREM

We summarize the results of the analysis of Secs. 3.3 and 3.4 in the following statement.

Theorem 3.5.1 (Existence). *For all piecewise-continuous f that satisfy*

$$|f(x)| \leq C_1 \exp\{C_2 |x|^{1+\alpha}\}, \qquad 0 \leq \alpha < 1, \qquad (3.5.1)$$

where C_1 *and* C_2 *are positive constants,*

$$u(x, t) = \int_{-\infty}^{\infty} K(x - \xi, t) f(\xi), \qquad t > 0, \qquad (3.5.2)$$

where K is defined by (3.2.1), *is a solution of the initial-value problem*

$$u_t = u_{xx}, \qquad -\infty < x < \infty, \qquad 0 < t,$$
$$u(x, 0) = f(x), \qquad -\infty < x < \infty. \qquad (3.5.3)$$

For all piecewise-continuous f that are asymptotic to $C_1 \exp\{C_2 x^2\}$ *at* $|x| = \infty$, *u is defined only for* $0 < t < (4C_2)^{-1}$ *and satisfies the initial-value problem*

$$u_t = u_{xx}, \qquad -\infty < x < \infty, \qquad 0 < t < (4C_2)^{-1},$$
$$u(x, 0) = f(x), \qquad -\infty < x < \infty. \qquad (3.5.4)$$

Proof. See Secs. 3.2, 3.3, and 3.4. ◻

3.6. UNIQUENESS

Recalling the estimate (3.4.11), we see that, for all piecewise-continuous f that satisfy $|f(x)| \leq C_1 \exp\{C_2 x^2\}$, an estimate similar to that of (3.4.12) can be employed to show that, for $0 < t < (16C_2)^{-1}$,

$$|u(x,t)| \leq C_1\{\exp 2C_2 x^2\} \pi^{-1/2} \int_{-\infty}^{\infty} \exp\{-\tfrac{1}{2}\rho^2\} d\rho$$

$$\leq \sqrt{2}\, C_1 \exp\{2C_2 x^2\}. \tag{3.6.1}$$

Hence, we shall begin our study of uniqueness by showing that there can be only one solution of the initial-value problem that satisfies the growth condition of the form

$$|u(x,t)| \leq C_1 \exp\{C_2 x^2\} \tag{3.6.2}$$

for $t > 0$.

Suppose that there are two solutions of the initial-value problem that satisfy (3.6.2). Let v denote their difference. Then we can assume that v satisfies

$$\begin{cases} v_t = v_{xx}, & -\infty < x < \infty, \quad 0 < t < T, \\ v(x,0) = 0, & -\infty < x < \infty, \\ |v(x,t)| \leq C_1 \exp\{C_2 x^2\}, & -\infty < x < \infty, \quad 0 < t < T, \end{cases} \tag{3.6.3}$$

where we have relabeled $2C_1$ as C_1. We shall show that $v = 0$.

We obtain a particular solution of $u_t = u_{xx}$ for comparison purposes from $K(x,t)$ via a multiplication by a constant, a translation, and an affine transformation. Set

$$W(x,t) = \sqrt{4\pi}\, K(\alpha x, 1 + \alpha^2 t), \tag{3.6.4}$$

where $\alpha = i\sqrt{4C_3}$ and $C_3 > 0$, $C_3 > C_2$, and $C_3 > (4T)^{-1}$. We see that

$$W(x,t) = (1 - 4C_3 t)^{-1/2} \exp\{C_3 x^2 (1 - 4C_3 t)^{-1}\} \tag{3.6.5}$$

is a solution of

$$W_t = W_{xx}, \quad -\infty < x < \infty, \quad 0 < t < (4C_3)^{-1} \tag{3.6.6}$$

$$W(x,0) = \exp\{C_3 x^2\}, \quad -\infty < x < \infty.$$

We select a point (x_1, t_1) with $0 < t_1 < (4C_3)^{-1}$ and consider the rectangle

$$R = \{(x,t) | -A \leq x \leq A, \quad 0 \leq t \leq (4C_3)^{-1}\}, \tag{3.6.7}$$

where A is any positive number such that $A > |x_1|$. We compare the functions $\pm v$ with

$$z = C_1 \exp\{(C_2 - C_3)A^2\} W \tag{3.6.8}$$

in R. For $t = 0$,

$$0 = |v(x,0)| \le z(x,0) = C_1 \exp\{(C_2 - C_3)A^2\} \cdot \exp\{C_3 x^2\}. \quad (3.6.9)$$

On the sides $x = \pm A$ of R,

$$|v(\pm A, t)| \le C_1 \exp\{C_2 A^2\}. \quad (3.6.10)$$

However, for $0 < t < (4C_3)^{-1}$,

$$z(\pm A, t) = C_1(1 - 4C_3 t)^{-1/2} \exp\{C_2 A^2\} \exp\{C_3 A^2 (1 - 4C_3 t)^{-1} - C_3 A^2\}. \quad (3.6.11)$$

Consequently,

$$|v(\pm A, t)| \le z(\pm A, t) \quad (3.6.12)$$

Applying the Extended Comparison Theorem 1.6.5, we see that

$$-z(x_1, t_1) \le v(x_1, t_1) \le z(x_1, t_1). \quad (3.6.13)$$

Hence, for all $A > 0$,

$$|v(x_1, t_1)| \le C_1 \exp\{(C_2 - C_3)A^2\} W(x_1, t_1), \quad (3.6.14)$$

Since $C_3 > C_2$, we can show that $|v(x_1, t_1)|$ is less than any positive number by selecting A sufficiently large. Thus,

$$v(x_1, t_1) = 0. \quad (3.6.15)$$

As (x_1, t_1) was any point such that $0 < t_1 < (4C_3)^{-1}$, it follows that $v \equiv 0$ for $-\infty < x < \infty$, $0 < t < (4C_3)^{-1}$. However, this implies that $v \equiv 0$ for $-\infty < x < \infty$ and $0 < t < T$. Otherwise, there would be a least time T_1, $(4C_3)^{-1} \le T_1 < T$, such that on a sequence $\{(x_n, t_n)\}$ with $t_n \downarrow T_1$, $v(x_n, t_n) \ne 0$. By repeating the above argument for the initial time $T_1 - \varepsilon$, where $\varepsilon > 0$ and sufficiently small, we obtain the contradiction

$$0 = v(x_n, t_n) \ne 0 \quad (3.6.16)$$

for all n sufficiently large. Hence, we have demonstrated the following result.

Theorem 3.6.1 (Uniqueness). *The solution u of Theorem 3.5.1 is unique within the class of solutions v of the initial-value problem that admit a finite number of bounded discontinuities on $t = 0$ and that satisfy a growth condition of the form*

$$|v(x, t)| \le C_1 \exp\{C_2 x^2\}, \quad (3.6.17)$$

where C_1 and C_2 are positive constants.

Proof. See the analysis preceding the theorem. $\qquad\square$

3.7. NONUNIQUENESS

We give an example here of a nontrivial solution to the initial-value problem

$$\begin{cases} u_t = u_{xx}, & -\infty < x < \infty, \quad 0 < t, \\ u(x,0) = 0, & -\infty < x < \infty. \end{cases} \tag{3.7.1}$$

The example here consists of the series

$$u(x,t) = \sum_{n=0}^{\infty} f^{(n)}(t) \frac{x^{2n}}{(2n)!}, \tag{3.7.2}$$

where

$$f(t) = \begin{cases} \exp\{-t^{-2}\}, & t > 0, \\ 0, & t = 0. \end{cases} \tag{3.7.3}$$

From Sec. 2.4 and Theorem 2.3.1, we see that the series converges and can be differentiated term by term. Hence, $u_t = u_{xx}$ and $u(x,0) \equiv 0$ since f and all of its derivatives vanish at $t = 0$. We note that, in view of Theorem 3.6.1, it is clear that u has a growth rate exceeding $C_1 \exp\{C_2 x^2\}$.

EXERCISES

3.1. Show that

$$\int_{-\infty}^{\infty} \exp\{-\rho^2\} \, d\rho = \pi^{1/2}.$$

Hint: Note that

$$\left[\int_{-\infty}^{\infty} \exp\{-\rho^2\} \, d\rho \right]^2 = \int_0^{\infty} \int_0^{2\pi} \exp\{-r^2\} r \, d\theta \, dr.$$

3.2. Show that

$$\frac{1}{2\pi} \int_{-\infty}^{\infty} \exp\{i\alpha(x-\zeta) - \alpha^2 t\} \, d\alpha = K(x-\zeta, t).$$

Hint: First, show that

$$\text{Im}\left\{ \frac{1}{2\pi} \int_{-\infty}^{\infty} \exp\{i\alpha(x-\zeta) - \alpha^2 t\} \, d\alpha \right\} = 0.$$

Next, show that

$$\exp\{i\alpha(x-\zeta) - \alpha^2 t\}$$

$$= \exp\left\{ -\frac{(x-\zeta)^2}{4t} \right\} \exp\left\{ -\left(\alpha t^{1/2} - i\frac{(x-\zeta)}{2t^{1/2}} \right)^2 \right\}.$$

Use the transformation $\eta = \alpha t^{1/2}$ to show that

$$\frac{1}{2\pi}\int_{-\infty}^{\infty}\exp\{i\alpha(x-\zeta)-\alpha^2 t\}\,d\alpha$$

$$=\pi^{-1/2}K(x-\zeta,t)\int_{-\infty}^{\infty}\exp\left\{-\left(\eta-i\frac{(x-\zeta)}{2t^{1/2}}\right)^2\right\}\,d\eta.$$

Next, integrate $\exp\{-z^2\}$ around the boundary of the finite rectangle formed by the lines $\operatorname{Im} z = 0$, $\operatorname{Im} z = -(x-\zeta)/(2t^{1/2})$, and $\operatorname{Re} z = \pm C$. Use Cauchy's Theorem. Let C tend to infinity to obtain

$$\int_{-\infty}^{\infty}\exp\left\{-\left(\eta-i\frac{(x-\zeta)}{2t^{1/2}}\right)^2\right\}\,d\eta=\int_{-\infty}^{\infty}\exp\{-\eta^2\}\,d\eta.$$

3.3. Show that the bounded solution of

$$v_t = v_{xx}, \qquad 0 < x < \infty, \qquad 0 < t,$$
$$v(0,t) = 0, \qquad 0 < t, \qquad\qquad\qquad \text{(A)}$$
$$v(x,0) = f(x), \qquad 0 < x < \infty,$$

is given by

$$v(x,t) = \int_0^{\infty} G(x,\zeta,t)f(\zeta)\,d\zeta, \qquad\qquad \text{(B)}$$

where

$$G(x,\zeta,t) = K(x-\zeta,t) - K(x+\zeta,t), \qquad t > 0. \qquad \text{(C)}$$

Hint: Make an odd extension of f to $-\infty < x < 0$ and solve the initial-value problem for the extended f.

3.4. Show that the bounded solution of

$$v_t = v_{xx}, \qquad 0 < x < \infty, \qquad 0 < t,$$
$$v_x(0,t) = 0, \qquad 0 < t, \qquad\qquad\qquad \text{(D)}$$
$$v(x,0) = f(x), \qquad 0 < x < \infty,$$

is given by

$$v(x,t) = \int_0^{\infty} N(x,\zeta,t)f(\zeta)\,d\zeta, \qquad\qquad \text{(E)}$$

where

$$N(x,\zeta,t) = K(x-\zeta,t) + K(x+\zeta,t), \qquad t > 0. \qquad \text{(F)}$$

Hint: Make an even extension of f to $-\infty < x < 0$ and solve the initial-value problem for the extended f.

3.5. Show that the bounded solution of

$$v_t = v_{xx}, \qquad 0 < x < 1, \qquad 0 < t,$$
$$v(x,0) = f(x), \qquad 0 < x < 1, \qquad\qquad \text{(G)}$$
$$v(0,t) = v(1,t) = 0, \qquad 0 < t,$$

is given by

$$v(x,t) = \sum_{n=1}^{\infty} a_n \exp\{-n^2\pi^2 t\} \sin n\pi x, \qquad\qquad \text{(H)}$$

where

$$a_n = 2\int_0^1 f(x)\sin n\pi x \, dx. \qquad\qquad \text{(I)}$$

3.6. Show that the bounded solution of Eq. (G) is also given by

$$v(x,t)$$
$$= \int_0^1 \left\{ \sum_{m=-\infty}^{+\infty} [K(x-(\zeta+2m),t) - K(x+(\zeta-2m),t)] \right\} f(\zeta) \, d\zeta.$$
$$\text{(J)}$$

Hint: Extend f as an odd function to $-1 < x < 0$. Then extend f as periodic with period 2 to $-\infty < x < \infty$ and solve the initial-value problem for the periodic extension. Note that $v(x,t)$ given by Eq. (H) is also a solution to the initial-value problem for this periodic extension of f.

3.7. Using the results of Exercises 3.5 and 3.6, show that for all x, ζ, and $t > 0$,

$$2 \sum_{n=1}^{\infty} \exp\{-n^2\pi^2 t\} \sin n\pi x \sin n\pi\zeta$$

$$= \sum_{m=-\infty}^{\infty} [K(x-(\zeta+2m),t) - K(x+(\zeta-2m),t)].$$

$$\text{(K)}$$

Hint: Use Eq. (I) to rewrite Eq. (H) as an integral. Compare this integral to Eq. (J) for arbitrary $f(\zeta)$.

Chapter 4

The Initial-Boundary-Value Problem for the Quarter Plane with Temperature-Boundary Specification

4.1. INTRODUCTION

We study in this chapter the problem of determining u satisfying

$$u_t = u_{xx}, \quad 0 < x < \infty, \quad 0 < t,$$
$$u(0, t) = g(t), \quad 0 < t, \qquad (4.1.1)$$
$$u(x, 0) = f(x), \quad 0 < x < \infty,$$

where f and g are known piecewise-continuous functions. (See Fig. 4.1.1.) By the linearity of $u_t = u_{xx}$, (4.1.1) can be replaced by two problems. (See Fig. 4.1.2.) Set $u = v + w$, where v satisfies

$$v_t = v_{xx}, \quad 0 < x < \infty, \quad 0 < t,$$
$$v(0, t) = 0, \quad 0 < t, \qquad (4.1.2)$$
$$v(x, 0) = f(x), \quad 0 < x < \infty,$$

and w satisfies

$$w_t = w_{xx}, \quad 0 < x < \infty, \quad 0 < t,$$
$$w(0, t) = g(t), \quad 0 < t, \qquad (4.1.3)$$
$$w(x, 0) = 0, \quad 0 < x < \infty.$$

Clearly, $v(x, t)$ is given by Eq. (B) in the Exercise section of Chapter 3,

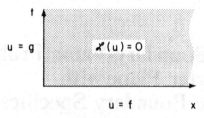

FIGURE 4.1.1

provided that f satisfies a growth condition that does not exceed one of the form $C_1\exp\{C_2 x^2\}$.

The problem (4.1.3) with $g \equiv 1$ has a simple solution that can be obtained via a Laplace transform. However, we have already displayed it in our study of the fundamental solution. Consider

$$W(x,t) = -2\int_0^t \frac{\partial K}{\partial x}(x,t-\tau)\,d\tau. \qquad (4.1.4)$$

From Sec. 1.5 we see that this convolution satisfies $W_t = W_{xx}$ for $x > 0$ and $t > 0$ since $\lim_{\tau \uparrow t}(\partial K/\partial x)(x, t - \tau) = 0$ for $x > 0$ and $t > 0$. From (3.2.8) in the proof of Property G of Theorem 3.2.1, we see that, by the change of variable $\rho = x/(2\sqrt{t-\tau})$,

$$W(x,t) = \frac{2}{\sqrt{\pi}} \int_{x/(2\sqrt{t})}^{\infty} \exp\{-\rho^2\}\,d\rho. \qquad (4.1.5)$$

Consequently,

$$\lim_{x \downarrow 0} W(x,t) = 1, \qquad t > 0, \qquad (4.1.6)$$

and

$$\lim_{t \downarrow 0} W(x,t) = 0, \qquad x > 0. \qquad (4.1.7)$$

We shall use the form (4.1.5) of W and the method of Duhamel to deduce the form of w. By setting

$$W(x,t) \equiv 0 \qquad (4.1.8)$$

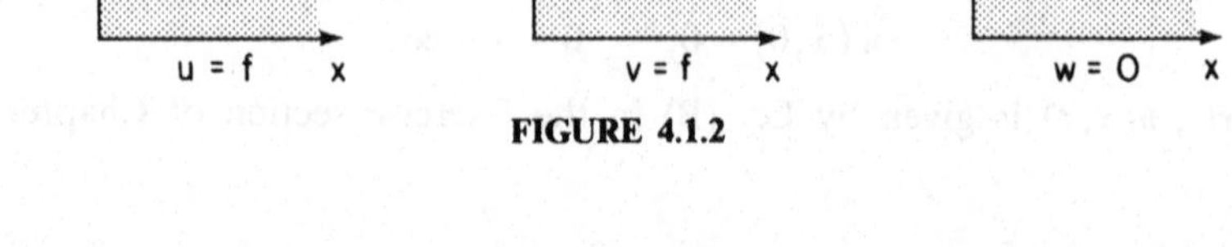

t ▲   u = g $\mathscr{L}(u) = 0$ =   v = 0 $\mathscr{L}(v) = 0$ +   w = g $\mathscr{L}(w) = 0$

u = f   x   v = f   x   w = 0   x

FIGURE 4.1.2

for $x > 0$ and $t \leq 0$, we see that for each τ, $W(x, t - \tau)$ satisfies

$$W_t = W_{xx}, \qquad 0 < x, \qquad \text{all } t,$$
$$W(0, t - \tau) = 1, \qquad \tau < t, \tag{4.1.9}$$
$$W(x, t - \tau) = 0, \qquad 0 < x, \qquad t \leq \tau.$$

For each positive integer n, consider

$$w_n(x, t) = \sum_{j=0}^{[nt]} \left\{ W(x, t - \tau_j) - W(x, t - \tau_{j+1}) \right\} \cdot g\left(\frac{\tau_{j+1} + \tau_j}{2} \right),$$

$$\tag{4.1.10}$$

where $\tau_j = jn^{-1}$. Clearly,

$$\frac{\partial w_n}{\partial t} = \frac{\partial^2 w_n}{\partial x^2}, \qquad w_n(x, 0) = 0 \qquad \text{for } x > 0,$$

and for almost all t,

$$w_n(0, t) = g\left(\frac{2[nt] + 1}{2n} \right). \tag{4.1.11}$$

Since

$$\lim_{n \to \infty} \frac{2[nt] + 1}{2n} = t, \tag{4.1.12}$$

the w_n are clearly approximations to w. As

$$W(x, t - \tau_j) - W(x, t - \tau_{j+1}) = \frac{\partial W}{\partial \tau}\left(x, t - \left(\frac{\tau_{j+1} + \tau_j}{2} \right) \right) \cdot \frac{1}{n} + o\left(\frac{1}{n} \right),$$

$$\tag{4.1.13}$$

the w_n are approximations to the Riemann integral

$$w(x, t) = \int_0^t \frac{\partial W}{\partial \tau}(x, t - \tau) g(\tau)\, d\tau = -2 \int_0^t \frac{\partial K}{\partial x}(x, t - \tau) g(\tau)\, d\tau.$$

$$\tag{4.1.14}$$

Remark. Before we study the properties of w, we should point out that the convolution expression can also be obtained from the use of the Laplace transform. Also, it should be emphasized again that the end result of a formalism can be studied independently of the formalism. The method of Duhamel is an interesting formalism with wide application.

4.2. PROPERTIES OF THE CONVOLUTION
$$w(x, t) = -2\int_0^t (\partial K/\partial x)(x, t - \tau) g(\tau)\, d\tau$$

For a fixed $x > 0$, we note that we can apply

$$\exp\{-\xi\} \leq p!\xi^{-p}, \qquad p = 1, 2, 3, \ldots, \tag{4.2.1}$$

to

$$\frac{\partial^{m+n}K}{\partial x^m \partial t^n}(x, t-\tau) = R_{m,n}\left(x, (t-\tau)^{1/2}\right) \cdot \exp\left\{-\frac{x^2}{4(t-\tau)}\right\}, \quad (4.2.2)$$

where $R_{m,n}$ is a rational function, to demonstrate that

$$\lim_{\tau \uparrow t} \frac{\partial^{m+n}K}{\partial x^m \partial t^n}(x, t-\tau) = 0 \quad (4.2.3)$$

and that $\partial^{m+n}K/(\partial x^m \partial t^n)$ are bounded respectively by constants that depend upon x, m, and n. Consequently, it is easy to specify the conditions for the existence of the convolution w. Namely, any integrable piecewise-continuous g will result in a well defined function w. At $\tau = 0$, we may permit g to grow as $\tau^{-\alpha}$, $0 < \alpha < 1$, since such a function is integrable.

From (4.2.2) and (4.2.3), we can apply Leibniz's rule any number of times to the integral. For $x > 0$ and $t > 0$, $w_t = w_{xx}$ follows from the fact that K_x satisfies the heat equation.

Clearly, for $x > 0$,

$$\lim_{t \downarrow 0} w(x, t) = 0. \quad (4.2.4)$$

In fact this limit is uniform for all $x \geq \delta > 0$. Applying (4.2.1) for $p = 2$ we see that

$$2\left|\frac{\partial K}{\partial x}(x, t-\tau)\right| \leq \frac{16\sqrt{t-\tau}}{\sqrt{\pi}\, x^3} \leq 16\pi^{-1/2}\delta^{-3}(t-\tau)^{1/2} \quad (4.2.5)$$

for all $x \geq \delta > 0$. Consequently, from the piecewise continuity and integrability of g, it follows that there exists a positive constant C such that for all $x \geq \delta > 0$ and $0 < t < T$,

$$|w(x, t)| \leq Ct^{(3-2\alpha)/2}, \quad 0 < \alpha < 1, \quad (4.2.6)$$

where here we have employed a growth of $\tau^{-\alpha}$ for g at $\tau = 0$. This implies the uniformity of the limit stated in (4.2.4) for all $x \geq \delta > 0$.

We turn now to the task of showing that, for $t > 0$,

$$\lim_{x \downarrow 0} w(x, t) = g(t) \quad (4.2.7)$$

when t is a point of continuity of g. We begin with $g(t)W(x, t)$ and note that, by (3.2.8) and Property F of Theorem 3.2.1,

$$\lim_{x \downarrow 0} g(t)W(x, t) = \lim_{x \downarrow 0} -2\int_0^t \frac{\partial K}{\partial x}(x, t-\tau)g(t)\, d\tau = g(t). \quad (4.2.8)$$

Moreover, from (4.1.10) it follows that, for $t \geq \gamma > 0$ and $x > 0$,

$$1 - \frac{2}{\sqrt{\pi}}\int_{x/(2\sqrt{t})}^{\infty} \exp\{-\rho^2\}\, d\rho = \frac{2}{\sqrt{\pi}}\int_0^{x/(2\sqrt{t})} \exp\{-\rho^2\}\, d\rho$$

$$\leq \frac{2}{\sqrt{\pi}}\int_0^{x/(2\sqrt{\gamma})} \exp\{-\rho^2\}\, d\rho$$

$$\leq \pi^{-1/2}\gamma^{-1/2}x. \quad (4.2.9)$$

Hence, for $t \geq \gamma > 0$, the limit in (4.2.8) is uniform. Thus, in order to show (4.2.7), it suffices to show that

$$\lim_{x \downarrow 0} -2 \int_0^t \frac{\partial K}{\partial x}(x, t - \tau)\{g(\tau) - g(t)\} \, d\tau = 0. \qquad (4.2.10)$$

Since g is continuous at t, for each $\varepsilon > 0$, there exists a $\delta_\varepsilon > 0$ such that $|g(t) - g(\tau)| < \varepsilon/2$ for all $|t - \tau| \leq \delta_\varepsilon$. We consider the integral

$$J = -2 \int_0^t \frac{\partial K}{\partial x}(x, t - \tau)\{g(\tau) - g(t)\} \, d\tau \qquad (4.2.11)$$

and write it in the form

$$J = J_1 + J_2, \qquad (4.2.12)$$

where

$$J_1 = -2 \int_0^{t - \delta_\varepsilon} \frac{\partial K}{\partial x}(x, t - \tau)\{g(\tau) - g(t)\} \, d\tau \qquad (4.2.13)$$

and

$$J_2 = -2 \int_{t - \delta_\varepsilon}^t \frac{\partial K}{\partial x}(x, t - \tau)\{g(\tau) - g(t)\} \, d\tau. \qquad (4.2.14)$$

As $-2(\partial K / \partial x) > 0$ for $x > 0$,

$$|J_2| \leq \frac{\varepsilon}{2}\left\{-2 \int_{t - \delta_\varepsilon}^t \frac{\partial K}{\partial x}(x, t - \tau) \, d\tau\right\} = \frac{\varepsilon}{\sqrt{\pi}} \int_{x/(2\sqrt{\delta_\varepsilon})}^\infty \exp\{-\rho^2\} \, d\rho < \frac{\varepsilon}{2}.$$
$$(4.2.15)$$

On the other hand,

$$|J_1| \leq 2^{-1} \pi^{-1/2} \delta_\varepsilon^{-3/2} x \int_0^t |g(\tau) - g(t)| \, d\tau$$

$$\leq 2^{-1} \pi^{-1/2} \delta_\varepsilon^{-3/2} x \int_0^T |g(\tau) - g(t)| \, d\tau, \qquad 0 < t < T. \quad (4.2.16)$$

Hence, for

$$0 < x < \pi^{1/2} \delta_\varepsilon^{3/2} \left\{\int_0^T |g(\tau) - g(t)| \, d\tau\right\}^{-1} \varepsilon, \qquad (4.2.17)$$

we see that

$$|J_1| < \frac{\varepsilon}{2}. \qquad (4.2.18)$$

Combining (4.2.15) with (4.2.18), we see that, given an $\varepsilon > 0$, there exists a $\Delta_\varepsilon > 0$ such that

$$|J| < \varepsilon \qquad (4.2.19)$$

for all x satisfying $0 < x < \Delta_\varepsilon$. Thus, we have shown the following result.

Lemma 4.2.1.   *At a point t of continuity of g,*

$$\lim_{x \downarrow 0} -2\int_0^t \frac{\partial K}{\partial x}(x, t-\tau)g(\tau)\,d\tau = g(t). \qquad (4.2.20)$$

Proof.   See the analysis preceding the statement of the lemma.   □

As a corollary of the preceding analysis, we have the following additional results.

Lemma 4.2.2.   *At a point t of continuity of g,*

$$\lim_{x \uparrow 0} -2\int_0^t \frac{\partial K}{\partial x}(x, t-\tau)g(\tau)\,d\tau = -g(t). \qquad (4.2.21)$$

Proof.   The analysis follows almost exactly as above.   □

Lemma 4.2.3.   *For piecewise-continuous g defined on $t > 0$,*

$$\lim_{x \downarrow 0} -2\int_0^t \frac{\partial K}{\partial x}(x, t-\tau)g(\tau)\,d\tau = g(t) \qquad (4.2.22)$$

uniformly for t belonging to a compact subset of an interval of continuity of g.

Proof.   For t in a closed, bounded subinterval of an interval of continuity of g, the δ_ε can be selected independent of t by the local uniform continuity of g.   □

4.3.   AN EXISTENCE THEOREM

We can summarize the results of the preceding sections in the following statement.

Theorem 4.3.1 (Existence).   *For piecewise-continuous f that satisfy*

$$|f(x)| \le C_1 \exp\{C_2|x|^{1+\alpha}\}, \qquad 0 \le \alpha < 1, \qquad (4.3.1)$$

where C_1 and C_2 are positive constants, and for piecewise-continuous g that satisfy

$$|g(t)| \le C_1 t^{-\alpha}, \qquad 0 < t < \varepsilon, \qquad (4.3.2)$$

the function

$$u(x,t) = -2\int_0^t \frac{\partial K}{\partial x}(x, t-\tau)g(\tau)\,d\tau + \int_0^\infty G(x, \xi, t)f(\xi)\,d\xi,$$

$$x > 0, \qquad t > 0, \qquad (4.3.3)$$

where

$$G(x, \xi, t) = K(x - \xi, t) - K(x + \xi, t), \qquad t > 0, \qquad (4.3.4)$$

and $K(x, t)$ is the fundamental solution defined by (3.1.8) or (3.2.1), is a *solution of the initial-boundary-value problem*

$$u_t = u_{xx}, \qquad 0 < x < \infty, \qquad 0 < t,$$
$$u(0, t) = g(t), \qquad 0 < t, \qquad (4.3.5)$$
$$u(x, 0) = f(x), \qquad 0 < x < \infty.$$

For piecewise-continuous f that are asymptotic to $C_1 \exp\{C_2 x^2\}$ at $x = +\infty$, u satisfies (4.3.5) only for $0 \le t < (4C_2)^{-1}$.

Proof. See Secs. 4.1 and 4.2. □

4.4. UNIQUENESS AND NONUNIQUENESS

We shall begin with nonuniqueness of the solution by giving two examples of nontrivial solutions to the problem

$$u_t = u_{xx}, \qquad 0 < x < \infty, \qquad 0 < t,$$
$$u(0, t) = 0, \qquad 0 < t, \qquad (4.4.1)$$
$$u(x, 0) = 0, \qquad 0 < x < \infty.$$

Our first example is

$$u = \frac{\partial K}{\partial x}. \qquad (4.4.2)$$

Clearly, $u_t = u_{xx}$, $\lim_{x \downarrow 0} u(x, t) = 0$ for $t > 0$, and $\lim_{t \downarrow 0} u(x, t) = 0$ for $x > 0$. Note that on the parabola $x = 2\sqrt{t}$, $t > 0$,

$$u(2\sqrt{t}, t) = Ct^{-1}, \qquad (4.4.3)$$

where

$$C = -(2e\pi^{1/2})^{-1}. \qquad (4.4.4)$$

Consequently, u is unbounded in each neighborhood of $x = t = 0$.

The second example consists of the series

$$u(x, t) = \sum_{n=0}^{\infty} f^{(n)}(t) \frac{x^{2n+1}}{(2n+1)!}, \qquad (4.4.5)$$

where

$$f(t) = \begin{cases} \exp\{-t^{-2}\}, & t > 0, \\ 0, & t \le 0. \end{cases} \qquad (4.4.6)$$

Trivial modifications of the discussion in Sec. 3.6 yield the following uniqueness result.

Theorem 4.5.1 (Uniqueness). *The solution u of Theorem 4.3.1 is unique within the class of solutions v that admit a finite number of bounded*

discontinuities on the parabolic boundary and that satisfy a growth condition of
the form

$$|v(x,t)| \leq C_1 \exp\{C_2 x^2\}, \tag{4.4.7}$$

where C_1 and C_2 are positive constants.

    Proof. The comparison function defined by Eq. (3.6.8) is positive
for $x = 0$ and $t > 0$. Hence, it can be used as a comparison function here
and the remainder of the argument is unchanged except that we select
$x_1 \geq 0$.   □

EXERCISES

4.1. Show that the problem of determining the unique bounded solution u
that satisfies

$$u_t = u_{xx}, \qquad 0 < x < \infty, \qquad 0 < t,$$

$$u(x,0) = f(x), \qquad 0 < x < \infty, \tag{A}$$

$$u_t(0,t) + \alpha(t)u_x(0,t) + \beta(t)u(0,t) = g(t), \qquad 0 < t,$$

and

$$|u(x,t)| \leq C_1 \exp\{C_2 x^2\}, \tag{B}$$

where C_i, $i = 1,2$, are positive constants, f is twice continuously
differentiable, and α and g are continuous, is equivalent to the
problem of determining the unique continuous solution φ to the
integral equation

$$\varphi(t) + \alpha(t)\int_0^\infty N(0,\xi,t)f'(\xi)\,d\xi - 2\alpha(t)\int_0^t K(0,t-\tau)\varphi(\tau)\,d\tau$$

$$+ \beta(t)f(0) + \beta(t)\int_0^t \varphi(\tau)\,d\tau = g(t), \qquad 0 < t, \tag{C}$$

where $K(x,t)$ is defined by Eq. (3.1.8) and $N(x,\xi,t)$ is defined by Eq.
(F) in the Exercise section of Chapter 3.

Hint: Define $u_t(0,t) = \varphi(t)$ and substitute $u(0,t) = f(0) + \int_0^t \varphi(\tau)\,d\tau$
into Eq. (4.3.3). The representation for u must be differentiated with
respect to x and an integration by parts performed after K_{xx} is
replaced by $-K_\tau$. Then, $u_x(0,t)$ is obtained by letting x tend to zero.
The derivation of Eq. (C) follows from the boundary condition in Eq.
(A), above.

4.2. Show that for continuous f, g, and s with $g(0) = \int_0^{s(0)} f(\xi)\, d\xi$, the solution u of

$$u_t = u_{xx}, \qquad 0 < x < \infty, \qquad 0 < t,$$
$$u(x,0) = f(x), \qquad 0 < x < \infty, \qquad \qquad \text{(D)}$$
$$\int_0^{s(t)} u(x,t)\, dx = g(t), \qquad 0 < t, \qquad 0 < s(t),$$

which satisfies a growth condition of the form

$$C_1 \exp\{ C_2 |x|^{1+\alpha} \}, \qquad 0 \le \alpha < 1,$$

where C_1 and C_2 are positive constants, has the representation

$$u(x,t) = \int_0^\infty G(x,\xi,t) f(\xi)\, d\xi - 2 \int_0^t \frac{\partial K}{\partial x}(x, t-\tau)\phi(\tau)\, d\tau, \quad \text{(E)}$$

where $K(x,t)$ is defined by (3.1.8) and $G(x,\xi,t)$ is defined by Eq. (C) in the Exercise section of Chapter 3, if and only if ϕ is a piecewise-continuous solution of

$$g(t) - \int_0^{s(t)} \left\{ \int_0^\infty G(x,\xi,t) f(\xi)\, d\xi \right\} dx$$
$$= 2 \int_0^t K(0, t-\tau)\phi(\tau)\, d\tau - 2\int_0^t K(s(t), t-\tau)\phi(\tau)\, d\tau. \quad \text{(F)}$$

Hint: Integrate Eq. (E) above, with respect to x, from 0 to $s(t)$, and apply Fubini's theorem.

Chapter 5

The Initial-Boundary-Value Problem for the Quarter Plane with Heat-Flux-Boundary Specification

5.1. EXISTENCE

The initial boundary value problem of the second kind is the problem of determining u satisfying

$$u_t = u_{xx}, \qquad 0 < x < \infty, \qquad 0 < t,$$
$$u_x(0, t) = g(t), \qquad 0 < t, \qquad (5.1.1)$$
$$u(x,0) = f(x), \qquad 0 < x < \infty,$$

where f and g are known continuous functions on the closures of their respective domains of definition. (See Fig. 5.1.1.) As we did in Sec. 4.1, we use linearity to replace (5.1.1) by two problems. (See Fig. 5.2.2.) Set $u = v + w$, where v satisfies

$$v_t = v_{xx}, \qquad 0 < x < \infty, \qquad 0 < t,$$
$$v_x(0, t) = 0, \qquad 0 < t, \qquad (5.1.2)$$
$$v(x,0) = f(x), \qquad 0 < x < \infty,$$

and w satisfies

$$w_t = w_{xx}, \qquad 0 < x < \infty, \qquad 0 < t,$$
$$w_x(0, t) = g(t), \qquad 0 < t, \qquad (5.1.3)$$
$$w(x,0) = 0, \qquad 0 < x < \infty.$$

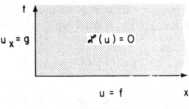

FIGURE 5.1.1

Theorem 5.1.1 (Existence).   *For all continuous g on $0 \le t$ and for all continuous f on $0 \le x$ that satisfy*

$$|f(x)| \le C_1 \exp\{C_2|x|^{1+\alpha}\}, \qquad 0 \le \alpha < 1, \qquad (5.1.4)$$

where C_1 and C_2 are positive constants,

$$u(x,t) = \int_0^\infty N(x, \xi, t) f(\xi)\, d\xi - 2\int_0^t K(x, t-\tau) g(\tau)\, d\tau,$$

$$x > 0, \qquad t > 0, \quad (5.1.5)$$

where $N(x, \xi, t)$ is defined by Eq. (F) of the exercise section in Chapter 3, and $K(x, t)$ is defined by (3.1.8), is a solution of the initial-boundary-value problem (5.1.1). For all continuous f on $0 \le x$ that are asymptotic to $C_1\exp\{C_2x^2\}$ at $x = +\infty$, u is defined and satisfies (5.1.10) only for $0 \le t \le (4C_2)^{-1}$.

   Proof.   The proof consists of straightforward modifications of the analysis preceding Theorem 4.3.1. The details are left as an exercise.   □

5.2.   UNIQUENESS

We demonstrate the following result.

   Theorem 5.2.1 (Uniqueness).   *The solution u of Theorem 5.1.1 is unique within the class of solutions v that satisfy a growth condition of the form*

$$|v(x,t)| \le C_1 \exp\{C_2 x^2\}, \qquad (5.2.1)$$

where C_1 and C_2 are positive constants.

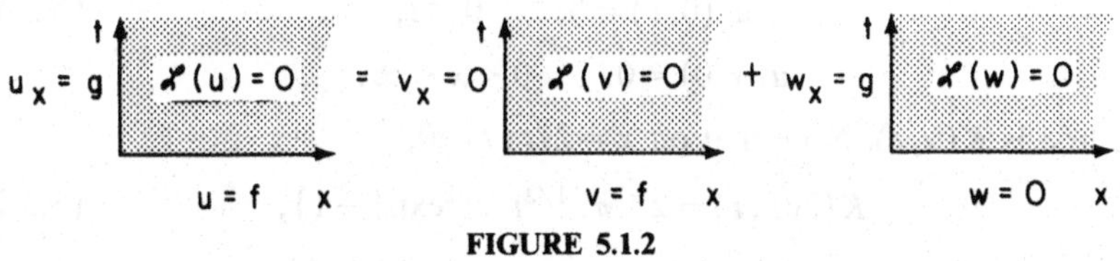

FIGURE 5.1.2

Proof. The comparison function z defined by (3.6.8) is an even function and hence satisfies $z_x(0, t) = 0$ as well as $z(x, 0) > 0$. For $\varepsilon > 0$, consider the function

$$\eta_\varepsilon(x, t) = z(x, t) + \varepsilon x - u(x, t), \qquad (5.2.2)$$

where u satisfies

$$u_t = u_{xx}, \qquad 0 < x < \infty, \qquad 0 < t,$$
$$u_x(0, t) = 0, \qquad 0 < t, \qquad (5.2.3)$$
$$u(x, 0) = 0, \qquad 0 < x < \infty,$$

and the growth condition (5.2.1). Clearly, η_ε satisfies the heat equation $\eta_\varepsilon(x, 0) > 0$, and $(\partial \eta_\varepsilon / \partial x)(0, t) = \varepsilon > 0$. Consequently, η_ε cannot assume its maximum on $x = 0$, $t > 0$, since otherwise $(\partial \eta_\varepsilon / \partial x)(0, t_0) \leq 0$, for some $t_0 > 0$. As $\eta_\varepsilon(A, t) > 0$ for $0 < t < (4C_3)^{-1}$ via the argument for Eqs. (3.6.10) through (3.6.12), it follows from the weak maximum principle that for $0 \leq x \leq A$ and $0 \leq t \leq (4C_3)^{-1}$,

$$u(x, t) \leq z(x, t) + \varepsilon x. \qquad (5.2.4)$$

Similarly, for $0 \leq x \leq A$ and $0 \leq t \leq (4C_3)^{-1}$,

$$-u(x, t) \leq z(x, t) + \varepsilon x. \qquad (5.2.5)$$

Selecting a point (x_1, t_1) with $x_1 > 0$ and $0 < t_1 < (4C_3)^{-1}$, from (5.2.4) and (5.2.5) we see that for all A sufficiently large and each $\varepsilon > 0$,

$$|u(x_1, t_1)| \leq z(x_1, t_1) + \varepsilon x_1. \qquad (5.2.6)$$

However, from (3.6.14) we see that $z(x_1, t_1)$ can be made as small as we wish by selecting A sufficiently large. Likewise, εx_1 can be made as small as we wish via the choice of ε. Thus, we see that

$$u(x_1, t_1) = 0. \qquad (5.2.7)$$

The remainder of the argument is a minor modification of that just preceding Theorem 3.6.1. $\qquad \square$

5.3. NONUNIQUENESS

As in Sec. 4.5, we can give here two examples of nontrivial solutions to the problem

$$u_t = u_{xx}, \qquad 0 < x < \infty, \qquad 0 < t,$$
$$u_x(0, t) = 0, \qquad 0 < t, \qquad (5.3.1)$$
$$u(x, 0) = 0, \qquad 0 < x < \infty.$$

The first is $K(x, t)$. Note that on $x = 2\sqrt{t}$, $t > 0$,

$$K(2\sqrt{t}, t) = 2^{-1} \pi^{-1/2} t^{-1/2} \exp\{-1\}, \qquad (5.3.2)$$

which becomes unbounded as t becomes small. The second has been given in Sec. 2.4. See Eq. (2.4.14).

EXERCISES

5.1. Show that the problem of determining the unique solution u that satisfies

$$u_t = u_{xx}, \qquad 0 < x < \infty, \qquad 0 < t,$$
$$u(x,0) = f(x), \qquad 0 < x < \infty, \tag{A}$$
$$u_x(0,t) + \alpha(t)u(0,t) = g(t), \qquad 0 < t,$$

and

$$|u(x,t)| \le C_1 \exp\{C_2 x^2\}, \tag{B}$$

where C_i, $i = 1, 2$, are positive constants, and where f, α, and g are piecewise-continuous functions, is equivalent to the problem of determining the unique piecewise solution φ to the integral equation

$$\varphi(t) + \alpha(t) \int_0^\infty N(0, \xi, t) f(\xi) \, d\xi$$
$$- 2\alpha(t) \int_0^t K(0, t - \tau) \varphi(\tau) \, d\tau = g(t), \qquad 0 < t, \tag{C}$$

where $K(x, t)$ is defined by Eq. (3.1.8) and $N(x, \xi, t)$ is defined by Eq. (F) of the Exercise section of Chapter 3.

5.2. Derive the equivalent integral equation for the boundary condition

$$u_x(0,t) + \alpha(t)u(0,t) + \int_0^t F(t,\tau)u(0,\tau) \, d\tau = g(t), \qquad 0 < t. \tag{D}$$

5.3. Derive the equivalent integral equation for the boundary condition

$$u_x(0,t) = F(t, u(0,t)), \qquad 0 < t. \tag{E}$$

5.4. Prove Theorem 5.1.1.

Chapter 6

The Initial-Boundary-Value Problem for the Semi-Infinite Strip with Temperature-Boundary Specification and Heat-Flux-Boundary Specification

6.1. INTRODUCTION

By linearity the problem (see Fig. 6.1.1)

$$
\begin{aligned}
u_t &= u_{xx}, && 0 < x < 1, && 0 < t, \\
u(x,0) &= f(x), && 0 < x < 1, \\
u(0,t) &= g(t), && 0 < t, \\
u(1,t) &= h(t), && 0 < t,
\end{aligned}
\tag{6.1.1}
$$

where f, g, and h are piecewise-continuous functions, can be partitioned into three problems. (See Fig. 6.1.2.) This is accomplished by setting $u = v + w + z$, where v satisfies

$$
\begin{aligned}
v_t &= v_{xx}, && 0 < x < 1, && 0 < t, \\
v(x,0) &= f(x), && 0 < x < 1, \\
v(0,t) &= 0, && 0 < t, \\
v(1,t) &= 0, && 0 < t,
\end{aligned}
\tag{6.1.2}
$$

w satisfies

$$
\begin{aligned}
w_t &= w_{xx}, && 0 < x < 1, && 0 < t, \\
w(x,0) &= 0, && 0 < x < 1, \\
w(0,t) &= g(t), && 0 < t, \\
w(1,t) &= 0, && 0 < t,
\end{aligned}
\tag{6.1.3}
$$

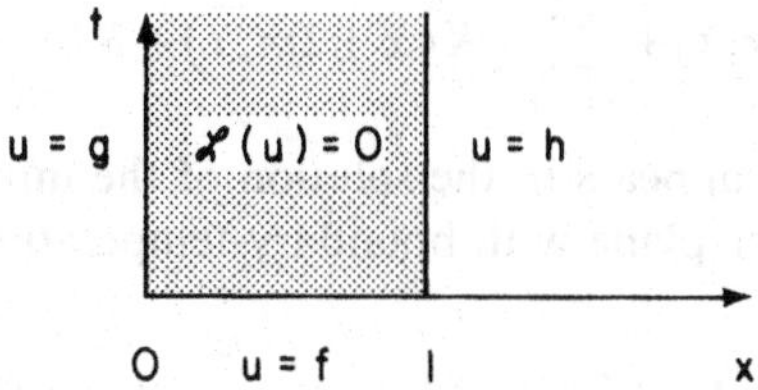

FIGURE 6.1.1

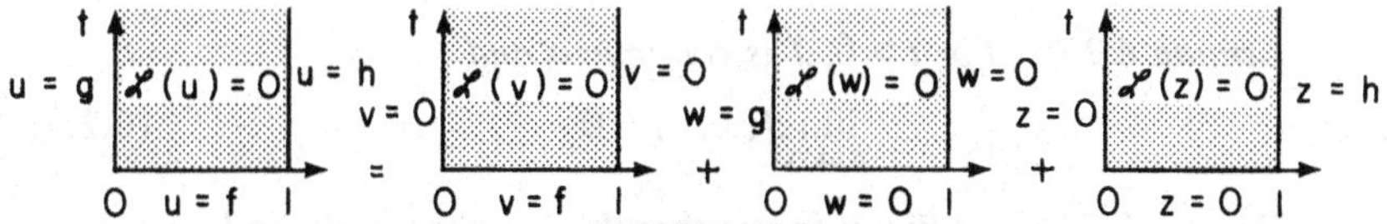

FIGURE 6.1.2

and z satisfies

$$z_t = z_{xx}, \qquad 0 < x < 1, \qquad 0 < t,$$
$$z(x,0) = 0, \qquad 0 < x < 1,$$
$$z(0,t) = 0, \qquad 0 < t,$$
$$z(1,t) = h(t), \qquad 0 < t.$$

(6.1.4)

From Exercises 3.3, 3.4, and 3.5, we see that for $0 < x < 1, 0 < t$,

$$v(x,t) = \int_0^1 \left\{ \sum_{m=-\infty}^{\infty} K(x-(\xi+2m),t) - K(x+(\xi-2m),t) \right\} f(\xi) \, d\xi.$$

(6.1.5)

The function

$$\theta(x,t) = \sum_{m=-\infty}^{\infty} K(x+2m,t), \qquad t > 0,$$

(6.1.6)

appears quite often in what follows and is called the *theta function*.

6.2. SOME PROPERTIES OF $\theta(x,t) = \sum_{m=-\infty}^{\infty} K(x+2m,t)$

First, from the uniform absolute convergence of the series for θ and its partial derivatives, it is clear that, for $t > 0$, $\theta(x,t) > 0$, the θ function is continuous and all of its partial derivatives are continuous. Next, it is easy to see that

$$\theta(x, t) = K(x, t) + \sum_{m=1}^{\infty} \{K(x+2m, t) + K(x-2m, t)\}, \quad (6.2.1)$$

Recalling that $\partial K/\partial x$ appears in the solution of the initial-boundary-value problem for the quarter plane with boundary-temperature specification, we consider

$$\frac{\partial \theta}{\partial x} = \frac{\partial K}{\partial x} + \sum_{m=1}^{\infty} \left\{ -\frac{(x+2m)}{2t} K(x+2m, t) - \frac{(x-2m)}{2t} K(x-2m, t) \right\}$$

$$= \frac{\partial K}{\partial x} + J(x, t). \quad (6.2.2)$$

Lemma 6.2.1. *For $t > 0$, J is continuous and*

$$\lim_{x \to 0} J(x, t) = 0. \quad (6.2.3)$$

Proof. It is not difficult to see that J is an absolutely and uniformly convergent series of continuous functions. Considering $x = 0$, we see that the mth term is

$$-\frac{m}{t} K(2m, t) + \frac{m}{t} K(-2m, t)\} = 0, \quad (6.2.4)$$

since $K(2m, t) = K(-2m, t)$. Consequently, $J(0, t) = 0$. □

From Lemma 6.2.1, we obtain the following results.

Lemma 6.2.2. *For $t > 0$,*

$$\lim_{x \downarrow 0} -2 \int_0^t \frac{\partial \theta}{\partial x}(x, t-\tau) g(\tau) \, d\tau = g(t) \quad (6.2.5)$$

at each point t of continuity of g.

Proof. The result (6.2.5) is a corollary of Lemma 6.2.1 and Lemma 4.2.1. □

Also, we can state the following result.

Lemma 6.2.3. *For $t > 0$ and piecewise-continuous g,*

$$\lim_{x \downarrow 0} -2 \int_0^t \frac{\partial \theta}{\partial x}(x, t-\tau) g(\tau) \, d\tau = g(t) \quad (6.2.6)$$

is uniform for t belonging to a compact subset of an interval of continuity of g.

Proof. This result is a corollary of Lemma 6.2.1, Lemma 6.2.2, and Lemma 4.2.3. □

Now that we have some information about the behavior of $\partial \theta/\partial x$ at $x = 0$, we consider $\partial \theta/\partial x$ at $x = 1$. From (6.2.2), it is not difficult to see that, for $t > 0$,

$$\frac{\partial \theta}{\partial x}(1,t) = \sum_{m=1}^{\infty} -\frac{(2m+1)}{2t} K(2m+1,t)$$

$$+ \sum_{m=1}^{\infty} \frac{2m-1}{2t} K(2m-1,t) - \frac{1}{2t} K(1,t)$$

$$= 0. \tag{6.2.7}$$

We can summarize this in the following statement.

Lemma 6.2.4. *For $t > 0$,*

$$\lim_{x \uparrow 1} \frac{\partial \theta}{\partial x}(x,t) = 0. \tag{6.2.8}$$

Proof. The result follows from the continuity of $\partial \theta / \partial x$ and (6.2.7).
□

Consequently, from Lebesgue's dominated-convergence theorem we have the following result.

Lemma 6.2.5. *For $t > 0$,*

$$\lim_{x \uparrow 1} -2 \int_0^t \frac{\partial \theta}{\partial x}(x,t-\tau) g(\tau)\, d\tau = 0 \tag{6.2.9}$$

for any Lebesgue-integrable g. Moreover, this limit is taken on uniformly with respect to t contained in compact sets.

Proof. The result (6.2.9) follows from (6.2.8) and the uniform continuity of $(\partial \theta / \partial x)(x,t)$ at $x = 1$ for $T \geq t \geq 0$. Note that the only singularity of $\partial \theta / \partial x$ occurs at $x = t = 0$. For the uniformity of the limit, note that

$$\left| 2 \int_0^t \frac{\partial \theta}{\partial x}(x,t-\tau) g(\tau)\, d\tau \right| \leq 2 \left\{ \sup_{0 \leq t \leq T} \left| \frac{\partial \theta}{\partial x}(x,t) \right| \right\} \int_0^T |g(\tau)|\, d\tau. \tag{6.2.10}$$

From the continuity of $\partial \theta / \partial x$ at $x = 1$, we see that the righthand side of (6.2.10) can be made as small as we like by selecting x sufficiently near 1. □

Since θ is a sum of solutions to the heat equation, it follows that θ, $\partial \theta / \partial x$, and

$$w(x,t) = -2 \int_0^t \frac{\partial \theta}{\partial x}(x,t-\tau) g(\tau)\, d\tau \tag{6.2.11}$$

are solutions to the heat equation. In particular, w is a solution for $x \neq 0$

and $t > 0$. Since $w(x, 0) = 0$, it follows from Lemma 6.2.3 and Lemma 6.2.5 that w is a solution of Problem (6.1.3).

It is not difficult to see from (6.2.2) and (6.2.7) that

$$\frac{\partial \theta}{\partial x}(-1, t) = 0. \tag{6.2.12}$$

Consequently, by an analysis similar to that above, we obtain for problem (6.1.4) the solution

$$z(x, t) = 2 \int_0^t \frac{\partial \theta}{\partial x}(x - 1, t - \tau) h(\tau) \, d\tau, \tag{6.2.13}$$

which is obtained from $-w$ for $-1 < x < 0$ by replacing g by h and translating by one to the right.

6.3. AN EXISTENCE AND UNIQUENESS THEOREM

We summarize the results of the preceding sections in the following statement.

Theorem 6.3.1 (Existence and Uniqueness). *For piecewise-continuous f and for piecewise-continuous g and h, the function*

$$u(x, t) = \int_0^1 \{\theta(x - \xi, t) - \theta(x + \xi, t)\} f(\xi) \, d\xi$$

$$- 2 \int_0^t \frac{\partial \theta}{\partial x}(x, t - \tau) g(\tau) \, d\tau + 2 \int_0^t \frac{\partial \theta}{\partial x}(x - 1, t - \tau) h(\tau) \, d\tau,$$

$$\tag{6.3.1}$$

where $\theta(x, t)$ is defined by (6.1.6) and the fundamental solution $K(x, t)$ is defined by (3.2.1), is a solution of

$$\begin{aligned}
u_t &= u_{xx}, & 0 < x < 1, & \quad 0 < t, \\
u(x, 0) &= f(x), & 0 < x < 1, & \\
u(0, t) &= g(t), & 0 < t, & \\
u(1, t) &= h(t), & 0 < t. &
\end{aligned} \tag{6.3.2}$$

Moreover, u is the only bounded solution of (6.3.2).

Proof. That u satisfies (6.3.2) is obvious from the analysis in Secs. 6.1 and 6.2. The uniqueness follows immediately from an application of the Extended Uniqueness Theorem 1.6.6. □

6.4. SPECIFICATION OF HEAT FLUX ON THE BOUNDARY

As a corollary of the preceding analysis, we can state the following result.

Theorem 6.4.1 (Existence and Uniqueness). *For piecewise-continuous f and for piecewise-continuous g and h, the function*

$$u(x,t) = \sum_{n=0}^{\infty} b_n \exp\{-n^2\pi^2 t\}\cos n\pi x$$
$$-2\int_0^t \theta(x,t-\tau)g(\tau)\,d\tau + 2\int_0^t \theta(x-1,t-\tau)h(\tau)\,d\tau,$$

$$(6.4.1)$$

where

$$b_n = 2\int_0^1 f(x)\cos n\pi x\,dx, \qquad n = 1,2,\ldots, \qquad (6.4.2)$$

$$b_0 = \int_0^1 f(x)\,dx, \qquad (6.4.3)$$

$\theta(x,t)$ *is defined by* (6.1.6) *and the fundamental solution* $K(x,t)$ *is defined by* (3.2.1), *is a solution of*

$$u_t = u_{xx}, \qquad 0 < x < 1, \qquad 0 < t,$$
$$u(x,0) = f(x), \qquad 0 < x < 1,$$
$$u_x(0,t) = g(t), \qquad 0 < t, \qquad (6.4.4)$$
$$u_x(1,t) = h(t), \qquad 0 < t.$$

Moreover, u is the only solution* of (6.4.4).

Remark. Recalling Exercises 3.3, 3.4, and 3.5 and noting that f here is extended as an even function, we can replace the series in (6.4.1) with

$$\int_0^1 \{\theta(x-\xi,t) + \theta(x+\xi,t)\}f(\xi)\,d\xi. \qquad (6.4.5)$$

Proof. We need concern ourselves here only with uniqueness. We give here a proof based upon integrals of squares, which is the so-called energy method. let z denote a solution to

$$z_t = z_{xx}, \qquad 0 < x < 1, \qquad 0 < t,$$
$$z(x,0) = z_x(0,t) = z_x(1,t) = 0. \qquad (6.4.6)$$

Since

$$zz_t = \tfrac{1}{2}(z^2)_t, \qquad (6.4.7)$$

---

*The class of solutions here is different from that of Definition 1.5.4. Here, as seen in the proof, we require square integrability of u_t, u_x, u_{xx} in order to present the energy method of proof.

and

$$zz_{xx} = (zz_x)_x - (z_x)^2, \tag{6.4.8}$$

we have for every $t > 0$,

$$0 = \int_0^t \int_0^1 \{ zz_t - zz_{xx} \} \, dx \, d\tau$$

$$= \frac{1}{2} \int_0^1 \{ z(x, t) \}^2 \, dx + \int_0^t \int_0^1 (z_x)^2 \, dx \, d\tau. \tag{6.4.9}$$

Consequently, $z \equiv 0$. As the difference of any two solutions of (6.4.4) is such a z, we conclude that the solution u given by (6.4.1) is unique.   □

6.5. REDUCTION TO A SYSTEM OF INTEGRAL EQUATIONS

A powerful technique for deducing the existence of the solution of a large number of problems is that of deriving an equivalent system of integral equations to which a single existence and uniqueness theory can be applied. We begin our study of these problems with a detailed account of the derivation of an equivalent system of integral equations for the initial-boundary-value problem (6.1.1).

We first make a smooth extension of f outside of $0 \le x \le 1$ so that the extended f is bounded and has compact support. The solution u is now assumed to possess the form

$$u(x, t) = \int_{-\infty}^{\infty} K(x - \xi, t) f(\xi) \, d\xi$$

$$- 2 \int_0^t \frac{\partial K}{\partial x}(x, t - \tau) \phi_1(\tau) \, d\tau + 2 \int_0^t \frac{\partial K}{\partial x}(x - 1, t - \tau) \phi_2(\tau) \, d\tau,$$

$$\tag{6.5.1}$$

where ϕ_1 and ϕ_2 are unknown piecewise-continuous functions that are to be determined. Note that the initial condition and the differential equation are satisfied. Also, note that the function

$$v(x, t) = \int_{-\infty}^{\infty} K(x - \xi, t) f(\xi) \, d\xi \tag{6.5.2}$$

is continuous at $x = 0$ and $x = 1$ when $t = 0$, which explains the purpose of the smooth extension of f. The system of integral equations is derived from the boundary conditions, Lemma 4.2.3 and Lemma 4.2.4. Allowing x to tend to zero and one, we obtain

$$g(t) = v(0, t) + \phi_1(t) + 2 \int_0^t \frac{\partial K}{\partial x}(-1, t - \tau) \phi_2(\tau) \, d\tau,$$

$$h(t) = v(1, t) + \phi_2(t) - 2 \int_0^t \frac{\partial K}{\partial x}(1, t - \tau) \phi_1(\tau) \, d\tau. \tag{6.5.3}$$

Consequently, if u possesses the form (6.5.1), then ϕ_1 and ϕ_2 must satisfy

(6.5.3). On the other hand, if the piecewise-continuous ϕ_1 and ϕ_2 satisfy (6.5.3) for all but a finite number of t values, then one can consider the form (6.5.1), which satisfies the differential equation and the initial condition. From Lemma 4.2.3 we see that the

$$\lim_{x \downarrow 0} u(x, t) = v(C, t) + \phi_1(t) + 2 \int_0^t \frac{\partial K}{\partial x}(-1, t - \tau)\phi_2(\tau)\, d\tau \quad (6.5.4)$$

is uniform for t belonging to compact subsets of the intervals of continuity of ϕ_1. Hence, u is two-dimensionally continuous at such points and from (6.5.3), $u(0, t) = g(t)$. Clearly, from (6.5.3), g and ϕ_1 have the same points of discontinuity since $v(0, t)$ and $2\int_0^t (\partial K/\partial x)(-1, t - \tau)\phi_2(\tau)\, d\tau$ are continuous. Likewise, u assumes the value h as $x \uparrow 1$ and is two-dimensionally continuous except for a finite number of discontinuities on $x = 1$, $t \geq 0$.

We can gather these results into the following statement.

Theorem 6.5.1 (Equivalence). *For piecewide-continuous f, g, and h, the solution u of*

$$\begin{aligned}
u_t &= u_{xx}, && 0 < x < 1, && 0 < t, \\
u(x, 0) &= f(x), && 0 < x < 1, \\
u(0, t) &= g(t), && 0 < t, \\
u(1, t) &= h(t), && 0 < t
\end{aligned} \qquad (6.5.5)$$

has the form

$$u(x, t) = v(x, t) - 2 \int_0^t \frac{\partial K}{\partial x}(x, t - \tau)\phi_1(\tau)\, d\tau$$

$$+ 2 \int_0^t \frac{\partial K}{\partial x}(x - 1, t - \tau)\phi_2(\tau)\, d\tau, \qquad (6.5.6)$$

where

$$v(x, t) = \int_{-\infty}^{\infty} K(x - \xi, t) f(\xi)\, d\xi, \qquad (6.5.7)$$

$K(x, t)$ *is defined by (3.2.1), and f here is a smooth, bounded extension of the f above, if and only if ϕ_1 and ϕ_2 are piecewise continuous solutions of*

$$g(t) = v(0, t) + \phi_1(t) + 2 \int_0^t \frac{\partial K}{\partial x}(-1, t - \tau)\phi_2(\tau)\, d\tau,$$

$$h(t) = v(1, t) + \phi_2(t) - 2 \int_0^t \frac{\partial K}{\partial x}(1, t - \tau)\phi_1(\tau)\, d\tau. \qquad (6.5.8)$$

Proof. See the analysis preceding the theorem statement. □

Remark. The uniqueness of the solution of the system (6.5.8) does not imply the uniqueness of the solution u, since we assumed the form here. The uniqueness must be obtained independently. However, if it is known that a given representation of a solution is unique, then the equivalence of uniqueness is valid. Here, the uniqueness of the representation of the solution means that $u(x, t)$ is given explicitly in terms of $u(x, 0)$, $u(0, t)$ (or $u_x(0, t)$), and $u(1, t)$ (or $u_x(1, t)$). The argument goes like this: If the initial-boundary-value problem has a unique solution, then the integral equations must possess a unique solution. Otherwise, the uniqueness of the solution of the initial-boundary-value problem is contradicted. Likewise, if the integral equations possess a unique solution, then the initial-boundary-value problem must possess a unique solution. Otherwise, the uniqueness of the solution of the integral equations is contradicted.

NOTES ON CHAPTERS 3–6

The Fourier-transform method applied to the initial-value problem is common to most texts. See Weinberger [6]. The Appell transformation can be found in Appell [1]. The method of Duhamel can be found in Weber [5]. The remaining material in these chapters is standard. For example, see Courant and Hilbert [2, 3], Goursat [4], Weinberger [6], Widder [7], Young [8], and Zachmanoglou and Thoe [9].

REFERENCES FOR CHAPTERS 3–6

1. Appell, P. Sur l'équation $(\partial^2 z / \partial x^2) - (\partial z / \partial y) = 0$ et la théorie de la chaleur, *J. Math. Pures Appl.* **8** (1892), 187–216.
2. Courant, R., and Hilbert, D., *Methods of Mathematical Physics*, Vol. I. Interscience Pub. Inc. (John Wiley & Sons), New York, 1953.
3. Courant, R., and Hilbert, D., *Methods of Mathematical Physics*, Vol. II, Interscience Pub. Inc. (John Wiley & Sons), New York-London-Sydney, 1962.
4. Goursat, E., *Cours d'Analyse Mathématique*, Vol. 3, Chapter 29; Gauthier-Villar, Paris, 1923.
5. Weber, H., *Die Partiellen-Differential Gleichungen der Mathematischen Physik*, Vol. 2, 101–105, 111–117. Vieweg, Braunschweig, Germany, 1912.
6. Weinberger, H. F., *A First Course in Partial-Differential Equations*, Blaisdell Pub. Co. (Ginn & Co.), Waltham, Massachusetts-Toronto-London, 1965.
7. Widder, D. V., *The Heat Equation*. Academic Press, New York, 1975.
8. Young, E. C., *Partial Differential Equations: An Introduction*. Allyn and Bacon, Inc., Boston, 1972.
9. Zachmanoglou, E. C., and Thoe, D., *Introduction to Partial Differential Equations*. The Williams & Wilkins Company, Baltimore, 1976.

Chapter 7

The Reduction of Some Initial-Boundary-Value Problems for the Semi-Infinite Strip, to Integral Equations: Some Exercises

EXERCISE 7.1. THE BOUNDARY CONDITION $u(0, t)$ WITH $u_x(1, t)$

The results of Sec. 6.5 enable us to state the following result.

Theorem 7.1.1. *For piecewise-continuous f, g, and h, the solution u of*

$$u_t = u_{xx}, \qquad 0 < x < 1, \qquad 0 < t,$$
$$u(x, 0) = f(x), \qquad 0 < x < 1,$$
$$u(0, t) = g(t), \qquad 0 < t, \qquad (7.1.1)$$
$$u_x(1, t) = h(t), \qquad 0 < t,$$

has the form

$$u(x, t) = v(x, t) - 2 \int_0^t \frac{\partial K}{\partial x}(x, t - \tau) \varphi_1(\tau) \, d\tau$$

$$+ 2 \int_0^t K(x - 1, t - \tau) \varphi_2(\tau) \, d\tau, \qquad (7.1.2)$$

where

$$v(x, t) = \int_{-\infty}^{\infty} K(x - \xi, t) f(\xi) \, d\xi, \qquad (7.1.3)$$

$K(x, t)$ is defined by (3.2.1), and f here is a smooth, bounded extension of the f

in (7.1.1), *if and only if* φ_1 *and* φ_2 *are piecewise-continuous solutions of*

$$g(t) = v(0,t) + \varphi_1(t) + 2\int_0^t K(-1,t-\tau)\varphi_2(\tau)\,d\tau,$$

$$h(t) = v_x(1,t) - 2\int_0^t \frac{\partial^2 K}{\partial x^2}(+1,t-\tau)\varphi_1(\tau)\,d\tau + \varphi_2(t). \tag{7.1.4}$$

Proof.   The proof is left to the reader.   □

Remark.   Note that the uniqueness of the solution φ_1 and φ_2 of (7.1.4) does not imply the uniqueness of u. The uniqueness of u for (7.1.1) can be obtained via the energy argument following Theorem 6.4.1. Note that the requirement of the square integrability of u_t, u_x, and u_{xx} changes the class of solutions from that of Definition 1.5.4.

EXERCISE 7.2.   THE BOUNDARY CONDITION
$u_x(0,t) + \alpha(t)u(0,t)$ WITH $u(1,t)$

As in Ex. 7.1, the results of Sec. 6.5 enable us to state the following result.

   Theorem 7.2.1.   *For piecewise-continuous f, α, g, and h, the solution u of*

$$\begin{aligned}
u_x &= u_{xx}, \qquad 0 < x < 1, \qquad 0 < t, \\
u(x,0) &= f(x), \qquad 0 < x < 1, \\
u_x(0,t) + \alpha(t)u(0,t) &= g(t), \qquad 0 < t, \\
u(1,t) &= h(t), \qquad 0 < t,
\end{aligned} \tag{7.2.1}$$

has the form

$$u(x,t) = v(x,t) - 2\int_0^t K(x,t-\tau)\varphi_1(\tau)\,d\tau$$

$$+ 2\int_0^t \frac{\partial K}{\partial x}(x-1,t-\tau)\varphi_2(\tau)\,d\tau, \tag{7.2.2}$$

where

$$v(x,t) = \int_{-\infty}^{\infty} K(x-\xi,t)f(\xi)\,d\xi, \tag{7.2.3}$$

$K(x,t)$ *is defined by* (3.2.1), *and f here is a smooth, bounded extension of the f in* (7.2.1), *if and only if φ_1 and φ_2 are piecewise-continuous solutions of*

$$g(t) = v_x(0,t) + \varphi_1(t) + 2\int_0^t \frac{\partial^2 K}{\partial x^2}(-1,t-\tau)\varphi_2(\tau)\,d\tau$$

$$+ \alpha(t)v(0,t) - 2\alpha(t)\int_0^t K(0,t-\tau)\varphi_1(\tau)\,d\tau$$

$$+ 2\alpha(t)\int_0^t \frac{\partial K}{\partial x}(-1,t-\tau)\varphi_2(\tau)\,d\tau, \tag{7.2.4}$$

$$h(t) = v(1,t) - 2\int_0^t K(1,t-\tau)\varphi_1(\tau)\,d\tau + \varphi_2(t).$$

Proof. The proof is left to the reader.   □

Remark. The uniqueness of the solution φ_1 and φ_2 of (7.2.4) does not imply the uniqueness of the solution u of (7.2.1). Consider here the method of proof that is given in the uniqueness of Theorem 6.4.1. Let z denote the difference of two solutions to (7.2.1). Then, z satisfies

$$
\begin{aligned}
z_t &= z_{xx}, & 0 < x < 1, & \quad 0 < t, \\
z(x,0) &= 0, & 0 < x < 1, & \\
z_x(0, t) &+ \alpha(t)z(0, t) = 0, & 0 < t, & \\
z(1, t) &= 0, & 0 < t. &
\end{aligned}
\tag{7.2.5}
$$

Following the computations given in (6.4.7) and (6.4.8), we see that, for $t > 0$,

$$
\begin{aligned}
0 &= \int_0^t \int_0^1 \{ zz_t - zz_{xx} \} \, dx \, d\tau \\
&= \frac{1}{2} \int_0^1 \{ z(x,t) \}^2 \, dx + \int_0^t \int_0^1 (z_x)^2 \, dx \, d\tau - \int_0^t \alpha(\tau)(z(0,\tau))^2 \, d\tau.
\end{aligned}
\tag{7.2.6}
$$

As

$$
\int_0^t \alpha(\tau)(z(0,\tau))^2 \, d\tau = -\int_0^t \alpha(\tau) \left\{ \int_0^1 2z(x,\tau)z_x(x,\tau) \, dx \right\} d\tau,
\tag{7.2.7}
$$

we see that

$$
\begin{aligned}
\int_0^1 \{ z(x,t) \}^2 \, dx &+ 2\int_0^t \int_0^1 z_x^2 \, dx \, d\tau = 2\int_0^t \alpha(\tau)(z(0,\tau))^2 \, d\tau \\
&= -2\int_0^t \alpha(\tau) \left\{ 2\int_0^1 z(x,\tau)z_x(x,\tau) \, dx \right\} d\tau.
\end{aligned}
\tag{7.2.8}
$$

If $\alpha(t) < 0$, then uniqueness follows immediately as in (6.4.9). Otherwise, we must split $2z(x, \tau)z_x(x, \tau)$, using

$$
2ab \le \varepsilon a^2 + \varepsilon^{-1} b^2,
\tag{7.2.9}
$$

where ε is any positive number. Thus,

$$
\begin{aligned}
-2\int_0^t \alpha(\tau) &\left\{ 2\int_0^1 z(x,\tau)z_x(x,\tau) \, dx \right\} d\tau \\
&\le 2\int_0^t |\alpha(\tau)| \left\{ \varepsilon^{-1} \int_0^1 (z(x,\tau))^2 \, dx + \varepsilon \int_0^1 (z_x(x,\tau))^2 \, dx \right\} d\tau \\
&\le 2\varepsilon^{-1} \int_0^t |\alpha(\tau)| \int_0^1 (z(x,\tau))^2 \, dx + 2\varepsilon \max_{0 \le \tau \le t} |\alpha(\tau)| \int_0^t \int_0^1 z_x^2 \, dx \, d\tau,
\end{aligned}
\tag{7.2.10}
$$

whence follows

$$\int_0^1 \{z(x,t)\}^2 \, dx + 2\left(1 - \varepsilon \max_{0 \le \tau \le t} |\alpha(\tau)|\right) \int_0^t \int_0^1 z_x^2 \, dx \, d\tau$$

$$\le 2\varepsilon^{-1} \int_0^t |\alpha(\tau)| \int_0^1 (z(x,\tau))^2 \, dx \, d\tau. \qquad (7.2.11)$$

If $\alpha(t)$ is bounded or bounded on compact subsets of $0 < t$, then ε can be chosen sufficiently small so that (7.2.11) reduces to

$$\int_0^1 (z(x,t))^2 \, dx \le 2\varepsilon^{-1} \int_0^t |\alpha(\tau)| \int_0^1 (z(x,\tau))^2 \, dx, \qquad (7.2.12)$$

which, by multiplication by $|\alpha(t)|$, is equivalent to the differential inequality

$$\begin{cases} E'(t) \le 2\varepsilon^{-1}|\alpha(t)|E(t), \\ E(0) = 0, \end{cases} \qquad (7.2.13)$$

where

$$E(t) = \int_0^t |\alpha(\tau)| \int_0^1 (z(x,\tau))^2 \, dx \, d\tau. \qquad (7.2.14)$$

Using the integrating factor $\exp\{-2\varepsilon^{-1}\int_0^t |\alpha(\tau)| \, d\tau\}$, we obtain

$$E(t)\exp\left\{-2\varepsilon^{-1}\int_0^t |\alpha(\tau)| d\tau\right\} \le E(0) = 0, \qquad (7.2.15)$$

whence follows

$$E(t) = 0, \qquad (7.2.16)$$

which implies that $z(x,t) = 0$ for $0 \le x \le 1$ and each t where $\alpha(t) \ne 0$. For any interval where $\alpha(t) = 0$, the problem reduces to that discussed in Sec. 6.4. We see that these distinctions on α can be phrased in terms of Lebesgue measure; for example, $\alpha \ne 0$ except for a set of measure zero, vs. $\alpha = 0$ on a set of positive measure.

We summarize the above discussion with the following statement.

Theorem 7.2.2 (Uniqueness). *The problem* (7.2.1) *can have at most one solution in the class of solutions with square integrable* u_t, u_x, *and* u_{xx}.

Proof. See the analysis preceding the theorem statement. □

As corollaries of Theorem 7.2.1 and the argument of Theorem 7.2.2, we can state the following results.

Corollary 7.2.1. *For piecewise-continuous* f, α, g, *and* h, *the solution* u *of*

$$\begin{aligned} u_t &= u_{xx}, & 0 < x < 1, && 0 < t, \\ u(x,0) &= f(x), & 0 < x < 1, \\ u_x(0,t) + \alpha(t)u(0,t) &= g(t), & && 0 < t, \\ u_x(1,t) &= h(t), & 0 < t, \end{aligned} \qquad (7.2.17)$$

has the form

$$u(x, t) = v(x, t) - 2\int_0^t K(x, t - \tau)\varphi_1(\tau)\, d\tau$$

$$+ 2\int_0^t K(x - 1, t - \tau)\varphi_2(\tau)\, d\tau, \qquad (7.2.18)$$

where

$$v(x, t) = \int_{-\infty}^{\infty} K(x - \xi, t)f(\xi)\, d\xi, \qquad (7.2.19)$$

$K(x, t)$ *is defined by* (3.2.1), *and f here is a smooth, bounded extension of f in* (7.2.17), *if and only if* φ_1 *and* φ_2 *are piecewise continuous and satisfy*

$$g(t) = v_x(0, t) + \varphi_1(t) + 2\int_0^t \frac{\partial K}{\partial x}(-1, t - \tau)\varphi_2(\tau)\, d\tau$$

$$+ \alpha(t)v(0, t) - 2\alpha(t)\int_0^t K(0, t - \tau)\varphi_1(\tau)\, d\tau$$

$$+ 2\alpha(t)\int_0^t K(-1, t - \tau)\varphi_2(\tau)\, d\tau, \qquad (7.2.20)$$

$$h(t) = v_x(1, t) - 2\int_0^t \frac{\partial K}{\partial x}(1, t - \tau)\varphi_1(\tau)\, d\tau + \varphi_2(t).$$

Moreover, the solution u of (7.2.17) *is unique in the class of solutions with square-integrable* u_t, u_x, *and* u_{xx}.

    Proof.   The proof is left to the reader.   □

    Corollary 7.2.2.   *For piecewise-continuous* f, α, g, β, *and* h, *the solution u of*

$$u_t = u_{xx}, \qquad 0 < x < 1, \qquad 0 < t,$$

$$u(x, 0) = f(x), \qquad 0 < x < 1,$$

$$u_x(0, t) + \alpha(t)u(0, t) = g(t), \qquad 0 < t, \qquad (7.2.21)$$

$$u_x(1, t) + \beta(t)u(1, t) = h(t), \qquad 0 < t,$$

has the form

$$u(x, t) = v(x, t) - 2\int_0^t K(x, t - \tau)\varphi_1(\tau)\, d\tau$$

$$+ 2\int_0^t K(x - 1, t - \tau)\varphi_2(\tau)\, d\tau, \qquad (7.2.22)$$

where

$$v(x, t) = \int_{-\infty}^{\infty} K(x - \xi, t)f(\xi)\, d\xi, \qquad (7.2.23)$$

$K(x, t)$ is defined by (3.2.1), and f here is a smooth, bounded extension of f in (7.2.21), if and only if φ_1 and φ_2 are piecewise-continuous and satisfy

$$g(t) = v_x(0, t) + \varphi_1(t) + 2 \int_0^t \frac{\partial K}{\partial x}(-1, t - \tau) \varphi_2(\tau) \, d\tau$$

$$+ \alpha(t) v(0, t) - 2\alpha(t) \int_0^t K(0, t - \tau) \varphi_1(\tau) \, d\tau$$

$$+ 2\alpha(t) \int_0^t K(-1, t - \tau) \varphi_2(\tau) \, d\tau,$$

$$h(t) = v_x(1, t) - 2 \int_0^t \frac{\partial K}{\partial x}(1, t - \tau) \varphi_1(\tau) \, d\tau + \varphi_2(t)$$

$$+ \beta(t) v(1, t) - 2\beta(t) \int_0^t K(1, t - \tau) \varphi_1(\tau) \, d\tau$$

$$+ 2\beta(t) \int_0^t K(0, t - \tau) \varphi_2(\tau) \, d\tau. \tag{7.2.24}$$

Moreover, the solution u of (7.2.21) is unique in the class of solutions with square-integrable u_t, u_x, and u_{xx}.

Proof.   The proof is left to the reader.   □

Remark.   When the transfer of heat from liquids to solids is considered, the heat flux is often taken to be proportional to the difference in the boundary temperature of the solid and the temperature of the liquid. Here, $\alpha(t)$ and $\beta(t)$ represent those proportionality factors.

EXERCISE 7.3.   THE BOUNDARY CONDITION $u_x(0, t) = F(t, u(0, t))$ WITH $u_x(1, t)$

In this section we draw on the results of Sec. 6.4 and state the following result.

Theorem 7.3.1.   *For piecewise-continuous f and h and continuous F, the solution u of the problem*

$$u_t = u_{xx}, \qquad 0 < x < 1, \qquad 0 < t,$$
$$u(x, 0) = f(x), \qquad 0 < x < 1,$$
$$u_x(0, t) = F(t, u(0, t)), \qquad 0 < t, \tag{7.3.1}$$
$$u_x(1, t) = h(t), \qquad 0 < t,$$

has the form

$$u(x, t) = w(x, t) - 2 \int_0^t \theta(x, t - \tau) F(\tau, \phi(\tau)) \, d\tau$$

$$+ 2 \int_0^t \theta(x - 1, t - \tau) h(\tau) \, d\tau, \tag{7.3.2}$$

where

$$w(x,t) = \int_0^1 \{\theta(x-\xi,t) + \theta(x+\xi,t)\} f(\xi) \, d\xi, \qquad (7.3.3)$$

and $\theta(x,t)$ is defined by (6.1.6), if and only if ϕ is a piecewise continuous solution of

$$\phi(t) = w(0,t) - 2\int_0^t \theta(0, t-\tau) F(\tau, \phi(\tau)) \, d\tau + 2\int_0^t \theta(-1, t-\tau) h(\tau) \, d\tau.$$

$$(7.3.4)$$

Moreover, uniqueness of u is assured under the additional assumption that F is Lipschitz continuous.

    Proof. The equivalence of the integral equation (7.3.4) and problem (7.3.1) follows from arguments similar to those given in Sec. 6.5 through Sec. 7.2. The equivalence of uniqueness follows from the uniqueness of the representation of a solution in terms of its flux on each boundary. See the remark following Theorem 6.5.1. A direct argument of uniqueness of the solution of problem (7.3.1) can be given for u in the class of solutions with square-integrable u_t, u_x, and u_{xx}. Let z denote the difference of two solutions of (7.3.1). Then, from the Lipschitz continuity of F,

$$|z_x(0,t)| = |F(t, u_1(0,t)) - F(t, u_2(0,t))| \le C(t)|z(0,t)|. \quad (7.3.5)$$

The remainder of the argument follows that given by (7.2.5) through (7.2.16).   □

We list here several corollaries of this result.

    Corollary 7.3.1. *For piecewise-continuous f, h, and α and for continuous F, the solution u of the problem*

$$\begin{aligned}
u_t &= u_{xx}, & 0 < x < 1, & \quad 0 < t, \\
u(x,0) &= f(x), & 0 < x < 1, & \\
u_x(0,t) &= F(t, u(0,t)), & 0 < t, & \\
u_x(1,t) + \alpha(t)u(1,t) &= h(t), & 0 < t,
\end{aligned} \qquad (7.3.6)$$

has the form

$$u(x,t) = w(x,t) - 2\int_0^t \theta(x, t-\tau) F(\tau, \phi_1(\tau)) \, d\tau$$

$$+ 2\int_0^t \theta(x-1, t-\tau) \phi_2(\tau) \, d\tau, \qquad (7.3.7)$$

where

$$w(x,t) = \int_0^1 \{\theta(x-\xi,t) + \theta(x+\xi,t)\} f(\xi) \, d\xi \qquad (7.3.8)$$

and $\theta(x, t)$ is defined by (6.1.6), if and only if ϕ_1 and ϕ_2 are piecewise-continuous functions that satisfy

$$\phi_1(t) = w(0, t) - 2 \int_0^t \theta(0, t - \tau) F(\tau, \phi_1(\tau)) \, d\tau$$

$$+ 2 \int_0^t \theta(-1, t - \tau) \phi_2(\tau) \, d\tau,$$

$$h(t) = \phi_2(t) + \alpha(t) w(1, t) - 2\alpha(t) \int_0^t \theta(1, t - \tau) \cdot F(\tau, \phi_1(\tau)) \, d\tau$$ (7.3.9)

$$+ 2\alpha(t) \int_0^t \theta(0, t - \tau) \phi_2(\tau) \, d\tau.$$

Moreover, uniqueness of u is assured under the additional assumption that F is Lipschitz continuous.

      Proof. The proof is left to the reader. $\square$

      Corollary 7.3.2. *For piecewise-continuous f and for continuous F and G, the solution u of the problem*

$$u_t = u_{xx}, \qquad 0 < x < 1, \qquad 0 < t,$$

$$u(x, 0) = f(x), \qquad 0 < x < 1,$$

$$u_x(0, t) = F(t, u(0, t)), \qquad 0 < t,$$ (7.3.10)

$$u_x(1, t) = G(t, u(1, t)), \qquad 0 < t,$$

has the form

$$u(x, t) = w(x, t) - 2 \int_0^t \theta(x, t - \tau) F(\tau, \phi_1(\tau)) \, d\tau$$

$$+ 2 \int_0^t \theta(x - 1, t - \tau) G(\tau, \phi_2(\tau)) \, d\tau,$$ (7.3.11)

where

$$w(x, t) = \int_0^1 \{\theta(x - \xi, t) + \theta(x + \xi, t)\} f(\xi) \, d\xi$$ (7.3.12)

and $\theta(x, t)$ is defined by (6.1.6), if and only if ϕ_1 and ϕ_2 are piecewise-continuous functions that satisfy

$$\phi_1(t) = w(0, t) - 2 \int_0^t \theta(0, t - \tau) F(\tau, \phi_1(\tau)) \, d\tau$$

$$+ 2 \int_0^t \theta(-1, t - \tau) G(\tau, \phi_2(\tau)) \, d\tau,$$

$$\phi_2(t) = w(1, t) - 2 \int_0^t \theta(1, t - \tau) F(\tau, \phi_1(\tau)) \, d\tau$$ (7.3.13)

$$+ 2 \int_0^t \theta(0, t - \tau) G(\tau, \phi_2(\tau)) \, d\tau.$$

Moreover, uniqueness of u is assured under the additional assumption that F and G are Lipschitz continuous.

   Proof. The proof is left to the reader.   □

   Corollary 7.3.3. *For piecewise-continuous f and h and for continuous F, the solution u of the problem*

$$u_t = u_{xx}, \qquad 0 < x < 1, \qquad 0 < t,$$

$$u(x, 0) = f(x), \qquad 0 < x < 1,$$

$$u_x(0, t) = F(t, u(0, t)), \qquad 0 < t, \tag{7.3.14}$$

$$u(1, t) = h(t), \qquad 0 < t,$$

has the form

$$u(x, t) = w(x, t) - 2 \int_0^t \theta(x, t - \tau) F(\tau, \phi_1(\tau)) \, d\tau$$

$$+ 2 \int_0^t \theta(x - 1, t - \tau) \phi_2(\tau) \, d\tau, \tag{7.3.15}$$

where

$$w(x, t) = \int_0^1 \{\theta(x - \xi, t) + \theta(x + \xi, t)\} f(\xi) \, d\xi, \tag{7.3.16}$$

and $\theta(x, t)$ is defined by (6.1.6), if and only if ϕ_1 and ϕ_2 are piecewise-continuous functions that satisfy

$$\phi_1(t) = w(0, t) - 2 \int_0^t \theta(0, t - \tau) F(\tau, \phi_1(\tau)) \, d\tau$$

$$+ 2 \int_0^t \theta(-1, t - \tau) \phi_2(\tau) \, d\tau,$$

$$h(t) = w(1, t) - 2 \int_0^t \theta(1, t - \tau) F(\tau, \phi_1(\tau)) \, d\tau + 2 \int_0^t \theta(0, t - \tau) \phi_2(\tau) \, d\tau.$$

$$\tag{7.3.17}$$

Moreover, uniqueness of u is assured under the additional assumption that F is Lipschitz continuous.

   Proof. The proof is left to the reader.   □

Remark. When the radiation of heat from a solid is considered, the heat flux is often taken to be proportional to the fourth power of the difference of the boundary temperature of the solid with the temperature of the surroundings. Here, F represents a general radiation law.

EXERCISE 7.4.   THE BOUNDARY CONDITION
$$u_t(0,t)+\alpha(t)u_x(0,t)+\beta(t)u(0,t) \text{ WITH } u(1,t)$$

We begin with the computation of $u_x(0,t)$. From Theorem 6.3.1, we may consider the representation

$$u(x,t) = \int_0^1 \{\theta(x-\xi,t)-\theta(x+\xi,t)\} u(\xi,0)\, d\xi$$
$$-2\int_0^t \frac{\partial\theta}{\partial x}(x,t-\tau)u(0,\tau)\, d\tau +2\int_0^t \frac{\partial\theta}{\partial x}(x-1,t-\tau)u(1,\tau)\, d\tau,$$

$$(7.4.1)$$

where $\theta(x,t)$ is defined by (6.1.6). Differentiating with respect to x, we obtain

$$u_x(x,t) = \int_0^1 \left\{ \frac{\partial\theta}{\partial x}(x-\xi,t)-\frac{\partial\theta}{\partial x}(x+\xi,t) \right\} u(\xi,0)\, d\xi$$
$$-2\int_0^t \frac{\partial^2\theta}{\partial x^2}(x,t-\tau)u(0,\tau)\, d\tau$$
$$+2\int_0^t \frac{\partial^2\theta}{\partial x^2}(x-1,t-\tau)u(1,\tau)\, d\tau$$
$$= J_1 + J_2 + J_3. \qquad (7.4.2)$$

Starting with J_1, we see that

$$\frac{\partial\theta}{\partial x}(x-\xi,t) = -\frac{\partial\theta}{\partial \xi}(x-\xi,t) \qquad \text{and} \qquad \frac{\partial\theta}{\partial x}(x+\xi,t) = \frac{\partial\theta}{\partial \xi}(x+\xi,t).$$

Hence

$$J_1 = -\int_0^1 \frac{\partial}{\partial \xi}\{\theta(x-\xi,t)+\theta(x+\xi,t)\} u(\xi,0)\, d\xi$$
$$= 2\{\theta(x,t)\} u(0,0) - \{\theta(x-1,t)+\theta(x+1,t)\} u(1,t)$$
$$+\int_0^1 \{\theta(x-\xi,t)+\theta(x+\xi,t)\} u_\xi(\xi,0)\, d\xi. \qquad (7.4.3)$$

For J_2 and J_3, we note that

$$\frac{\partial^2\theta}{\partial x^2} = \frac{\partial\theta}{\partial t} = -\frac{\partial\theta}{\partial \tau} \qquad (7.4.4)$$

and that

$$\lim_{\tau \uparrow t} \theta(x,t-\tau) = 0, \qquad 0 < x < 1. \qquad (7.4.5)$$

Consequently, we can integrate by parts to obtain

$$J_2 = -2\theta(x,t)u(0,0)-2\int_0^t \theta(x,t-\tau)u_\tau(0,\tau)\, d\tau \qquad (7.4.6)$$

and

$$J_3 = +2\theta(x-1,t)u(1,0)+2\int_0^t\theta(x-1,t-\tau)u_\tau(1,\tau)\,d\tau. \quad (7.4.7)$$

Summing the results (7.4.3), (7.4.6), and (7.4.7), employing the Lebesgue dominated-convergence theorem, and observing that

$$\theta(x-1,t) = \theta(x+1,t), \qquad (7.4.8)$$

we obtain

$$\lim_{x\downarrow 0} u_x(x,t) = 2\int_0^1\theta(\xi,t)u_\xi(\xi,0)\,d\xi - 2\int_0^t\theta(0,t-\tau)u_\tau(0,\tau)\,d\tau$$

$$+2\int_0^t\theta(-1,t-\tau)u_\tau(1,\tau)\,d\tau, \qquad (7.4.9)$$

where we have also used the fact that

$$\theta(-\xi,t) = \theta(\xi,t). \qquad (7.4.10)$$

From

$$\theta(1-\xi,t) = \theta(1+\xi,t), \qquad (7.4.11)$$

it follows in a similar manner that

$$\lim_{x\uparrow 1} u_x(x,t) = 2\int_0^1\theta(1+\xi,t)u_\xi(\xi,0)\,d\xi - 2\int_0^t\theta(1,t-\tau)u_\tau(0,\tau)\,d\tau$$

$$+2\int_0^t\theta(0,t-\tau)u_\tau(1,\tau)\,d\tau. \qquad (7.4.12)$$

Using the arguments of Sec. 6.5 through Ex. 7.3, we can state the following result.

Theorem 7.4.1. *For continuously differentiable f and h with $f(1) = h(0)$ and for continuous g, α, and β, the solution of the problem*

$$u_t = u_{xx}, \qquad 0 < x < 1, \quad 0 < t,$$

$$u(x,0) = f(x), \qquad 0 < x < 1,$$

$$u_t(0,t)+\alpha(t)u_x(0,t)+\beta(t)u(0,t) = g(t), \qquad 0 < t, \qquad (7.4.13)$$

$$u(1,t) = h(t), \qquad 0 < t,$$

has the representation

$$u(x,t) = \int_0^1\{\theta(x-\xi,t)-\theta(x+\xi,t)\}f(\xi)\,d\xi$$

$$-2\int_0^t\frac{\partial\theta}{\partial x}(x,t-\tau)\left\{f(0)+\int_0^\tau\phi(\eta)\,d\eta\right\}d\tau$$

$$+2\int_0^t\frac{\partial\theta}{\partial x}(x-1,t-\tau)h(\tau)\,d\tau, \qquad (7.4.14)$$

where $\theta(x, t)$ is defined by (6.1.6), if and only if ϕ is a continuous solution of

$$g(t) = \phi(t) + 2\alpha(t)\int_0^1 \theta(\xi, t)f'(\xi)\,d\xi$$

$$-2\alpha(t)\int_0^t \theta(0, t - \tau)\phi(\tau)\,d\tau + 2\alpha(t)\int_0^t \theta(-1, t - \tau)h'(\tau)\,d\tau$$

$$+ \beta(t)f(0) + \beta(t)\int_0^t \phi(\tau)\,d\tau. \tag{7.4.15}$$

Proof. See the analysis preceding the theorem statement. $\qquad\square$

Remark. The equivalence of uniqueness carries over here, since u is uniquely determined by the representation (7.4.1).

As corollaries of the above Theorem 7.4.1, we can state the following results.

Corollary 7.4.1. *For continuously differentiable f and for continuous g, h, α, β, γ, and δ, the solution of the problem*

$$u_t = u_{xx}, \qquad 0 < x < 1, \qquad 0 < t,$$

$$u(x, 0) = f(x), \qquad 0 < x < 1,$$

$$u_t(0, t) + \alpha(t)u_x(0, t) + \beta(t)u(0, t) = g(t), \qquad 0 < t,$$

$$u_t(1, t) + \gamma(t)u_x(1, t) + \delta(t)u(1, t) = h(t), \qquad 0 < t, \tag{7.4.16}$$

has the representation

$$u(x, t) = \int_0^1 \{\theta(x - \xi, t) - \theta(x + \xi, t)\}f(\xi)\,d\xi$$

$$-2\int_0^t \frac{\partial\theta}{\partial x}(x, t - \tau)\left\{f(0) + \int_0^\tau \phi_1(\eta)\,d\eta\right\}d\eta$$

$$+2\int_0^t \frac{\partial\theta}{\partial x}(x - 1, t - \tau)\left\{f(1) + \int_0^\tau \phi_2(\eta)\,d\eta\right\}d\tau, \tag{7.4.17}$$

where $\theta(x, t)$ is defined by (6.1.6) if and only if ϕ_1 and ϕ_2 are continuous functions that satisfy

$$g(t) = \phi_1(t) + 2\alpha(t)\int_0^1 \theta(\xi, t)f'(\xi)\,d\xi - 2\alpha(t)\int_0^t \theta(0, t - \tau)\phi_1(\tau)\,d\tau$$

$$+ 2\alpha(t)\int_0^t \theta(-1, t - \tau)\phi_2(\tau)\,d\tau + \beta(t)f(0) + \beta(t)\int_0^t \phi_1(\tau)\,d\tau,$$

$$h(t) = \phi_2(t) + 2\gamma(t)\int_0^1 \theta(1 + \xi, t)f'(\xi)\,d\xi - 2\gamma(t)\int_0^t \theta(1, t - \tau)\phi_1(\tau)\,d\tau$$

$$+ 2\gamma(t)\int_0^t \theta(0, t - \tau)\phi_2(\tau)\,d\tau + \delta(t)f(1) + \delta(t)\int_0^\tau \phi_2(\tau)\,d\tau. \tag{7.4.18}$$

Proof. The proof is left to the reader. $\qquad\square$

Remark. The equivalence of uniqueness carries over here, since u is uniquely determined by the representation (7.4.1).

    Corollary 7.4.2. *For continuously differentiable f and for continuous g, h, α, β, and γ, the solution of the problem*

$$u_t = u_{xx}, \qquad 0 < x < 1, \qquad 0 < t,$$

$$u(x, 0) = f(x), \qquad 0 < x < 1,$$

$$u_t(0, t) + \alpha(t)u_x(0, t) + \beta(t)u(0, t) = g(t), \qquad 0 < t.$$

$$u_x(1, t) + \gamma(t)u(1, t) = h(t), \qquad 0 < t,$$

    (7.4.19)

has the representation

$$u(x, t) = \int_0^1 \{\theta(x - \xi, t) - \theta(x + \xi, t)\} f(\xi) \, d\xi$$

$$-2 \int_0^t \frac{\partial \theta}{\partial x}(x, t - \tau) \left\{ f(0) + \int_0^\tau \phi_1(\eta) \, d\eta \right\} d\tau$$

$$+2 \int_0^t \frac{\partial \theta}{\partial x}(x - 1, t - \tau) \left\{ f(1) + \int_0^\tau \phi_2(\eta) \, d\eta \right\} d\tau, \quad (7.4.20)$$

where $\theta(x, t)$ is defined by (6.1.6), if and only if ϕ_1 and ϕ_2 are continuous functions that satisfy

$$g(t) = \phi_1(t) + 2\alpha(t) \int_0^t \theta(\xi, t) f'(\xi) \, d\xi - 2\alpha(t) \int_0^t \theta(0, t - \tau) \phi_1(\tau) \, d\tau$$

$$+ 2\alpha(t) \int_0^t \theta(-1, t - \tau) \phi_2(\tau) \, d\tau + \beta(t) f(0) + \beta(t) \int_0^t \phi_1(\tau) \, d\tau,$$

$$h(t) = 2 \int_0^1 \theta(1 + \xi, t) f'(\xi) \, d\xi - 2 \int_0^t \theta(1, t - \tau) \phi_1(\tau) \, d\tau$$

$$+ 2 \int_0^t \theta(0, t - \tau) \phi_2(\tau) \, d\tau + \gamma(t) f(1) + \gamma(t) \int_0^t \phi_2(\tau) \, d\tau. \quad (7.4.21)$$

    Proof. The proof is left to the reader.   □

Remark. The equivalence of uniqueness carries over here, since u is uniquely determined by the representation (7.4.1).

Remark. Here, the physics of heat flow does not present an easy example. However, in diffusion of chemicals the boundary condition

$$u_t + \alpha(t)u_x + \beta(t)u = g$$

can easily represent a boundary reaction, where $\alpha(t)u_x$ represents the diffusive transport of materials to the boundary.

EXERCISE 7.5.   THE BOUNDARY CONDITION $\int_0^{s(t)} u(x,t)\,dx$ WITH $u(1,t)$

The analysis given in Ex. 7.4 enables us to state immediately the following result.

Theorem 7.5.1.   *For continuous f, g, h and s with $g(0) = \int_0^{s(0)} f(\xi)\,d\xi$, the solution of*

$$u_t = u_{xx}, \qquad 0 < x < 1, \qquad 0 < t,$$

$$u(x,0) = f(x), \qquad 0 < x < 1,$$

$$\int_0^{s(t)} u(x,t)\,dx = g(t), \qquad 0 < t, \qquad 0 < s(t) < 1, \tag{7.5.1}$$

$$u(1,t) = h(t), \qquad 0 < t,$$

has the representation

$$u(x,t) = \int_0^1 \{\theta(x - \xi, t) - \theta(x + \xi, t)\} f(\xi)\,d\xi$$

$$- 2\int_0^t \frac{\partial \theta}{\partial x}(x, t - \tau)\phi(\tau)\,d\tau + 2\int_0^t \frac{\partial \theta}{\partial x}(x - 1, t - \tau)h(\tau)\,d\tau,$$

$$\tag{7.5.2}$$

where $\theta(x,t)$ is defined by (6.1.6), if and only if ϕ is a piecewise-continuous solution of

$$g(t) - \int_0^{s(t)} \left\{ \int_0^1 \{\theta(x - \xi, t) - \theta(x + \xi, t)\} f(\xi)\,d\xi \right\} dx$$

$$= 2\int_0^t \theta(0, t - \tau)\phi(\tau)\,d\tau - 2\int_0^t \theta(s(t), t - \tau)\phi(\tau)\,d\tau$$

$$+ 2\int_0^t \theta(s(t) - 1, t - \tau)h(\tau)\,d\tau - 2\int_0^t \theta(-1, t - \tau)h(\tau)\,d\tau.$$

$$\tag{7.5.3}$$

Proof.   The proof is left to the reader.   □

Remark.   The equivalence of uniqueness follows from the representation (7.4.1).

Remark.   The condition $\int_0^{s(t)} u(x,t)\,dx$ represents the specification of a relative heat content of a portion of the conductor. For diffusion, the condition is equivalent to the specification of mass in a portion of the domain of diffusion.

NOTES

The material in Sec. 6.5 for the reduction of the initial-boundary-value problem in a semi-infinite strip to a system of integral equations was given by Gevrey [3]. The reduction of the problem with boundary condition $u_x = F(t, u(0, t))$ to an integral equation was studied by Adler [1], Mann and Wolf [4], Mann and Roberts [5], and Padmavally [6]. The specification of energy $\int_0^{s(t)} u(x, t)\, dx$ was studied by Cannon [2].

REFERENCES

1. Adler, G., Un type nouveau des problémes aux limites de la conduction de la chaleur, *Magyar Tud. Akad. Mat. Kutato Int. Kozl*, **4** (1959), 109–127.
2. Cannon, J. R., The solution of the heat equation subject to the specification of energy, *Quart. Appl. Math.*, **21** (1963), 155–160.
3. Gevrey, M., Sur les équations aux dérivées partielles du type parabolique, *J. Math. Pures Appl.*, **9** (1913), 305–471.
4 Mann, W. R., and Wolf, F., Heat transfer between solids and gases under nonlinear boundary conditions, *Quart. Appl. Math.*, **9** (1951), 163–184.
5. Mann, W. R., and Roberts, J. H., A nonlinear integral equation of Volterra type, *Pacific J. Math.*, **1** (1951), 431–445.
6. Padmavally, K., On a nonlinear integral equation, *J. Math. & Mech.*, **7** (1958), 533–555.

Chapter 8

Integral Equations

8.1. VOLTERRA INTEGRAL EQUATIONS OF THE FIRST KIND AND THE ABEL INTEGRAL EQUATION

The classical, linear Volterra integral equation of the first kind can be written in the form

$$g(t) = \int_0^t H(t, \tau)\phi(\tau)\, d\tau, \tag{8.1.1}$$

where g and H are known continuously differentiable functions of their arguments. Differentiating equation (8.1.1) with respect to t, we obtain

$$g'(t) = H(t, t)\phi(t) + \int_0^t \frac{\partial H}{\partial t}(t, \tau)\phi(\tau)\, d\tau, \tag{8.1.2}$$

which can be written in the form

$$\phi(t) = f(t) + \int_0^t F(t, \tau)\phi(\tau)\, d\tau, \tag{8.1.3}$$

where

$$f(t) = g'(t)[H(t, t)]^{-1} \tag{8.1.4}$$

and

$$F(t, \tau) = -\frac{\partial H}{\partial t}(t, \tau)[H(t, t)]^{-1}, \tag{8.1.5}$$

provided that

$$H(t, t) \neq 0. \tag{8.1.6}$$

The equation (8.1.3) is in the form of a Volterra integral equation of the second kind. Such equations are treated in Sec. 8.2, where a general theory is developed. When considering various problems for the heat equation it is quite often the case that $H(t, t) = 0$. Hence, special methods other than simple differentiation are required in order to treat some of these problems.

Consider first the Exercise 4.2. When $s(t) \equiv + \infty$ is considered, we are presented with the classical Abel integral equation

$$g(t) = \int_0^t \frac{\phi(\tau)}{\sqrt{t - \tau}} dt. \tag{8.1.7}$$

Note that $H(t, \tau) = (t - \tau)^{-1/2}$, $H(t, t)$ is undefined, and $(\partial H / \partial t)(t, t)$ is not integrable. Actually, we can solve a slightly more general problem.

Theorem 8.1.1.   *The generalized Abel equation*

$$g(t) = \int_0^t \frac{\phi(\tau)}{(t - \tau)^\alpha} d\tau, \qquad 0 < \alpha < 1, \tag{8.1.8}$$

has the solution

$$\phi(t) = \frac{\sin(\alpha\pi)}{\pi} \frac{d}{dt} \int_0^t \frac{g(\tau)}{(t - \tau)^{1 - \alpha}} d\tau$$

$$= \frac{\sin(\alpha\pi)}{\pi} \left\{ \frac{g(0)}{t^{1 - \alpha}} + \int_0^t \frac{g'(\tau)}{(t - \tau)^{1 - \alpha}} d\tau \right\} \tag{8.1.9}$$

for absolutely continuous g with bounded g'. In particular, for $\alpha = \frac{1}{2}$,

$$\phi(t) = \frac{1}{\pi} \frac{d}{dt} \int_0^t \frac{g(\tau) d\tau}{(t - \tau)^{1/2}}$$

$$= \frac{1}{\pi} \left\{ \frac{g(0)}{t^{1/2}} + \int_0^t \frac{g'(\tau) d\tau}{(t - \tau)^{1/2}} \right\}. \tag{8.1.10}$$

Proof.   The inversion technique involves the multiplication of (8.1.8) by $(\eta - t)^{\alpha - 1}$ and the integration with respect to t from 0 to η. Thus we obtain

$$\int_0^\eta \frac{g(t) dt}{(\eta - t)^{1 - \alpha}} = \int_0^\eta \frac{1}{(\eta - t)^{1 - \alpha}} \int_0^\tau \frac{\phi(\tau)}{(t - \tau)^\alpha} d\tau dt. \tag{8.1.11}$$

By Fubini's theorem,

$$\int_0^\eta \frac{1}{(\eta-t)^{1-\alpha}} \int_0^t \frac{\phi(\tau)}{(t-\tau)^\alpha} d\tau \, dt = \int_0^\eta \phi(\tau) \left\{ \int_\tau^\eta \frac{dt}{(\eta-t)^{1-\alpha}(t-\tau)^\alpha} \right\} d\tau.$$

$$(8.1.12)$$

Under the transformation of variables $t = \tau + (\eta - \tau)\xi$,

$$\int_\tau^\eta \frac{dt}{(\eta-t)^{1-\alpha}(t-\tau)^\alpha} = \int_0^1 \frac{d\xi}{(1-\xi)^{1-\alpha}\xi^\alpha}$$

$$= \Gamma(\alpha)\Gamma(1-\alpha) = \frac{\pi}{\sin(\alpha\pi)}. \qquad (8.1.13)$$

Consequently,

$$\int_0^\eta \phi(\tau) \, d\tau = \frac{\sin(\alpha\pi)}{\pi} \int_0^\eta \frac{g(t) \, dt}{(\eta-t)^{1-\alpha}}, \qquad (8.1.14)$$

from whence follows the result (8.1.9).   □

As an application of Theorem 8.1.1, we consider the integral Eq. (F) of the exercise section in Chapter 4 for the initial-boundary-value problem with boundary condition $\int_0^{s(t)} u(x,t) \, dx$ for $s(t) \neq \infty$. This equation and several others given in Chapter 7 have the form

$$g(t) + \int_0^t H(t,\tau)\phi(\tau) \, d\tau = \int_0^t \frac{\phi(\tau) \, d\tau}{(t-\tau)^\alpha} \qquad (8.1.15)$$

for $\alpha = \frac{1}{2}$. Here, we shall assume that $0 < \alpha < 1$. Formally, applying the inversion formula (8.1.9), we obtain

$$\phi(t) = \frac{\sin(\alpha\pi)}{\pi} \left\{ \frac{d}{dt} \int_0^t \frac{g(\tau) \, d\tau}{(t-\tau)^{1-\alpha}} \right.$$

$$\left. + \frac{d}{dt} \int_0^t \frac{1}{(t-\tau)^{1-\alpha}} \int_0^\tau H(\tau,\eta)\phi(\eta) \, d\eta \, d\tau \right\}. \quad (8.1.16)$$

Since

$$\frac{d}{dt} \int_0^t \frac{g(\tau) \, d\tau}{(t-\tau)^{1-\alpha}} = \frac{g(0)}{t^{1-\alpha}} + \int_0^t \frac{g'(\tau)}{(t-\tau)^{1-\alpha}} d\tau, \qquad (8.1.17)$$

it follows from Leibnitz's rule that

$$\frac{d}{dt} \int_0^t \frac{1}{(t-\tau)^{1-\alpha}} \int_0^\tau H(\tau,\eta)\phi(\eta) \, d\eta \, dt$$

$$= \int_0^t \frac{1}{(t-\tau)^{1-\alpha}} \left\{ H(\tau,\tau)\phi(\tau) + \int_0^\tau \frac{\partial H}{\partial \tau}(\tau,\eta)\phi(\eta) \, d\eta \right\} d\tau.$$

$$(8.1.18)$$

From Fubini's theorem, we see that

$$\int_0^t \frac{1}{(t-\tau)^{1-\alpha}} \int_0^\tau \frac{\partial H}{\partial \tau}(\tau, \eta) \phi(\eta) \, d\eta \, d\tau$$

$$= \int_0^t \phi(\eta) \left\{ \int_\eta^t (t-\tau)^{\alpha-1} \frac{\partial H}{\partial \tau}(\tau, \eta) \, d\tau \right\} d\eta. \qquad (8.1.19)$$

Consequently, from (8.1.16) through (8.1.19) we see that

$$\phi(t) = \frac{\sin(\alpha\pi)}{\pi} \left\{ \frac{g(0)}{t^{1-\alpha}} + \int_0^t \frac{g'(\tau)}{(t-\tau)^{1-\alpha}} d\tau \right.$$

$$+ \int_0^t \phi(\eta) \left\{ (t-\eta)^{\alpha-1} H(\eta, \eta) \right.$$

$$\left. + \int_\eta^t (t-\tau)^{\alpha-1} \frac{\partial H}{\partial \tau}(\tau, \eta) \, d\tau \right\} d\eta \right\}. \qquad (8.1.20)$$

We can summarize the preceding analysis in the following statement.

Theorem 8.1.2. *An integral equation of the form*

$$g(t) + \int_0^t H(t, \tau) \phi(\tau) \, d\tau = \int_0^t \frac{\phi(\tau) \, d\tau}{(t-\tau)^\alpha} \qquad (8.1.21)$$

for $0 < \alpha < 1$ is equivalent to the Volterra integral equation of the second kind,

$$\phi(t) = \frac{\sin(\alpha\pi)}{\pi} \left\{ \frac{g(0)}{t^{1-\alpha}} + \int_0^t \frac{g'(\tau)}{(t-\tau)^{1-\alpha}} d\tau \right.$$

$$+ \int_0^t \phi(\eta) \left\{ (t-\eta)^{\alpha-1} H(\eta, \eta) \right.$$

$$\left. + \int_\eta^t (t-\tau)^{\alpha-1} \frac{\partial H}{\partial \tau}(\tau, \eta) \, d\tau \right\} d\eta \right\}, \qquad (8.1.22)$$

provided that g is absolutely continuous with bounded g',

$$H(t, t) = \lim_{\tau \uparrow t} H(t, \tau) \qquad (8.1.23)$$

exists, and

$$\int_0^t (t-\eta)^{\alpha-1} |H(\eta, \eta)| \, d\eta < \infty, \qquad (8.1.24)$$

and $(\partial H/\partial \tau)(\tau, \eta)$ exists, and

$$F(t, \eta) = \int_\eta^t (t-\tau)^{\alpha-1} \frac{\partial H}{\partial \tau}(\tau, \eta) \, d\tau \qquad (8.1.25)$$

exists as an integrable function with respect to η for each t.

Proof. The analysis preceding the statement of the Theorem 8.1.2 shows that a solution ϕ of (8.1.21) must satisfy (8.1.22). Under the conditions given on H, the steps of that analysis are reversible. $\square$

Remark. For the applications of this result to the equations given in Exercises 4.2, 7.3, 7.4, and 7.5, $H(t, t) \equiv 0$ and $(\partial H/\partial\tau)(\tau, \eta)$ is a bounded continuous function of its arguments. Consequently, the equation (8.1.22) will satisfy the conditions of Theorem 8.2.1 given below.

8.2. VOLTERRA INTEGRAL EQUATIONS OF THE SECOND KIND

We begin with the study of the existence and uniqueness of a piecewise-continuous solution $\phi = \phi(t)$ of the equation

$$\phi(t) = g(t) + \int_0^t H(t, \tau, \phi(\tau)) \, d\tau, \qquad t > 0, \qquad (8.2.1)$$

where g is a piecewise-continuous function and $H = H(t, \tau, \phi)$ is continuous in its arguments for all ϕ and for $t > \tau$. Moreover, we assume that H satisfies the Lipschitz condition

$$|H(t, \tau, \phi_1) - H(t, \tau, \phi_2)| \leq L(t, \tau)|\phi_1 - \phi_2|, \qquad (8.2.2)$$

where L satisfies the integrability condition

$$\int_{t_1}^{t_2} L(t_2, \tau) \, d\tau \leq \alpha(t_2 - t_1), \qquad t_2 > t_1, \qquad (8.2.3)$$

for some monotone-increasing function α with

$$\lim_{\eta \downarrow 0} \alpha(\eta) = 0. \qquad (8.2.4)$$

Remark. For most applications here, $L(t, \tau) = C_1(t - \tau)^{-1/2}$ and $\alpha(\eta) = C_2\eta^{1/2}$, where C_1 and C_2 are positive constants.

> **Lemma 8.2.1.** *For any piecewise-continuous ψ,*
> $$\int_0^t |H(t, \tau, \psi(\tau))| \, d\tau < \infty \qquad (8.2.5)$$
> *provided that*
> $$\int_0^t |H(t, \tau, 0)| \, d\tau < \infty. \qquad (8.2.6)$$

Proof. We add and subtract $H(t, \tau, 0)$ and obtain

$$\int_0^t |H(t, \tau, \psi(\tau))| \, d\tau \leq \int_0^t |H(t, \tau, \psi(\tau)) - H(t, \tau, 0)| \, d\tau + \int_0^t |H(t, \tau, 0)| \, d\tau$$

$$\leq \int_0^t L(t, \tau)|\psi(\tau)| \, d\tau + \int_0^t |H(t, \tau, 0)| \, d\tau. \qquad (8.2.7)$$

Since ψ is bounded, the result follows from (8.2.3) and (8.2.6). $\square$

Lemma 8.2.2. *If there exists a nonnegative function β such that*

$$\int_{t_1}^{t_2} |H(t_2, \tau, 0)| \, d\tau \leq \beta(t_2 - t_1), \qquad t_2 \geq t_1, \qquad (8.2.8)$$

and that

$$\lim_{\eta \downarrow 0} \beta(\eta) = 0, \qquad (8.2.9)$$

then for any piecewise-continuous ψ

$$\int_0^t H(t, \tau, \psi(\tau)) \, d\tau \qquad (8.2.10)$$

is a continuous function of t.

Remark. For applications that are linear, $\beta \equiv 0$, while for most others, $\beta(\eta) = C_3 \eta^{1/2}$ where C_3 is a positive constant.

Proof. Since

$$\left| \int_0^{t_2} H(t_2, \tau, \psi(\tau)) \, d\tau - \int_0^{t_1} H(t_1, \tau, \psi(\tau)) \, d\tau \right|$$

$$\leq \int_{t_1}^{t_2} |H(t_2, \tau, \psi(\tau))| \, d\tau + \int_{t_1 - \delta}^{t_1} |H(t_2, \tau, \psi(\tau))| \, d\tau$$

$$+ \int_{t_1 - \delta}^{t_1} |H(t_1, \tau, \psi(\tau))| \, d\tau$$

$$+ \int_0^{t_1 - \delta} |H(t_2, \tau, \psi(\tau)) - H(t_1, \tau, \psi(\tau))| \, d\tau$$

$$= J_1 + J_2 + J_3 + J_4, \qquad (8.2.11)$$

where we have assumed $t_2 > t_1$, it suffices to show $\lim_{t_1 \downarrow t_1} J_i = 0$ for $i = 1, \ldots, 4$. Consider J_1. Now,

$$J_1 \leq \int_{t_1}^{t_2} |H(t_2, \tau, \psi(\tau)) - H(t_2, \tau, 0)| \, d\tau + \int_{t_1}^{t_2} |H(t_2, \tau, 0)| \, d\tau$$

$$\leq \int_{t_1}^{t_2} L(t_2, \tau) |\psi(\tau)| \, d\tau + \beta(t_2 - t_1)$$

$$\leq \left\{ \max_{0 \leq \tau \leq t} |\psi(\tau)| \right\} \alpha(t_2 - t_1) + \beta(t_2 - t_1), \qquad t > t_2. \qquad (8.2.12)$$

Consequently, $\lim_{t_2 \downarrow t_1} J_1 = 0$ follows from (8.2.4) and (8.2.9). In a similar manner, we see that

$$\lim_{\substack{t_2 \downarrow t_1 \\ \delta \downarrow 0}} J_i = 0, \qquad i = 2, 3.$$

All that remains now is to show that for fixed $\delta > 0$, $\lim_{t_2 \downarrow t_1} J_4 = 0$. Since

$H(t, \tau, \phi)$ is continuous, it follows that for, $0 \le \tau \le t_1 - \delta$,

$$t_1 - \frac{\delta}{2} \le t \le T \quad \text{and} \quad |\phi| \le \max_{0 \le \tau \le T} |\psi(\tau)|,$$

H is bounded, and

$$\lim_{t_2 \downarrow t_1} \left\{ H(t_2, \tau, \psi(\tau)) - H(t_1, \tau, \psi(\tau)) \right\} = 0. \qquad (8.2.13)$$

By the Lebesgue bounded-convergence theorem we obtain $\lim_{t_2 \downarrow t_1} J_4 = 0$. □

It is sufficient to consider the case of continuous g. Let $t_1 < t_2 < \cdots < t_m$ denote the bounded jumps of g in the interval $0 \le t \le T$, $0 \le T$. Suppose that ϕ has been uniquely determined up to t_k and satisfies (8.2.1). Then, for $t_k \le t \le t_{k+1}$, we consider

$$\phi(t) = g(t) + \int_0^{t_k} H(t, \tau, \phi(\tau)) \, d\tau + \int_{t_k}^t H(t, \tau, \phi(\tau)) \, d\tau$$

$$= G(t) + \int_{t_k}^t H(t, \tau, \phi(\tau)) \, d\tau, \qquad t \ge t_k, \qquad (8.2.14)$$

where G is continuous since g and $\int_0^{t_k} H(t, \tau, \phi(\tau)) \, d\tau$ are. The extension of ϕ to t_{k+1} is obtained by utilizing the values of the solution for (8.2.14) for $t_k \le t \le t_{k+1}$.

Lemma 8.2.3. *For continuous g and for H that is continuous in its arguments for all ϕ and $t > \tau$ and satisfies (8.2.2), (8.2.3), and (8.2.8), the mapping F defined by*

$$(F\phi)(t) = g(t) + \int_0^t H(t, \tau, \phi(\tau)) \, d\tau, \qquad 0 \le t \le T, \qquad (8.2.15)$$

maps the space of continuous functions on $0 \le t \le T$ into itself. Moreover,

$$\|F\phi_1 - F\phi_2\|_T \le \alpha(T) \|\phi_1 - \phi_2\|_T, \qquad (8.2.16)$$

where

$$\|\psi\|_t = \sup_{0 \le \tau \le t} |\psi(\tau)|. \qquad (8.2.17)$$

Proof. The continuity of the function $F\phi$ follows immediately from Lemma 8.2.2 and the continuity of g. For the estimate (8.2.16), we see that for each t in $0 \le t \le T$,

$$|(F\phi_1)(t) - (F\phi_2)(t)| \le \int_0^t |H(t, \tau, \phi_1(\tau)) - H(t, \tau, \phi_2(\tau))| \, d\tau$$

$$\le \int_0^t L(t, \tau) |\phi_1(\tau) - \phi_2(\tau)| \, d\tau$$

$$\le \alpha(t) \|\phi_1 - \phi_2\|_t \le \alpha(T) \|\phi_1 - \phi_2\|_T, \qquad (8.2.18)$$

whence follows (8.2.16).

Lemma 8.2.4. *For T sufficiently small, F is a contracting map of the Banach space[1] of continuous functions on $0 \leq t \leq T$ with the uniform norm $\| \ \|_T$ into itself; that is, for T sufficiently small, $\alpha(T) < 1$ in (8.2.16).*

Proof. This follows from Lemma 8.2.3 since α is monotone and $\lim_{t \downarrow 0} \alpha(t) = 0$. □

Lemma 8.2.5. *A contracting map F of the Banach space of continuous functions on $0 \leq t \leq T$ with uniform norm $\| \ \|_T$ into itself possesses a unique fixed point. In other words, there exists a unique function ϕ such that $\phi = F\phi$.*

Proof. This is a standard theorem in most texts on real and functional analysis. However, we shall give a proof here in this context for the sake of completeness.

Let ϕ_0 denote a continuous function on the interval $0 \leq t \leq T$ and consider the sequence of Picard iterates

$$\phi_n = F\phi_{n-1}, \qquad n = 1, 2, 3, \ldots \tag{8.2.19}$$

Clearly, the ϕ_n are continuous functions on $0 \leq t \leq T$. Moreover, since F is a contracting map, there exists a ρ, $0 < \rho < 1$, such that

$$\|\phi_n - \phi_{n-1}\|_T < \rho \|\phi_{n-1} - \phi_{n-2}\|_T. \tag{8.2.20}$$

Hence,

$$\|\phi_n - \phi_{n-1}\|_T \leq \rho^{n-1} \|\phi_1 - \phi_0\|_T. \tag{8.2.21}$$

Now, we can write

$$\phi_n = \phi_0 + \sum_{k=1}^{n} \{\phi_k - \phi_{k-1}\}. \tag{8.2.22}$$

Considering

$$\phi_n - \phi_m = \sum_{k=m+1}^{n} \{\phi_k - \phi_{k-1}\}, \tag{8.2.23}$$

it follows from (8.2.21) that

$$\|\phi_n - \phi_m\|_T \leq \sum_{k=m+1}^{n} \rho^{k-1} \|\phi_1 - \phi_0\|_T$$

$$\leq \|\phi_1 - \phi_0\|_T \sum_{k=m}^{\infty} \rho^k$$

$$= \|\phi_1 - \phi_0\|_T \frac{\rho^m}{1 - \rho}. \tag{8.2.24}$$

---

[1]For the term Banach space, the reader can substitute the notion that uniformly Cauchy sequences of functions from the space have a unique limit function in the space.

Thus,

$$\lim_{m,n\to\infty} \|\phi_n - \phi_m\|_T = 0. \tag{8.2.25}$$

Hence, the sequence of Picard iterates converges uniformly to a continuous function ϕ. Since

$$\|\phi - F\phi\|_T \le \|\phi - \phi_n\|_T + \|F\phi_{n-1} - F\phi\|_T$$
$$\le \|\phi - \phi_n\|_T + \rho\|\phi - \phi_{n-1}\|_T \tag{8.2.26}$$

holds for all n, it follows from

$$\lim_{n\uparrow\infty} \|\phi - \phi_n\| = 0 \tag{8.2.27}$$

that

$$\phi = F\phi. \tag{8.2.28}$$

For unicity we assume that ϕ_i, $i = 1,2$, are two fixed points and observe that

$$\phi_1 - \phi_2 = F\phi_1 - F\phi_2 \tag{8.2.29}$$

and that

$$\|\phi_1 - \phi_2\|_T < \rho\|\phi_1 - \phi_2\|_T, \tag{8.2.30}$$

which implies that $\phi_1 = \phi_2$. □

We can summarize the above results in the following statement.

Theorem 8.2.1. *For piecewise-continuous g, the integral equation*

$$\phi(t) = g(t) + \int_0^t H(t,\tau,\phi(\tau))\, d\tau, \tag{8.2.31}$$

possesses a unique, piecewise-continuous solution ϕ for all $t > 0$ if $H(t,\tau,\phi)$ is continuous for all ϕ and for $t > \tau$, if

$$|H(t,\tau,\phi_1) - H(t,\tau,\phi_2)| \le L(t,\tau)|\phi_1 - \phi_2|, \tag{8.2.32}$$

where

$$\int_{t_1}^{t_2} L(t_2,\tau)\, d\tau \le \alpha(t_2 - t_1), \qquad t_2 > t, \tag{8.2.33}$$

for some monotone-increasing function α, with

$$\lim_{\eta\downarrow 0} \alpha(\eta) = 0, \tag{8.2.34}$$

and if

$$\int_{t_1}^{t_2} |H(t_2,\tau,0)|\, d\tau \le \beta(t_2 - t_1), \qquad t_2 > t_1, \tag{8.2.35}$$

for some nonnegative function β, with

$$\lim_{\eta\downarrow 0} \beta(\eta) = 0. \tag{8.2.36}$$

Proof. From Lemmas 8.2.1 through 8.2.5, we see that it suffices to consider g continuous and that there exists a $T > 0$ such that (8.2.31) possesses a unique solution on $0 \leq t \leq T$. Let us assume that (8.2.31) possesses a unique solution for $0 \leq t \leq nT$, where n is a positive integer. We shall show that (8.2.31) possesses a unique solution for $0 \leq t \leq (n+1)T$. Consider the problem

$$\phi(t) = g(t) + \int_0^{nT} H(t, \tau, \phi(\tau)) \, d\tau + \int_{nT}^t H(t, \tau, \phi(\tau)) \, d\tau \quad (8.2.37)$$

for $nT \leq t \leq (n+1)T$. As ϕ is known for $0 \leq t \leq nT$, the second term on the righthand side of (8.2.36) can be combined with g to form a known continuous function G defined over $nT \leq t \leq (n+1)T$. Consequently, (8.2.36) reduces to the problem

$$\phi(t) = G(t) + \int_{nT}^t H(t, \tau, \phi(\tau)) \, d\tau, \qquad nT \leq t \leq (n+1)T. \quad (8.2.38)$$

It is easy to see that Lemmas 8.2.1 through 8.2.5 apply to Eq. (8.2.37). Hence, (8.2.37) possesses a unique continuous solution ϕ defined on $nT \leq t \leq (n+1)T$. From the definition of G we see that ϕ, regarded as a function on $0 \leq t \leq (n+1)T$, satisfies (8.2.31). Moreover, from the unicity of ϕ on each interval $nT \leq t \leq (n+1)T$, $n = 0, 1, 2, \ldots$, it follows that ϕ is unique. $\square$

As a corollary of the arguments given above, we can state the following generalization.

Corollary 8.2.1. *For piecewise-continuous g_i, $i = 1, \ldots, m$, the system of integral equations*

$$\phi_i(t) = g_i(t) + \int_0^t H_i(t, \tau, \phi_1(\tau), \ldots, \phi_m(\tau)) \, d\tau, \qquad i = 1, \ldots, m,$$

$$(8.2.39)$$

possesses unique solution $(\phi_1, \ldots, \phi_m)$ for all $t > 0$ if $H_i(t, \tau, \phi_1, \ldots, \phi_m)$, $i = 1, 2, \ldots, m$, are continuous for all ϕ_i, $i = 1, 2, \ldots, m$ and for $t > \tau$, if

$$\left| H_i(t, \tau, \phi_1^1, \ldots, \phi_m^1) - H_i(t, \tau, \phi_1^2, \ldots, \phi_m^2) \right| \leq L(t, \tau) \left\{ \sum_{i=1}^m |\phi_i^1 - \phi_i^2| \right\},$$

$$i = 1, \ldots, m, \quad (8.2.40)$$

where

$$\int_{t_1}^{t_2} L(t_2, \tau) \, d\tau \leq \alpha(t_2 - t_1), \qquad t_2 > t_1, \quad (8.2.41)$$

for some monotone-increasing function α with

$$\lim_{\eta \downarrow 0} \alpha(\eta) = 0, \quad (8.2.42)$$

and if

$$\int_{t_1}^{t_2} |H_i(t_2,\tau,0,\dots,0)| \, d\tau \le \beta(t_2 - t_1), \qquad t_2 > t_1, \qquad (8.2.43)$$

for some nonnegative function β with

$$\lim_{\eta \downarrow 0} \beta(\eta) = 0. \tag{8.2.44}$$

8.2.1. Applications

In Sec. 6.5 through 7.5, numerous initial-boundary-value problems were reduced to equivalent systems of integral equations. Most of these systems are already in the form of (8.2.1) or (8.2.39). The rest can be reduced to one of these forms by applying the inversion method for the Abel integral equation, as shown in Sec. 8.1. We shall leave to the reader the application of Theorem 8.2.1 and Corollary 8.2.1 to those systems, as well as any needed additional assumptions on the initial and/or boundary data in those problems.

8.3. CONTINUOUS DEPENDENCE OF THE SOLUTIONS OF INTEGRAL EQUATIONS UPON THE DATA

We need consider here only Volterra integral equations of the second kind since equations of the first kind, which we are able to analyze, are either explicitly solvable or reducible to the second kind.

We begin by supposing that ϕ_i, $i = 1, 2$, are the solutions of the equations

$$\phi_i(t) = g_i(t) + \int_0^t H_i(t, \tau, \phi_i(\tau)) \, d\tau, \qquad t > 0, \qquad i = 1, 2, \quad (8.3.1)$$

where g_i and H_i, $i = 1, 2$, satisfy the properties set forth in Theorem 8.2.1. Setting $\psi(t) = \phi_1(t) - \phi_2(t)$ and differencing the two equations in (8.3.1), we can write the equation

$$\psi(t) = \{ g_1(t) - g_2(t) \} + \int_0^t \{ H_1(t, \tau, \phi_1(\tau)) - H_2(t, \tau, \phi_1(\tau)) \} \, d\tau$$

$$+ \int_0^t \{ H_2(t, \tau, \phi_1(\tau)) - H_2(t, \tau, \phi_2(\tau)) \} \, d\tau, \tag{8.3.2}$$

whence follows

$$|\psi(t)| \le |g_1(t) - g_2(t)| + \int_0^t |H_1(t, \tau, \phi_1(\tau)) - H_2(t, \tau, \phi_1(\tau))| \, d\tau$$

$$+ \int_0^t L(t, \tau) |\psi(\tau)| \, d\tau. \tag{8.3.3}$$

Let $\delta > 0$ denote a fixed number. We can write

$$\int_0^t |H_1(t,\tau,\phi_1(\tau)) - H_2(t,\tau,\phi_1(\tau))| \, d\tau$$

$$= \int_0^{t-\delta} |H_1(t,\tau,\phi_1(\tau)) - H_2(t,\tau,\phi_1(\tau))| \, d\tau$$

$$+ \int_{t-\delta}^t |H_1(t,\tau,\phi_1(\tau)) - H_2(t,\tau,\phi_1(\tau))| \, d\tau$$

$$= J_1 + J_2. \tag{8.3.4}$$

Now, in the case that $t \geq \delta$, it is easy to see that

$$J_1 \leq (t-\delta) \, \Delta H(\delta), \tag{8.3.5}$$

where

$$\Delta H(\delta) = \sup_{S_\delta} |H_1(t,\tau,\phi) - H_2(t,\tau,\phi)|, \tag{8.3.6}$$

and

$$S_\delta = \{(t,\tau,\phi): -\infty < \phi < \infty, \quad t-\delta \geq \tau \geq 0, \quad \delta \leq t \leq T\}.$$

For J_2 we employ the analysis of Lemma 8.2.2 to obtain

$$J_2 \leq 2\|\phi_1\|_T \alpha(\delta) + 2\beta(\delta). \tag{8.3.7}$$

Hence, for all t in $0 \leq t \leq T$, we see that

$$J_1 + J_2 \leq T\Delta H(\delta) + 2\|\phi_1\|_T \alpha(\delta) + 2\beta(\delta). \tag{8.3.8}$$

Consequently, the righthand side of (8.3.3) can be estimated, yielding, for all t in $0 \leq t \leq T$,

$$|\psi(t)| \leq \|g_1 - g_2\|_T + T\Delta H(\delta) + 2\|\phi_1\|_T \alpha(\delta) + 2\beta(\delta) + \alpha(T)\|\psi\|_T. \tag{8.3.9}$$

Thus,

$$\|\psi\|_T \leq D + \alpha(T)\|\psi\|_T, \tag{8.3.10}$$

where

$$D = \|g_1 - g_2\|_T + T\Delta H(\delta) + 2\|\phi_1\|_T \alpha(\delta) + 2\beta(\delta). \tag{8.3.11}$$

If $\alpha(T) < 1$, then

$$\|\psi\|_T \leq (1 - \alpha(T))^{-1} D. \tag{8.3.12}$$

Suppose now that T in each part of D has been replaced by nT for some large positive integer n. Then, for $T \leq t \leq 2T$, (8.3.3) can be written as

$$|\psi(t)| \leq D + \int_0^T L(t,\tau)|\psi(\tau)| \, d\tau + \int_T^t L(t,\tau)|\psi(\tau)| \, d\tau$$

$$\leq D + \alpha(2T)\|\psi\|_T + \alpha(T)\|\psi\|_{2T}. \tag{8.3.13}$$

As α is monotone increasing, it follows, from (8.3.10) and (8.3.13) that

$$\|\psi\|_{2T} \leq D + \alpha(2T)\|\psi\|_T + \alpha(T)\|\psi\|_{2T}. \tag{8.3.14}$$

Consequently, from $\alpha(T) < 1$ and (8.3.12), we obtain

$$\|\psi\|_{2T} \leq \left\{(1 - \alpha(T))^{-1} + \alpha(2T)(1 - \alpha(T))^{-2}\right\} D. \tag{8.3.15}$$

In a similar manner we see that

$$\|\psi\|_{3T} \leq D + \alpha(3T)\|\psi\|_{2T} + \alpha(T)\|\psi\|_{3T}, \tag{8.3.16}$$

whence it follows that

$$\|\psi\|_{3T} \leq \left\{(1 - \alpha(T))^{-1} + \alpha(3T)(1 - \alpha(T))^{-2} + \alpha(3T)\alpha(2T)(1 - \alpha(T))^{-3}\right\} D.$$

$$\leq \left[\sum_{k=1}^{3} \{\alpha(3T)\}^{k-1}(1 - \alpha(T))^{-k} \right] D, \tag{8.3.17}$$

since α is monotone increasing. Assume now that

$$\|\psi\|_{nT} \leq \left[\sum_{k=1}^{n} \{\alpha(nT)\}^{k-1}(1 - \alpha(T))^{-k} \right] D. \tag{8.3.18}$$

Then, in a similar manner, we see that

$$\|\psi\|_{(n+1)T} \leq D + \alpha((n+1)T)\|\psi\|_{nT} + \alpha(T)\|\psi\|_{(n+1)T}, \tag{8.3.19}$$

whence it follows that

$$\|\psi\|_{(n+1)T} \leq \left\{ (1 - \alpha(T))^{-1} + \alpha((n+1)T) \right.$$

$$\left. \cdot \sum_{k=1}^{n} \{\alpha(nT)\}^{k-1}(1 - \alpha(T))^{-(k+1)} \right\} D$$

$$\leq \left\{ \sum_{k=1}^{(n+1)} \{\alpha((n+1)T)\}^{k-1}(1 - \alpha(T))^{-k} \right\} D, \tag{8.3.20}$$

since α is monotone increasing. Hence, the induction is complete and (8.3.18) provides us with the estimate for arbitrary $n = 1, 2, 3, \ldots$, and thus for any time interval we wish.

We gather the above into the following statement.

Theorem 8.3.1. *Suppose that* ϕ_i, $i = 1, 2$, *are, respectively, solutions of*

$$\phi_i(t) = g_i(t) + \int_0^t H_i(t, \tau, \phi_i(\tau)) \, d\tau, \qquad t > 0, \qquad i = 1, 2, \tag{8.3.21}$$

where g_i *and* H_i, $i = 1, 2$, *satisfy the hypotheses of Theorem 8.2.1. Then, for each* $n = 1, 2, 3, \ldots$,

$$\|\phi_1 - \phi_2\|_{nT} \leq \left\{ \sum_{k=1}^{n} (\alpha(nT))^{k-1}(1 - \alpha(T))^{-k} \right\} D, \tag{8.3.22}$$

where $0 < \alpha(T) < 1$,

$$
\begin{cases}
D = \|g_1 - g_2\|_{nT} + (nT)\,\Delta H(\delta) + 2\|\phi_1\|_{nT}\alpha(\delta) + 2\beta(\delta), \\
\Delta H(\delta) = \sup_{S_\delta} |H_1(t,\tau,\phi) - H_2(t,\tau,\phi)|, \\
S_\delta = \{(t,\tau,\phi): t-\delta \geq \tau \geq 0, \quad \delta \leq t \leq nT, \quad |\phi| \leq \|\phi_1\|_{nT}\},
\end{cases}
$$

$$(8.3.23)$$

δ is an arbitrarily chosen positive number, and α and β are functions related to H in Theorem 8.2.1. Also, the result remains valid if $\Delta H(\delta)$ in (8.2.23) is replaced by

$$
\Delta H(\delta) = \sup_{\delta \leq t \leq nT} \int_0^{t-\delta} |H_1(t,\tau,\phi_1(\tau)) - H_2(t,\tau,\phi_2(\tau))|\,d\tau.
$$

Proof. See the analysis preceding the statement of the theorem. □

Remark. As a consequence of Theorem 8.3.1, we see that if a sequence $\{g_n\}$ converges uniformly to a function g and a sequence $\{H_n\}$ converges uniformly to H on every compact subset of the set $t > \tau$ and $-\infty < \phi < \infty$, then the solutions ϕ_n of

$$
\phi_n(t) = g_n(t) + \int_0^t H_n(t,\tau,\phi_n(\tau))\,d\tau, \qquad n = 1,2,\ldots, \quad (8.3.24)
$$

converge uniformly to the solution ϕ of

$$
\phi(t) = g(t) + \int_0^t H(t,\tau,\phi(\tau))\,d\tau, \tag{8.3.25}
$$

provided that g, H, g_n, and H_n, $n = 1,2,\ldots$, satisfy the hypotheses of Theorem 8.2.1. Recalling the remark just before Lemma 8.2.1, the assumption of a common α and β for H, and H_n, $n = 1,2,\ldots$, is not restrictive for most applications.

As a corollary of the argument given above, we can state the following result.

Corollary 8.3.1. *Suppose that* $(\phi_1^{(i)},\ldots,\phi_m^{(i)})$, $i = 1,2$, *are, respectively, solutions of the systems*

$$
\phi_k^{(i)}(t) = g_k^{(i)}(t) + \int_0^t H_k^{(i)}(t,\tau,\phi_1^{(i)}(\tau),\ldots,\phi_m^{(i)}(\tau))\,d\tau,
$$

$$
t > 0, \qquad k = 1,\ldots,m, \qquad i = 1,2, \quad (8.3.26)
$$

where the $g_k^{(i)}$ *and* $H_k^{(i)}$, $k = 1,\ldots,m$, $i = 1,2$, *satisfy the hypotheses of Corollary 8.2.1. Then, for each* $n = 1,2,3,\ldots$,

$$
\|\phi^{(1)} - \phi^{(2)}\|_{nT} \leq \left\{ \sum_{l=1}^n (m\alpha(nT))^{l-1}(1 - m\alpha(T))^{-l} \right\} D, \quad (8.3.27)
$$

where $m\alpha(T) < 1$,

$$\|\phi^{(1)} - \phi^{(2)}\|_{nT} = \max_{1 \le k \le m} \left\{ \sup_{0 \le t \le nT} |\phi_k^{(1)}(t) - \phi_k^{(2)}(t)| \right\}, \quad (8.3.28)$$

and

$$\begin{cases} D = \|g^{(1)} - g^{(2)}\|_{nT} + nT\Delta H(\delta) + 2m\|\phi^{(1)}\|_{nT}\alpha(\delta) + 2m\beta(\delta), \\ \Delta H(\delta) = \max_{1 \le k \le m} \left\{ \sup_{S_\delta} |H_k^{(1)}(t, \tau, \phi_1, \ldots, \phi_m) - H_k^{(2)}(t, \tau, \phi_1, \ldots, \phi_m)| \right\} \\ S_\delta = \{(t, \tau, \phi_1, \ldots, \phi_m) : t - \delta \ge \tau \ge 0, \quad \delta \le t \le nT, \quad |\phi_k| \le \|\phi^{(1)}\|_{nT}, \\ k = 1, \ldots, m \}, \end{cases}$$

$$(8.3.29)$$

δ *is an arbitrarily chosen positive number, and* α *and* β *are the functions related to the* $H_k^{(i)}$ *in Corollary 8.2.1. Here, we assume that* α *and* β *work for both sets of* H_k. *The result remains valid if* $\Delta H(\delta)$ *in (8.3.29) is replaced by*

$$\Delta H(\delta) = \sup_{\delta \le t \le nT} \int_0^{t-\delta} \sum_{k=1}^m |H_1(t, \tau, \phi_1^{(1)}(\tau), \ldots, \phi_m^{(1)}(\tau))$$

$$- H_2(t, \tau, \phi_1^{(1)}(\tau), \ldots, \phi_m^{(1)}(\tau))| \, d\tau.$$

Proof.   The proof is left to the reader.   □

Remark.   The remark following Theorem 8.3.1 can be modified to the situation of systems. In other words, if the data converges uniformly, then the solutions converge uniformly to the solution of the system with the limit data.

8.4.   *A PRIORI* BOUNDS ON SOLUTIONS OF INTEGRAL EQUATIONS

We shall use the continuous-dependence estimate (8.3.22)–(8.3.23) of Theorem 8.3.1 to derive an *a priori* bound on $\|\phi\|_{nT}$, where ϕ is the solution of

$$\phi(t) = g(t) + \int_0^t H(t, \tau, \phi(\tau)) \, d\tau, \quad t > 0, \quad (8.4.1)$$

where g and H satisfy the hypotheses of Theorem 8.2.1. Note that in the estimate (8.3.22)–(8.3.23), there appears the term $\|\phi_1\|_{nT}$. We proceed now to introduce a special ϕ_1 into this estimate. Namely, $\phi_1 \equiv 0$. By Theorem 8.2.1, we see that $\phi \equiv 0$ is the unique solution of

$$\phi(t) = -\int_0^t H(t, \tau, 0) \, d\tau + \int_0^t H(t, \tau, \phi(\tau)) \, d\tau. \quad (8.4.2)$$

Consequently, in Theorem 8.3.1, we may take

$$g_1(t) = -\int_0^t H(t, \tau, 0)\, d\tau, \tag{8.4.3}$$

$g_2 = g$, $H_1 = H_2$, $\phi_1 \equiv 0$, and $\phi_2 = \phi$. A direct application of the theorem yields

$$\|\phi\|_{nT} \leq \left\{ \sum_{k=1}^n (\alpha(nT))^{k-1}(1-\alpha(T))^{-k} \right\} D, \tag{8.4.4}$$

where

$$D = \{ \|g\|_{nT} + \|\beta\|_{nT} \}, \tag{8.4.5}$$

where here $\beta = \beta(t)$ is the function related to H in Theorem 8.2.1. The terms $\alpha(\delta)$ and $\beta(\delta)$ disappear from D, since δ may be chosen arbitrarily small.

We gather this analysis into the following statement.

Theorem 8.4.1. *For the solution ϕ of*

$$\phi(t) = g(t) + \int_0^t H(t, \tau, \phi(\tau))\, d\tau, \qquad t > 0, \tag{8.4.6}$$

we see that

$$\|\phi\|_{nT} \leq \left\{ \sum_{k=1}^n (\alpha(nT))^{k-1}(1-\alpha(T))^{-k} \right\} D, \tag{8.4.7}$$

where $0 < \alpha(T) < 1$,

$$D = \|g\|_{nT} + \|\beta\|_{nT}, \tag{8.4.8}$$

and $\alpha = \alpha(t)$ and $\beta = \beta(t)$ are functions related to H in Theorem 8.2.1.

Proof. See the analysis preceding the statement of the theorem. ☐

As a corollary to the argument above we can apply the Corollary 8.3.1 to obtain the following result.

Corollary 8.4.1. *For the solution $\phi = (\phi_1, \ldots, \phi_m)$ of the system*

$$\phi_k(t) = g_k(t) + \int_0^t H_k(t, \tau, \phi_1(\tau), \ldots, \phi_m(\tau))\, d\tau, \qquad t > 0, \qquad k = 1, \ldots, m, \tag{8.4.9}$$

we see that

$$\|\phi\|_{nT} \leq \left\{ \sum_{l=1}^n (m\alpha(nT))^{l-1}(1-m\alpha(T))^{-l} \right\} D, \tag{8.4.10}$$

where, for vectors

$$\|\phi\|_{nT} = \max_{1 \leq k \leq m} \{ \|\phi_k\|_{nT} \}, \qquad 0 < m\alpha(T) < 1, \tag{8.4.11}$$

$$D = \|g\|_{nT} + \|\beta\|_{nT}, \tag{8.4.12}$$

and α and β are the functions related to the H_k in Corollary 8.2.1.

Proof. The proof is left to the reader.   □

If often occurs, in the estimation of a solution of an integral equation, that the Lipschitz function $L(t, \tau)$ is bounded. When this occurs, the estimation of the solution of the integral equation reduces to the estimation of an inequality of the form

$$\|\phi\|_t \leq C_1 + C_2 \int_0^t \|\phi\|_\tau \, d\tau, \tag{8.4.13}$$

which is a special case of Gronwall's inequality.

Lemma 8.4.1 (Gronwall). *Let* $b = b(t)$ *be a continuous solution of*

$$0 \leq b(t) \leq \int_0^t \{a(\tau)b(\tau) + c(\tau)\} \, d\tau + f(t) \tag{8.4.14}$$

for $0 \leq t \leq T$, *where* $a \geq 0$, $c \geq 0$, *and f are continuous functions on the interval* $0 \leq t \leq T$. *Then,*

$$0 \leq b(t) \leq \left\{ \int_0^T c(t) \, dt + \|f\|_T \right\} \exp\left\{ \int_0^t a(\tau) \, d\tau \right\}. \tag{8.4.15}$$

Proof. Clearly, we can multiply (8.4.14) through by $a(t)\exp\{ - \int_0^t a(\tau) \, d\tau \}$ and obtain:

$$\frac{d}{dt}\left[\left\{ \int_0^t a(\tau)b(\tau) \, d\tau \right\} \exp\left\{ - \int_0^t a(\varsigma) \, d\varsigma \right\} \right]$$

$$\leq \left[\int_0^t c(\tau) \, d\tau + f(\tau) \right] a(t)\exp\left\{ - \int_0^t a(\varsigma) \, d\varsigma \right\}$$

$$\leq \left[\int_0^T c(\tau) \, d\tau + \|f\|_T \right] a(t)\exp\left\{ - \int_0^t a(\varsigma) \, d\varsigma \right\}. \tag{8.4.16}$$

Integrating (8.4.16), we get:

$$\left\{ \int_0^t a(\tau)b(\tau) \, d\tau \right\} \leq \left[\int_0^T c(\tau) \, d\tau + \|f\|_T \right] \left\{ \exp\left\{ \int_0^t a(\tau) \, d\tau \right\} - 1 \right\}. \tag{8.4.17}$$

Substituting (8.4.17) into (8.4.14), we see that

$$0 \leq b(t) \leq \left[\int_0^T c(\tau) \, d\tau + \|f\|_T \right] \exp\left\{ \int_0^t a(\tau) \, d\tau \right\}$$

$$+ \int_0^t c(\tau) \, d\tau + f(t) - \left[\int_0^T c(\tau) \, d\tau + \|f\|_T \right], \tag{8.4.18}$$

whence follows the result (8.4.15).   □

Applying Lemma 8.4.1 to (8.4.13) yields the following estimate.

Lemma 8.4.2. *For*

$$\|\phi\|_t \le C_1 + C_2 \int_0^t \|\phi\|_\tau \, d\tau, \qquad\qquad (8.4.19)$$

we have

$$\|\phi\|_t \le C_1 \exp\{C_2 t\}. \qquad\qquad (8.4.20)$$

Proof.   The proof is left to the reader.   □

Remark.   The estimate (8.4.20) is valid for systems also, since $\|\phi\|_t$ can represent a vector norm.

EXERCISES

8.1.   Consider the integral equation (F) in the Exercise section of Chapter 4. Using the Abel inversion technique, reduce (F) in the Exercise section of Chapter 4 to a Volterra integral equation of the second kind. Give conditions on $f(x)$, $g(t)$ and $s(t)$ to validate the reduction.

8.2.   Consider the integral equation (7.5.3). Using the Abel inversion technique, reduce (7.5.3) to a Volterra integral equation of the second kind. Give conditions on $f(x)$, $g(t)$, $h(t)$, and $s(t)$ to validate the reduction.

8.3.   Show that if the mth iteration of a map F of the Banach space of continuous functions on $0 \le t \le T$ with uniform norm $\| \ \|_T$ into itself is a contraction, then F possesses a unique fixed point. Here, $F^m \phi = F(F^{m-1}\phi)$, $F^2\phi = F(F\phi)$.

8.4.   For F defined by (8.2.15), assume that $|L(t, \tau)| \le$ Constant. Show that for any $T > 0$, there exists an $m > 0$ such that F^m is a contraction.

8.5.   Do Exercise 8.4 with $|L(t, \tau)| \le C_1 (t - \tau)^{-1/2}$.

8.6.   Show that the integral Eq. (C) in the Exercise section of Chapter 4 possesses a unique solution.

8.7.   Using Gronwall's inequality (8.4.15) or (8.4.20), estimate the solution to the integral equation (C) in the Exercise section of Chapter 4.

8.8.   Show that the system of integral equations (6.5.8) possesses a unique solution.

8.9.   Using Gronwall's inequality (8.4.15) or (8.4.20), estimate the solution to the system (6.5.8).

NOTES

The material on integral equations here can be found also in Courant and Hilbert [1], Tricomi [2], Volterra [3].

REFERENCES

1. Courant, R., and Hilbert, D., *Methods of Mathematical Physics*, Vol. I, Interscience Pub. Inc. (John Wiley & Sons), New York, 1953.
2. Tricomi, F. G., *Integral Equations*. Interscience Pub. Inc. (John Wiley & Sons), New York, 1957.
3. Volterra, V., *Leçons sur les équations intégrales et des équations intégro-différentielles*, Chapter II. Gauthier Villars, Paris, 1913.

Chapter 9

Solutions of Boundary-Value Problems for All Times and Periodic Solutions

9.1. THE INFINITE STRIP WITH TEMPERATURE-BOUNDARY CONDITIONS

We consider here the problem (see Fig. 9.1.1)

$$u_t = u_{xx}, \qquad 0 < x < 1, \qquad -\infty < t < \infty,$$
$$u(0, t) = g(t), \qquad -\infty < t < \infty, \qquad (9.1.1)$$
$$u(1, t) = h(t), \qquad -\infty < t < \infty,$$

where g and h are known piecewise-continuous functions.

As a first observation we see that the solution

$$v(x, t) = \exp\{-\pi^2 t\} \sin \pi x \qquad (9.1.2)$$

provides us with an example of nonuniqueness of the solution of (9.1.1) unless some growth restriction is placed upon u as t tends to minus infinity. The essence of the growth condition is that the solution cannot grow as fast as the fundamental mode exhibited in (9.1.2). Set

$$M(T) = \sup_{0 < x < 1} |u(x, T)| \qquad (9.1.3)$$

and suppose that

$$\lim_{T \to -\infty} M(T) \exp\{\pi^2 T\} = 0. \qquad (9.1.4)$$

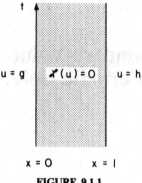

FIGURE 9.1.1

We shall demonstrate that under the growth condition (9.1.4), uniqueness of the solution holds. Suppose that u_1 and u_2 are solutions of (9.1.1) that satisfy (9.1.4). Let $z = u_1 - u_2$. Then, z satisfies

$$z_t = z_{xx}, \qquad 0 < x < 1, \qquad -\infty < t < \infty,$$
$$z(0, t) = z(1, t) = 0, \qquad -\infty < t < \infty, \qquad (9.1.5)$$
$$|z(x, T)| \le \tilde{M}(T) = M_1(T) + M_2(T),$$

where M_1 and M_2 are defined by (9.1.3) with u replaced respectively by u_1 and u_2. From Sec. 6.1 we see that, for fixed x and t,

$$z(x, t) = \sum_{n=1}^{\infty} a_n \exp\{-n^2\pi^2(t - T)\} \sin n\pi x, \qquad (9.1.6)$$

where

$$a_n = 2 \int_0^1 z(x, T) \sin n\pi x \, dx \qquad (9.1.7)$$

and $t > T$. Clearly,

$$|a_n| \le 2\tilde{M}(T). \qquad (9.1.8)$$

Consequently,

$$|z(x, t)| \le 2\tilde{M}(T) \sum_{n=1}^{\infty} \exp\{-n^2\pi^2(t - T)\}$$

$$\le 2\tilde{M}(T)\left\{ \exp\{-\pi^2(t - T)\} + \int_1^\infty 2x \exp\{-\pi^2(t - T)x^2\} \, dx \right\}$$

$$\le 2\{1 + \pi^{-2}(t - T)^{-1}\} \exp\{-\pi^2 t\} \tilde{M}(T) \exp\{\pi^2 T\}. \qquad (9.1.9)$$

For any $\varepsilon > 0$, we see that

$$|z(x, t)| < \varepsilon \qquad (9.1.10)$$

follows from (9.1.4) by taking T sufficiently negative. Hence, $z(x, t) = 0$, and $u_1 \equiv u_2$.

For existence, one uses

$$u(x, t) = -2 \int_{-\infty}^{t} \frac{\partial \theta}{\partial x}(x, t - \tau) g(\tau) \, d\tau + 2 \int_{-\infty}^{t} \frac{\partial \theta}{\partial x}(x - 1, t - \tau) h(\tau) \, d\tau,$$

$$(9.1.11)$$

with g and h satisfying the growth condition $g = h = \mathcal{O}(\exp\{\beta|t|\})$, $\beta < \pi^2$, as t tends to minus infinity, since $(\partial \theta / \partial x)(x, t - \tau) = \mathcal{O}(\exp\{-\pi^2|\tau|\})^*$ as τ tends to minus infinity. (See [2, 3].)

We can summarize the above results in the following statement.

Theorem 9.1.1. *The problem*

$$u_t = u_{xx}, \quad 0 < x < 1, \quad -\infty < t < \infty,$$
$$u(0, t) = g(t), \quad -\infty < t < \infty, \quad (9.1.12)$$
$$u(1, t) = h(t), \quad -\infty < t < \infty,$$

possesses a solution u provided that g and h are piecewise continuous functions that satisfy

$$g = \mathcal{O}(\exp\{\beta|t|\}), \quad -\infty < t \le T, \quad (9.1.13)$$

and

$$h = \mathcal{O}(\exp\{\beta|t|\}), \quad -\infty < t \le T, \quad (9.1.14)$$

for some constant $\beta < \pi^2$ and some constant T. Moreover, u satisfies

$$|u(x, t)| = \mathcal{O}(\exp\{\beta|t|\}), \quad (9.1.15)$$

and u is unique within the class of solutions v such that

$$\lim_{t \to -\infty} \left\{ \sup_{0 < x < 1} |v(x, t)| \right\} \exp\{\pi^2 t\} = 0. \quad (9.1.16)$$

Proof. See the analysis preceding the theorem statement. $\square$

As a corollary of the theorem we can state the following result.

Corollary 9.1.1. *If g and h are piecewise-continuous functions that are periodic with period P, then the solution u of (9.1.12) is periodic in t with period P.*

Proof. The function $U(x, t) = u(x, t + P)$ is clearly a solution of (9.1.12), since g and h are periodic with period P. By the uniqueness of the solution, it follows that $U(x, t) \equiv u(x, t)$ or, that is, $u(x, t + P) \equiv u(x, t)$. Consequently, u is periodic in t with period P. $\square$

---

*Use $\theta(x, t) = 1 + 2 \sum_{k=1}^{\infty} \exp\{-k^2\pi^2 t\} \cos k\pi x.$ $(9.1.17)$

9.2. THE INFINITE STRIP WITH FLUX-BOUNDARY CONDITIONS

In this section we discuss the problem

$$u_t = u_{xx}, \qquad 0 < x < 1, \qquad -\infty < t < \infty,$$
$$u_x(0, t) = g(t), \qquad -\infty < t < \infty, \qquad (9.2.1)$$
$$u_x(1, t) = h(t), \qquad -\infty < t < \infty,$$

where g and h are known piecewise-continuous functions.

As in Sec. 9.1, we see that the solution

$$v(x, t) = b_0 + \exp\{-\pi^2 t\}\cos \pi x \qquad (9.2.2)$$

provides us with an example of nonuniqueness of the solution of (9.2.1) unless some growth condition is placed upon u. The growth condition would eliminate the second term on the right side of (9.2.2) but can't handle the term b_0. From Sec. 6.4, the solution u given in (6.4.2) for $g = h = 0$ is

$$u(x, t) = \sum_{n=0}^{\infty} b_n \exp\{-n^2\pi^2 t\}\cos n\pi x, \qquad (9.2.3)$$

where

$$b_n = 2\int_0^1 u(x, 0)\cos n\pi x\, dx \qquad (9.2.4)$$

and

$$b_0 = \int_0^1 u(x, 0)\, dx. \qquad (9.2.5)$$

Consequently, we see a direct relationship between b_0 and the initial heat content of the solution u. Here the term "initial" refers to $t = -\infty$. Consequently, the solution here must satisfy two additional requirements. These are that

$$\lim_{t \to -\infty} \left\{ \sup_{0 < x < 1} u(x, t) \right\}\exp\{+\pi^2 t\} = 0 \qquad (9.2.6)$$

and that there exists a known constant Q such that

$$\lim_{t \to -\infty} \int_0^1 u(x, t)\, dx = Q. \qquad (9.2.7)$$

Suppose now that u_1 and u_2 are two solutions of (9.2.1) that satisfy (9.2.6) and (9.2.7). Consider the difference $z = u_1 - u_2$. We shall show that $z \equiv 0$. From (9.2.3) it follows that, for fixed x and t,

$$z(x, t) = \sum_{n=0}^{\infty} b_n \exp\{-n^2\pi^2(t - T)\}\cos n\pi x, \qquad (9.2.8)$$

where $t > T$,

$$b_n = 2\int_0^1 z(x, T)\cos n\pi x \, dx, \tag{9.2.9}$$

and

$$b_0 = \int_0^1 z(x, T) \, dx. \tag{9.2.10}$$

The same type of estimate given in Sec. 9.1 takes care of the terms $n \geq 1$. But for b_0, we have

$$\lim_{T \to -\infty} \int_0^1 z(x, T) \, dx = 0, \tag{9.2.11}$$

since u_1 and u_2 satisfy (9.2.7). Consequently, $z \equiv 0$, which implies $u_1 \equiv u_2$.
For existence, consider

$$u(x, t) = Q - 2\int_{-\infty}^t \theta(x, t - \tau)g(\tau) \, d\tau + 2\int_{-\infty}^t \theta(x - 1, t - \tau)h(\tau) \, d\tau. \tag{9.2.12}$$

Elementary estimates of the series expression for θ given by (9.1.17) yields

$$|\theta(x, t - \tau)| = \mathcal{O}(1) \tag{9.2.13}$$

and

$$|\theta(x - 1, t - \tau)| = \mathcal{O}(1). \tag{9.2.14}$$

Consequently, the integrals in (9.2.12) will converge if in a neighborhood of $-\infty$,

$$g(\tau) = \mathcal{O}(|\tau|^\alpha) \tag{9.2.15}$$

and

$$h(\tau) = \mathcal{O}(|\tau|^\alpha), \qquad \alpha < -1. \tag{9.2.16}$$

As the singularity of θ at $\tau = -\infty$ dominates those of the derivatives of θ, (9.2.15) and (9.2.16) are sufficient for the convergence of the derivatives of u. Consequently, u satisfies the heat equation. From the convergence of the integrals, it is clear that u satisfies (9.2.7). Since, for any $T < t$,

$$\lim_{x \downarrow 0} -2\int_T^t \frac{\partial \theta}{\partial x}(x, t - \tau)g(\tau) \, d\tau = g(t) \tag{9.2.17}$$

can be demonstrated by using the methods of Sec. 4.2 and Sec. 6.2, the convergence of the integrals implies that u satisfies the boundary conditions $u_x(0, t) = g(t)$ and $u_x(1, t) = h(t)$. Thus, we see that u given by (9.2.12) is a solution of (9.2.1) that satisfies the conditions (9.2.6) and (9.2.7).
We summarize the preceding analysis in the following statement.

Theorem 9.2.1. *For the problem*

$$u_t = u_{xx}, \qquad 0 < x < 1, \qquad -\infty < t < \infty,$$

$$u_x(0, t) = g(t), \qquad -\infty < t < \infty,$$

$$u_x(1, t) = h(t), \qquad -\infty < t < \infty,$$

$$\lim_{t \to -\infty} \int_0^1 u(x, t)\, dx = Q, \qquad (9.2.18)$$

where g and h are piecewise-continuous functions and Q is a given constant, the solution is given by

$$u(x, t) = Q - 2 \int_{-\infty}^t \theta(x, t - \tau) g(\tau)\, d\tau + 2 \int_{-\infty}^t \theta(x - 1, t - \tau) h(\tau)\, d\tau,$$

$$(9.2.19)$$

provided that, in a neighborhood of $\tau = -\infty$,

$$g(\tau) = \mathcal{O}(|\tau|^\alpha) \qquad (9.2.20)$$

and

$$h(t) = \mathcal{O}(|\tau|^\alpha), \qquad \alpha < -1. \qquad (9.2.21)$$

Moreover, the solution u is unique among the solutions v that satisfy the growth condition

$$\lim_{t \to -\infty} \left\{ \sup_{0 < x < 1} v(x, t) \right\} \exp\{\pi^2 t\} = 0. \qquad (9.2.22)$$

Proof. See the analysis preceding the theorem statement. □

In this case there is no simple application of the above theorem to produce a periodic solution. So, we consider the problem

$$u_t = u_{xx}, \qquad 0 < x < 1, \qquad 0 < t \le P,$$

$$u_x(0, t) = g(t), \qquad 0 < t \le P,$$

$$u_x(1, t) = h(t), \qquad 0 < t \le P,$$

$$u(x, 0) = u(x, P), \qquad 0 < x < 1,$$

where

$$\int_0^1 u(x, 0)\, dx = Q.$$

We shall show by direct methods that it possesses a unique solution if the data g and h satisfy a certain condition.

We begin by integrating $u_t = u_{xx}$ over the region $0 < x \le 1, 0 < t \le P$, and obtain

$$\int_0^1 u(x, P)\, dx - \int_0^1 u(x, 0)\, dx = \int_0^P h(t)\, dt - \int_0^P g(t)\, dt. \quad (9.2.23)$$

Since $u(x,0) = u(x, P)$, it follows necessarily that we assume

$$\int_0^P \{h(t) - g(t)\} \, dt = 0. \qquad (9.2.24)$$

Let $w = w(x, t)$ denote the solution of the problem

$$\begin{aligned}
w_t &= w_{xx}, & 0 < x < 1, & \quad 0 < t \le P, \\
w_x(0, t) &= g(t), & 0 < t \le P, & \\
w_x(1, t) &= h(t), & 0 < t \le P, & \qquad (9.2.25) \\
w(x, 0) &= 0, & 0 < x < 1. &
\end{aligned}$$

The existence and uniqueness of w follows from the analysis given in Sec. 6.4. We note that

$$\int_0^1 w(x, P) \, dx = \int_0^P \{h(t) - g(t)\} \, dt = 0. \qquad (9.2.26)$$

Set $u = v + w$, where v satisfies

$$\begin{aligned}
v_t &= v_{xx}, & 0 < x < 1, & \quad 0 < t \le P, \\
v_x(0, t) &= v_x(1, t) = 0, & 0 < t \le P, & \qquad (9.2.27) \\
v(x, 0) &= u(x, 0). &
\end{aligned}$$

Clearly,

$$\int_0^1 v(x, P) \, dx = \int_0^1 u(x, 0) \, dx = Q. \qquad (9.2.28)$$

In fact, if we let

$$b_n = 2 \int_0^1 u(x, 0) \cos n\pi x \, dx, \qquad n = 1, 2, \ldots, \qquad (9.2.29)$$

then v has the representation

$$v(x, t) = Q + \sum_{n=1}^{\infty} b_n \exp\{-n^2\pi^2 t\} \cos n\pi x \qquad (9.2.30)$$

and $u(x, 0)$ has the representation

$$u(x, 0) = Q + \sum_{n=1}^{\infty} b_n \cos n\pi x. \qquad (9.2.31)$$

Using the fact that $u(x, 0) = u(x, P) = v(x, P) + w(x, P)$, we obtain the equation

$$Q + \sum_{n=1}^{\infty} b_n \cos n\pi x = w(x, P) + Q + \sum_{n=1}^{\infty} b_n \exp\{-n^2\pi^2 P\} \cos n\pi x.$$

$$(9.2.32)$$

Setting

$$c_n = 2 \int_0^1 w(x, P) \cos n\pi x \, dx, \qquad (9.2.33)$$

we use the orthogonality of the $\cos n\pi x$, $n=1,2,3,\ldots$, to obtain the elementary system of equations

$$b_n = c_n + b_n \exp\{-n^2\pi^2 P\}, \qquad n=1,2,3,\ldots, \tag{9.2.34}$$

which can be solved for the unknown b_n to obtain

$$b_n = \left(1 - \exp\{-n^2\pi^2 P\}\right)^{-1} c_n, \qquad n=1,2,3,\ldots \tag{9.2.35}$$

Note that (9.2.26) implies that $c_0 = 0$. Hence, we see that the solution u is uniquely determined from the data g, h, and Q.

We summarize the above analysis in the following statement.

Theorem 9.2.2. *In order for the problem*

$$\begin{aligned} u_t &= u_{xx}, & 0 < x < 1, & \quad 0 < t \le P, \\ u_x(0,t) &= g(t), & 0 < t \le P, \\ u_x(1,t) &= h(t), & 0 < t \le P, \\ u(x,0) &= u(x,P), & 0 < x \le 1, \end{aligned} \tag{9.2.36}$$

where

$$\int_0^1 u(x,0)\, dx = Q,$$

where Q is a constant, P is a positive constant, and g and h are piecewise-continuous functions, to possess a unique solution, u, it is necessary and sufficient that

$$\int_0^P \{h(t) - g(t)\}\, dt = 0. \tag{9.2.37}$$

Proof. See the analysis preceding the theorem statement.   □

9.3. THE INFINITE STRIP WITH TEMPERATURE- AND FLUX-BOUNDARY CONDITIONS: SOME EXERCISES

In this section we shall discuss the problem

$$\begin{aligned} u_t &= u_{xx}, & 0 < x < 1, & \quad -\infty < t < \infty, \\ u(0,t) &= g(t), & -\infty < t < \infty, \\ u_x(1,t) &= h(t), & -\infty < t < \infty, \end{aligned} \tag{9.3.1}$$

where g and h are known piecewise-continuous functions.

As in Sec. 9.1 we see that the solution

$$v(x,t) = \exp\left\{-\frac{\pi^2}{4}t\right\}\sin\frac{\pi x}{2} \tag{9.3.2}$$

provided us with an example of the nonuniqueness of the solution of (9.3.1) unless some growth restriction is placed upon u as t tends to minus infinity. As in Sec. 9.1, the essence of the growth condition is that the solution cannot grow as fast as the fundamental mode exhibited in (9.2.2). Set

$$M(T) = \sup_{0 \le x \le 1} |u(x, T)| \qquad (9.3.3)$$

and suppose that

$$\lim_{T \to -\infty} M(T) \exp\left\{\frac{\pi^2}{4} T\right\} = 0. \qquad (9.3.4)$$

Exercise 9.1. Demonstrate that under the growth condition (9.3.4) uniqueness of the solution holds.

Theorem 9.3.1. *The problem*

$$
\begin{aligned}
u_t &= u_{xx}, & 0 < x < 1, &\quad -\infty < t < \infty, \\
u(0, t) &= g(t), & -\infty < t < \infty, & \qquad (9.3.5) \\
u_x(1, t) &= 0, & -\infty < t < \infty,
\end{aligned}
$$

possesses a solution u provided that g is a piecewise-continuous function that satisfies

$$|g(t)| = \mathcal{O}(\exp\{\beta|t|\}) \qquad (9.3.6)$$

$$\beta < \frac{\pi^2}{4}$$

for some positive constant C and some constant T. Moreover, u satisfies

$$|u(x, t)| = \mathcal{O}(\exp\{\beta|t|\}) \qquad (9.3.7)$$

and u is a unique within the class of solutions v such that

$$\lim_{t \to -\infty} \left\{\sup_{0 < x < 1} |v(x, t)|\right\} \exp\left\{\frac{\pi^2 t}{4}\right\} = 0. \qquad (9.3.8)$$

Exercise 9.2. Prove Theorem 9.3.1.

Exercise 9.3. Generalize Theorem 9.3.1 for $h \neq 0$.

As a corollary of the theorem we can state the following result.

Corollary 9.3.1. *If g is a piecewise-continuous function that is periodic with period P, then the solution u of (9.3.5) is periodic in t with period P.*

Exercise 9.4. Prove Corollary 9.3.1.

NOTES

The presentation of solutions for all times given in this chapter was motivated by Fife's [1] study for more general parabolic equations.

REFERENCES

1. Fife, P., Solutions of parabolic boundary problems existing for all time, *Arch. Rat. Mech. & Anal.*, **16** (1964), 155–168.
2. Hartman, P., and Wintner, A., On the solutions of the equation of heat conduction, *Am. J. Math.*, **72**, 367–395.
3. Widder, D. V., *The Heat Equation*. Academic Press, Inc., New York, 1975, p. 90.

Chapter 10

Analyticity of Solutions

10.1. INTRODUCTION

The theory of analytic functions of a complex variable is extensive. Its theorems have many applications in other areas of mathematics as well as other fields of science. In order to utilize many results from this theory, it is of interest to study the analyticity of various solutions of the heat equations. From Sec. 1.3, we see that all of those solutions are analytic functions of both variables. In fact they are entire analytic in both variables. The fundamental solution presents us with a different structure. At $t = 0$ we have a branch point for $t^{1/2}$ and an essential singularity for $\exp\{-x^2/4t\}$ as a function of t. Add to this the complication that the various solutions to initial- and/or boundary-value problems are convolutions, and the reader should see the utility of the following result.

   Theorem 10.1.1.   *Let D denote a domain in the complex $z = x + iy$, $i = \sqrt{-1}$, plane and let Γ denote an interval in $\mathbf{R}^1$. If $F = F(z, \xi)$ is defined and continuous on the Cartesian product $D \times \Gamma$, if for each $\xi \in \Gamma$, $F(z, \xi)$ is an analytic function of the complex variable z, and if for each compact $\mathcal{K} \subset D$, the integral*

$$\int_{\Gamma} |F(z, \xi)| \, d\xi \qquad (10.1.1)$$

converges uniformly for $z \in \mathscr{K}$, then the functoin

$$G(z) = \int_{\Gamma} F(z, \xi)\, d\xi \qquad (10.1.2)$$

is analytic in D. Moreover, if $M_{\mathscr{K}}$ is a positive constant such that

$$\int_{\Gamma} |F(z, \xi)|\, d\xi < M_{\mathscr{K}} \qquad (10.1.3)$$

for all $z \in \mathscr{K}$, then

$$|G(z)| < M_{\mathscr{K}} \qquad (10.1.4)$$

for all $z \in \mathscr{K}$.

   Proof. Let Γ' denote a closed and bounded subinterval of Γ. Consider a compact set $\mathscr{K} \subset D$. Then, the continuity of F on $\mathscr{K} \times \Gamma'$ implies that the Riemann integral

$$G_{\Gamma'}(z) = \int_{\Gamma'} F(z, \xi)\, d\xi \qquad (10.1.5)$$

exists. Moreover, for $z \in \mathscr{K}$, $G_{\Gamma'}(z)$ is the uniform limit of Riemann sums. As the Riemann sums are analytic functions of the complex variable z, $G_{\Gamma'}(z)$ is an analytic function in $\mathscr{K}$. Choosing now a sequence of $\Gamma' \uparrow \Gamma$, we see that the uniform convergence of the integral

$$\int_{\Gamma} |F(z, \xi)|\, d\xi$$

for $z \in \mathscr{K}$ implies that the $G_{\Gamma'}(z)$ converge uniformly to $G(z)$. Hence, $G(z)$ is analytic in $\mathscr{K}$. The bound (10.1.4) follows trivially from (10.1.2) and (10.1.3). $\qquad\qquad\qquad\qquad\qquad\qquad\qquad\qquad\qquad\qquad\qquad\qquad\qquad\quad$ □

10.2. THE ANALYTICITY OF THE SOLUTION OF THE INITIAL-VALUE PROBLEM IN THE SPACE VARIABLE

We shall apply Theorem 10.1.1 here to find conditions on f so that

$$u(z, t) = \int_{-\infty}^{\infty} K(z - \xi, t) f(\xi)\, d\xi \qquad (10.2.1)$$

is an analytic function of the complex variable $z = x + iy$, $i = \sqrt{-1}$.

   Consider the function $K(z - \xi, t)$. Obviously, the continuity and analyticity requirements of Theorem 10.1.1 are satisfied. Since

$$\exp\left\{ -\frac{(z-\xi)^2}{4t} \right\} = \exp\left\{ \frac{y^2}{4t} \right\} \exp\left\{ -\frac{(x-\xi)^2}{4t} \right\} \exp\left\{ -\frac{2i(x-\xi)y}{4t} \right\},$$

$$(10.2.2)$$

we see that for $|y| \le Y$,

$$|K(z-\xi, t)| \le \exp\left\{\frac{Y^2}{4t}\right\} K(x-\xi, t). \tag{10.2.3}$$

Consequently, for $|y| \le Y$, it follows from (10.2.1) and the positivity of K that

$$|u(z, t)| \le \exp\left\{\frac{Y^2}{4t}\right\} \int_{-\infty}^{\infty} K(x-\xi, t)|f(\xi)| \, d\xi. \tag{10.2.4}$$

Hence, the uniform absolute convergence of (10.2.1) depends upon the convergence of

$$\int_{-\infty}^{\infty} K(x-\xi, t)|f(\xi)| \, d\xi. \tag{10.2.5}$$

Recalling Sec. 3.3, we can state the following result.

Theorem 10.2.1. *If, for all x,*

$$|f(x)| \le C_1 \exp\left\{C_2 |x|^{1+\alpha}\right\}, \tag{10.2.6}$$

where C_1 is a positive constant, C_2 is a nonnegative constant, and α is a positive constant such that $0 \le \alpha \le 1$, then

$$u(z, t) = \int_{-\infty}^{\infty} K(z-\xi, t)f(\xi) \, d\xi \tag{10.2.7}$$

is an entire analytic function in z for all $t > 0$ when $\alpha < 1$ and for each t in $0 < t < (4C_2)^{-1}$ when $\alpha = 1$. Moreover, for $C_2 = 0$ and $|\operatorname{Im} z| \le Y$,

$$|u(z, t)| \le C_1 \exp\left\{\frac{Y^2}{4t}\right\}. \tag{10.2.8}$$

For $C_2 > 0$ and $0 \le \alpha < 1$, there exists a positive-valued function $C_3 = C_3(t, C_1, C_2, \alpha, \varepsilon)$ for $0 < \varepsilon < 1$ such that

$$|u(z, t)| \le C_3 \exp\left\{\frac{(1-\varepsilon)\varepsilon^{-1} X^2 + Y^2}{4t}\right\} \tag{10.2.9}$$

holds for all z such that $|\operatorname{Re} z| \le X$ and $|\operatorname{Im} z| \le Y$. Finally, for $C_2 > 0$, $\alpha = 1$, $0 < \varepsilon < 1$, and $0 < t < (1-\varepsilon)(4C_2)^{-1}$, there exists a positive-valued function $C_4 = C_4(t, C_1, C_2, \varepsilon)$ such that

$$|u(z, t)| \le C_4 \exp\left\{\frac{(1-\varepsilon)\varepsilon^{-1} X^2 + Y^2}{4t}\right\} \tag{10.2.10}$$

holds for all z such that $|\operatorname{Re} z| \le X$ and $|\operatorname{Im} z| \le Y$.

Proof. The analyticity of $u(x, t)$ follows immediately from the discussion of the convergence of the convolution given in Sec. 3.3. The estimate (10.2.8) follows trivially from (10.2.3) and Property E of Theorem 3.2.1. For the estimate (10.2.9), we recall (3.3.2) and (3.3.14), and combine them with (10.2.3), to obtain

$$|u(x,t)| \le C_1 \exp\left\{ \frac{(1-\varepsilon)\varepsilon^{-1}X^2 + Y^2}{4t} \right\} \int_{-\infty}^{\infty} (4\pi t)^{-1/2}$$

$$\cdot \exp\left\{ C_2|\xi|^{1+\alpha} - (4t)^{-1}(1-\varepsilon)\xi^2 \right\} d\xi, \qquad (10.2.11)$$

provided $0 < \varepsilon < 1$, $|\operatorname{Re} z| \le X$, and $|\operatorname{Im} z| \le Y$. Under the transformation $\rho = \xi(4t)^{-1/2}$, we need only estimate the integral

$$J = \frac{2}{\sqrt{\pi}} \int_0^{\infty} \exp\left\{ C_2 (4t)^{(1+\alpha)/2} \rho^{1+\alpha} - (1-\varepsilon)\rho^2 \right\} d\rho. \quad (10.2.12)$$

Since

$$C_2 (4t)^{(1+\alpha)/2} - (1-\varepsilon)\rho^{(1-\alpha)} \le -1 \qquad (10.2.13)$$

for all

$$\rho \ge \mu = \left[(1-\varepsilon)^{-1} \left(C_2 (4t)^{(1+\alpha)/2} + 1 \right) \right]^{(1-\alpha)^{-1}}, \qquad (10.2.14)$$

we see that

$$J \le J_1 + J_2, \qquad (10.2.15)$$

where

$$J_1 = \frac{2}{\sqrt{\pi}} \int_0^{\mu} \exp\left\{ C_2 (4t)^{(1+\alpha)/2} \rho^{1+\alpha} \right\} d\rho < \infty, \qquad (10.2.16)$$

and

$$J_2 = \frac{2}{\sqrt{\pi}} \int_{\mu}^{\infty} \exp\left\{ -\rho^{1+\alpha} \right\} d\rho < \infty. \qquad (10.2.17)$$

Combining (10.2.16) and (10.2.17), we see that the existence of $C_3 = C_3(t, C_1, C_2, \alpha, \varepsilon)$ is assured. The estimate (10.2.10) follows in a similar manner.   □

10.3. THE ANALYTICITY OF THE SOLUTION OF THE INITIAL-VALUE PROBLEM IN THE TIME VARIABLE

We shall apply Theorem 10.1.1 here to show that

$$u(x, \eta) = \int_{-\infty}^{\infty} K(x - \xi, \eta) f(\xi) \, d\xi \qquad (10.3.1)$$

is an analytic function of the complex variable $\eta = t + i\tau$, $i = \sqrt{-1}$, in the domains

$$D_\mu = \{ \eta \mid t > 0 \quad \text{and} \quad |\tau| \le \mu t \}, \quad \mu > 0, \qquad (10.3.2)$$

where the branch of $t^{1/2}$ in K is chosen so that $\eta^{1/2}$ reduces to $t^{1/2}$ when $\tau = 0$. Since

$$|\eta|^{-1/2} \leq t^{-1/2} \tag{10.3.3}$$

and

$$\left|\exp\left\{-\frac{(x-\xi)^2}{4\eta}\right\}\right| \leq \exp\left\{-\frac{(x-\xi)^2}{4(1+\mu^2)t}\right\} \tag{10.3.4}$$

for $\eta \in D_\mu$, we see that for $\eta \in D_\mu$,

$$|K(x-\xi,\eta)| \leq (1+\mu^2)^{1/2} K(x-\xi,(1+\mu^2)t). \tag{10.3.5}$$

Consequently, the analyticity of $u(x,\eta)$ in D_μ for each $\mu > 0$ follows from the uniform convergence of the integral

$$\int_{-\infty}^{\infty} K(x-\xi,(1+\mu)t)|f(\xi)|\,d\xi. \tag{10.3.6}$$

Recalling again Sec. 3.3, we can state the following result.

Theorem 10.3.1. *If, for all x,*

$$|f(x)| \leq C_1 \exp\{C_2|x|^{1+\alpha}\}, \tag{10.3.7}$$

where C_1 is a positive constant, C_2 is a nonnegative constant, and α is a positive constant such that $0 \leq \alpha \leq 1$, then

$$u(x,\eta) = \int_{-\infty}^{\infty} K(x-\xi,\eta)f(\xi)\,d\xi \tag{10.3.8}$$

is an analytic function in D_μ for each $\mu > 0$ when $\alpha < 1$. When $\alpha = 1$, $u(x,\eta)$ is an analytic function of η for

$$\eta \in D_{\mu,\beta} = \{\eta | 0 < t < \beta \quad \text{and} \quad |\tau| \leq \mu t\}$$

for each $\mu > 0$ and each $\beta, 0 < \beta < (4(1+\mu^2)C_2)^{-1}$. Moreover, for $C_2 = 0$ and $\eta \in D_\mu$,

$$|u(x,\eta)| \leq (1+\mu^2)^{1/2}C_1. \tag{10.3.9}$$

Note that this estimate is uniform in x. For $C_2 > 0$, $0 \leq \alpha < 1$, $0 < \varepsilon < 1$, and

$$\eta \in D_{\mu,\beta,\gamma} = \{\eta | 0 < \gamma \leq t \leq \beta \quad \text{and} \quad |\tau| \leq \mu t\},$$

there exists a positive-valued function $C_5 = C_5(\mu, \beta, \gamma, C_1, C_2, \varepsilon, \alpha)$ such that

$$|u(x,\eta)| \leq C_5 \exp\left\{\frac{\varepsilon^{-1}(1-\varepsilon)x^2}{4(1+\mu^2)\gamma}\right\}. \tag{10.3.10}$$

For $C_2 > 0$, $\alpha = 1$, $0 < \varepsilon < 1$, and $0 < \beta < (1 - \varepsilon)(4(1 + \mu^2)C_2)^{-1}$, there exists a positive-valued function $C_6 = C_6(\mu, \beta, \gamma, C_1, C_2, \varepsilon)$ such that

$$|u(x, \eta)| \leq C_6 \exp\left\{ \frac{\varepsilon^{-1}(1 - \varepsilon)x^2}{4(1 + \mu^2)\gamma} \right\} \tag{10.3.11}$$

holds for all $\eta \in D_{\mu, \beta, \gamma}$.

Proof.   The analyticity of $u(x, \eta)$ follows immediately from the discussion of the convergence of the convolution given in Sec. 3.3. Since

$$|u(x, \eta)| \leq (1 + \mu^2)^{1/2} \int_{-\infty}^{\infty} K(x - \xi, (1 + \mu^2)t)|f(\xi)| \, d\xi, \tag{10.3.12}$$

the estimate (10.3.9) follows from Property E of Theorem 3.2.1. By an argument similar to that given in the proof of Theorem 10.2.1., we obtain the estimates (10.3.10) and (10.3.11).   □

10.4.   THE ANALYTICITY OF THE SOLUTION OF THE INITIAL-BOUNDARY-VALUE PROBLEM FOR THE QUARTER PLANE IN THE SPACE VARIABLE

For the solution of the initial-boundary-value problem for the quarter plane with temperature-boundary condition, we need only consider the function

$$w(x, t) = -2 \int_0^t \frac{\partial K}{\partial x}(x, t - \tau)g(\tau) \, d\tau \tag{10.4.1}$$

since the function

$$v(x, t) = \int_0^{\infty} G(x, \xi, t)f(\xi) \, d\xi \tag{10.4.2}$$

can be rewritten in the form of a solution to an initial-value problem to which Sec. 10.2. applies.

Theorem 10.4.1.   *For z in the domain*

$$D^+ = \{z \mid x > 0 \quad \text{and} \quad |y| < x\}, \tag{10.4.3}$$

the function

$$w(z, t) = -2 \int_0^t \frac{\partial K}{\partial x}(z, t - \tau)g(\tau) \, d\tau \tag{10.4.4}$$

is analytic in the variable z. Moreover, for z in the domains

$$D_\mu^+ = \{z \mid x > 0 \quad \text{and} \quad |y| \leq \mu x\}, \quad 0 < \mu < 1, \tag{10.4.5}$$

$$|w(z, t)| \leq (1 + \mu^2)^{1/2}(1 - \mu^2)^{-1/2}\|g\|_t, \tag{10.4.6}$$

where

$$\|g\|_t = \sup_{0 \le \tau \le t} |g(\tau)|. \tag{10.4.7}$$

Proof. Clearly, the analyticity and continuity properties of $(\partial K / \partial x)(z, t - \tau)$ are satisfied. Let z belong to D_μ^+. Then,

$$|z| \le (1 + \mu^2)^{1/2} x \tag{10.4.8}$$

and

$$\left| \exp\left\{ -\frac{z^2}{4(t-\tau)} \right\} \right| = \exp\left\{ -\frac{(x^2 - y^2)}{4(t-\tau)} \right\} \le \exp\left\{ -\frac{(1-\mu^2)x^2}{4(t-\tau)} \right\}. \tag{10.4.9}$$

From (10.4.8) and (10.4.9), it follows that for $z \in D_\mu^+$,

$$\left| \frac{\partial K}{\partial x}(z, t - \tau) \right| \le -\frac{(1 + \mu^2)^{1/2}}{(1 - \mu^2)^{1/2}} \frac{\partial K}{\partial x} \left((1 - \mu^2)^{1/2} x, t - \tau \right). \tag{10.4.10}$$

Under the transformation $\rho = 2^{-1}(t - \tau)^{-1/2}(1 - \mu^2)^{1/2} x$,

$$\left| -2 \int_0^t \frac{\partial K}{\partial x} \left((1 - \mu^2)^{1/2} x, t - \tau \right) g(\tau) \, d\tau \right| \le \|g\|_t. \tag{10.4.11}$$

Hence, for all $z \in D_\mu^+$,

$$|w(z, t)| \le (1 + \mu^2)^{1/2} (1 - \mu^2)^{-1/2} \|g\|_t, \tag{10.4.12}$$

demonstrating (10.4.6) and the uniform absolute convergence of the integral (10.4.4), whence the analyticity follows.   □

As a corollary of the argument, we state the following result.

Corollary 10.4.1. *For z in the domain*

$$D^- = \{ z \mid x < 0 \quad and \quad |y| < |x| \}, \tag{10.4.13}$$

the function

$$w(z, t) = -2 \int_0^t \frac{\partial K}{\partial x}(z, t - \tau) g(\tau) \, d\tau \tag{10.4.14}$$

is analytic in the variable z. Moreover, for all z in the domains

$$D_\mu^- = \{ z \mid x < 0 \quad and \quad |y| \le \mu|x| \}, \qquad 0 < \mu < 1, \tag{10.4.15}$$

$$|w(z, t)| \le (1 + \mu^2)^{1/2} (1 - \mu^2)^{-1/2} \|g\|_t, \tag{10.4.16}$$

where $\|g\|_t$ is defined by (10.4.7).

Proof. The proof is left to the reader.   □

As a further corollary of the method of analysis we can state the following result.

Corollary 10.4.2. *For z in D^+ as defined by (10.4.3), the function*

$$w(z,t) = -2\int_0^t K(z,t-\tau)g(\tau)\,d\tau \qquad (10.4.17)$$

is analytic. Moreover, for z in D^+,

$$|w(z,t)| \le 2\pi^{-1/2}t^{1/2}\|g\|_t, \qquad (10.4.18)$$

where $\|g\|_t$ is defined by (10.4.7). Moreover, the same results hold in D^-, which is defined by (10.4.13).

  Proof. For $z \in D^+$, we note that

$$|K(z,t-\tau)| \le 2^{-1}\pi^{-1/2}(t-\tau)^{-1/2}. \qquad (10.4.19) \quad \square$$

10.5. THE ANALYTICITY OF THE SOLUTION OF THE INITIAL-BOUNDARY-VALUE PROBLEM FOR THE SEMI-INFINITE STRIP IN THE SPACE VARIABLE

Recalling Theorem 6.5.1, we can write the solution u of

$$\begin{aligned}
u_t &= u_{xx}, & 0 < x < 1, & \quad 0 < t, \\
u(x,0) &= f(x), & 0 < x < 1, & \\
u(0,t) &= g(t), & 0 < t, & \\
u(1,t) &= h(t), & 0 < t, &
\end{aligned} \qquad (10.5.1)$$

in the form

$$u(x,t) = v(x,t) - 2\int_0^t \frac{\partial K}{\partial x}(x,t-\tau)\phi_1(\tau)\,d\tau$$

$$+ 2\int_0^t \frac{\partial K}{\partial x}(x-1,t-\tau)\phi_2(\tau)\,d\tau, \qquad (10.5.2)$$

where

$$v(x,t) = \int_{-\infty}^{\infty} K(x-\xi,t)f(\xi)\,d\xi, \qquad (10.5.3)$$

f is a smooth bounded extension of f, and the pair, ϕ_1 and ϕ_2, is the solution of the system of integral equations

$$\phi_1(t) = g(t) - v(0,t) - 2\int_0^t \frac{\partial K}{\partial x}(-1,t-\tau)\phi_2(\tau)\,d\tau,$$

$$\phi_2(t) = h(t) - v(1,t) + 2\int_0^t \frac{\partial K}{\partial x}(1,t-\tau)\phi_1(\tau)\,d\tau. \qquad (10.5.4)$$

  From Sec. 10.2, we see that $v(x,t)$ is analytic in the space variable and, for $|f(x)| < C_1$,

$$|v(z,t)| \le C_1\exp\left\{\frac{Y^2}{4t}\right\} \qquad (10.5.5)$$

for all z such that $|\operatorname{Im} z| \le Y$. In order to employ the results of Sec. 10.4, we need estimates of ϕ_1 and ϕ_2. Consider the first equation in (10.5.4). Taking absolute values of the terms and using the inequality $e^{-x} \le p! x^{-p}$ with $p = 2$, we see that

$$|\phi_1(t)| \le \|g\|_t + C_1 + 16\pi^{-1/2} \int_0^t (t-\tau)^{1/2} |\phi_2(\tau)| \, d\tau, \quad (10.5.6)$$

where $\|g\|_t = \sup_{0 \le \tau \le t} |g(\tau)|$. Consequently, for $0 \le t \le T$,

$$|\phi_1(t)| \le \|g\|_T + C_1 + 16\pi^{-1/2} T^{1/2} \int_0^t \|\phi\|_\tau d\tau, \quad (10.5.7)$$

where

$$\|\phi\|_\tau = \max\{ \|\phi_1\|_\tau, \|\phi_2\|_\tau \}. \quad (10.5.8)$$

Similarly,

$$|\phi_2(t)| \le \|h\|_T + C_1 + 16\pi^{-1/2} T^{1/2} \int_0^t \|\phi\|_\tau d\tau. \quad (10.5.9)$$

Combining (10.5.7) and (10.5.9), it follows that

$$\|\phi\|_t \le \{ \|g\|_T + \|h\|_T + C_1 \} + 16\pi^{-1/2} T^{1/2} \int_0^t \|\phi\|_\tau d\tau. \quad (10.5.10)$$

With an application of Lemma 8.4.1, Gronwall's Inequality, we see that, for $0 \le t \le T$,

$$\|\phi\|_t \le \{ \|g\|_T + \|h\|_T + C_1 \} \exp\{ 16\pi^{-1/2} T^{1/2} t \}. \quad (10.5.11)$$

Now, we can apply the results of Sec. 10.2, Theorem 10.4.1, and Corollary 10.4.1, to obtain the following result.

Theorem 10.5.1. *For z in the domain*

$$D_\mu = \{ z | 0 < x < 1, |y| \le \mu|x|, |y| \le \mu|1 - x| \}, \qquad 0 < \mu < 1,$$

$$(10.5.12)$$

(see Fig. 10.5.1) the solution u of

$$u_t = u_{xx}, \qquad 0 < x < 1, \qquad 0 < t,$$
$$u(x, 0) = f(x), \qquad 0 < x < 1,$$
$$u(0, t) = g(t), \qquad 0 < t, \qquad (10.5.13)$$
$$u(1, t) = h(t), \qquad 0 < t,$$

is an analytic function of z. Moreover, if $|f| < C_1$, $C_1 > 0$, then for $z \in D_\mu$ and $0 < t \le T$,

$$|u(z, t)| \le C_1 \exp\{ (32)^{-1} t^{-1} \} + 2(1 + \mu^2)^{1/2} (1 - \mu^2)^{-1/2}$$
$$\cdot \{ \|g\|_T + \|h\|_T + C_1 \} \exp\{ 16\pi^{-1/2} T^{1/2} t \}, \quad (10.5.14)$$

where $\|g\|_T$ and $\|h\|_T$ are the uniform norms over $0 < t \le T$.

Proof. Take $Y = \sqrt{2}/4$ in (10.5.5). The remainder is a straightforward application of Theorem 10.4.1 and Corollary 10.4.1.   □

10.6.   SOME PROBLEMS IN ESTIMATING THE MODULUS OF ANALYTIC FUNCTIONS

In the next chapter we shall consider the continuous dependence of solutions of some state estimation problems upon the data. Most of those problems are not well posed, in the sense of Hadamard. We shall have more to say about this in the next chapter. In order to demonstrate the estimates of continuous dependence, we shall use the analyticity of the solution in the space and time variable and the following estimates of the modulus of analytic functions of a complex variable.

We begin with the statement of a general problem and its solution.

Problem 10.6.1. Let D denote a domain in the complex z-plane with boundary ∂D, that consists of a finite number of piecewise-smooth closed Jordan curves. Suppose that $f = f(z)$ is an analytic function defined in D, continuous in $\bar{D}$, and $|f| < C$ throughout $\bar{D}$, where C is positive constant. Moreover, suppose that $|f| < \varepsilon$, $\varepsilon < C$, on an arc $\Gamma \in \partial D$. Can one find a known function $F(z, \varepsilon, C, D)$ that is independent of f such that, for all $z \in D$, $\lim_{\varepsilon \to 0} F(z, \varepsilon, C, D) = 0$, and

$$|f(z)| < f(z, \varepsilon, C, D)? \tag{10.6.1}$$

The answer is yes! We shall discuss here the following solution.

Solution. Let $h = h(z)$ denote the bounded harmonic function (solution of Laplace's equation: $h_{xx} + h_{yy} = 0$) in D that is zero on $\partial D - \Gamma$

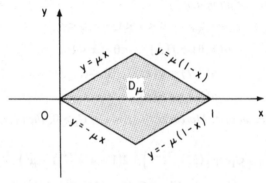

FIGURE 10.5.1

and one on Γ. It is well known [5] that such a function h exists. Consider the function $\log C^{-1}f(z)$. Its real part, $\log C^{-1}|f(z)|$, is harmonic in D. Moreover, on $\partial D - \Gamma$,

$$\log C^{-1}|f(z)| < \log C^{-1}C = 0 = h(z)\log C^{-1}\varepsilon, \qquad (10.6.2)$$

while on Γ

$$\log C^{-1}|f(z)| < \log C^{-1}\varepsilon = h(z)\log C^{-1}\varepsilon. \qquad (10.6.3)$$

By the maximum principle for harmonic functions, we see that, for $z \in \bar{D}$,

$$\log C^{-1}|f(z)| < h(z)\log C^{-1}\varepsilon < \log C^{-h(z)}\varepsilon^{h(z)}, \qquad (10.6.4)$$

whence it follows that

$$|f(z)| < C^{1-h(z)}\varepsilon^{h(z)}. \qquad (10.6.5)$$

Hence, the function

$$F(z, \varepsilon, C, D) = C^{1-h(z)}\varepsilon^{h(z)} \qquad (10.6.6)$$

provides an answer to the question.   □

Remark. Note that from the strong maximum principle for harmonic functions for $z \in D$, $0 < h(z) < 1$. Consequently, for each $z \in D$, we see that if ε tends to zero, then the analytic functions f that satisfy $|f| < C$ throughout D with $|f| < \varepsilon$ on Γ must tend to zero in modulus via the estimate (10.6.5).

We proceed now with some examples of some simple regions for which simple estimates for $h(z)$ can be obtained. Hence, we shall have some explicit tools to use in the next chapter.

> **Lemma 10.6.1** (Carleman).   *Let $f(z)$ be an analytic function in the sector*

$$D = \left\{ z = (r, \theta) | 0 < r < R, \quad -\frac{\alpha\pi}{2} < \theta < \frac{\alpha\pi}{2} \right\} \qquad (10.6.7)$$

which is continuous in the closure $\bar{D}$ of D. For $\varepsilon < C$, suppose that

$$|f(z)| < \varepsilon, \quad z = (R, \theta), \quad -\frac{\alpha\pi}{2} < \theta < \frac{\alpha\pi}{2}, \quad 0 < \alpha < 2,$$

$$(10.6.8)$$

and that

$$|f(z)| < C, \quad z = \left(r, -\frac{\alpha\pi}{2}\right), \quad and \quad z = \left(r, \frac{\alpha\pi}{2}\right) \quad for\ 0 \le r \le R.$$

$$(10.6.9)$$

Then, for $z = (x, 0)$,

$$|f(x)| < C^{1-(x/R)^{1/\alpha}}\varepsilon^{(x/R)^{1/\alpha}}. \qquad (10.6.10)$$

Proof.   The $h(z)$ here would be zero on the linear sides of the sector and one on the arc of the circle. Note that $\{\log C^{-1}\varepsilon\} < 0$. Hence, if $h_1(z)$ is a harmonic function such that $h_1(z) < h(z)$, then $\{\log C^{-1}\varepsilon\} h(z) < \{\log C^{-1}\varepsilon\} h_1(z)$, and $\{\log C^{-1}\varepsilon\} h_1(z)$ can be used in (10.6.4) to yield

$$|f(z)| < C^{1-h_1(z)} \varepsilon^{h_1(z)}. \tag{10.6.11}$$

Here an elementary harmonic lower bound is provided by

$$h_1(z) = \left(\frac{r}{R}\right)^{1/\alpha} \cos\left(\frac{\theta}{\alpha}\right) = \mathrm{Re}\left(\frac{z}{R}\right)^{1/\alpha}. \tag{10.6.12}$$

Clearly, $h_1(z) = 0$ on the linear sides of the sector D while

$$h_1(Re^{i\theta}) = \cos\left(\frac{\theta}{\alpha}\right), \qquad -\frac{\alpha\pi}{2} < \theta < \frac{\alpha\pi}{2},$$

on the circular arc. Since $\cos(\theta/\alpha) \le 1$ for $-(\alpha\pi/2) < \theta < (\alpha\pi/2)$, it follows that $h_1(z) \le h(z)$. This, combined with (10.6.11) and (10.6.12), yields the estimate (10.6.10).

Lemma 10.6.2. *Let $f(z)$ be a founded analytic function in the semi-infinite strip*

$$D = \{z \,|\, x > 0 \quad and \quad |y| < Y\}, \tag{10.6.13}$$

which is continuous in the closure $\bar{D}$ of D. For $\varepsilon < C$, suppose that

$$|f(iy)| < \varepsilon, \qquad |y| < Y, \qquad x = 0, \tag{10.6.14}$$

and that

$$|f(z)| < C, \qquad z = x \pm iY, \qquad x \ge 0. \tag{10.6.15}$$

Then, for $z = x$,

$$|f(x)| < C^{1-\exp\{-2^{-1}Y^{-1}\pi x\}} \varepsilon^{\exp\{-2^{-1}Y^{-1}\pi x\}}. \tag{10.6.16}$$

Proof. We need only find a lower estimate for the harmonic function h that is zero when $z = x \pm iY$, $x \ge 0$, and one when $z = iy$, $|y| \le Y$. The harmonic function

$$h_1(z) = \exp\left\{-\frac{\pi x}{2Y}\right\} \cos\frac{\pi y}{2Y} \tag{10.6.17}$$

provides us with an explicit lower bound for h. Consequently, the result (10.6.16) follows from the combination of (10.6.17) with (10.6.11). □

Lemma 10.6.3. *Let $f(z)$ be an analytic function in the rectangle*

$$D = \{z \,|\, a < x < b \quad and \quad c < y < d\}, \tag{10.6.18}$$

which is continuous in the closure $\bar{D}$ of D. For $\varepsilon < C$, suppose that

$$|f(x + ic)| < \varepsilon, \qquad a < x < b, \tag{10.6.19}$$

and that $|f| < C$ on the remaining three sides of D. Then for $z = ((a+b)/2)+$

iy, c ≤ y ≤ d,

$$\left| f\left(\frac{a+b}{2} + iy \right) \right| < C^{1-S(y)} \varepsilon^{S(y)}, \tag{10.6.20}$$

where

$$S(y) = \frac{\sinh[\pi(y-d)/(b-a)]}{\sinh[\pi(c-d)/(b-a)]}. \tag{10.6.21}$$

Proof. As in the proofs above, we need only find a lower harmonic estimate for h that is one for $z = x + ic$, $a < x < b$, and zero on the remaining three sides of D. Such an estimate is provided by the harmonic function

$$h_1(z) = S(y)\sin\left[\frac{(x-a)\pi}{(b-a)} \right]. \tag{10.6.22}$$

As above, the result (10.6.20) follows from the combination of (10.6.22) with (10.6.11). □

Another useful estimate is given in the following statement.

Lemma 10.6.4 (Lindelöf). *Let $f(z)$ be an analytic function in an n-sided regular polygon D. Suppose that $|f(z)| < C$ for z on $(n-1)$ sides of ∂D, while $|f(z)| < \varepsilon$ for z on the remaining side. Then at the center z_0 of D,*

$$|f(z_0)| < C^{(n-1)/n}\varepsilon^{1/n}. \tag{10.6.23}$$

Proof. By a translation and a rotation, it can be assumed that $z_0 = 0$ and one of the vertices of D lies on the real axis. Consider the function

$$F(z) = f(z)f(\zeta z)\cdots f(\zeta^{n-1}z),$$

where $\zeta = \exp\{i(2\pi/n)\}$. The arguments $\zeta z, \ldots, \zeta^{n-1}z$ generate successive rotations of the regular polygon D onto itself. It is not difficult to see that on the boundary ∂D, $|F(z)| < C^{n-1}\varepsilon$. At $z = 0$, $F(0) = f(0)^n$. Hence, from the maximum modulus theorem for analytic functions of a complex variable, we see that $|f(0)|^n < C^{n-1}\varepsilon$, from whence follows the result (10.6.23). □

Remark. Especially useful below is the application of this result for equilateral triangles.

We now apply the above Lemmas to obtain estimates in two situations that will occur frequently in the following chapter.

Lemma 10.6.5. *Let $f(z)$ be an analytic function in the infinite strip*

$$D = \{z \mid -\infty < x < \infty \quad and \quad |y| < Y\}, \tag{10.6.24}$$

which is continuous in the closure $\overline{D}$ of D. Suppose that $|f| < C$ throughout $\overline{D}$. In addition, suppose that for real $z = x$, $a \le x \le b$, $|f(x)| < \varepsilon$, where $\varepsilon < C$.

Then, for all real $z = x \geq (a+b)/2$,

$$|f(x)| < C^{1-S(2^{-1}Y)\exp\{-Y^{-1}\pi(x-[(a+b)/2])\}} \cdot \varepsilon^{S(2^{-1}Y)\exp\{-Y^{-1}\pi(x-[(a+b)/2])\}}$$

$$(10.6.25)$$

where

$$S(2^{-1}Y) = \frac{\sinh[\pi Y/(2(b-a))]}{\sinh[\pi Y/(b-a)]},$$

$$(10.6.26)$$

and for all real $z = x \leq (a+b)/2$,

$$|f(z)| < C^{1-S(2^{-1}Y)\exp\{Y^{-1}\pi(x-[(a+b)/2])\}}$$

$$\cdot \varepsilon^{S(2^{-1}Y)\exp\{Y^{-1}\pi(x-[(a+b)/2])\}}.$$

$$(10.6.27)$$

Proof. First, we apply Lemma 10.6.3 with $c = 0$ and $d = Y$ to obtain for $|y| \leq 2^{-1}Y$,

$$\left| f\left(\frac{a+b}{2} + iy\right) \right| < C^{1-S(2^{-1}Y)} \varepsilon^{S(2^{-1}Y)}.$$

$$(10.6.28)$$

Next, we apply Lemma 10.6.2 with Y replaced by $2^{-1}Y$, and obtain for $z = x \geq (a+b)/2$,

$$|f(x)| \leq C^{1-\exp\{-Y^{-1}\pi(x-[(a+b)/2])\}} \cdot \left\{ C^{1-S(2^{-1}Y)} \varepsilon^{S(2^{-1}Y)} \right\}^{\exp\{-Y^{-1}\pi(x-[(a+b)/2])\}},$$

$$(10.6.29)$$

from which (10.6.25) follows. In a similar manner, we obtain (10.6.27). $\square$

Lemma 10.6.6. *Let $f(z)$ be an analytic function in the infinite sector*

$$D_\mu^+ = \{ z \mid x > 0 \quad and \quad |y| \leq \mu|x| \}, \qquad 0 < \mu, \qquad (10.6.30)$$

which is continuous in the closure $\overline{D}_\mu^+$ of D_μ^+. Suppose that $|f| < C$ throughout D_μ^+. In addition, suppose that for real $z = x$, $0 < a \leq x \leq b$, $|f(x)| < \varepsilon$, where $\varepsilon < C$. Then, for all real $z = x$, $0 < x < a$,

$$|f(x)| < C^{1-(1/3)(x/R)^{1/\alpha}} \varepsilon^{(1/3)(x/R)^{1/\alpha}},$$

$$(10.6.31)$$

where if $\sqrt{3}(b-a) \leq \mu(a+b)$, then

$$R = \left\{ 4^{-1}(a+b)^2 + 12^{-1}(b-a)^2 \right\}^{1/2}$$

$$(10.6.32)$$

and

$$\alpha = \frac{2}{\pi} \text{Arctan}\left(\frac{\sqrt{3}(b-a)}{3(b+a)} \right)$$

$$(10.6.33)$$

while if $\sqrt{3}(b-a) > \mu(a+b)$, then

$$R = 3^{-1/2}10^{1/2}(3^{1/2} - \mu)^{-2} a$$

$$(10.6.34)$$

and

$$\alpha = \frac{2}{\pi} \text{Arctan}\left(\frac{\mu}{3}\right). \tag{10.6.35}$$

Proof. If $\sqrt{3}(b-a) \le \mu(a+b)$, then we can construct the equilateral triangle with base $a \le x \le b$ and vertex $(2^{-1}(a+b), 2^{-1}\sqrt{3}(b-a))$ and be assured that it lies in $\bar{D}_\mu^+$. (See Fig. 10.6.1.) Since $|f| < C$ on the sides of the triangle and $|f| < \varepsilon$ on its base, we can apply Lemma 10.6.4 to obtain $|f(z_0)| < C^{2/3}\varepsilon^{1/3}$ at the center $z_0 = 2^{-1}(a+b) + i(6)^{-1}\sqrt{3}(b-a)$. Consider the line segment through $z = a$ and z_0 extended to the side opposite $z = a$. Considering the family of equilateral triangles formed by moving the side opposite $z = a$ down the line segment through z_0 to $z = a$, we see that the centers of those triangles constitute the line segment between $z = a$ and z_0. From Lemma 10.6.4, we see that $|f| < C^{2/3}\varepsilon^{1/3}$ on that segment. By a similar argument, $|f| < C^{2/3}\varepsilon^{1/3}$ on the line segments z_0 to $z = b$, $z = b$ to $\bar{z}_0$, and $\bar{z}_0$ to $z = a$. By the maximum modulus theorem for analytic functions, $|f| < C^{2/3}\varepsilon^{1/3}$ throughout the diamond-shaped region with vertices $z = a$, z_0, $z = b$, and $\bar{z}_0$. We consider now the sector of the circle centered at $z = 0$ which has the linear sides $z = 0$ to z_0 and $z = 0$ to $\bar{z}_0$, and whose circular arc is contained in the diamond-shaped region and connects z_0 to $\bar{z}_0$. On this

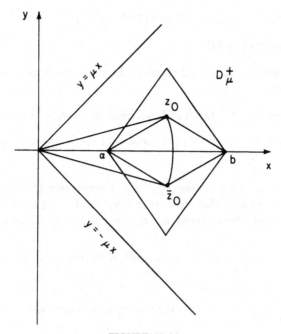

FIGURE 10.6.1

circular arc, $|f| < C^{2/3}\varepsilon^{1/3}$ while $|f| < C$ on the linear sides of the sector. We can apply the Lemma 10.6.1 here with

$$\alpha = \frac{2}{\pi} \text{Arctan}\left(\frac{\sqrt{3}\,(b-a)}{3(b+a)} \right) \tag{10.3.36}$$

and

$$R = \left\{ 4^{-1}(a+b)^2 + 12^{-1}(b-a)^2 \right\}^{1/2} \tag{10.6.37}$$

to obtain the result (10.6.31).

On the other hand if $\sqrt{3}\,(b-a) > \mu(a+b)$, then the line $y = \mu x$ intersects the line $y = \sqrt{3}\,(x-a)$ at the point

$$z_1 = \sqrt{3}\,\left(\sqrt{3} - \mu \right)^{-1} a + i\sqrt{3}\,\mu\left(\sqrt{3} - \mu \right)^{-1} a$$

and, using z_1 as the vertex off of $a \le x \le b$, we construct an equilateral triangle with its base on $a \le x \le b$ and another of its vertices at $z = a$. As above, this triangle lies in $\overline{D_\mu^+}$. We can repeat the argument above, using first Lemma 10.6.4, followed by the application of Lemma 10.6.1, with

$$\alpha = \frac{2}{\pi} \text{Arctan}\left(\frac{\mu}{3} \right) \tag{10.6.38}$$

and

$$R = 3^{-1/2} 10^{1/2} \left(3^{1/2} - \mu \right)^{-2} a \tag{10.6.39}$$

to obtain the result (10.6.31). $\qquad\qquad\qquad\qquad\qquad\qquad\qquad\qquad\square$

As a corollary of the argument of Lemma 10.6.6, we can state the following result.

Corollary 10.6.1.   *Let $f(z)$ be an analytic function in the region*

$$D_\mu = \{ z \,|\, 0 < x < 1, \quad |y| \le \mu|x| \quad \text{and} \quad |y| \le \mu|1 - x| \}, \qquad \mu > 0,$$

$$\tag{10.6.40}$$

which is continuous in the closure $\overline{D}_\mu$ of D_μ. Suppose that $|f| < C$ throughout $\overline{D}_\mu$. In addition, suppose that, for real $z = x$, $0 < a \le x \le b < 1$, $|f(x)| < \varepsilon$, where $\varepsilon < C$. Then, there exists $\alpha = \alpha(a, b, \mu)$, $0 < \alpha < 1$, such that, for $0 < x < a$,

$$|f(x)| < C^{1 - (1/3)(x/b)^{1/\alpha}} \varepsilon^{(1/3)(x/b)^{1/\alpha}}, \tag{10.6.41}$$

while for $b < x < 1$,

$$|f(x)| < C^{1 - (1/3)((1-x)/(1-a))^{1/\alpha}} \varepsilon^{(1/3)((1-x)/(1-a))^{1/\alpha}}. \tag{10.6.42}$$

Proof.   The proof is left to the reader. $\qquad\qquad\qquad\qquad\qquad\qquad\qquad\square$

EXERCISES

The application of Theorem 10.1.1 to the integrals in Secs. 10.2, 10.3, and 10.4 is incorrect when the data f and g are piecewise continuous.

10.1. Generalize Theorem 10.1.1 for $F(z, \xi)$ piecewise continuous in ξ.
10.2. Generalize Theorem 10.1.1 for $F(z, \xi)$ with an integrable singularity at an endpoint of Γ.
10.3. Generalize Theorem 10.1.1 for $F(z, \xi)$ with both piecewise continuity in ξ and an integrable singularity at an endpoint of Γ.
10.4. Show that

$$w(z, t) = \int_0^t K(z - s(\tau), t - \tau) \varphi(\tau) \, d\tau \qquad \text{(A)}$$

is analytic in the domain

$$D_\mu^+(t) = \{ z \mid x > s(t), \quad |y| < \mu |x - s(t)| \}, \qquad 0 < \mu < 1, \quad \text{(B)}$$

for piecewise-continuous φ and smooth s.

10.5. Show that

$$v(z, t) = \int_0^t \frac{\partial K}{\partial x} (z - s(\tau), t - \tau) \varphi(\tau) \, d\tau \qquad \text{(C)}$$

is analytic in $D_\mu^+(t)$ for piecewise-continuous φ and smooth s.

10.6. Consider $w(x, t) = -2 \int_0^t (\partial K / \partial x)(x, t - \tau) g(\tau) \, d\tau$. Investigate the analyticity of $w(x, t)$ with respect to t.

NOTES

The analyticity of solutions to various initial-boundary-value problems was presented by Gevrey [4]. The material on the estimation of the modulus of analytic functions of a complex variable can be found in Behnke and Sommer [1], Cannon and Miller [2], and Carleman [3]. Theorem 10.1.1 was a private communication to the author from Gerald R. MacLane.

REFERENCES

1. Behnke, H., and Sommer, F., *Theorie der Analytischen Funktionen einer Komplexen Veranderlichen.* Springer-Verlag, Berlin, 1962, 128.
2. Cannon, J. R., and Miller, K., Some problems in numerical analytic continuation, *SIAM J. Num. Anal.*, 2 (1965), 87–98.
3. Carleman, T., *Fonctions Quasi Analytiques.* Gauthier-Villars, Paris, 1926, 3–5.
4. Gevrey, M., Sur les équations aux dérivées partielles du type parabolique, *J. Math. Pures Appl.*, 9 (1913), 305–471.
5. Kellogg, O. D., *Foundations of Potential Theory.* Dover Publications, Inc., New York, 1953.

Chapter 11

Continuous Dependence upon the Data for Some State-Estimation Problems

11.1. INTRODUCTION

The determination of the temperature in a conductor from initial and/or boundary data is a classical state-estimation problem, with the temperature representing the state that is being estimated. For the problem

$$u_t = u_{xx}, \qquad 0 < x < 1, \qquad 0 < t,$$
$$u(x,0) = f(x), \qquad 0 < x < 1,$$
$$u(0,t) = g(t), \qquad 0 < t, \tag{11.1.1}$$
$$u(1,t) = h(t), \qquad 0 < t,$$

it is an easy consequence of the Extended Comparison Theorem 1.6.5 that, for $0 \le x \le 1$ and $0 \le t$,

$$|u(x,t)| \le \max\{\|f\|_1, \|g\|_t, \|h\|_t\}, \tag{11.1.2}$$

where

$$\|\psi\|_a = \sup_{0 \le \tau \le a} |\psi(\tau)| \tag{11.1.3}$$

for arbitrary ψ and constant a. We see that as f, g, and h tend to zero uniformly, so does u. From the linearity of (11.1.1), it follows that the difference of two solutions tends to zero when the difference of the data tends to zero. Consequently, the solution of Problem (11.1.1) depends continuously upon the data.

Likewise, we see that for the problem

$$u_t = u_{xx}, \qquad 0 < x < \infty, \qquad 0 < t,$$
$$u(x,0) = f(x), \qquad 0 < x < \infty, \qquad (11.1.4)$$
$$u(0,t) = g(t), \qquad 0 < t,$$

with bounded f, the bounded solution satisfies

$$|u(x,t)| \le \max\{\|f\|_\infty, \|g\|_t\} \qquad (11.1.5)$$

for $0 \le x < \infty$ and $0 \le t$. Also, we see that the bounded solution of

$$u_t = u_{xx}, \qquad -\infty < x < \infty, \qquad 0 < t,$$
$$u(x,0) = f(x), \qquad -\infty < x < \infty, \qquad (11.1.6)$$

with bounded f satisfies

$$|u(x,t)| \le \sup_{-\infty < x < \infty} |f(x)| \qquad (11.1.7)$$

for $-\infty < x < \infty$ and $0 \le t$. Consequently, these solutions depend continuously upon the data too.

Recalling Sec. 8.3 and Sec. 8.4, it was shown that solutions of integral equations depend continuously upon the data. Consequently, it follows that the solutions of all of the problems discussed in Chapters 4, 5, 6, and 7 depend continuously upon the data. We shall leave the details of each specific case in those chapters for the reader.

All the problems above required the specification of the initial temperature. Such a specification is possible only in certain circumstances. For example, one might place a conductor in an ice bath for quite a long time and make the reasonable assumption that $u(x,0) \equiv 0$. Oftentimes, one of the boundaries of the conductor is not accessible. Consequently, still more of the data needed in the above problems might not be measurable. The Cauchy problem of Chapter 2 is exactly the treatment of one way to get around the lack of initial temperature and one of the boundary temperatures. Namely, on a boundary one measures the temperature and the heat flux in order to determine the temperature in the interior of the conductor. Section 2.7 presents an example of the lack of continuous dependence upon the data for the Cauchy problem. This problem is an example of a *not-well-posed problem*, in the sense of Hadamard.

Well-posed problems are those that have solutions that are unique and depend continuously upon the data. *A problem is not-well-posed if it does not possess solutions, or if its solutions are not uniquely determined by the data, or if its unique solutions do not depend continuously upon the data.* The Cauchy problem here is an example of the latter case. Often, the specification of additional information restricts the class of solutions sufficiently so that it is possible to derive continuous dependence upon data for solutions in the restricted class. One of the most physically realistic assumptions that can be made is that of an *a priori* known bound upon the solution u. For instance, if no change of phase has occurred, then a knowledge of the melting point coupled with absolute zero yields such a constant. In all of the sections that follow, we shall assume that there exists a known positive constant C_1 such that

$$|u(x,t)| < C_1. \qquad (11.1.8)$$

11.2. AN INTERIOR TEMPERATURE MEASUREMENT IN A FINITE CONDUCTOR WITH ONE UNKNOWN BOUNDARY CONDITION

We shall consider here the problem

$$\begin{aligned} u_t &= u_{xx}, & 0 < x < 1, & \quad 0 < t, \\ u(x,0) &= f(x), & 0 < x < 1, \\ u(a,t) &= g(t), & 0 < t, & \quad 0 < a < 1, \\ u(1,t) &= h(t), & 0 < t, \\ |u(x,t)| &< C_1, & 0 \le x \le 1, & \quad 0 \le t, \end{aligned} \qquad (11.2.1)$$

where f, g, and h are piecewise-continuous functions. See Fig. 11.2.1. We shall assume that a solution to (11.2.1) exists. This implies that g must be infinitely differentiable and that $u(0, t)$ is a bounded, piecewise-continuous function. From the maximum principle, it follows that for $a \le x \le 1$ and $0 \le t$,

$$|u(x,t)| \le \max\{\|f\|_1, \|g\|_t, \|h\|_t\}. \qquad (11.2.2)$$

The purpose of the discussion here shall be the derivation of an *a priori* estimate for u in the region $0 < x < a$, $0 < t$, in terms of f, g, h, and C_1. It is no loss of generality to assume that $\max\{\|f\|_1, \|g\|_t, \|h\|_t\} < C_1$.

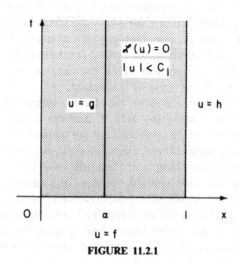

FIGURE 11.2.1

We begin by recalling Theorem 10.5.1. Since $|u(0, t)| < C_1$, a direct application of Theorem 10.5.1 yields the fact that, for z in the domains

$$D_\mu = \{z \,|\, 0 < x < 1, \quad |y| \le \mu |x|, \quad |y| \le \mu|1 - x|\}, \qquad 0 < \mu < 1,$$

(11.2.3)

$u = u(z, t)$ is an analytic function of z, and for $z \in D_\mu$ and $0 \le t \le T$,

$$|u(z, t)| < C_1 \exp\{(32)^{-1} t^{-1}\}$$
$$+ 6(1 + \mu^2)^{1/2}(1 - \mu^2)^{-1/2} C_1 \exp\{16\pi^{-1/2} T^{1/2} t\}.$$

(11.2.4)

Recalling (11.2.2) we see that we can make a direct application of Corollary 10.6.1 to obtain the following *a priori* estimate.

Theorem 11.2.1. *For $0 < x < a$ and $0 < t \le T$,*

$$|u(x, t)| \le C^{1 - (1/3)(x)^{(1/a)}} \varepsilon^{(1/3)(x)^{(1/a)}},$$

(11.2.5)

where $\alpha = \alpha(a, \mu)$, $0 < \alpha < 1$,

$$\varepsilon = \max\{\|f\|_1, \|g\|_t, \|h\|_t\},$$

(11.2.6)

and

$$C = C_1 \exp\{(32)^{-1} t^{-1}\} + 6(1 + \mu^2)^{1/2}(1 - \mu^2)^{-1/2} C_1 \exp\{16\pi^{-1/2} T^{1/2} t\}.$$

(11.2.7)

Proof.   See the analysis preceding the theorem statement.   □

Remark.   From Theorem 11.2.1, it follows that the solution u of problem (11.2.1) depends continuously upon the data. In fact, the dependence is Hölder. Note that the Hölder exponent $(1/3)(x)^{1/\alpha}$ tends to zero as x tends to zero. This is a standard occurrence for continuation estimates; that is, the estimate becomes worse as you move away from the data.

11.3. AN INTERIOR TEMPERATURE MEASUREMENT ON A CHARACTERISTIC IN A FINITE CONDUCTOR WITH NO INITIAL DATA

We shall consider here the problem

$$u_t = u_{xx}, \qquad 0 < x < 1, \qquad 0 < t,$$
$$u(x, T) = f(x), \qquad 0 < T, \qquad 0 \le a < x < b \le 1,$$
$$u(0, t) = g(t), \qquad 0 < t,$$
$$u(1, t) = h(t), \qquad 0 < t,$$
$$|u(x, t)| < C_1, \qquad 0 \le x \le 1, \qquad 0 \le t,$$

(11.3.1)

where T is a positive constant and f, g, and h are piecewise-continuous

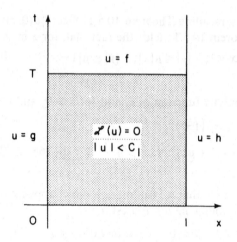

FIGURE 11.3.1

functions. See Fig. 11.3.1. We shall assume that a solution to (11.3.1) exists. This implies that f is an analytic function of x and that $u(x,0)$ is a bounded, piecewise-continuous function. The purpose of the discussion here is to derive an *a priori* estimate of u in the region $0 < x < 1$, $0 < t$, in terms of f, g, h, and C_1. It is no loss of generality to assume that

$$\|f\|_{[a,b]} + \max\{\|g\|_T, \|h\|_T\} < C_1,$$

where

$$\|f\|_{[a,b]} = \sup_{a \le x \le b} |f(x)|. \tag{11.3.2}$$

We begin by setting $u = v + w$, where

$$\begin{aligned} v_t &= v_{xx}, & 0 < x < 1, \quad 0 < t, \\ v(x,0) &= 0, & 0 < x < 1, \\ v(0,t) &= g(t), & 0 < t, \\ v(1,t) &= h(t), & 0 < t, \end{aligned} \tag{11.3.3}$$

and

$$\begin{aligned} w_t &= w_{xx}, & 0 < x < 1, \quad 0 < t, \\ w(x,T) &= f(x) - v(x,T), & 0 \le a \le x \le b \le 1, \quad 0 < T, \\ w(0,t) &= w(1,t) = 0, & 0 < t, \\ w(x,0) &= u(x,0) \\ |w(x,t)| &< C_1, \end{aligned} \tag{11.3.4}$$

where $|w(x,t)| < C_1$ follows from $|u(x,0)| < C_1$ and the boundary conditions via the maximum principle. Since

$$|v(x,t)| \le \max\{\|g\|_t, \|h\|_t\} \qquad (11.3.5)$$

holds, from an application of the maximum principle, it suffices to concentrate on w.

Now, an application of (11.3.5) for $t = T$ yields

$$|w(x,T)| \le \|f\|_{[a,b]} + \max\{\|g\|_T, \|h\|_T\} \qquad (11.3.6)$$

for $a \le x \le b$. Taking $w(x,0) = u(x,0)$ and extending it as an odd function over $-1 \le x < 0$ and extending the resulting function over $-1 < x < 1$ as a periodic function with period 2, we have seen that $w(x,t)$ can be represented as a solution of an initial-value problem with initial data given by the forementioned extension of $u(x,0)$. As $|u(x,0)| < C_1$, we see from Theorem 10.2.1 that $w(z,t)$ is an analytic function of the complex variable z. Moreover, for $|\operatorname{Im} z| \le Y$,

$$|w(z,t)| \le C_1 \exp\left\{\frac{Y^2}{4t}\right\}. \qquad (11.3.7)$$

Recalling (11.3.6), we are able to make a direct application of Lemma 10.6.5 to obtain for $x \ge (a+b)/2$

$$|w(x,T)| < C^{1-\alpha(x)} \varepsilon^{\alpha(x)}, \qquad (11.3.8)$$

where

$$\varepsilon = \{\|f\|_{[a,b]} + \max\{\|g\|_T, \|h\|_T\}\}, \qquad (11.3.9)$$

$$C = C_1 \exp\{4^{-1} T^{-1} Y^2\}, \qquad (11.3.10)$$

$$\alpha(x) = S(2^{-1}Y) \exp\left\{-Y^{-1}\pi\left(x - \frac{(a+b)}{2}\right)\right\}, \qquad (11.3.11)$$

and

$$S(2^{-1}Y) = \frac{\sinh\left(\frac{\pi Y}{2(b-a)}\right)}{\sinh\left(\frac{Y}{(b-a)}\right)}. \qquad (11.3.12)$$

For $x \le (a+b)/2$, we get the same estimate except for a change of the minus sign to a plus sign in front of Y^{-1} within the exponential in the definition of $\alpha(x)$. Since the estimate (11.3.8) can be written in the form

$$|w(x,T)| < C\left(\frac{\varepsilon}{C}\right)^{\alpha(x)}, \qquad (11.3.13)$$

where $\varepsilon < C_1$ and (11.3.10) imply that $\varepsilon < C$, it then follows that, for $0 \le x \le 1$,

$$|w(x,T)| < C^{1-\beta} \varepsilon^{\beta}, \qquad (11.3.14)$$

where

$$\beta = \min\{\alpha(0), \alpha(1)\} \tag{11.3.15}$$

and $\alpha(0)$ employs the $+$ sign in front of Y^{-1} in the exponential in $\alpha(x)$ while $\alpha(1)$ employs the minus sign. From the periodicity of w with respect to x, we see that, for all $-\infty < x < \infty$,

$$|w(x, T)| < C^{1-\beta}\varepsilon^{\beta}. \tag{11.3.16}$$

Consequently, for all x and all $t \geq T$,

$$|w(x, t)| < C^{1-\beta}\varepsilon^{\beta}. \tag{11.3.17}$$

We use again the fact that w is the solution of an initial-value problem with bounded initial data. Recall Theorem 10.3.1. We see that for each x, w is an analytic function $w(x, \eta)$ of the complex variable $\eta = t + i\tau$ in the domains

$$D_{\mu} = \{\eta \mid t > 0 \quad \text{and} \quad |\tau| \leq \mu t\}, \qquad \mu > 0. \tag{11.3.18}$$

Moreover, for $\eta \in D_{\mu}$,

$$|w(x, \eta)| \leq (1+\mu^2)^{1/2}C_1. \tag{11.3.19}$$

Coupling (11.3.19) together with (11.3.17) for $T \leq t \leq 2T$, we find ourselves in the setting for an application of Lemma 10.6.6. For ease in presentation, we select $\mu = \sqrt{3}/3$, so that the condition $\sqrt{3}(b-a) \leq (a+b)\mu$ of the Lemma 10.6.6 is satisfied. Here, $a = T$ and $b = 2T$. Consequently, from Lemma 10.6.6 we see that, for $0 < t < T$,

$$|w(x, t)| < C_1^{1-(\beta/3)(t/R)^{1/\alpha}}\varepsilon^{(\beta/3)(t/R)^{1/\alpha}}, \tag{11.3.20}$$

where ε is given by (11.3.9), and β is given by (11.3.15), where also

$$C_2 = \max\left(C, \left(\tfrac{4}{3}\right)^{1/2}C_1\right), \tag{11.3.21}$$

C is given by (11.3.10),

$$R = \left(\tfrac{7}{3}\right)^{1/2}T, \tag{11.3.22}$$

and where

$$\alpha = \frac{2}{\pi}\operatorname{Arctan}\left(\frac{\sqrt{3}}{9}\right). \tag{11.3.23}$$

We can now combine (11.3.5), (11.3.17), and (11.3.20) into an estimate of $u(x, t)$.

Theorem 11.3.1. *For the solution u of the problem*

$$u_t = u_{xx}, \qquad 0 < x < 1, \qquad 0 < t,$$
$$u(x, T) = f(x), \qquad 0 \leq a < x < b \leq 1, \qquad 0 < T,$$
$$u(0, t) = g(t), \qquad 0 < t,$$
$$u(1, t) = h(t), \qquad 0 < t,$$
$$|u(x, t)| < C_1, \qquad 0 \leq x \leq 1, \qquad 0 \leq t, \tag{11.3.24}$$

where T and C_1 are positive constants and f, g, and h are piecewise continuous, we see that, for $0 \le x \le 1$ and $t \ge T$, there exists a positive constant β, $0 < \beta \le 1$, and a positive $C = C(C_1, T)$ such that

$$|u(x,t)| \le \max\{\|g\|_t, \|h\|_t\} + C^{1-\beta}\{\|f\|_{[a,b]} + \max\{\|g\|_T, \|h\|_T\}\}^{\beta},$$

while for $0 \le x \le 1$ and $0 < t < T$, there exists another positive $C = C(C_1, T)$ such that

$$|u(x,t)| \le \max\{\|g\|_t, \|h\|_t\} + C^{1-(\beta/3)(t/R)^{1/\alpha}}$$
$$\cdot \{\|f\|_{[a,b]} + \max\{\|g\|_T, \|h\|_T\}\}^{(\beta/3)(t/R)^{1/\alpha}}, \qquad (11.3.25)$$

where

$$R = \left(\tfrac{7}{3}\right)^{1/2} T, \qquad (11.3.26)$$

$$\alpha = \frac{2}{\pi}\text{Arctan}\frac{\sqrt{3}}{9}, \qquad (11.3.27)$$

for any function ψ,

$$\|\psi\|_{[a,b]} = \sup_{a \le x \le b} |\psi(x)|, \qquad (11.3.28)$$

and

$$\|\psi\|_t = \sup_{0 \le \tau \le t} |\psi(\tau)|. \qquad (11.3.29)$$

Proof. See the analysis preceding the theorem statement. $\qquad\square$

Remark. It can be seen clearly from the estimates presented in Theorem 11.3.1 that for $0 \le x \le 1$ and $0 < t$, the solution u of (11.3.1) depends continuously upon the data f, g, and h. Moreover as we move down away from $t = T$, the estimate becomes worse. However, for each t, the estimate is Hölder.

We conclude this section with a discussion of whether the boundary data are necessary in order to obtain an estimate of a solution for a finite region from data on a characteristic. In other words, we consider the problem

$$u_t = u_{xx}, \qquad 0 < x < 1, \qquad 0 < t,$$
$$u(x,T) = f(x), \qquad 0 < x < 1, \qquad (11.3.30)$$
$$|u(x,t)| \le C_1,$$

where f is piecewise continuous and C_1 and T are positive constants. The assumption of the existence of a solution u implies that f is necessarily analytic. However, even this is not enough to obtain an estimate of u in terms of f that would tend to zero as f tends to zero. To see this, we consider the problem

$$u_t = u_{xx}, \qquad 0 < x < 1, \qquad 0 < t,$$
$$u(x, T) \equiv 0, \qquad 0 < x < 1, \qquad (11.3.31)$$
$$|u(x, t)| \le C_1.$$

We shall construct a nonzero solution of it.

Recall Sec. 2.4 and consider the function

$$f(t) = \begin{cases} \exp\left\{ -\dfrac{1}{(T-t)^2} \right\}, & t < T \\ 0, & t \ge T. \end{cases} \qquad (11.3.32)$$

For each $\gamma_1 > 0$ it is clear that there exists a $C = C(\gamma_1) > 0$ such that $f \in H(\gamma_1, \gamma_2, C(\gamma_1), 0)$ for each fixed $\gamma_2 > 0$. Consequently, for $-\infty < x < \infty$ and $0 \le t \le T$, the series

$$u(x, t) = \sum_{j=0}^{\infty} f^{(j)}(t) \frac{x^{2j}}{(2j)!} \qquad (11.3.33)$$

converges and is a solution of the heat equation. Moreover, $u(x, T) \equiv 0$. Let

$$C_2 = \max_{\substack{0 \le x \le 1 \\ 0 \le t \le T}} |u(x, t)|. \qquad (11.3.34)$$

Then, the solution

$$v(x, t) = C_1 C_2^{-1} u(x, t) \qquad (11.3.35)$$

provides us with a solution of (11.3.31) that is not identically zero. Hence, no estimate is possible in the absence of the boundary data. Moreover, the function

$$u(x, t) = \sum_{j=0}^{\infty} f^{(j)}(t) \frac{x^{2j+1}}{(2j+1)!} \qquad (11.3.36)$$

provides us with the example that shows that the inclusion of one-boundary data is still not sufficient to conclude that an estimate can be obtained. Hence, the necessity of both pieces of boundary data.

Theorem 11.3.2. *For a solution u of*

$$u_t = u_{xx}, \qquad 0 < x < 1, \qquad 0 < t,$$
$$u(x, T) = f(x), \qquad 0 < x < 1, \qquad (11.3.37)$$
$$|u(x, t)| \le C_1, \qquad 0 \le x \le 1, \qquad 0 \le t,$$

or

$$u_t = u_{xx}, \qquad 0 < x < 1, \qquad 0 < t,$$
$$u(x, T) = f(x), \qquad 0 < x < 1,$$
$$u(0, t) = g(t), \qquad 0 < t, \qquad (11.3.38)$$
$$|u(x, t)| \le C_1, \qquad 0 \le x \le 1, \qquad 0 \le t,$$

where C_1 and T are positive constants and f and g are piecewise continuous functions, an *a priori* estimate for u in terms of the data cannot be derived. In other words, there is no continuous dependence upon the data for either problem.

Proof. See the analysis preceding the statement of the theorem.   □

11.4. THE CAUCHY PROBLEM: THE SPECIFICATION OF TEMPERATURE AND HEAT FLUX ON A BOUNDARY

We shall consider here the problem

$$u_t = u_{xx}, \qquad 0 < x < 1, \qquad 0 < t,$$
$$u(1, t) = g(t), \qquad 0 < t,$$
$$u_x(1, t) = h(t), \qquad 0 < t, \qquad (11.4.1)$$
$$|u(x, t)| < C_1, \qquad 0 \le x \le 1, \qquad 0 \le t,$$

where g and h are assumed to be piecewise-continuous functions. (See Fig. 11.4.1.) We shall assume that a solution of (11.4.1) exists. This implies that $u(0, t)$ and $u(x, 0)$ are bounded, piecewise-continuous functions. Our purpose here is to derive an *a priori* estimate of u in terms of g, h, and C_1.

We begin here by recalling the Green's function

$$G(x, t, \xi, \tau) = K(x - \xi, t - \tau) - K(x + \xi, t - \tau) \qquad (11.4.2)$$

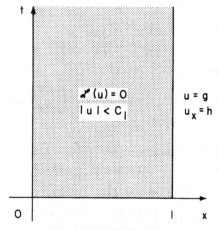

FIGURE 11.4.1

and the Newmann's function

$$N(x,t,\xi,\tau) = K(x-\xi,t-\tau) + K(x+\xi,t-\tau), \qquad (11.4.3)$$

where $K = K(x,t)$, the Fundamental Solution, is defined by (3.2.1). By differentiation it is clear that, with respect to ξ and τ,

$$G_{\xi\xi} + G_\tau = 0. \qquad (11.4.4)$$

Hence, for $u = u(\xi,\tau)$,

$$\{u_\xi G - uG_\xi\}_\xi = (uG)_\tau. \qquad (11.4.5)$$

Integrating this identity over $0 \le \xi \le 1$ and $0 \le \tau \le t - \varepsilon$, we obtain

$$\int_0^1 G(x,\xi,t,t-\varepsilon)u(\xi,t-\varepsilon)\,d\xi$$

$$= \int_0^1 G(x,\xi,t,0)u(\xi,0)\,d\xi + \int_0^{t-\varepsilon} G(x,1,t,\tau)u_\xi(1,\tau)\,d\tau$$

$$- \int_0^{t-\varepsilon} G_\xi(x,1,t,\tau)u(1,\tau)\,d\tau$$

$$- \int_0^{t-\varepsilon} G(x,0,t,\tau)u_\xi(0,\tau)\,d\tau$$

$$+ \int_0^{t-\varepsilon} G_\xi(x,0,t,\tau)u(0,\tau)\,d\tau. \qquad (11.4.6)$$

For $\xi = 0$, $G(x,0,t,\tau) = 0$. Also, letting $\varepsilon \to 0$, we see that for $0 < x < 1$, $0 < t$, we obtain the representation

$$u(x,t) = \int_0^1 G(x,\xi,t,0)u(\xi,0)\,d\xi - 2\int_0^t \frac{\partial K}{\partial x}(x,t-\tau)u(0,\tau)\,d\tau$$

$$+ \int_0^t G(x,1,t,\tau)h(\tau)\,d\tau + \int_0^t \frac{\partial N}{\partial x}(x,1,t,\tau)g(\tau)\,d\tau,$$

$$(11.4.7)$$

where we have used

$$\frac{\partial G}{\partial \xi} = -\frac{\partial N}{\partial x}. \qquad (11.4.8)$$

Consequently, we see that

$$u(x,t) = v(x,t) + \int_0^t G(x,1,t,\tau)h(\tau)\,d\tau + \int_0^t \frac{\partial N}{\partial x}(x,1,t,\tau)g(\tau)\,d\tau,$$

$$(11.4.9)$$

where v is the bounded solution of

$$\begin{aligned} v_t &= v_{xx}, & 0 < x < \infty, & \quad 0 < t, \\ v(0,t) &= u(0,t), & 0 < t, \\ v(x,0) &= \begin{cases} u(x,0), & 0 < x < 1, \\ 0, & 1 \le x < \infty. \end{cases} \end{aligned} \qquad (11.4.10)$$

From (11.4.1), we see that

$$|v(x,t)| < C_1. \tag{11.4.11}$$

First, we shall estimate the two integrals in (11.4.9) for $0 < x < 1$, $0 < t$. Then, we will use the data in (11.4.1), coupled with estimates of the limits of the two integrals as x tends to 1, to obtain an estimate for $v(1,t)$. From the maximum principle for bounded solutions in a quarter plane, the bound on v will hold for all $x \geq 1$ and $0 \leq t$. We will then estimate v in $0 < x < 1$, $0 < t$, via the analyticity of v in the spatial variable coupled with Corollary 10.6.1.

We begin our estimate of $u(x,t)$ with the integral

$$J_1 = \int_0^t G(x,1,t,\tau)h(\tau)\,d\tau. \tag{11.4.12}$$

Now, from (3.2.1),

$$|K(x,t)| \leq 2^{-1}\pi^{-1/2}t^{-1/2}. \tag{11.4.13}$$

Recalling (11.4.2), we see that

$$|G(x,t,\xi,\tau)| \leq \pi^{-1/2}(t-\tau)^{-1/2}. \tag{11.4.14}$$

Consequently, for $0 \leq x \leq 1$, $0 \leq t$,

$$|J_1| \leq 2\pi^{-1/2}t^{1/2}\|h\|_t. \tag{11.4.15}$$

Turning now to the second integral

$$J_2 = \int_0^t \frac{\partial N}{\partial x}(x,1,t,\tau)g(\tau)\,d\tau, \tag{11.4.16}$$

we see that J_2 satisfies the heat equation in $0 < x < 1$, $0 < t$. Initially, $J_2 = 0$. Hence, to obtain an estimate of J_2 in $0 \leq x \leq 1$, $0 \leq t$, it suffices to estimate it for $x = 0$ and $x = 1$. We consider first $x = 0$. Here,

$$\frac{\partial N}{\partial x}(0,1,t,\tau) = \frac{\partial K}{\partial x}(-1,t-\tau) + \frac{\partial K}{\partial x}(1,t-\tau). \tag{11.4.17}$$

Employing the estimate $\exp\{-x\} \leq p!x^{-p}$ for any $p = 1,2,3,\dots$ with $p = 1$, we see that

$$\left|\frac{\partial K}{\partial x}(\pm 1, t-\tau)\right| = \frac{1}{4\pi^{1/2}(t-\tau)^{3/2}}\exp\left\{\frac{-1}{4(t-\tau)}\right\} \leq \pi^{-1/2}(t-\tau)^{-1/2}. \tag{11.4.18}$$

Hence,

$$\left|\frac{\partial N}{\partial x}(0,1,t,\tau)\right| \leq 2\pi^{-1/2}(t-\tau)^{-1/2}. \tag{11.4.19}$$

Thus,

$$|J_2(0,t)| \leq 4\pi^{-1/2}t^{1/2}\|g\|_t. \tag{11.4.20}$$

For $x = 1$, we must use Lemma 4.2.2. Its application here yields

$$\lim_{x \uparrow 1} J_2(x, t) = \tfrac{1}{2} g(t) + \int_0^t \frac{\partial K}{\partial x} (2, t - \tau) g(\tau) \, d\tau. \qquad (11.4.21)$$

Using $\exp\{-x\} \le p! x^{-p}$ for $p = 1$ again, we obtain

$$\left| \frac{\partial K}{\partial x} (2, t - \tau) \right| \le 2^{-1} \pi^{-1/2} (t - \tau)^{-1/2}, \qquad (11.4.22)$$

whence it follows that

$$\left| \int_0^t \frac{\partial K}{\partial x} (2, t - \tau) g(\tau) \, d\tau \right| \le \pi^{-1/2} t^{1/2} \| g \|_t. \qquad (11.4.23)$$

Combining (11.4.23) with (11.4.21) yields

$$|J_2(1, t)| \le \{ 2^{-1} + \pi^{-1/2} t^{1/2} \} \| g \|_t. \qquad (11.4.24)$$

From (11.4.20) and (11.4.24) we see, from the maximum principle, that for $0 \le x \le 1$ and $0 \le t$,

$$|J_2(x, t)| \le \{ 2^{-1} + 5\pi^{-1/2} t^{1/2} \} \| g \|_t. \qquad (11.4.25)$$

Note that we have added (11.4.20) and (11.4.24) to simplify the bounding expression.

We turn now to v. Recalling now that v is the solution of the quarter-plane problem (11.4.11) and that

$$v(x, t) = u(x, t) - J_1(x, t) - J_2(x, t),$$

we see that

$$v(1, t) = g(t) - J_1(1, t) - \tfrac{1}{2} g(t) - \int_0^t \frac{\partial K}{\partial x} (2, t - \tau) g(\tau) \, d\tau$$

$$= - J_1(1, t) + \tfrac{1}{2} g(t) - \int_0^t \frac{\partial K}{\partial x} (2, t - \tau) g(\tau) \, d\tau. \qquad (11.4.26)$$

Hence, to the estimate of $J_2(1, t)$ we need only add the estimate of $J_1(x, t)$ in order to bound $v(1, t)$. This results in

$$|v(1, t)| \le 2\pi^{-1/2} t^{1/2} \| h \|_t + \{ 2^{-1} + \pi^{-1/2} t^{1/2} \} \| g \|_t. \qquad (11.4.27)$$

From (11.4.11) we see that, from the maximum principle for bounded solutions of quarter-plane problems, for $1 \le x$ and $0 \le t$,

$$|v(x, t)| \le 2\pi^{-1/2} t^{1/2} \| h \|_t + \{ 2^{-1} + \pi^{-1/2} t^{1/2} \} \| g \|_t. \qquad (11.4.28)$$

From Sec. 10.2 and 10.4, we see that $v = v(z, t)$ is an analytic function of the complex variable z. Moreover, in the domain

$$D = \{ z \mid x > 0, \quad |y| < \mu x \quad \text{and} \quad |y| \le 1 \}, \qquad 0 < \mu < 1, \qquad (11.4.29)$$

it follows from those sections that for $z \in D$ and $t > 0$,

$$|v(z, t)| < C_1 \exp\{ 4^{-1} t^{-1} \} + (1 + \mu^2)^{1/2} (1 - \mu^2)^{-1/2} C_1. \qquad (11.4.30)$$

As an application of the argument of Lemma 10.6.6 it follows that there exists a positive constant $C = C(C_1, \mu, t)$ and a positive constant α, $0 < \alpha < 1$, such that for $0 < t$ and $0 < x < 1$,

$$|v(x,t)| \leq C^{1-(1/3)(x/2)^{1/\alpha}} \varepsilon^{(1/3)(x/2)^{1/\alpha}}, \qquad (11.4.31)$$

where

$$\varepsilon = 2\pi^{-1/2}t^{1/2}\|h\|_t + \{2^{-1} + \pi^{-1/2}t^{1/2}\}\|g\|_t.$$

Combining (11.4.15), (11.4.25), and (11.4.31) we obtain the following estimate for u.

Theorem 11.4.1. *For the solution u of the problem*

$$u_t = u_{xx}, \qquad 0 < x < 1, \qquad 0 < t,$$
$$u(1,t) = g(t), \qquad 0 < t,$$
$$u_x(1,t) = h(t), \qquad 0 < t, \qquad (11.4.32)$$
$$|u(x,t)| < C_1, \qquad 0 \leq x \leq 1, \qquad 0 \leq t,$$

where C_1 is a positive constant and g and h are piecewise-continuous functions, there exist positive constants $C = C(C_1, t)$ and α, $0 < \alpha < 1$ such that, for $0 < x < 1$, $0 < t$,

$$|u(x,t)| \leq 2\pi^{-1/2}t^{1/2}\|h\|_t + \{2^{-1} + 5\pi^{-1/2}t^{1/2}\}\|g\|_t$$
$$+ C^{1-(1/3)(x/2)^{1/\alpha}}\{2\pi^{-1/2}t^{1/2}\|h\|_t$$
$$+ \{2^{-1} + \pi^{-1/2}t^{1/2}\}\|g\|_t\}^{(1/3)(x/2)^{1/\alpha}}, \qquad (11.4.33)$$

where

$$\|\psi\|_t = \sup_{0 \leq \tau \leq t} |\psi(\tau)| \qquad (11.4.34)$$

for any function ψ.

    Proof.   See the analysis preceding the theorem statement.   □

Remark. Note, as in previous sections, that the estimate for u grows worse as you move away from the data at $x = 1$.

Corollary 11.4.1. *For the solution u of the problem*

$$u_t = u_{xx}, \qquad 0 < x < 1, \qquad 0 < t,$$
$$u(x,0) = f(x), \qquad 0 < x < 1,$$
$$u(1,t) = g(t), \qquad 0 < t, \qquad (11.4.35)$$
$$u_x(1,t) = h(t), \qquad 0 < t,$$
$$|u(x,t)| < C_1, \qquad 0 \leq x \leq 1, \qquad 0 \leq t,$$

where C_1 is a positive constant and f, g, and h are piecewise-continuous *functions, there exist positive constants $C = C(C_1)$ and α, $0 < \alpha < 1$, such that for $0 < x < 1$, $0 < t$,*

$$|u(x,t)| \leq \|f\|_1 + 2\pi^{-1/2}t^{1/2}\|h\|_t + \{2^{-1} + 5\pi^{-1/2}t^{1/2}\}\|g\|_t + C^{1-(1/3)(x/2)^{1/\alpha}}$$

$$\times \{\|f\|_1 + 2\pi^{-1/2}t^{1/2}\|h\|_t + \{2^{-1} + \pi^{-1/2}t^{1/2}\}\|g\|_t\}^{(1/3)(x/2)^{1/\alpha}},$$

$$(11.4.36)$$

where $\| \ \|_t$ is defined by (11.4.34).

    Proof. In this case the function v reduces to the quarter-plane problem with zero initial value and boundary value $u(0, t)$. The integral

$$\int_0^1 G(x, \xi, t, 0) f(\xi) \, d\xi \qquad (11.4.37)$$

is bounded uniformly by $\|f\|_1$, which must be incorporated into the estimate of $v(1, t)$ and thus into (11.4.36).   □

    Corollary 11.4.2. *For the solution u of the problem*

$$u_t = u_{xx}, \qquad 0 < x < 1, \qquad 0 < t,$$

$$u(0, t) = \varphi(t), \qquad 0 < t,$$

$$u(1, t) = g(t), \qquad 0 < t, \qquad (11.4.38)$$

$$u_x(1, t) = h(t), \qquad 0 < t,$$

$$|u(x, t)| \leq C_1, \qquad 0 \leq x \leq 1, \qquad 0 \leq t,$$

where C_1 is a positive constant and φ, g, and h are piecewise-continuous functions, there exist positive constants $C = C(C_1, t)$ and α, $0 < \alpha < 1$, such that for $0 < x < 1$ and $0 < t$,

$$|u(x,t)| \leq \|\varphi\|_t + 2\pi^{-1/2}t^{1/2}\|h\|_t + \{2^{-1} + 5\pi^{-1/2}t^{1/2}\}\|g\|_t$$

$$+ C^{1-\alpha}\{\|\varphi\|_t + 2\pi^{-1/2}t^{1/2}\|h\|_t + \{2^{-1} + \pi^{-1/2}t^{1/2}\}\|g\|_t\}^{\alpha},$$

$$(11.4.39)$$

where $\| \ \|_t$ is defined by (11.4.34).

    Proof. In this case the function v reduces to the integral

$$\int_0^1 G(x, \xi, t, 0) u(\xi, 0) \, d\xi, \qquad (11.4.40)$$

which is equivalent to an initial-value problem for an odd extension of $u(\xi, 0)$. The integral involving φ is simply the solution to a quarter-plane problem with zero initial value and boundary value $\varphi(t)$. This integral is dominated by $\|\varphi\|_t$, which must be incorporated into the estimate of $v(1, t)$. The application of Lemma 11.6.5 results in an estimate for v of the form $C^{1-\theta(x)}\varepsilon^{\theta(x)}$ that is good for $-\infty < x < 1$. Since we only require an estimate over $0 < x < 1$, we can select $\alpha = \min_{0 \leq x \leq 1}\theta(x)$.   □

Remark. Here it appears that the estimate (11.4.39) yields a uniform estimate. However, $C = C_1 \exp\{4^{-1}t^{-1}\}$ is the form of the constant in (11.4.39). This means that as t tends to zero, C tends exponentially to infinity. So we see again that, as we move away from the data, the estimate grows worse.

11.5. SPECIFICATION OF AN INTERIOR TEMPERATURE AND FLUX ALONG WITH BOUNDARY TEMPERATURES: AN EXERCISE

Exercise 11.1. Show that for the solution u of the problem

$$u_t = u_{xx}, \qquad 0 < x < 1, \qquad 0 < t,$$

$$u(0, t) = g(t), \qquad 0 < t,$$

$$u(1, t) = h(t), \qquad 0 < t, \qquad (11.5.1)$$

$$u(a, t) = \varphi(t), \qquad 0 < a < 1, \qquad 0 < t_0 < t < t_1,$$

$$u_x(a, t) = \psi(t), \qquad 0 < t_0 < t < t_1,$$

$$|u(x, t)| \le C_1, \qquad 0 \le x \le 1, \qquad 0 \le t,$$

where a, $0 < a < 1$, t_0, t_1, and C_1 are positive constants and g, h, φ, and ψ are piecewise-continuous functions, there exist positive constants $C = C(C_1, t_1, t_1 - t_0, a)$ and $\alpha = \alpha(a)$, $0 < \alpha < 1$, such that for $0 \le x \le 1$ and $0 < t < t_1$,

$$|u(x, t)| \le \max\{\|g\|_t, \|h\|_t\} + C^{1 - (\alpha/3)(t/R)^{1/\gamma}}$$

$$\cdot \left\{ \|\varphi\|_{[t_0, t_1]} + \|\psi\|_{[t_0, t_1]} + \max\{\|g\|_{t_1}, \|h\|_{t_1}\} \right\}^{(\alpha/3)(t/R)^{1/\gamma}},$$

$$(11.5.2)$$

where

$$R = \left(\tfrac{7}{3}\right)^{1/2} t_1 \qquad (11.5.3)$$

and

$$\gamma = \frac{2}{\pi} \arctan \frac{\sqrt{3}}{9}, \qquad (11.5.4)$$

while for $0 \le x \le 1$ and $t_1 \le t$,

$$|u(x, t)| \le \max\{\|g\|_t, \|h\|_t\}$$

$$+ C\left\{ \|\varphi\|_{[t_0, t_1]} + \|\psi\|_{[t_0, t_1]} + \max\{\|g\|_{t_1}, \|h\|_{t_1}\} \right\}^\alpha,$$

$$(11.5.5)$$

where

$$\|g\|_{[t_0, t_1]} = \sup_{t_0 \le \tau \le t_1} |g(\tau)| \qquad (11.5.6)$$

for any t_0, t_1, and function g, and $\|g\|_t = \|g\|_{[0, t]}$.

11.6. THE SPECIFICATION OF AN INTERIOR TEMPERATURE IN A FINITE CONDUCTOR ALONG WITH THE TEMPERATURE ON THE BOUNDARIES: AN EXERCISE

Here, we shall consider the problem

$$u_t = u_{xx}, \qquad 0 < x < 1, \qquad 0 < t,$$
$$u(0, t) = g(t), \qquad 0 < t,$$
$$u(1, t) = h(t), \qquad 0 < t, \qquad (11.6.1)$$
$$u(a, t) = \varphi(t), \qquad 0 < t_0 < t < t_1, \qquad 0 < a < 1,$$
$$|u(x, t)| \le C_1, \qquad 0 \le x \le 1, \qquad 0 \le t,$$

where a, t_0, t_1, and C_1 are positive constants and g, h, and φ are piecewise-continuous functions.

Exercise 11.2. Show that for $a = p/q$, a rational number, Problem (11.6.1) has infinitely many solutions.

Exercise 11.3. Show that for a an irrational number, Problem (11.6.1) can have at most one solution.

Exercise 11.4. Show that for a solution of problem (11.6.1), there exists a positive constant $C = C(C_1, t_0)$ and a positive constant α, $0 < \alpha < 1$, such that for each N for which $\sin k\pi a \ne 0$, $\|\psi\|_t = \|\psi\|_{[0, t]}$, and $k = 1, \dots, N$,

$$|u(x, t)| \le C \left\{ \sum_{n=1}^{N} (\sin n\pi a)^{-1} \exp\left\{ \pi^2 n^2 \left[\frac{t_0 + t_1}{2} - t \right] \right\} \right\}$$

$$\cdot \left\{ \|\varphi\|_{[t_0, t_1]} + \max\{ \|g\|_{t_1}, \|h\|_{t_1} \} \right\}^\alpha \qquad (11.6.2)$$

$$+ 2C_1 \pi^{-2} t^{-1} \exp\{ -(N+1)^2 \pi^2 t \} + \max\{ \|g\|_t, \|h\|_t \}$$

holds for $0 \le x \le 1$, $0 < t$, where

$$\|\psi\|_{[t_0, t_1]} = \sup_{t_0 \le \tau \le t_1} |\psi(\tau)| \qquad (11.6.3)$$

for any function ψ. *Hint:* Begin by writing $u = v + w$, where

$$w_t = w_{xx}, \qquad 0 < x < 1, \qquad 0 < t,$$
$$w(0, t) = w(1, t) = 0, \qquad 0 < t, \qquad (11.6.4)$$
$$w(a, t) = \varphi(t) - v(a, t), \qquad 0 < t_0 < t < t_1,$$
$$w(x, 0) = u(x, 0).$$

Show that

$$|w(a, t)| < \|\varphi\|_{[t_0, t_1]} + \max\{ \|g\|_{t_1}, \|h\|_{t_1} \}. \qquad (11.6.5)$$

Note that $w(a, t)$ is an analytic function of the complex variable $\eta = t + i\tau$, which is periodic with period $2\pi i$. Show that

$$|w(a, \eta)| \leq \pi^{-2} t_0^{-1} C_1 \exp\{-\pi^2 t_0\} = C. \quad (11.6.6)$$

Applying Lemma 10.6.3 with $a = t_0$, $b = t_1$, $c = 0$, $d = 4\pi$,

$$C = \pi^{-2} t_0^{-1} C_1 \exp\{-\pi^2 t_0\}, \quad (11.6.7)$$

and

$$\varepsilon = \|\varphi\|_{[t_0, t_1]} + \max\{\|g\|_{t_1}, \|h\|_{t_1}\}, \quad (11.6.8)$$

where it is no loss of generality to assume that C_1 is large enough so that $\varepsilon < C$, it follows that there exists a positive constant α, $0 < \alpha < 1$, such that

$$\left| w\left(a, \frac{t_0 + t_1}{2} + i\tau\right) \right| \leq C^{1-\alpha} \varepsilon^\alpha \quad (11.6.9)$$

holds for $0 \leq \tau \leq 2\pi$. From the periodicity of $w(a, \eta)$, it follows that (11.6.9) holds for all τ. Under the transformation $\zeta = \exp\{-\pi^2 \eta\}$, $w(a, \eta)$ becomes the analytic function

$$H(\zeta) = \sum_{n=1}^{\infty} (a_n \sin n\pi a) \zeta^{n^2}. \quad (11.6.10)$$

Consequently,

$$(a_k \sin k\pi a) = \frac{H^{(k^2)}(0)}{k^2!} \quad (11.6.12)$$

and

$$a_k = \frac{H^{(k^2)}(0)}{(\sin k\pi a) k^2!}, \quad k = 1, 2, 3, \ldots \quad (11.6.13)$$

Estimate $H^{(k^2)}(0)$ from the Cauchy–Riemann formula and finish the estimate (11.6.2).

Remark.   For a an irrational number, the estimate (11.6.2) implies continuous dependence of u upon the data g, h, and φ since N may be selected arbitrarily. In other words, given a positive number ε, the term $2C_1 \pi^{-2} t^{-1} \exp\{-(N+1)^2 \pi t\}$ can be made less than $\varepsilon/3$ by selecting an N sufficiently large. Upon fixing N, the first and third terms can be made less than $\varepsilon/3$ each if the data g, h, and φ are taken uniformly small enough. Hence, if g, h, and φ tend to zero uniformly, then u tends to zero for each (x, t) such that $0 \leq x \leq 1$ and $0 < t$.

For a rational number, the estimate (11.6.2) does not provide continuous dependence of u upon the data. However, a finite number of harmonics of u do depend continuously upon the data. Moreover, as t increases, the higher harmonics damp out and the estimate (11.6.2) may become practical.

11.7. INTERIOR TEMPERATURE SPECIFICATION ON A CHARACTERISTIC FOR AN INFINITE CONDUCTOR WITH UNKNOWN INITIAL TEMPERATURE

We shall conclude this chapter with the estimation of the solution to the problem

$$u_t = u_{xx}, \qquad -\infty < x < \infty, \qquad 0 < t,$$

$$u(x, T) = f(x), \qquad a < x < b, \qquad 0 < T, \qquad (11.7.1)$$

$$|u(x, t)| \le C_1, \qquad -\infty < x < \infty, \qquad 0 \le t,$$

where a, b are constants, C_1 and T are positive constants, and f is a piecewise-continuous function. (See Fig. 11.7.1.) The assumption of the existence of u implies that $u(x, 0)$ is piecewise continuous and f is analytic. However, we shall only use the fact that

$$u(x, t) = \int_{-\infty}^{\infty} K(x - \xi, t) u(\xi, 0) \, d\xi \qquad (11.7.2)$$

and the analyticity of such representations, as described in Secs. 10.2 and 10.3.

In the previous sections with the notable exception of Sec. 11.6, the estimates of the solutions were Hölder with respect to the norms of the data. In other words, for each (x, t), there existed an $\alpha = \alpha(x, t)$, $0 < \alpha < 1$, such

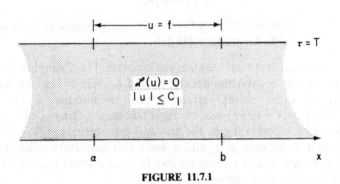

FIGURE 11.7.1

that

$$|u(x,t)| = \mathcal{O}(\|\text{Data}\|^{\alpha}). \tag{11.7.3}$$

Here the expanse of the domain forces a different estimate of the solution. Namely, we shall show that

$$|u(x,t)| = \mathcal{O}\big([\log\|\text{Data}\|^{-1}]^{-\alpha}\big). \tag{11.7.4}$$

This type of estimate is typical of infinite-conductor problems, which is the reason for its inclusion here.

First, we shall obtain an estimate for u in the compact set $|x| \leq X_1$, $T < t \leq 2T$, using the analyticity of u with respect to the space variable. Then, we shall estimate u in $|x| \leq X_1$, $0 < t \leq T$, via the analyticity of u with respect to the time variable.

It is no loss of generality to assume that the interval $a < x < b$ is contained in the interval $|x| \leq X_1$. From Sec. 10.2 we see that u is an analytic function of the complex variable $z = x + iy$ and for $z \in D = \{z \mid |y| \leq Y\}$,

$$|u(z,T)| \leq C_1 \exp\left\{\frac{Y^2}{4T}\right\}. \tag{11.7.5}$$

From (11.7.1), for $y = 0$, $a < x < b$,

$$|u(x,T)| \leq \|f\|_{[a,b]}, \tag{11.7.6}$$

where

$$\|f\|_{[a,b]} = \sup_{a \leq x \leq b} |f(x)|. \tag{11.7.7}$$

We can now apply Lemma 10.6.5 to obtain, for $|x| \leq X_2$,

$$|u(x,T)| < C\{\|f\|_{[a,b]}\}^{\alpha \exp\{-\beta X_2\}}, \tag{11.7.8}$$

where $C = C(C_1, T)$, $\alpha = \alpha(a,b)$, and β are positive constants. As

$$u(x,t) = \int_{-\infty}^{\infty} K(x-\xi, t-T) u(\xi, T) \, d\xi \tag{11.7.9}$$

for $t > T$, we can write

$$|u(x,t)| \leq \int_{|\xi| \leq X_2} K(x-\xi, t-T)|u(\xi, T)| \, d\xi$$

$$+ \int_{|\xi| > X_2} K(x-\xi, t-T)|u(\xi, T)| \, d\xi$$

$$\leq C\{\|f\|_{[a,b]}\}^{\alpha \exp\{-\beta X_2\}} + C_1 \int_{|\xi| > X_2} K(x-\xi, t-T) \, d\xi.$$

$$\tag{11.7.10}$$

For $|x| \le X_1$ and $T < t \le 2T$,

$$\int_{|\xi| > X_2} K(x - \xi, t - T) \, d\xi < \frac{2}{\sqrt{\pi}} \int_{|X_2 - X_1|/(2\sqrt{2} \, T^{1/2})}^{\infty} \exp\{-\rho^2\} \, d\rho.$$

$$(11.7.11)$$

As we shall take $X_2 > X_1$, it is no loss of generality to assume that $|X_2 - X_1| > 2\sqrt{2} \, T^{1/2}$. Thus, $\rho > 1$ and the right side of (11.7.11) is less than

$$\frac{2}{\sqrt{\pi}} \int_{|X_2 - X_1|/(2\sqrt{2} \, T^{1/2})}^{\infty} \rho \exp\{-\rho^2\} \, d\rho = \pi^{-1/2} \exp\left\{-\frac{(X_2 - X_1)^2}{8T}\right\}.$$

$$(11.7.12)$$

Hence, for $|x| \le X_1$ and $T < t < 2T$,

$$|u(x, t)| \le C\{\|f\|_{[a, b]}\}^{\alpha \exp\{-\beta X_2\}} + C_1 \pi^{-1/2} \exp\left\{-\frac{(X_2 - X_1)^2}{8T}\right\}$$

$$(11.7.13)$$

holds for all $X_2 > X_1$ with $|X_2 - X_1| > 2\sqrt{2} \, T^{1/2}$.

From Sec. 10.3, we have that $u(x, \eta)$ is an analytic function of the complex variable $\eta = t + i\tau$ in the domain

$$D_\mu = \{\eta \, | \, 0 < t, \quad |\tau| \le \mu t\}.$$

Moreover, for $\eta \in D_\mu$,

$$|u(x, \eta)| \le (1 + \mu^2)^{1/2} C_1. \tag{11.7.14}$$

Given the estimate (11.7.13) of u for $|x| \le X_1$ and $T < t \le 2T$, we can apply Lemma 10.6.6 with $\mu = \sqrt{3}/3$, so that the condition $\sqrt{3} \, (b - a) \le \mu(a + b)$ is satisfied. Here, $a = T$ and $b = 2T$. Thus, for $|x| \le X_1$ and $0 < t \le T$,

$$|u(x, t)| \le C_2^{1 - (1/3)(t/R)^{1/\gamma}} \varepsilon^{(1/3)(t/R)^{1/\gamma}}, \tag{11.7.15}$$

where $C_2 = C_2(C_1)$,

$$R = \left(\tfrac{7}{3}\right)^{1/2} T, \tag{11.7.16}$$

$$\gamma = \frac{2}{\pi} \text{Arctan} \frac{\sqrt{3}}{9}, \tag{11.7.17}$$

and

$$\varepsilon = C\{\|f\|_{[a, b]}\}^{\alpha \exp\{-\beta X_2\}} + C_1 \pi^{-1/2} \exp\left\{-\frac{(X_2 - X_1)^2}{8T}\right\}.$$

$$(11.7.18)$$

We can summarize the results in the following statement.

Theorem 11.7.1. *For the solution u of the problem*

$$u_t = u_{xx}, \qquad -\infty < x < \infty, \qquad 0 < t,$$

$$u(x, T) = f(x), \qquad a < x < b, \qquad 0 < T, \qquad (11.7.19)$$

$$|u(x, t)| \le C_1, \qquad -\infty < x < \infty, \qquad 0 \le t,$$

where a and b are constants, C_1 and T are positive constants, and f is a piecewise-continuous function, there exist positive constants $C = C(C_1, T)$, $\alpha = \alpha(a, b)$, β, and $C_2 = C_2(C_1)$, such that, for $|x| \le X_1$ and $0 < t \le T$,

$$|u(x, t)| \le C_2^{1 - (1/3)(t/R)^{1/\gamma}} \varepsilon^{(1/3)(t/R)^{1/\gamma}}, \qquad (11.7.20)$$

where

$$R = \left(\tfrac{7}{3}\right)^{1/2} T, \qquad (11.7.21)$$

$$\gamma = \frac{2}{\pi} \text{Arctan} \frac{\sqrt{3}}{9}, \qquad (11.7.22)$$

$$\varepsilon = C\left\{\|f\|_{[a, b]}\right\}^{\alpha \exp\{-\beta X_2\}} + C_1 \pi^{-1/2} \exp\left\{\frac{-(X_2 - X_1)^2}{8T}\right\}, \qquad (11.7.23)$$

and where $\|f\|_{[a, b]} = \sup_{a \le x \le b} |f(x)|$, and X_2 is any positive constant such that $X_2 > X_1$ and

$$|X_2 - X_1| > 2\sqrt{2}\, T. \qquad (11.7.24)$$

For $|x| \le X_1$ and $T \le t \le 2T$,

$$|u(x, t)| < \varepsilon, \qquad (11.7.25)$$

where ε is given by (11.7.23).

Proof. See the analysis preceding the theorem statement. $\qquad\qquad\square$

Remark. Given a function

$$f(x) = \varepsilon^{\exp\{-x\}} + \exp\{-x\}, \qquad x > 0, \qquad (11.7.26)$$

$$\inf_{x > 0} f(x) \le 2 \exp\{-x_0\}, \qquad (11.7.27)$$

where x_0 is the solution of

$$\varepsilon^{\exp\{-x_0\}} = \exp\{-x_0\}, \qquad 0 < \varepsilon < 1. \qquad (11.7.28)$$

Now, taking the log of both sides,

$$\exp\{-x_0\} \log \varepsilon = -x_0. \qquad (11.7.29)$$

Again, we see

$$-x_0 + \log\log \varepsilon^{-1} = \log x_0, \qquad (11.7.30)$$

whence it follows that

$$\log\log \varepsilon^{-1} = x_0 + \log x_0. \tag{11.7.31}$$

As ε tends to zero,

$$x_0 \sim \log\log \varepsilon^{-1}. \tag{11.7.32}$$

Hence, as ε tends to zero

$$\inf_{x>0} f(x) \le 2\exp\{-\log\log \varepsilon^{-1}\} = 2(\log \varepsilon^{-1})^{-1}. \tag{11.7.33}$$

Applying a similar analysis to the expression in (11.7.23), we see that as $\|f\|_{[a,b]}$ tends to zero, our estimate (11.7.20) is

$$\mathcal{O}\left(\left[\log\{\|f\|_{[a,b]}^{-1}\}\right]^{-\alpha}\right) \tag{11.7.34}$$

where $\alpha = \alpha(x, t)$, $0 < \alpha < 1$.

NOTES

Another discussion of the material of Sec. 11.1 on continuous dependence upon the data can be found in Payne's [10] discussion of improperly posed problems (not-well-posed). There the reader can find several hundreds of references to the literature on such problems. The remainder of the material in this chapter is the joint work of Cannon with Douglas, Hill, and Klein [1,...,9].

REFERENCES

1. Cannon, J. R., "The Backward Continuation in Time by Numerical Means of the Solution of the Heat Equation in a Rectangle." M.A. Thesis, Rice University, 1960.
2. Cannon, J. R., Error estimates for some unstable continuation problems, *SIAM Journal*, **12** (1964), 270–284.
3. Cannon, J. R., *A priori* estimate for the continuation of the solution of the heat equation in the space variable, *Annali di Mat. Pura ed Appl.* (IV) **65** (1964), 377–388.
4. Cannon, J. R., A Cauchy problem for the heat equation, *Annali di Mat. Pura ed Appl.* (IV) **66** (1964), 155–165.
5. Cannon, J. R., and Douglas, J., Jr., "The approximation of harmonic and parabolic functions on half spaces from interior data." C.I.M.E., *Numerical Analysis of Partial Differential Equations*, Ispra, Varese, 3–11 Iuglio 1967. Editore, Cremonese Roma, 1968.
6. Cannon, J. R., and Douglas, J., Jr., The Cauchy problem for the heat equation, *SIAM J. Numer. Anal.*, **4** (1967), 317–336.

7. Cannon, J. R., and Hill, C. D., Continuous dependence of bounded solutions of a linear parabolic partial differential equation upon interior Cauchy data, *Duke Math. J.*, **35** (1968), 217–230.

8. Cannon, J. R., and Klein, R. E., "Optimal selection of measurement location in a conductor for approximate determination of temperature distributions." Proceedings of 1970 Joint Automatic Control Conference, 750–756.

9. Cannon, J. R., and Klein, R. E., Optimal selection of measurement locations in a conductor for approximate determination of temperature distributions, Trans. of ASME, *J. Dyn. Sys. Meas. and Cont.*, **93** (1971), 193–199.

10. Payne, L. E., Improperly Posed Problems in Partial Differential Equations. CBMS Regional Conference Series in Applied Math., *Soc. Ind. Appl. Math.*, Philadelphia, PA, 1975.

Chapter 12

Some Numerical Methods for Some State-Estimation Problems

12.1. SOME NUMERICAL RESULTS FOR THE SOLUTION OF THE HEAT EQUATION BACKWARDS IN TIME*

Introduction

Assume the existence of the bounded solution $v = v(x, t)$ of the problem (see Fig. 11.7.1)

$$v_{xx} = v_t, \qquad -\infty < x < \infty, \qquad 0 < t,$$

$$v(x, T) = g(x), \qquad -q < x < q, \tag{12.1.1}$$

$$|v(x, 0)| < M, \qquad -\infty < x < \infty,$$

where g is a given function and M, q, and T are given positive constants. For the time being, suppose that it is possible to construct a function $\bar{v}$ that

*The author wishes to thank John Wiley & Sons, Inc., for permission to reprint in this section his article: J. R. Cannon, Some Numerical Results for the Solution of the Heat Equation Backwards in Time, *Numerical Solutions of Nonlinear Differential Equations*, Edited by Donald Greenspan, John Wiley & Sons, Inc., N.Y., 1966, pp. 21–54.

satisfies

$$\bar{v}_{xx} = \bar{v}_t, \qquad -\infty < x < \infty, \qquad 0 < t,$$

$$|\bar{v}(x,0)| \le M, \qquad -\infty < x < \infty, \qquad (12.1.2)$$

$$\|g(x) - \bar{v}(x,T)\|_{[-q,q]} \le \varepsilon,$$

where ε is a known positive number and, for any function f defined on $a \le x \le b$,

$$\|f(x)\|_{[a,b]} = \sup_{a \le x \le b} |f(x)|. \qquad (12.1.3)$$

When ε is small, it is natural to consider $\bar{v}$ as an approximation to v for $0 < t < T$ and $-\infty < x < \infty$. The following theorem indicates just how good an approximation to v that $\bar{v}$ really is.

Theorem.   *For any compact subset D of the half-plane $t > 0$, there exist computable positive constants k_1, k_2, and k_3, $0 < k_3 < 1$, which depend only upon D, T, and M, and there exists a function $\alpha(X)$, $0 < \alpha(X) < 1$, which tends to zero when $|X|$ tends to infinity such that for any (x,t) in D,*

$$|v(x,t) - \bar{v}(x,t)| \le k_1 \left(\varepsilon^{\alpha(X)} + \exp\left\{ -k_2(X-c)^2 \right\} \right)^{k_3}, \qquad (12.1.4)$$

where c is a positive constant chosen so that the projection of D onto the x-axis is contained in the interval $-c \le x \le c$.

Proof.   For the complete details, see Theorem 11.7.1. However, for the completeness of this section, a brief description of the method of estimation will be included here. Set $w(x,t) = v(x,t) - \bar{v}(x,t)$. Then, w satisfies

$$w_{xx} = w_t, \qquad -\infty < x < \infty, \qquad 0 < t,$$

$$\|w(x,T)\|_{[-q,q]} \le \varepsilon, \qquad (12.1.5)$$

$$\|w(x,0)\|_{[-\infty,\infty]} \le 2M.$$

From the representation

$$w(x,t) = \frac{1}{(4\pi t)^{1/2}} \int_{-\infty}^{\infty} \exp\left\{ \frac{-(x-\xi)^2}{4t} \right\} w(\xi,0) \, d\xi, \qquad (12.1.6)$$

it is easy to see that $w(x,t)$ is analytic in both t and x. Using first the analyticity with respect to x, $w(x,t)$ can be estimated on the interval $[-X, X]$ by first using Lindelof's Lemma 10.6.4 to obtain $|w(x,T)| \le k_4 \varepsilon^{k_5}$, $0 < k_5 < 1$, in a full complex neighborhood of a point x_0 in $(-q, q)$, and then using Carleman's Lemma 10.6.1 to obtain the estimate $|w(x,T)| \le k_6 \varepsilon^{\alpha(X)}$ for x in $[-X, X]$. It follows from the exponential decay of solutions of the initial-value problem for the heat equation that for $-c \le x \le c$ and $T \le t \le 2T$,

$$|w(x,t)| \le k_7 \left(\varepsilon^{\alpha(X)} + \exp\left\{ -k_2(X-c)^2 \right\} \right). \qquad (12.1.7)$$

For (x, t) in the rectangle defined by $0 < t_0 \le t \le T$ and $-c \le x \le c$, the result (12.1.4) follows from the analyticity of w with respect to t, the estimate (12.1.5) and an application of Lindelof's Lemma 10.6.4 and Carleman's Lemma 10.6.1 for complex t. Note that it has been assumed that D lies in the rectangle defined by $-c \le x \le c$ and $t_0 \le t \le 2T$.   □

In general, $g(x)$ would be known only approximately as $G(x)$, where

$$\|G(x) - g(x)\|_{[-q, q]} \le \mu, \qquad \mu > 0, \qquad (12.1.8)$$

and where μ may or may not be known. By the theorem, a method for approximating v would be to find a solution $\bar{v}$ of the heat equation that satisfies

$$|\bar{v}(x, 0)| \le M, \qquad -\infty < x < \infty,$$

and that makes the number $\|G(x) - \bar{v}(x, T)\|_{[-q, q]}$ approximately as small as possible. Such a $\bar{v}$ can be determined via the method of linear programming.

Let $[-Q, Q]$ be an interval on the x-axis that contains or is contained in $[-q, q]$. Let N_1 and N_2 be positive integers and set

$$h_1 = \frac{2Q}{N_1} \qquad \text{and} \qquad h_2 = \frac{2q}{N_2}. \qquad (12.1.9)$$

Let

$$x_i = -Q + ih_1, \qquad i = 0, \dots, N_1, \qquad (12.1.10)$$

and

$$\eta_j = -q + jh_2, \qquad j = 0, \dots, N_2. \qquad (12.1.11)$$

Next, set

$$K_i(x, t) = \frac{1}{(4\pi t)^{1/2}} \int_{x_{i-1}}^{x_i} \exp\left\{-\frac{(x - \xi)^2}{4t}\right\} d\xi. \qquad (12.1.12)$$

For the set of real numbers b_i, $i = 1, \dots, N_1$, set

$$\bar{v}(x, t, \mathbf{b}) = \sum_{i=1}^{N_1} b_i K_i(x, t). \qquad (12.1.13)$$

Let B denote the set of vectors $\mathbf{b}$ such that $|b_i| \le M$, $i = 1, \dots, N_i$. Since it is desired to find a $\bar{v}$ that makes $\|G(x) - \bar{v}(x, T)\|_{[-q, q]}$ as small as possible, it is natural to consider the following approximation to the minimization problem

$$\inf_{\mathbf{b} \in B} \|G(x) - \bar{v}(x, T, \mathbf{b})\|_{[-q, q]}. \qquad (12.1.14)$$

Let

$$\delta = \inf_{\mathbf{b} \in B} \max_{0 \le j \le N_2} |G(\eta_j) - \bar{v}(\eta_j, T, \mathbf{b})|. \qquad (12.1.15)$$

This problem possesses a solution. Indeed, δ is equal to the minimum of the linear function

$$L(b_1,\ldots,b_{N_1},\zeta) = \zeta \qquad (12.1.16)$$

subject to the linear constraints

$$|G(\eta_j) - \bar{v}(\eta_j, T, b)| \leq \zeta, \qquad j = 0, \ldots, N_2, \qquad (12.1.17)$$

$$|b_i| \leq M, \qquad i = 1, \ldots, N_1, \qquad (12.1.18)$$

and

$$\zeta \geq 0. \qquad (12.1.19)$$

The problem of determining the minimum of L, subject to (12.1.17), (12.1.18), and (12.1.14), is a feasible linear-programming problem. The solution of this linear-programming problem defines a solution $\bar{v}(x,t)$ that makes $\|G(x) - \bar{v}(x,T)\|_{[-q,q]}$ approximately as small as possible. *A priori* and *a posteriori* estimates of $\|g(x) - \bar{v}(x,T)\|_{[-q,q]}$ can be derived.

With the preceding discussion as a theoretical and notational basis, results obtained from some numerical experiments of continuing a solution of the heat equation backwards in time will be discussed in the remainder of this paper.

Outline of the Numerical Experiment

The solution of the heat equation

$$v(x,t) = \frac{10}{2\sqrt{\pi}} \left[\frac{1}{\sqrt{t+.01}} \exp\left\{ -\frac{(x+.5)^2}{4(t+.01)} \right\} \right.$$

$$\left. + \frac{1}{\sqrt{t+.01}} \exp\left\{ -\frac{(x-.5)^2}{4(t+.01)} \right\} \right] \qquad (12.1.20)$$

was selected as the theoretical solution for a series of numerical tests. This solution generates the initial data

$$v(x,0) = \frac{100}{2\sqrt{\pi}} \left[\exp\left\{ -\frac{(x+.5)^2}{.04} \right\} + \exp\left\{ -\frac{(x-.5)^2}{.04} \right\} \right], \qquad (12.1.21)$$

which is graphed in Fig. 1. *All figures are displayed at the end of this section.* The parameters q, Q, N_1, and N_2 were chosen as follows:

$$q = 2.5, \qquad Q = 1, \qquad N_1 = 16, \qquad \text{and} \qquad N_2 = 10. \qquad (12.1.22)$$

Two cases of T were considered.

The first case was $T = .1$ (Figs. 1–10). The case was broken down into two parts: Figs. 1–5 and Figs. 6–10. For the first part (Figs. 1–5) of the first case,

$$G(x) = v(x,.1), \qquad (12.1.23)$$

which means that the approximation function $\bar{v}$ was determined by the linear-programming problem with no error in the data. In the second part

(Figs. 6–10) of the first case,

$$G(\eta_j) = v(\eta_j,.1) + (.01)(E(\eta_j))v(\eta_j,.1), \qquad j = 0,\dots,N_2,$$

$$(12.1.24)$$

where $E(\eta_j)$, $j = 0,\dots,N_2$, are the percentage error values (Fig. 24), which were determined from the distribution of error shown in Fig. 23.

The second case was $T = 1$ (Figs. 11–22). This case was broken down into three parts: Figs. 11–14, Figs. 15–18, and Figs. 19–22. Parts 1 and 2 (Figs. 11–14 and Figs. 15–18) were conducted like Parts 1 and 2 of Case 1 with the exception that $v(x,.1)$ in (2.4) and 2.5) was replaced by $v(x,1)$ for this case. Some of the results of Part 2 of Case 2 were not as good as the results of Part 2 of Case 1. The parameter N_2 was increased from 10 to 20, and Part 2 was rerun, with better results (Part 3: Figs. 19–22). The increase of data points for Part 3 of Case 2 necessitated the choosing of a new set of percentage errors from the error distribution of Fig. 23. These percentage errors are displayed in Fig. 25.

On each figure is a short legend:

T = .1 (or 1),
NO ERROR IN G (or ERROR IN G),
$(N_2 + 1) \times N_1(N_2 + 1$ data points and N_1b's),
ζ = (the minimum value from the linear programming problem),
t = 0 (or .001, or .01, or .1, or .5, or 1),
v(x, t):_____,
$\bar{v}$(x, t):_____,   (12.1.25)

which summarizes what each graph describes.

Discussion of the Numerical Results

In summary, the preceding section states that Parts 1 of Case 1 and Case 2 are a study of the truncation error (Figs. 1–5 and Figs. 11–14) and that the remaining parts of Case 1 and Case 2 are a study of the continuation in the presence of data error as well as truncation error.

Figures 1, 6, 11, 15, and 19 compare the solution v at $t = 0$ with the respective approximations $\bar{v}$. It is apparent that the approximations do not fit v at $t = 0$ very well, which should remind the reader that the theorem in the preceding section does not guarantee that the initial condition of v will be approximated by the initial condition of the approximation.

Figures 2 and 7 show the difference at $t = T$ between v and $\bar{v}$. As expected $v - \bar{v}$ can be fairly large in the intervals between consecutive η_j's even though $v - \bar{v}$ is reasonably small at η_j, $j = 0,\dots,N_2$, $t = T$.

Figures 3, 8, 12, 16, and 20 compare the solution v with the respective approximation $\bar{v}$ at a point in time 50% of the way back from T to $t = 0$. The results of all parts of both cases are quite good.

Figures 4, 9, 13, 17, and 21 compare the solution v with the respective approximations $\bar{v}$ at a point in time 90% of the way back from T to $t = 0$. With the exception of Fig. 17, the results displayed in Figs. 4, 9, 13, and 21 are reasonably good. Figure 17 represents the attempt to continue backwards in time, ten times as far with the same amount of data as for Part 2 of Case 1. This continuation in the presence of truncation and data error did not yield an adequate approximation of the solution 90% of the way back to $t = 0$. Figure 21 represents the author's correction to Figure 17. The approximation in Figure 21 was obtained from 21 data points η_j instead of the 11 points used for the determination of the approximation displayed in Fig. 17.

Figures 5, 10, 14, 18, and 22 demonstrate that the approximation of v becomes pretty poor at 99% of the way back to $t = 0$ from $t = T$. However, with the exception of Fig. 18 (see the remarks for Fig. 17 in the preceding paragraph), the approximations still retain some of the qualitative behavior of v.

Graphs of the Results of Numerical Experiments of Continuing a Solution of the Heat Equation Backwards in Time

Figures 1 to 25 follow.

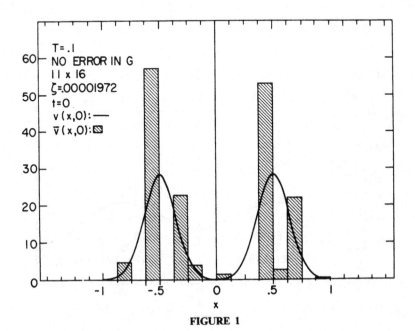

FIGURE 1

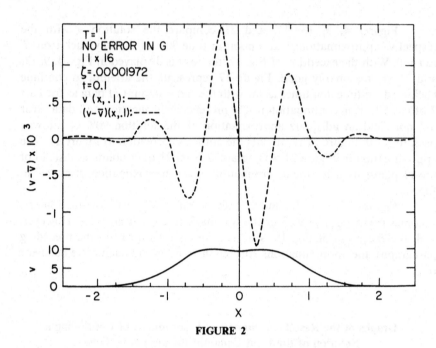

FIGURE 2

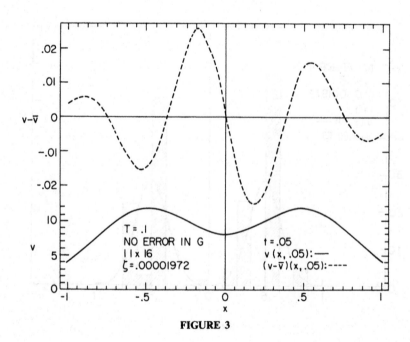

FIGURE 3

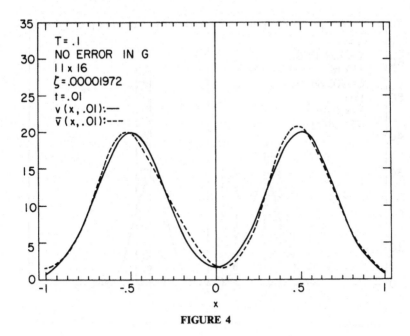

FIGURE 4

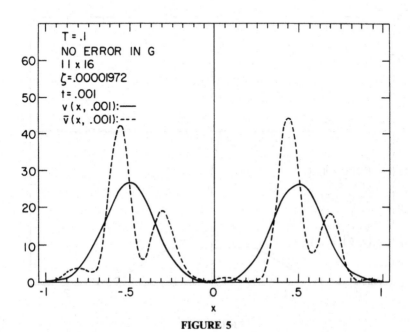

FIGURE 5

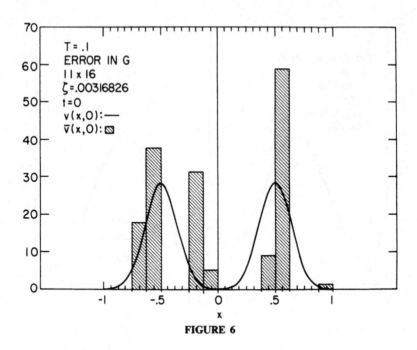

FIGURE 6

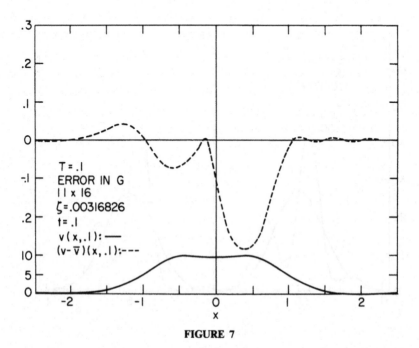

FIGURE 7

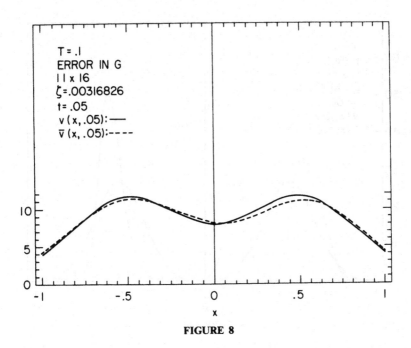

FIGURE 8

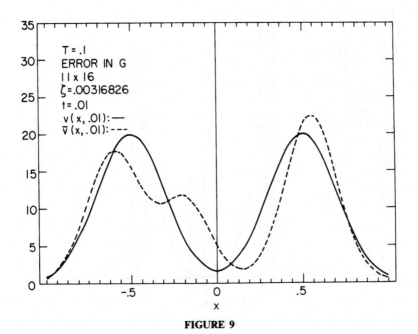

FIGURE 9

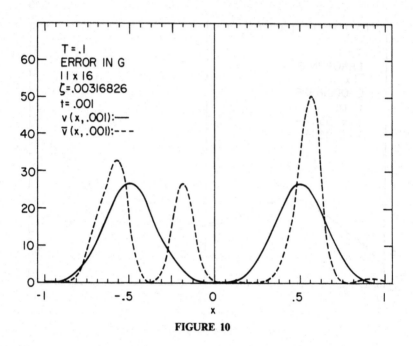

FIGURE 10

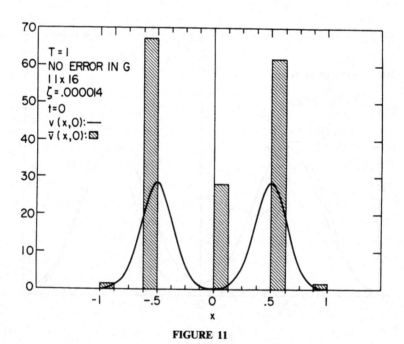

FIGURE 11

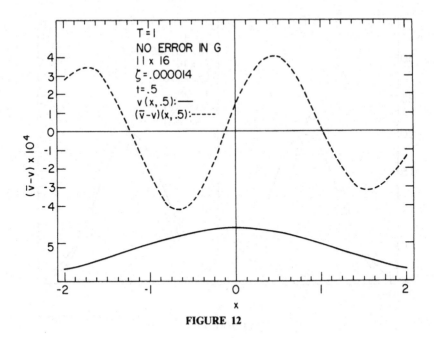

FIGURE 12

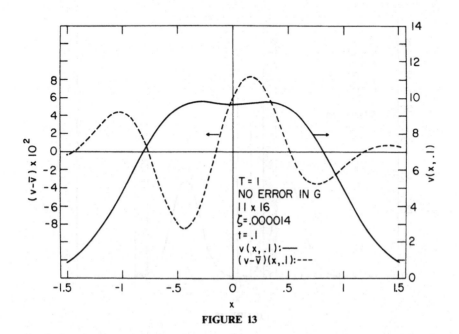

FIGURE 13

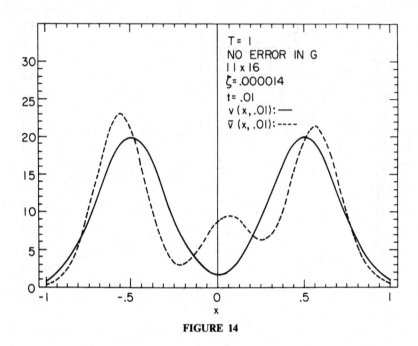

FIGURE 14

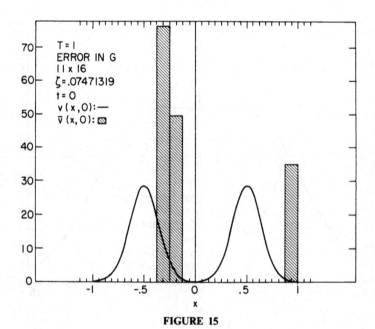

FIGURE 15

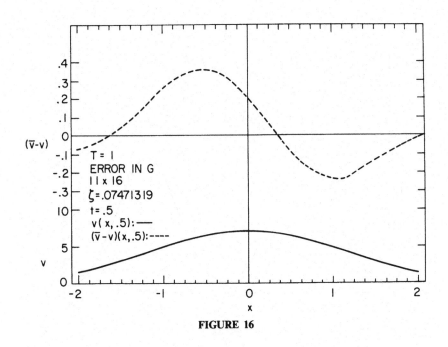

FIGURE 16

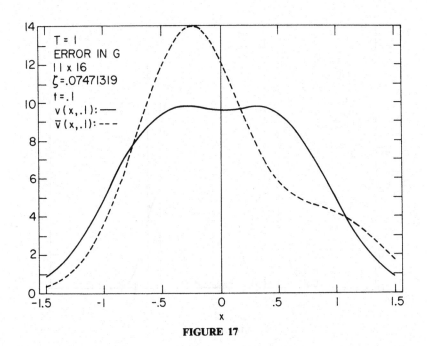

FIGURE 17

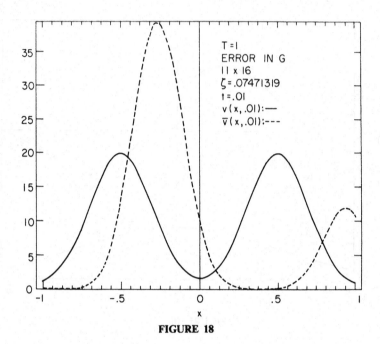

FIGURE 18

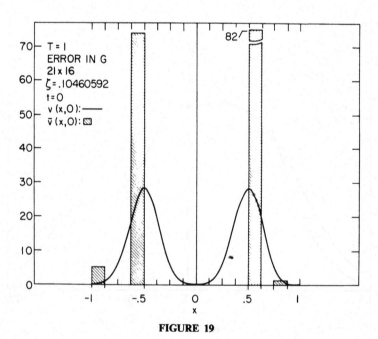

FIGURE 19

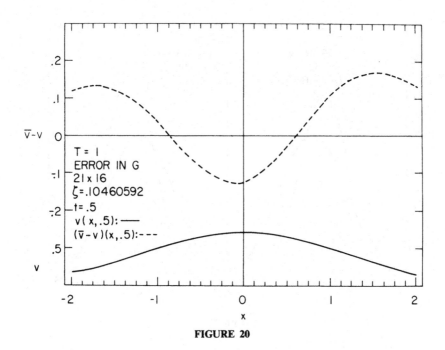

FIGURE 20

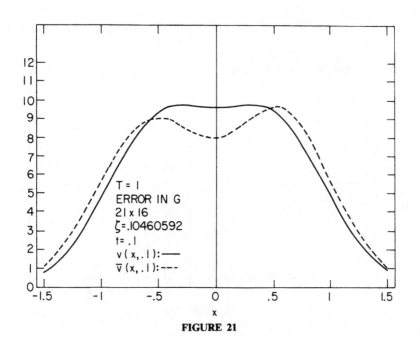

FIGURE 21

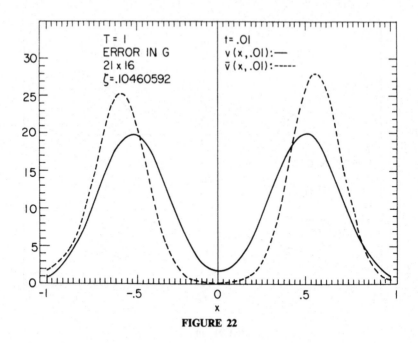

FIGURE 22

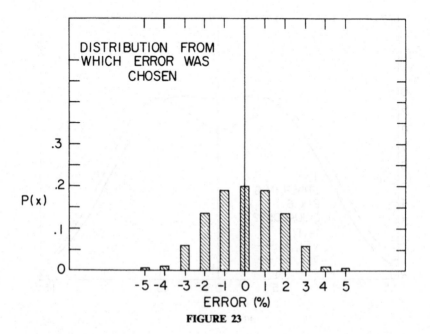

FIGURE 23

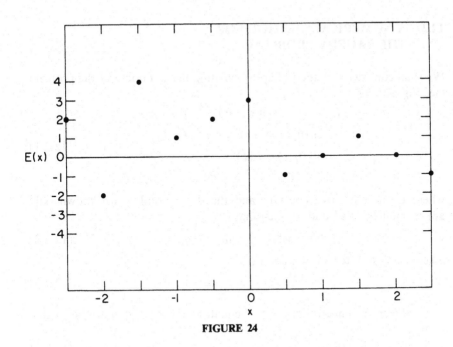

FIGURE 24

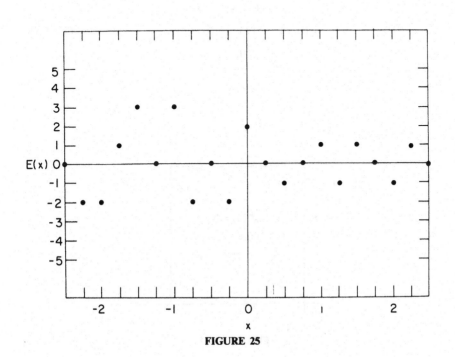

FIGURE 25

12.2. A NUMERICAL METHOD FOR THE CAUCHY PROBLEM

We shall consider a method of approximating the solution u of the problem (see Fig. 12.2.1)

$$u_t = u_{xx}, \qquad 0 < x < 1, \qquad 0 < t \le T,$$
$$u(0,t) = g(t), \qquad 0 < t \le T,$$
$$u_x(0,t) = h(t), \qquad 0 < t \le T, \qquad (12.2.1)$$
$$|u(x,t)| \le C_1, \qquad 0 \le x \le 1, \qquad 0 \le t \le T,$$

where C_1 is a positive constant and the data g and h are known only approximately as g^* and h^* such that

$$\|g - g^*\|_T, \qquad \|h - h^*\|_T < \varepsilon, \qquad (12.2.2)$$

where $\varepsilon > 0$ and, for any function ψ,

$$\|\psi\|_T = \sup_{0 < t \le T} |\psi(t)|. \qquad (12.2.3)$$

From the linearity of the heat equation, the solution u can be written as

$$u = v + w, \qquad (12.2.4)$$

where v satisfies

$$v_t = v_{xx}, \qquad 0 < x, \qquad 0 < t \le T,$$
$$v_x(0,t) = h(t), \qquad 0 < t \le T, \qquad (12.2.5)$$
$$v(x,0) = 0, \qquad 0 < x,$$

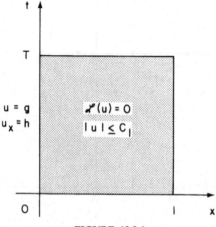

FIGURE 12.2.1

and w satisfies

$$
\begin{aligned}
w_t &= w_{xx}, && 0 < x < 1, && 0 < t \le T, \\
w(0, t) &= g(t) - v(0, t), && 0 < t \le T, \\
w_x(0, t) &= 0, && 0 < t \le T, \\
|w(x, 0)| &\le C_1, && 0 < x < 1.
\end{aligned}
\tag{12.2.6}
$$

Since the bounded solution v of (12.2.5) is given by

$$
v(x, t) = -2 \int_0^t K(x, t - \tau) h(\tau) \, d\tau,
\tag{12.2.7}
$$

where $K = K(x, t)$ is the fundamental solution defined by (3.2.1), it follows from elementary estimates that

$$
|v(x, t)| \le 2\pi^{-1/2} T^{1/2} \|h\|_T \le 2\pi^{-1/2} T^{1/2} C_1.
\tag{12.2.8}
$$

By constructing an even extension of w, we see that w satisfies

$$
\begin{aligned}
w_t &= w_{xx}, && -1 < x < 1, && 0 < t \le T, \\
w(0, t) &= g(t) - v(0, t), && 0 < t \le T, \\
w_x(0, t) &= 0, && 0 < t \le T, \\
|w(x, 0)| &< C_1, && -1 \le x \le 1, \\
|w(-1, t)| &< C_1(1 + 2\pi^{-1/2} T^{1/2}), && 0 < t \le T, \\
|w(+1, t)| &< C_1(1 + 2\pi^{-1/2} T^{1/2}), && 0 < t \le T.
\end{aligned}
\tag{12.2.9}
$$

From Theorem 10.5.1, we see that w can be extended into the complex $z = x + iy$ plane as an analytic function in the domains

$$
D_\mu = \{ z \mid -1 < x < 1, \quad |y| < \mu(x + 1), \quad |y| < \mu(1 - x) \}
\tag{12.2.10}
$$

for any μ satisfying $0 < \mu < 1$. Moreover, there exists a constant $C_2 = C_2(C_1, \mu, T, \delta)$ such that for $0 < \delta \le t \le T$ and $z \in D_\mu$,

$$
|w(z, t)| \le C_2.
\tag{12.2.11}
$$

There is a $\mu_0 > 0$ such that the disc $|z| < 2/3$ is contained in D_{μ_0}. Since w can be represented as a power series in that disc, we can estimate the coefficients of that power series from the Cauchy–Riemann integral formula. Hence,

$$
\frac{1}{n!} \left| \frac{\partial^n w}{\partial x^n}(0, t) \right| \le C_2 \left(\frac{3}{2} \right)^n, \qquad n = 1, 2, \ldots, \quad \text{for } 0 < \delta \le t \le T.
\tag{12.2.12}
$$

Let $v^{**}(x, t)$ denote a numerical approximation to the quadrature

$$
v^*(x, t) = -2 \int_0^t K(x, t - \tau) h^*(\tau) \, d\tau
\tag{12.2.13}
$$

and let

$$w_n^{**}(x, t) = w^{**}(0, t) + \sum_{i=1}^{[(n-1)]/2} \frac{\Delta_i w^{**}(0, t)}{k_i^i} \frac{x^{2i}}{(2i)!}, \quad (12.2.14)$$

where

$$w^{**}(0, t) = \{ g^*(t) - v^{**}(0, t) \}, \quad (12.2.15)$$

$[(n-1)/2]$ denotes the largest integer in $(n-1)/2$ and $\Delta_i w^{**}(0, t)/k_i^i$ denotes a standard forward difference approximation to $(\partial^i w/\partial t^i)(0, t)$ with mesh size k_i. As our approximation to u, we set

$$u_n^{**}(x, t) = v^{**}(x, t) + w_n^{**}(x, t). \quad (12.2.16)$$

In what follows, we shall discuss how to select k_i and n in order to obtain a good approximation to u. These selections will arise in the discussion of the error in the approximation of u.

First, it is easy to see that there exists a constant C_3 such that for $0 \le x < \infty$, $0 < t \le T$,

$$|v(x, t) - v^{**}(x, t)| < C_3(\varepsilon + r), \quad (12.2.17)$$

where r denotes the maximum evaluation and rounding error. Next, we shall estimate the tail of the power-series representation for w. Since $w_x(0, t) = 0$ implies, from the differential equation, that every odd x-derivation of w vanishes at $x = 0$, it follows that for n odd

$$\sum_{i=n+1}^{\infty} \frac{\partial^i w}{\partial x^i}(0, t) \frac{x^i}{(i)!} = \sum_{i=(n+1)/2}^{\infty} \frac{\partial^{2i} w}{\partial x^{2i}}(0, t) \frac{x^{2i}}{(2i)!}. \quad (12.2.18)$$

From (12.2.12), it follows that for $0 < \delta \le t \le T$ and $|x| \le \frac{1}{2}$,

$$\left| \sum_{i=(n+1)/2}^{\infty} \frac{\partial^{2i} w}{\partial x^{2i}}(0, t) \frac{x^{2i}}{(2i)!} \right| \le \frac{C_2 \left[\left(\frac{3}{2} x \right)^2 \right]^{(n+1)/2}}{1 - \left(\frac{3}{2} x \right)^2} \le \frac{16}{7} C_2 \left(\frac{3}{2} x \right)^{n+1}.$$

$$(12.2.19)$$

Finally, we shall estimate the first n terms in the power series for w. First, for $0 < \delta \le t \le T$.

$$|w(0, t) - w^{**}(0, t)| \le |g(t) - g^*(t)| + |v(0, t) - v^{**}(0, t)|$$
$$\le \varepsilon + C_3(\varepsilon + r) = C_4(\varepsilon + r). \quad (12.2.20)$$

Next, for each $i = 1, 2, \ldots, (n-1)/2$, we see that

$$\left| \frac{\partial^{2i} w}{\partial x^{2i}}(0, t) - \frac{\Delta_i w^{**}(0, t)}{k_i^i} \right| \le \left| \frac{\partial^{2i} w}{\partial x^{2i}}(0, t) - \frac{\Delta_i w(0, t)}{k_i^i} \right|$$
$$+ |\Delta_i w(0, t) - \Delta_i w^{**}(0, t)| k_i^{-i},$$

$$(12.2.21)$$

where $\Delta_i w(0, t)/k_i^i$ is the finite difference approximation of $(\partial^i w/\partial t^i)(0, t)$. From Taylor's theorem with remainder, we see that

$$\left| \frac{\partial^{2i} w}{\partial x^{2i}}(0, t) - \frac{\Delta_i w(0, t)}{k_i^i} \right| = \left| \frac{\partial^i w}{\partial t^i}(0, t) - \frac{\Delta_i w(0, t)}{k_i^i} \right|$$

$$\leq \sup_{\delta \leq \tau \leq T} \left| \frac{\partial^{i+1} w}{\partial t^{i+1}}(0, \tau) \right| \frac{i}{2} k_i$$

$$\leq \sup_{\delta \leq \tau \leq T} \left| \frac{\partial^{2i+2} w}{\partial x^{2i+2}}(0, \tau) \right| \frac{i}{2} k_i$$

$$\leq 2^{-1} C_2 (2i+2)! \left(\frac{3}{2} \right)^{2i+2} i k_i, \quad (12.2.22)$$

while employing the representation

$$|\Delta_i w(0, t) - \Delta_i w^{**}(0, t)| k_i^{-i}$$

$$= i! \sum_{s=0}^{i} \frac{w(0, t_s) - w^{**}(0, t_s)}{(t_s - t_0) \cdots (t_s - t_{s-1})(t_s - t_{s+1}) \cdots (t_s - t_i)},$$

where $t_s = t + sk_i$, $s = 0, 1, 2, \ldots, i$, we obtain from (12.2.20)

$$|\Delta_i w(0, t) - \Delta_i w^{**}(0, t)| k_i^{-i} \leq \frac{i!}{k_i^i} \sum_{s=0}^{i} |w(0, t_s) - w^{**}(0, t_s)|$$

$$\leq \frac{C_4 (i+1)!}{k_i^i} (\varepsilon + r). \quad (12.2.23)$$

Combining (12.2.21) throughout (12.2.23), we see that for each $i = 1, 2, \ldots, (n-1)/2$,

$$\left| \frac{\partial^{2i} w}{\partial x^{2i}}(0, t) - \frac{\Delta_i w^{**}(0, t)}{k_i^i} \right| \leq 2^{-1} C_2 (2i+2)! \left(\frac{3}{2} \right)^{2i+2} i k_i$$

$$+ \frac{C_4 (i+1)!}{k_i^i} (\varepsilon + r). \quad (12.2.24)$$

We now select

$$k_i = \left\{ \frac{2C_4 (i+1)! (\varepsilon + r)}{C_2 (2i+2)! (3/2)^{(2i+2)}} \right\}^{1/(i+1)} \quad (12.2.25)$$

and substitute into (12.2.24) to obtain, for $0 < \delta \leq t \leq T$,

$$\left| \frac{\partial^{2i} w}{\partial x^{2i}}(0,t) - \frac{\Delta_i w^{**}(0,t)}{k_i^i} \right|$$

$$\leq C_2 (2i+2)! \left(\frac{3}{2}\right)^{2i+2} i \{ C_4(i+1)!(\varepsilon+r) \}^{1/(i+1)}$$

$$\leq C_2 (2i+2)! \left(\frac{3}{2}\right)^{2i+2} i(i+1) \{ C_4(\varepsilon+r) \}^{1/(i+1)}. \qquad (12.2.26)$$

We now employ the preceding estimates in the estimate of the error in u. Let w_n denote the first n terms of w. Then,

$$|u(x,t) - u_n^{**}(x,t)| = |v(x,t) + w(x,t) - v^{**}(x,t) - w_n^{**}(x,t)|$$

$$\leq |v(x,t) - v^{**}(x,t)| + |w_n(x,t) - w_n^{**}(x,t)|$$

$$+ \left| \sum_{i=(n+1)/2}^{\infty} \frac{\partial^{2i} w}{\partial x^{2i}}(0,t) \frac{x^{2i}}{(2i)!} \right|. \qquad (12.2.27)$$

Using (12.2.17), (12.2.19), (12.2.20), and (12.2.26), we see that, for $0 \leq x \leq \frac{1}{2}$ and $0 < \delta \leq t \leq T$,

$$|u(x,t) - u_n^{**}(x,t)| \leq C_3(\varepsilon+r) + \tfrac{16}{7} C_2 (\tfrac{3}{2}x)^{n+1} + C_4(\varepsilon+r)$$

$$+ \sum_{i=1}^{(n-1)/2} C_2(2i+2)! \left(\frac{3}{2}\right)^{2i+2} i(i+1) \{ C_4(\varepsilon+r) \}^{1/(i+1)} \frac{x^{2i}}{(2i)!}$$

$$\leq C_5(\varepsilon+r) + C_6(\tfrac{3}{2}x)^{n+1} + C_7(n+1)^4(\varepsilon+r)^{1/(n+1)}$$

$$\leq C_8(n+1)^4(\varepsilon+r)^{1/(n+1)} + C_6(\tfrac{3}{2}x)^{n+1}, \qquad (12.2.28)$$

where C_i, $i = 5,6,7,8$ are positive constants that depend only upon C_1, T, μ_0, and δ. Taking the case of $x = \frac{1}{2}$, we see that for $0 \leq x \leq \frac{1}{2}$ and $0 < \delta \leq t \leq T$,

$$|u(x,t) - u_n^{**}(x,t)| \leq C_8(n+1)^4(\varepsilon+r)^{1/(n+1)} + C_6(\tfrac{3}{4})^{n+1}. \qquad (12.2.29)$$

Selecting

$$n = \left[\left\{ \frac{\log(\varepsilon+r)^{-1}}{\log(4/3)} \right\}^{1/2} \right] - 1, \qquad (12.2.30)$$

it follows that

$$|u(x,t) - u_n^{**}(x,t)| \le C_9 \left\{ \frac{\log(\varepsilon + r)^{-1}}{\log(4/3)} \right\}^2$$
$$\cdot \exp\left\{ -\left(\log(\varepsilon + r)^{-1} \right)^{1/2} [\log(\tfrac{4}{3})]^{1/2} \right\},$$

$$(12.2.31)$$

where $C_9 = C_9(C_1, T, \mu_0, \delta)$ is a positive constant. Using the simple estimate

$$\exp\{-x\} \le \frac{n!}{x^n} \qquad (12.2.32)$$

for any positive integer n and selecting $n \ge 5$, we see that for $0 \le x \le \tfrac{1}{2}$ and $\delta \le t \le T$, u_n^{**} tends to u as ε and r tend to zero.

We can summarize the above analysis in the following statement.

Theorem 12.2.1. *For the solution u of the problem*

$$u_t = u_{xx}, \qquad 0 < x < 1, \qquad 0 < t \le T,$$
$$u(0,t) = g(t), \qquad 0 < t \le T,$$
$$u_x(0,t) = h(t), \qquad 0 < t \le T,$$
$$|u(x,t)| \le C_1, \qquad 0 \le x \le 1, \qquad 0 \le t \le T,$$

$$(12.2.33)$$

where C_1 is a positive constant and the data g and h are known only approximately as g^* and h^* such that

$$\|g - g^*\|_T, \quad \|h - h^*\|_T < \varepsilon, \qquad (12.2.34)$$

where $\varepsilon > 0$ and for any function ψ,

$$\|\psi\|_T = \sup_{0 \le t \le T} |\psi(t)|, \qquad (12.2.35)$$

we define the approximation

$$u_n^{**}(x,t) = v^{**}(x,t) + w_n^{**}(x,t), \qquad (12.2.36)$$

where n is a positive odd integer, $v^{**}(x,t)$ is a numerical approximation of the quadrature

$$v^*(x,t) = -2 \int_0^t K(x,t-\tau) h^*(\tau) \, d\tau, \qquad (12.2.37)$$

where $K = K(x,t)$ is the fundamental solution defined by (3.2.1), and

$$w_n^{**}(x,t) = w^{**}(0,t) + \sum_{i=1}^{[(n-1)/2]} \frac{\Delta_i w^{**}(0,t)}{k_i^i} \frac{x^{2i}}{(2i)!}, \qquad (12.2.38)$$

where

$$w^{**}(0,t) = \{ g^*(t) - v^{**}(0,t) \}, \qquad (12.2.39)$$

$[(n-1)/2]$ denotes the largest integer in $(n-1)/2$, and $\Delta_i w^{**}(0,t)/k_i^i$

denotes a standard forward difference approximation to $(\partial^i w/\partial t^i)(0, t)$ with mesh size k_i. Then if, for each i,

$$k_i = \mathcal{O}\left((\varepsilon + r)^{1/(i+1)}\right) \tag{12.2.40}$$

where r denotes the maximum evaluation and rounding error, and if n is selected via

$$n = \left[\left\{\left(\frac{\log(\varepsilon + r)^{-1}}{\log(4/3)}\right)^{-1}\right\}^{1/2}\right] - 1, \tag{12.2.41}$$

then there exists a constant $C = C(C_1, T, \delta)$ such that for $0 \le x \le \frac{1}{2}$ and $0 < \delta \le t \le T$,

$$|u(x, t) - u_n^{**}(x, t)| \le C\left\{\frac{\log(\varepsilon + r)^{-1}}{\log(4/3)}\right\}^2$$
$$\cdot \exp\left\{-\left(\log(\varepsilon + r)^{-1}\right)^{1/2}\left[\log(\tfrac{4}{3})\right]^{1/2}\right\}, \tag{12.2.42}$$

for $(\varepsilon + r) < 1$.

    Proof.   See the analysis preceding the statement of the theorem.   □

Remark 1. Using $\exp\{-x\} \le n! x^{-n}$, the convergence is faster than any positive power of $[\log(\varepsilon + r)^{-1}]^{-1}$ as $\varepsilon + r$ tends to zero.

Remark 2. In practice one does not have to use a different k_i for each $i = 1, \ldots, (n-1)/2$. In practice one can get by with about three choices of k_i for $n = 21$. For example, $k_1 = 10^{-3}$, $k_2 = k_3 = 5 \times 10^{-3}$, and $k_4 = \cdots = k_{10} = 10^{-2}$ were found to be adequate when $2C_4(\varepsilon + r)/C_2 = .0001$.

    Some numerical experiments were conducted on the problem

$$u_t = u_{xx}, \qquad 0 < x < 1, \qquad 0 < t \le T,$$
$$u(0, t) = \exp\{-t\}, \qquad 0 < t \le T, \tag{12.2.43}$$
$$u_x(0, t) = 0, \qquad 0 < t \le T,$$

which has the unique solution

$$u(x, t) = \exp\{-t\}\cos x. \tag{12.2.44}$$

Since $u_x(0, t) = 0$, the approximation u_n^{**} involved only the polynomial approximation to the series representation for u:

$$u(x, t) = \sum_{j=0}^{\infty} (-1)^j \exp\{-t\}\frac{x^{2j}}{(2j)!} = \sum_{j=0}^{\infty} \frac{\partial^j u}{\partial t^j}(0, t)\frac{x^{2j}}{(2j)!}. \tag{12.2.45}$$

The values $n = 2, 4, 6, 8, 10, 12$, for the degree of the polynomial approximation u_n^{**} were chosen for study. The derivatives $\partial^j u / \partial t^j$, $j = 1, 2, 3, 4, 5, 6$ were approximated using centered difference formulae. This was a minor departure from the analysis given above. Also, only one value of the time step was used for computing the centered differences.

The numerical experiments were begun with the consideration of $u - u_n^{**}$ for $n = 2, 4, 6, 8, 10, 12$, $t = .5$, $x = j/10$, $j = 1, \ldots, 10$, six-digit accuracy in $u(0, t)$ in the calculation of the centered differences, and time step $k_i = k$, where three values of the time step k were selected. The best overall approximation to u was obtained for $n = 6$, and $k = .01$, as illustrated in Table 12.2.1.

The results of Table 12.2.1 illustrate the effects of finite significance in computations. We note that

$$\frac{\partial u}{\partial t}(0, .5) = \frac{u(0, .5 + k) - u(0, .5 - k)}{2k} + \mathcal{O}(k^2) \qquad (12.2.46)$$

for perfect arithmetic, while

$$\frac{\partial u}{\partial t}(0, .5) = \frac{u^*(0, .5 + k) - u^*(0, .5 - k)}{2k} + \mathcal{O}\left(k^2 + \frac{r}{k}\right) \quad (12.2.47)$$

where u^* denotes the approximate evaluation of u to machine limits and r

TABLE 12.2.1

x \ k	$k = .1$	$k = .01$	$k = .001$
$x = .1$	5.0664E − 06	5.96046E − 08	5.96046E − 08
$x = .2$	2.0206E − 05	2.38419E − 07	1.84774E − 06
$x = .3$	4.52995E − 05	3.57268E − 07	2.70009E − 05
$x = .4$	8.04663E − 05	5.36442E − 07	1.61231E − 04
$x = .5$	1.2517E − 04	0	6.31809E − 04
$x = .6$	1.79708E − 04	1.49012E − 06	1.91253E − 03
$x = .7$	2.43902E − 04	5.1558E − 06	4.86153E − 03
$x = .8$	3.1811E − 04	1.24574E − 05	.0108881
$x = .9$	4.03404E − 04	2.50638E − 05	.0221496
$x = 1$	5.01603E − 04	4.52399E − 05	.0417797

The values of $|u - u_n^{**}|$ for $t = .5$ and $x = j/10$, $j = 1, \ldots, 10$, degree of the approximation polynomial $n = 6$, and six-digit accuracy in the centered difference approximations of $(\partial^j u / \partial t^j)(0, .5)$, $j = 1, 2, 3$, are displayed for time steps $k = .1$, .01, and .001.

denotes the rounding error. Assuming u is of the order of one while r is 10^{-6}, we can estimate the minimum of $k^2 + (r/k)$, $k > 0$. We find that the minimum is $k = \sqrt[3]{.5} \times 10^{-2}$, which explains rather well the results in Table 12.2.1.

Selecting $n = 6$ and $k = .01$, error in u in the computation of the centered differences was introduced via the factor $f = (1 + (2\text{RND}(0) - 1)E)$, where $\text{RND}(0)$ denotes a random number selected from a uniform distribution over the interval $(0,1)$ and E denotes a positive maximum choice of error. For example, assuming perfect arithmetic,

$$u^*(0, .5 + k) = u(0, .5 + k)(1 + (2\text{RND}(0) - 1)E). \quad (12.2.48)$$

The effects of $E = .0001$, $.001$, and $.01$ are displayed in Tables 12.2.2, 12.2.3, and 12.2.4.

The results of Tables 12.2.2, 12.2.3, and 12.2.4 display the well known phenomenon of not-well-posed problems, in the sense of Hadamard. That is, the error increases as one moves away from the data location, and the error increases significantly for small changes in the data error. Some improvement in results was obtained for time step $k = .1$. This follows from the discussion following Table 12.2.1, since r now contains data error as well as finite significance. However, the overall behavior is similar and does not warrant display via tables.

TABLE 12.2.2

x	$u(x,.5)$	$u_6^{**}(x,.5)$	$\|u - u_6^{**}\|$
$x = .1$	.603501	.603492	8.88109E − 06
$x = .2$	.594441	.594431	1.00136E − 05
$x = .3$	.579441	.579345	9.63807E − 05
$x = .4$	.558652	.558357	2.94387E − 04
$x = .5$	.532281	.531841	4.39763E − 04
$x = .6$	.500591	.500614	2.21729E − 05
$x = .7$	.4639	.466196	2.29585E − 03
$x = .8$	.422574	.431124	8.55017E − 03
$x = .9$	.377026	.399317	.0222911
$x = 1$	.32771	.376502	.0487921

Comparison of $u(x, .5) = \exp\{-.5\}\cos x$ with $u_6^{**}(x, .5) = \sum_{j=0}^{3} c_j(x^{2j}/(2j)!)$, where c_j is the jth central difference for u at $x = 0$ and $t = .5$ with time step $k = .01$. The central differences were computed with the error factor $f = (1 + (2\text{RND}(0) - 1)E)$ applied to each u where $\text{RND}(0)$ denotes a random number chosen from a uniform distribution over the interval $(0,1)$ and $E = .0001$.

TABLE 12.2.3

x	$u(x,.5)$	$u_6^{**}(x,.5)$	$\lvert u - u_6^{**}\rvert$
$x = .1$	.603501	.603714	2.13146E−04
$x = .2$	.594441	.594248	1.9294E−04
$x = .3$	.579441	.579115	3.25739E−04
$x = .4$	.558652	.560216	1.5645E−03
$x = .5$	.532281	.542077	9.79638E−03
$x = .6$	.500591	.533715	.033124
$x = .7$	.4639	.55104	.0871401
$x = .8$	.422574	.619785	.197211
$x = .9$	.377026	.778973	.401947
$x = 1$	.32771	1.08491	.757202

Comparison of $u(x,.5) = \exp\{-.5\}\cos x$ with $u_6^{**}(x,.5) = \sum_{j=0}^{3} c_j (x^{2j}/(2j)!)$, where c_j is the jth central difference for u at $x = 0$ and $t = .5$ with time step $k = .01$. The central differences were computed with the error factor $f = (1 + (2\,\mathrm{RND}(0) - 1)E)$ applied to each u, where $\mathrm{RND}(0)$ denotes a random number chosen from a uniform distribution over the interval $(0,1)$ and $E = .001$.

TABLE 12.2.4

x	$u(x,.5)$	$u_6^{**}(x,.5)$	$\lvert u - u_6^{**}\rvert$
$x = .1$	.603501	.599608	3.89212E−03
$x = .2$	.594441	.589183	5.25755E−03
$x = .3$	.579441	.558689	.0207517
$x = .4$	.558652	.488824	.069828
$x = .5$	.532281	.353309	.178972
$x = .6$	.500591	.11964	.380952
$x = .7$	.4639	−.249956	.713857
$x = .8$	.422574	−.797345	1.21992
$x = .9$	.377026	−1.5671	1.94412
$x = 1$	.32771	−2.60489	2.9326

Comparison of $u(x,.5) = \exp\{-.5\}\cos x$ with $u_6^{**}(x,.5) = \sum_{j=0}^{3} c_j (x^{2j}/(2j)!)$, where c_j is the jth central difference for u at $x = 0$ and $t = .5$ with time step $k = .01$. The central differences were computed with the error factor $f = (1 + (2\,\mathrm{RND}(0) - 1)E)$ applied to each u, where $\mathrm{RND}(0)$ denotes a random number chosen from a uniform distribution over the interval $(0,1)$ and $E = .01$.

12.3. A LEAST-SQUARES METHOD

To illustrate the utility of the method of least squares, we shall consider here
the problem

$$u_t = u_{xx}, \qquad 0 < x < 1, \qquad 0 < t,$$
$$u(0, t) = u(1, t) = 0, \qquad 0 < t,$$
$$u(a, t) = \varphi(t), \qquad 0 < t_0 < t < t_1, \qquad (12.3.1)$$
$$u_x(a, t) = \psi(t), \qquad 0 < t_0 < t < t_1,$$
$$|u(x, t)| \le C_1, \qquad 0 \le x \le 1, \qquad 0 \le t,$$

where C_1 and a, $0 < a < 1$, are positive constants and the functions φ and ψ
are known only approximately as φ^* and ψ^*, where

$$\|\varphi - \varphi^*\|_{[t_0, t_1]} < \varepsilon \qquad (12.3.2)$$

and

$$\|\psi - \psi^*\|_{[t_0, t_1]} < \varepsilon. \qquad (12.3.3)$$

Here,

$$\|f\|_{[a, b]} = \sup_{a \le t \le b} |f(t)| \qquad (12.3.4)$$

and the positive number ε incorporates measurement error as well as any
error arising from the handling of boundary conditions other than the
homogeneous ones given above. (See Fig. 12.3.1.)

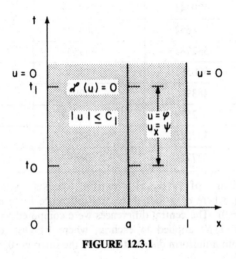

FIGURE 12.3.1

Now, the solution u has the representation

$$u(x,t) = \sum_{n=1}^{\infty} a_n \exp\{-n^2\pi^2 t\} \sin n\pi x, \qquad (12.3.5)$$

where the

$$a_n = 2\int_0^1 u(x,0)\sin n\pi x \, dx \qquad (12.3.6)$$

are unknown constants that satisfy

$$|a_n| < 2C_1. \qquad (12.3.7)$$

For our purpose here, we see that

$$\int_0^1 |u(x,0)|^2 \, dx \le C_1^2. \qquad (12.3.8)$$

As

$$u(x,0) = \sum_{n=1}^{\infty} a_n \sin n\pi x, \qquad (12.3.9)$$

we see that (12.3.8) yields

$$\sum_{n=1}^{\infty} a_n^2 \le 2C_1^2. \qquad (12.3.10)$$

We shall use (12.3.10) as the constraint in what follows.

For a positive integer N, let $\mathbf{b} = (b_1,\ldots,b_N)$ denote a vector in N-dimensional Euclidean space $\mathbf{R}^N$. Set

$$v(x,t;\mathbf{b}) = \sum_{n=1}^{N} b_n \exp\{-n^2\pi^2 t\}\sin n\pi x. \qquad (12.3.11)$$

We are particularly interested in the family F of solutions v that is generated by all $\mathbf{b} \in B$ where

$$B = \left\{ \mathbf{b} \in \mathbf{R}^N \mid \sum_{n=1}^{N} b_n^2 \le 2C_1^2 \right\}.$$

Since B is a compact set in $\mathbf{R}^N$, there exists a point $\mathbf{b} \in B$ at which the continuous function

$$J(\mathbf{b}) = \int_{t_0}^{t_1} \big(\varphi^*(t) - v(a,t;\mathbf{b})\big)^2 \, dt + \int_{t_0}^{t_1}\big(\psi^*(t) - v_x(a,t;\mathbf{b})\big)^2 \, dt$$

$$(12.3.12)$$

assumes its minimum value. Let

$$\eta^* = J(\mathbf{b}^*) = \inf_{\mathbf{b} \in B} J(\mathbf{b}). \qquad (12.3.13)$$

Lemma 12.3.1. *For each $N > 0$,*

$$0 \le \eta^* \le C_2\{\epsilon^2 + \exp\{-2N^2\pi^2 t_0\}\}, \qquad (12.3.14)$$

where $C_2 = C_2(C_1, t_0, t_1)$ is a positive constant.

Proof. Note that $J(\mathbf{b}^*) \le J(\mathbf{a})$, where $\mathbf{a} = (a_1, \ldots, a_N)$ and a_n is defined by (12.3.6). We estimate $J(\mathbf{a})$. Consider the first integral in J. Adding and subtracting φ in the integrand, it follows that we need only estimate

$$I_1 = \int_{t_0}^{t_1} \left(\varphi^*(t) - \varphi(t) \right)^2 dt \qquad (12.3.15)$$

and

$$I_2 = \int_{t_0}^{t_1} \left(\sum_{n=N+1}^{\infty} a_n \exp\{ -n^2\pi^2 t \} \sin n\pi a \right)^2 dt. \qquad (12.3.16)$$

From (12.3.2)

$$I_1 \le (t_1 - t_0)\varepsilon^2. \qquad (12.3.17)$$

Now, the series in the integrand of I_2 is dominated by

$$2C_1 \sum_{n=N+1}^{\infty} \exp\{ -n^2\pi^2 t_0 \} \le C_1 \int_N^{\infty} 2n \exp\{ -\pi^2 t_0 n^2 \} \, dn$$

$$\le C_1 \pi^{-2} t_0^{-1} \exp\{ -N^2\pi^2 t_0 \}. \qquad (12.3.18)$$

Consequently,

$$I_2 \le (t_1 - t_0) C_1^2 \pi^{-4} t_0^{-2} \exp\{ -2N^2\pi^2 t_0 \}. \qquad (12.3.19)$$

Combining (12.3.17) and (12.3.19) and using the fact that the first integral in J is bounded by $2(I_1 + I_2)$, we see that

$$2(I_1 + I_2) \le 2(t_1 - t_0)\left[\varepsilon^2 + C_1^2 \pi^{-4} t_0^{-2} \cdot \exp\{ -2N^2\pi^2 t_0 \} \right]. \qquad (12.3.20)$$

As the second integral in J can be handled in a similar manner, we see that there exists a constants $C_2 = C_2(C_1, t_0, t_1)$ such that

$$J(\mathbf{a}) \le C_2\{ \varepsilon^2 + \exp\{ -2N^2\pi^2 t_0 \} \}. \qquad (12.3.21) \quad \square$$

We define

$$v^*(x, t) = v(x, t, \mathbf{b}^*) \qquad (12.3.22)$$

as our approximation to u and consider the error in using $v^*(x, t)$. We shall employ Exercise 11.1 for this purpose. In order to apply the theorem, we need to estimate $|u(a, t)) - v^*(a, t)|$ and $|u_x(a, t) - v_x^*(a, t)|$ for $a < t < b$, where $t_0 < a < b < t_1$.

Lemma 12.3.2. *Let $\zeta(t)$ denote a smooth function defined on $t_0 \le t \le t_1$ such that $\zeta(t_0) = \zeta(t_1) = 0$ while $\zeta(t) \equiv 1$ for $a \le t \le b$, where $t_0 < a < b < t_1$. For any continuously differentiable function f defined on $t_0 \le t \le t_1$, we have, for $a \le t \le b$,*

$$f(t)^2 \le 2\left(\int_{t_0}^{t_1} (f(t))^2 \, dt \right)^{1/2} \left(\int_{t_0}^{t_1} \{ \zeta(t)(f(t)\zeta(t))' \}^2 \, dt \right)^{1/2}.$$

$$(12.3.23)$$

Proof. For $a \leq t \leq b$,

$$(f(t))^2 = (f(t)\zeta(t))^2 = 2\int_{t_0}^t f(\tau)\zeta(\tau)(f(\tau)\zeta(\tau))' d\tau, \quad (12.3.24)$$

whence (12.3.23) follows via Schwarz's inequality. $\qquad\square$

Since the series

$$\sum_{n=1}^{\infty} n^p \exp\{-\alpha n^2\} < \infty \quad (12.3.25)$$

holds for any p and $\alpha > 0$, it follows from (12.3.7) and $\mathbf{b^*} \in B$ that there exists a positive constant $C_3 = C_3(C_1, t_0)$ such that, for $t \geq t_0$,

$$|u_t(a, t)| \leq C_3, \quad (12.3.26)$$

$$|u_{xt}(a, t)| \leq C_3, \quad (12.3.27)$$

$$|v_t^*(a, t)| \leq C_3, \quad (12.3.28)$$

and

$$|v_{xt}^*(a, t)| \leq C_3. \quad (12.3.29)$$

Setting

$$a = \frac{3t_0 + t_1}{4} \quad \text{and} \quad b = \frac{t_0 + 3t_1}{4},$$

it follows, from (12.3.26) through (12.3.29) and Lemma 12.3.2, that there exists a constant $C_4 = C_4(C_1, t_0, t_1)$ such that for $a \leq t \leq b$,

$$|u(a, t) - v^*(a, t)| \leq C_4 \left(\int_{t_0}^{t_1} (\varphi(t) - v^*(a, t))^2 dt \right)^{1/4}, \quad (12.3.30)$$

and that, for $a \leq t \leq b$,

$$|u_x(a, t) - v_x^*(a, t)| \leq C_4 \left(\int_{t_0}^{t_1} (\psi(t) - v_x^*(a, t))^2 dt \right)^{1/4}. \quad (12.3.31)$$

Now, from (12.3.2), (12.3.13), and (12.3.22),

$$\int_{t_0}^{t_1} (\varphi(t) - v^*(a, t))^2 dt$$

$$\leq 2 \int_{t_0}^{t_1} (\varphi(t) - \varphi^*(t))^2 dt + 2 \int_{t_0}^{t_1} (\varphi^*(t) - v^*(a, t))^2 dt$$

$$\leq 2(t_1 - t_0)\varepsilon^2 + 2\eta^*. \quad (12.3.32)$$

Likewise,

$$\int_{t_0}^{t_1} (\psi(t) - v_x^*(a, t))^2 dt \leq 2(t_1 - t_0)\varepsilon^2 + 2\eta^*. \quad (12.3.33)$$

Hence, we can apply Lemma 12.3.1 to obtain the following estimates.

Lemma 12.3.3. *For $a \leq t \leq b$, we have, a posteriori, that*

$$|u(a,t) - v^*(a,t)|, \quad |u_x(a,t) - v_x^*(a,t)| \leq C_4 \left(2(t_1 - t_0)\varepsilon^2 + 2\eta^*\right)^{1/4}$$
$$\text{(12.3.34)}$$

while a priori

$$|u(a,t) - v^*(a,t)|, \quad |u_x(a,t) - v_x^*(a,t)|$$
$$\leq C_4 \left\{ 2\left[(t_1 - t_0) + C_2\right]\varepsilon^2 + 2C_2\exp\{-2N^2\pi t_0\} \right\}^{1/4}, \quad \text{(12.3.35)}$$

where $C_2 = C_2(C_1, t_0, t_1)$ and $C_4 = C_4(C_1, t_0, t_1)$ are positive constants.

Proof. See the analysis preceding the statement of the lemma.   □

We are now in a position to make a direct application of Exercise 11.1 to obtain the following estimate of the error.

Theorem 12.3.1. *Let u denote the solution of the problem*

$$u_t = u_{xx}, \qquad 0 < x < 1, \qquad 0 < t,$$
$$u(0,t) = u(1,t) = 0, \qquad 0 < t,$$
$$u(a,t) = \varphi(t), \qquad 0 < t_0 < t < t_1, \qquad 0 < a < 1, \qquad \text{(12.3.36)}$$
$$u_x(a,t) = \psi(t), \qquad 0 < t_0 < t < t_1,$$
$$|u(x,t)| \leq C_1, \qquad 0 \leq x \leq 1, \qquad 0 \leq t,$$

where a, $0 < a < 1$, t_0, t_1, and C_1 are positive constants and the functions φ and ψ are known only approximately as φ^* and ψ^* with

$$\sup_{t_0 \leq t \leq t_1} |\varphi(t) - \varphi^*(t)| < \varepsilon \qquad \text{(12.3.37)}$$

and

$$\sup_{t_0 \leq t \leq t_1} |\psi(t) - \psi^*(t)| < \varepsilon, \qquad \text{(12.3.38)}$$

where ε is a positive constant. Let

$$v^*(x,t) = v(x,t,\mathbf{b}^*) = \sum_{n=1}^{N} b_n^*\exp\{-n^2\pi^2 t\}\sin n\pi x, \quad \text{(12.3.39)}$$

where $\mathbf{b}^* = (b_1^*,\ldots,b_N^*)$ is a point in

$$B = \left\{ \mathbf{b} \in \mathbb{R}^N \middle| \sum_{n=1}^{N} b_n < 2C_1^2 \right\} \qquad \text{(12.3.40)}$$

such that

$$\eta^* = J(\mathbf{b}^*) = \inf_{\mathbf{b} \in B} J(\mathbf{b}), \qquad \text{(12.3.41)}$$

where

$$J(\mathbf{b}) = \int_{t_0}^{t_1} \left(\varphi^*(t) - v(a, t, \mathbf{b}) \right)^2 dt + \int_{t_0}^{t_1} \left(\psi^*(t) - v_x(a, t, \mathbf{b}) \right)^2 dt.$$

$$(12.3.42)$$

Then, there exist positive constants $C = C(C_1, t_1, t_0, t_1 - t_0, a)$ and $\alpha = \alpha(a)$, $0 < \alpha < 1$, such that $0 \le x \le 1$ and $0 < t < t_1$, we have a posteriori that

$$\left| u(x, t) - v^*(x, t) \right| \le C^{1 - (\alpha/12)(t/R)^{1/\gamma}} \left(\varepsilon^2 + \eta^* \right)^{(\alpha/12)(t/R)^{1/\gamma}},$$

$$(12.3.43)$$

where

$$R = \left(\tfrac{7}{3} \right)^{1/2} t,$$

$$(12.3.44)$$

and

$$\gamma = \frac{2}{\pi} \text{Arctan} \frac{\sqrt{3}}{9},$$

$$(12.3.45)$$

while a priori we have

$$\left| u(x, t) - v^*(x, t) \right| \le C^{1 - (\alpha/12)(t/R)^{1/\gamma}}$$

$$\cdot \left(\varepsilon^2 + \exp\left\{ -2N^2 \pi t_0 \right\} \right)^{(\alpha/12)(t/R)^{1/\gamma}}.$$

$$(12.3.46)$$

On the other hand, for $0 \le x \le 1$ and $t_1 \le t$, we have a posteriori that

$$\left| u(x, t) - v^*(x, t) \right| \le C \left\{ \varepsilon^2 + \eta^* \right\}^{\alpha/4} \qquad (12.3.47)$$

while a priori

$$\left| u(x, t) - v^*(x, t) \right| \le C \left\{ \varepsilon^2 + \exp\left\{ -N^2 \pi^2 t_0 \right\} \right\}^{\alpha/4}. \qquad (12.3.48)$$

Proof. See the analysis preceding the statement of the theorem. □

Remark 1. The ε in the statement above can also include evaluation errors as well as measurement errors.

Remark 2. The solution to the minimization problem

$$\inf_{\mathbf{b} \in B} J(\mathbf{b})$$

needs some discussion. The existence of a $\mathbf{b}^*$ is clear from the continuity of J and the compactness of B. Its computation is another matter, since it amounts to a least-squares problem subject to a quadratic constraint. This can be accomplished by the addition of a Lagrange multiplier term. Set

$$J_\lambda(\mathbf{b}) = J(\mathbf{b}) + \lambda \left(\sum_{n=1}^{N} b_n^2 \right). \qquad (12.3.49)$$

Since this is a well-posed, least-squares problem, there exists a $\mathbf{b}_\lambda$ such that

$$J_\lambda(\mathbf{b}_\lambda) = \inf J_\lambda(\mathbf{b}). \tag{12.3.50}$$

We substitute $\mathbf{b}_\lambda$ into $\sum_{n=1}^N b_n^2$ and obtain

$$C_\lambda = \sum_{n=1}^N \left\{ (b_\lambda)_n \right\}^2. \tag{12.3.51}$$

We start with $\lambda = 0$; if $C_0 \le 2C_1^2$, then we have located $\mathbf{b}^* = \mathbf{b}_0$. On the other hand, if $C_0 > 2C_1^2$, we increase λ until $C_\lambda < 2C_1^2$. By the method of interval halving, we can locate λ^* for which $C_{\lambda^*} = 2C_1^2$. Setting $\mathbf{b}^* = \mathbf{b}_{\lambda^*}$, we conclude that $\mathbf{b}^*$ can be obtained via a sequence of ordinary least-squares problems.

12.4. AN EXERCISE

Exercise 12.1.   In Problem (12.3.1), select a function $u(x,0)$. Generate $\varphi(t)$ and $\psi(t)$. Conduct numerical experiments for the least-squares method.

NOTES

Section 12.1 is a reprinting of an article by Cannon [1] on the backward heat equation. The application of linear programming (see Gass [5]) as a numerical procedure for improperly (not-well) posed problems was first introduced by Douglas [4]. Miller [6] analyzed the method of least squares subject to a quadratic constraint as a method for numerically approximating solutions to improperly (not-well) posed problems. A Lagrange multiplier method for this can be found in Cannon and Cecchi [2]. The material in Sec. 12.2 on power series is the work of Cannon and Ewing [3].

REFERENCES

1. Cannon, J. R., Some numerical results for the solution of the heat equation backwards in time, *Numerical Solutions of Nonlinear Differential Equations*, ed., Donald Greenspan, John Wiley and Sons, Inc., 1966, 21–54.
2. Cannon, J. R., and Cecchi, M. M., The numerical solution of some biharmonic problems by mathematical programming techniques, *J. SIAM Numer. Anal.*, 3 (1966), 451–466.
3. Cannon, J. R., and Ewing, R. E., A direct numerical procedure for the Cauchy problem for the heat equation, *J. Math. Anal. Appl.*, **56** (1976), 7–17.
4. Douglas, J., Jr., "A numerical method for analytic continuation," in *Boundary Problems in Differential Equations* (Edited by R. Langer), 179–189, U. of Wisc. Press, Madison, 1960.
5. Gass, S. I., *Linear Programming Methods and Applications*. McGraw-Hill Book Co., Inc., New York-Toronto-London, 1958.

Chapter 13

Determination of an Unknown Time-Dependent Diffusivity $a(t)$ from Overspecified Data

13.1. INTRODUCTION

In this chapter we shall consider the problem of determining the thermal diffusivity of a homogeneous conductor, that is changing with time. A physical example of such a problem arises from heat conduction in a material that is undergoing radioactive decay or damage. The thermal conductivity varies with the degree of decay, which can be related to time. Hence, the equation of heat conduction in such a material can be written

$$u_t = a(t)u_{xx}, \tag{13.1.1}$$

where $a(t) > 0$ is the time-dependent thermal diffusivity.

The study in this chapter is based upon the representations of solutions of the heat equation to which equation (13.1.1) can be reduced via the transformation

$$\theta(t) = \int_0^t a(y)\,dy, \qquad 0 \le t \le T. \tag{13.1.2}$$

Since

$$\theta'(t) = a(t) > 0, \qquad 0 \le t \le T, \tag{13.1.3}$$

there exists a unique function $\varphi(\theta)$ such that

$$\theta(\varphi(\tau)) = \tau, \qquad 0 \le \tau \le \theta(T), \tag{13.1.4}$$

$$\varphi(\theta(t)) = t, \qquad 0 \le t \le T, \tag{13.1.5}$$

$$\varphi'(\tau) = \theta'(\varphi(\tau))^{-1} = (a(\varphi(\tau)))^{-1}, \qquad 0 \le \tau \le \theta(T). \tag{13.1.6}$$

Let

$$U(x, \eta) = u(x, \varphi(\eta)).$$ (13.1.7)

Then,

$$U_\eta(x, \eta) = u_t(x, \varphi(\eta))\varphi'(\eta) = u_t(x, \varphi(\eta))(a(\varphi(\eta)))^{-1}$$
$$= u_{xx}(x, \varphi(\eta)) = U_{xx}(x, \eta).$$ (13.1.8)

Consequently, to obtain the representation for $u(x, t)$, we substitute $\eta = \theta(t)$ into the representation for $U(x, \eta)$.

For example, consider the initial-value problem

$$u_t = a(t)u_{xx}, \qquad -\infty < x < \infty, \qquad 0 < t,$$
$$u(x, 0) = f(x), \qquad -\infty < x < \infty,$$ (13.1.9)

which reduces to

$$U_\eta = U_{xx}, \qquad -\infty < x < \infty, \qquad 0 < \eta,$$
$$U(x, 0) = f(x), \qquad -\infty < x < \infty,$$ (13.1.10)

with bounded solution

$$U(x, \eta) = \int_{-\infty}^{\infty} K(x - \xi, \eta)f(\xi)\,d\xi.$$ (13.1.11)

Substituting in $\eta = \theta(t)$, we obtain

$$u(x, t) = \int_{-\infty}^{\infty} K(x - \xi, \theta(t))f(\xi)\,d\xi$$

$$= \left(4\pi \int_0^t a(y)\,dy\right)^{-1/2} \int_{-\infty}^{\infty} \exp\left\{-\frac{(x - \xi)^2}{4\int_0^t a(y)\,dy}\right\}f(\xi)\,d\xi.$$

(13.1.12)

Next, the problem

$$u_t = a(t)u_{xx}, \qquad 0 < x < \infty, \qquad 0 < t,$$
$$u(x, 0) = f(x), \qquad 0 < x < \infty,$$ (13.1.13)
$$u(0, t) = \psi(t), \qquad 0 < t,$$

reduces to

$$U_\eta = U_{xx}, \qquad 0 < x < \infty, \qquad 0 < \eta,$$
$$U(x, 0) = f(x), \qquad 0 < x < \infty,$$ (13.1.14)
$$U(0, \eta) = \psi(\varphi(\eta)), \qquad 0 < \eta,$$

with bounded solution

$$U(x, \eta) = -2\int_0^\eta \frac{\partial K}{\partial x}(x, \eta - \tau)\psi(\varphi(\tau))\,d\tau + \int_0^\infty G(x, \xi, \eta)f(\xi)\,d\xi,$$

(13.1.15)

where $K(x, t)$ is the fundamental solution defined by (3.2.1) and $G(x, \xi, \eta)$ is the Green's function defined by (4.3.4). Substituting $\eta = \theta(t)$ into (13.1.15), we obtain

$$u(x, t) = -2 \int_0^{\theta(t)} \frac{\partial K}{\partial x}(x, \theta(t) - \tau) \psi(\varphi(\tau)) \, d\tau$$

$$+ \int_0^\infty G(x, \xi, \theta(t)) f(\xi) \, d\xi. \qquad (13.1.16)$$

In the first integral, we change the variable to $\tau = \theta(y)$, and via $d\tau = a(y) \, dy$, $\theta(t) = \theta(y)$ implies $y = t$ and $\varphi(\theta(y)) = y$, we obtain

$$u(x, t) = -2 \int_0^t \frac{\partial K}{\partial x}\left(x, \int_y^t a(\eta) \, d\eta\right) \psi(y) a(y) \, dy$$

$$+ \int_0^\infty G\left(x, \xi, \int_0^t a(\eta) \, d\eta\right) f(\xi) \, d\xi. \qquad (13.1.17)$$

In a similar manner, the problem

$$\begin{aligned}
u_t &= a(t) u_{xx}, & 0 < x < 1, & \quad 0 < t, \\
u(x, 0) &= f(x), & 0 < x < 1, & \\
u(0, t) &= \psi(t), & 0 < t, & \\
u(1, t) &= \zeta(t), & 0 < t, &
\end{aligned} \qquad (13.1.18)$$

possesses the unique bonded solution

$$u(x, t) = \sum_{k=1}^\infty A_k \exp\left\{ -k^2 \pi^2 \int_0^t a(y) \, dy \right\} \sin k\pi x$$

$$- \int_0^t \frac{\partial M\left(x, \int_\tau^t a(y) \, dy\right)}{\partial x} \psi(\tau) a(\tau) \, d\tau$$

$$+ \int_0^t \frac{\partial M\left(x - 1, \int_\tau^t a(y) \, dy\right)}{\partial x} \zeta(\tau) a(\tau) \, d\tau \qquad (13.1.19)$$

where

$$M(\xi, \eta) = \pi^{-1/2} \eta^{-1/2} \sum_{k=-\infty}^{+\infty} \exp\left\{ -\frac{(\xi + 2k)^2}{4\eta} \right\}, \qquad \eta > 0,$$

$$(13.1.20)$$

and

$$A_k = 2 \int_0^t f(x) \sin k\pi x \, dx, \qquad k = 1, 2, \ldots. \qquad (13.1.21)$$

To determine the $a(t)$, we shall overspecify the data. Two types of data can be considered. The first type involves interior measurements of temperature at a point x_0. Consequently, for this type of measurement we need only substitute x_0 into the appropriate formula and consider the resulting equation for $a(t)$. The second type involves the measurement of flux at the boundary. Fourier's law states that the measured quantity at $x = 0$ is

$$-a(t)u_x(0, t) = g(t), \tag{13.1.22}$$

where $g(t)$ is the known measurement. The calculation of $u_x(0, t)$ follows. Consider the solution u given in (13.1.19). Differentiating with respect to x, using

$$\frac{\partial^2 M\left(x - \xi, \int_\tau^t a(y)\,dy\right)}{\partial x^2} = -\frac{1}{a(\tau)}\frac{\partial M\left(x - \xi, \int_\tau^t a(y)\,dy\right)}{\partial \tau}, \qquad x \neq \xi, \tag{13.1.23}$$

and

$$\lim_{\tau \uparrow t} M\left(x - \xi, \int_\tau^t a(y)\,dy\right) = 0, \qquad x \neq \xi, \tag{13.1.24}$$

and integrating by parts yields

$$u_x(x, t) = \sum_{k=1}^\infty k\pi A_k \exp\left\{-k^2\pi^2\int_0^t a(y)\,dy\right\}\cos k\pi x$$

$$+ M\left(x - 1, \int_0^t a(y)\,dy\right)\zeta(0) - M\left(x, \int_0^t a(y)\,dy\right)\psi(0)$$

$$- \int_0^t M\left(x, \int_\tau^t a(y)\,dy\right)\psi'(\tau)\,d\tau$$

$$+ \int_0^t M\left(x - 1, \int_\tau^t a(y)\,dy\right)\zeta'(\tau)\,d\tau. \tag{13.1.25}$$

Taking the limit as x tends to zero results in

$$u_x(0, t) = \sum_{k=1}^\infty k\pi A_k \exp\left\{-k^2\pi^2\int_0^t a(y)\,dy\right\}$$

$$+ M\left(-1, \int_0^t a(y)\,dy\right)\zeta(0) - M\left(0, \int_0^t a(y)\,dy\right)\psi(0)$$

$$- \int_0^t M\left(0, \int_\tau^t a(y)\,dy\right)\psi'(\tau)\,d\tau + \int_0^t M\left(-1, \int_\tau^t a(y)\,dy\right)\zeta'(\tau)\,d\tau. \tag{13.1.26}$$

Substituting (13.1.26) into (13.1.22) results in the equation

$$a(t) = g(t)\left\{ M\left(0, \int_0^t a(y)\, dy\right)\psi(0) - M\left(-1, \int_0^t a(y)\, dy\right)\varsigma(0)\right.$$

$$+ \int_0^t M\left(0, \int_\tau^t a(y)\, dy\right)\psi'(\tau)\, d\tau - \int_0^t M\left(-1, \int_\tau^t a(y)\, dy\right)\varsigma'(\tau)\, d\tau$$

$$\left. - \sum_{k=1}^\infty k\pi A_k \exp\left\{ - k^2\pi^2 \int_0^t a(y)\, dy\right\}\right\}^{-1} \qquad (13.1.27)$$

for the finite conductor. This is a nonlinear equation for $a(t)$.

In a similar manner, we obtain from (13.1.22) and (13.1.17) the equation

$$a(t) = g(t)\left\{ 2K\left(0, \int_0^t a(y)\, dy\right)\psi(0) + 2\int_\tau^t K\left(0, \int_\tau^t a(y)\, dy\right)\psi'(\tau)\, d\tau\right.$$

$$\left. - \int_0^\infty G_x\left(0, \xi, \int_0^t a(y)\, dy\right)f(\xi)\, d\xi\right\}^{-1} \qquad (13.1.28)$$

for the semi-infinite conductor.

The representations of the solutions for arbitrary $a(t) > 0$ allows us to reduce the problem for determining $a(t)$ and u down to a single nonlinear equation for $a(t)$. Each problem for determining a unique pair, $a(t) > 0$ and u, is equivalent to the determination of a unique positive solution $a(t)$ of the corresponding nonlinear equation for $a(t)$. We shall leave the equivalence proof for each particular problem to the reader.

13.2. SOME PROBLEMS WITH EXPLICIT SOLUTIONS

Problem 1. Interior Measurement for a Finite Conductor

We shall consider first the problem of determining a positive function $a(t)$ and a function $u(x, t)$ such that the pair (a, u) satisfies

$$\begin{aligned} u_t &= a(t)u_{xx}, & 0 &< x < 1, & 0 &< t,\\ u(0, t) &= u(1, t) = 0, & 0 &< t,\\ u(x, 0) &= \sin \pi x, & 0 &< x < 1,\\ u(\tfrac{1}{2}, t) &= h(t), & 0 &< t, \end{aligned} \qquad (13.2.1)$$

where the last condition in (13.2.1) is an interior temperature measurement, which is the necessary overspecification of data.

From (13.1.19) we see that for any positive $a(t)$,

$$u(x, t) = \exp\left\{ - \pi^2 \int_0^t a(y)\, dy\right\}\sin \pi x. \qquad (13.2.2)$$

From this and the last equation in (13.2.1) it follows that $a(t)$ must satisfy

$$h(t) = \exp\left\{ -\pi^2 \int_0^t a(y)\, dy \right\}. \tag{13.2.3}$$

From the equation, we can read off the necessary conditions on the data $h(t)$. Namely, h is positive, continuously differentiable, h' is negative, and $h(0) = 1$. For such a function, it follows that

$$a(t) = -\frac{h'(t)}{\pi^2 h(t)}, \qquad 0 < t. \tag{13.2.4}$$

Clearly, the pair (a, u) defined by (13.2.4) and (13.2.2) constitute the unique solution to problem (13.2.1).

Problem 2.   Boundary Flux Measurement for a Finite Conductor

We consider next another variation of (13.2.1). In other words, we consider the determination of the pair $a(t) > 0$ and u such that the pair (a, u) satisfies

$$
\begin{aligned}
u_t &= a(t)u_{xx}, & 0 < x < 1, & \quad 0 < t, \\
u(0, t) &= u(1, t) = 0, & 0 < t, & \\
u(x, 0) &= \sin \pi x, & 0 < x < 1, & \\
-a(t)u_x(0, t) &= g(t), & 0 < t, &
\end{aligned}
\tag{13.2.5}
$$

where the last equation in (13.2.5) is a boundary-flux measurement, which is the overspecification of the data here.

From (13.1.27), we obtain immediately the equation

$$a(t) = g(t)\left\{ -\pi \exp\left\{ -\pi^2 \int_0^t a(y)\, dy \right\} \right\}^{-1}. \tag{13.2.6}$$

Consequently, the first necessary condition on g is that $g < 0$. For such a g, we rewrite (13.2.6) as

$$-\pi a(t)\exp\left\{ -\pi^2 \int_0^t a(y)\, dy \right\} = g(t). \tag{13.2.7}$$

Integrating the equation (13.2.7), we obtain

$$+\exp\left\{ -\pi^2 \int_0^t a(y)\, dy \right\} = 1 + \pi \int_0^t g(\tau)\, d\tau, \tag{13.2.8}$$

which we substitute into (13.2.6) to obtain

$$a(t) = -g(t)\pi^{-1}\left\{ 1 + \pi \int_0^t g(\tau)\, d\tau \right\}^{-1}, \tag{13.2.9}$$

which is positive only if

$$1 + \pi \int_0^t g(\tau)\, d\tau > 0; \tag{13.2.10}$$

that is,

$$-\int_0^t g(\tau)\, d\tau < \pi^{-1}.$$

(13.2.11)

The statement (13.2.11) is a condition on the accumulated energy loss from the conductor—namely, that you cannot have more energy flowing out than you have initially present.

Consequently, under the conditions g continuous, $g < 0$, and (13.2.11), $a(t)$ given by (13.2.9) is continuous and positive. This $a(t)$, together with u defined by (13.2.2), constitute the unique solution to (13.2.5).

Problem 3. Boundary-Flux Measurement for a Semi-Infinite Conductor

In the conclusion of this section we study the case of a semi-infinite conductor with zero initial condition and constant positive boundary temperature.

Consider the determination of the pair $a(t) > 0$ and u such that (a, u) satisfies

$$u_t = a(t)u_{xx}, \qquad 0 < x < \infty, \qquad 0 < t,$$
$$u(x,0) = 0, \qquad 0 < x < \infty,$$
$$u(0, t) = \psi_0, \qquad 0 < t,$$
$$-a(t)u_x(0, t) = g(t), \qquad 0 < t.$$

(13.2.12)

From (13.1.28) we obtain the equation

$$a(t) = \pi^{1/2} g(t)\psi_0^{-1}\left(\int_0^t a(y)\, dy\right)^{1/2}.$$

(13.2.13)

From this it follows that

$$\frac{d}{dt}\left(\int_0^t a(y)\, dy\right)^{1/2} = \frac{\pi^{1/2} g(t)}{2\psi_0}.$$

(13.2.14)

Integrating and solving for $a(t)$ by differentiating, we obtain

$$a(t) = \frac{\pi}{2\psi_0} g(t)\int_0^t g(\tau)\, d\tau.$$

(13.2.15)

Consequently, under the condition that g is continuous and positive, Eq. (13.2.15) yields a positive and continuous $a(t)$ for $0 < t \le T$. If $a(0)$ is to exist, we must have

$$\ell = \lim_{t\downarrow 0} g(t)\int_0^t g(\tau)\, d\tau > 0.$$

(13.2.16)

Under such a condition $a(t)$ is positive and continuous on $0 \le t \le T$. This $a(t)$ coupled with the solutions u defined by (13.1.17) constitute the unique solution to Problem (13.2.12).

13.3. SOME PROBLEMS THAT REDUCE
TO THE SOLUTION OF $F(\int_0^t a(y)\,dy) = h(t)$

Problem 1.   Interior Temperature Measurement for an Infinite Conductor

We begin here with the consideration of the initial-value problem

$$u_t = a(t)u_{xx}, \qquad -\infty < x < \infty, \qquad 0 < t,$$

$$u(x,0) = \begin{cases} 1, & -b \le x \le b, \\ 0, & |x| > b, \end{cases} \tag{13.3.1}$$

$$u(0, t) = h(t),$$

for the determination of $a(t) > 0$ and u from the interior temperature measurement $h(t)$. From (13.1.12) we see that for arbitrary $a(t) > 0$,

$$u(x,t) = \left(4\pi \int_0^t a(y)\,dy\right)^{-1/2} \int_{-b}^b \exp\left\{ -\frac{(x-\xi)^2}{4\int_0^t a(y)\,dy} \right\} d\xi. \tag{13.3.2}$$

from whence it follows that

$$h(t) = \left(4\pi \int_0^t a(y)\,dy\right)^{-1/2} \int_{-b}^b \exp\left\{ -\frac{\xi^2}{4\int_0^t a(y)\,dy} \right\} d\xi. \tag{13.3.3}$$

Defining

$$F(\eta) = (4\pi\eta)^{-1/2} \int_{-b}^b \exp\left\{ -\frac{\xi^2}{4\eta} \right\} d\xi, \tag{13.3.4}$$

we see that (13.3.3) is in the form $F(\int_0^t a(y)\,dy) = h(t)$. Now, using the fact that $\exp\{-x^2\}$ is even and the transformation of variable $\rho = \xi/2\eta^{1/2}$, we obtain

$$F(\eta) = 2\pi^{-1/2} \int_0^{b/2\sqrt{\eta}} \exp\{-\rho^2\}\,d\rho, \tag{13.3.5}$$

whence it follows that $F(0) = 1$, $\lim_{\eta \to \infty} F(\eta) = 0$, and $F'(\eta) < 0$, $\eta > 0$. Consequently, $F(\eta)$ is a monotone-decreasing function. Let G denote the inverse of F. Then

$$\int_0^t a(y)\,dy = G(h(t)), \tag{13.3.6}$$

and

$$a(t) = \{ F'(G(h(t))) \}^{-1} h'(t). \tag{13.3.7}$$

From (13.3.6) and (13.3.7), we must require that h be continuous, $h' < 0$, $h(0) = 1$, and $\lim_{t \uparrow \infty} h(t) = 0$. From these requirements on h it follows that $a(t)$ is positive, integrable, and continuous for $0 < t$. For $a(t)$ to be continuous at $t = 0$ with $a(0) > 0$, we must have

$$\ell = \lim_{t \downarrow 0} \left\{ F'(G(h(t))) \right\}^{-1} h'(t) > 0. \qquad (13.3.8)$$

Under all of the above requirements, it follows that $a(t)$ defined by (13.3.7) and u defined by (13.3.2) constitute the unique solution to (13.3.1).

Problem 2. Interior Temperature Measurement for a Finite Conductor with Zero Boundary Temperatures

Consider the problem of determining $a(t) > 0$ and $u = u(x, t)$ such that the pair (a, u) satisfies

$$\begin{aligned}
u_t &= a(t) u_{xx}, & 0 < x < 1, & \quad 0 < t, \\
u(0, t) &= u(1, t) = 0, & 0 < t, & \\
u(x, 0) &= f(x), & 0 < x < 1, & \\
u(x_0, t) &= h(t), & 0 < x_0 < 1, &
\end{aligned} \qquad (13.3.9)$$

where x_0 is a given constant and f and h are given functions.

From (13.1.19), we see that for arbitrary $a(t) > 0$,

$$u(x, t) = \sum_{k=1}^{\infty} A_k \exp\left\{ -k^2 \pi^2 \int_0^t a(y)\, dy \right\} \sin k\pi x, \qquad (13.3.10)$$

where

$$A_k = 2 \int_0^t f(\xi) \sin k\pi \xi\, d\xi. \qquad (13.3.11)$$

Hence, we have the equation

$$h(t) = \sum_{k=1}^{\infty} A_k \exp\left\{ -k^2 \pi^2 \int_0^t a(y)\, dy \right\} \sin k\pi x_0 \qquad (13.3.12)$$

from which we must determine $a(t)$.

Now, the function

$$v(x, \eta) = \sum_{k=1}^{\infty} A_k \exp\{ -k^2 \pi^2 \eta \} \sin k\pi x \qquad (13.3.13)$$

satisfies

$$\begin{aligned}
v_\eta &= v_{xx}, & 0 < x < 1, & \quad 0 < \eta, \\
v(0, \eta) &= v(1, \eta) = 0, & 0 < \eta, & \\
v(x, 0) &= f(x), & 0 < x < 1. &
\end{aligned} \qquad (13.3.14)$$

Now, if f is nonnegative continuous in $0 \le x \le 1$, $f(0) = f(1) = 0$, and $f(x_0) > 0$, and if f is twice continuously differentiable in $0 < x < 1$ such that $f'' < 0$ and f'' is bounded for $0 < x < 1$, then, from the maximum principle and (13.3.13), it follows that

$$v_\eta(x_0, \eta) < 0, \qquad 0 \le \eta, \tag{13.3.15}$$

$$v(x_0, 0) = f(x_0) > 0, \tag{13.3.16}$$

and

$$\lim_{\eta \to \infty} v(x_0, \eta) = 0. \tag{13.3.17}$$

Set

$$F(\eta) = v(x_0, \eta) \tag{13.3.18}$$

and consider the equation

$$F\left(\int_0^t a(y)\, dy \right) = h(t). \tag{13.3.19}$$

If $0 < h(t) < f(x_0)$, then

$$\int_0^t a(y)\, dy = G(h(t)), \tag{13.3.20}$$

where G is the inverse of F. Differentiating (13.3.20), we get

$$a(t) = \left\{ F'(G(h(t))) \right\}^{-1} h'(t). \tag{13.3.21}$$

Hence, if h' is continuous and $h' < 0$, then $a(t)$ given by (13.3.21) is continuous and positive. In order to obtain the continuity of $a(t)$ at $t = 0$ and $a(0) > 0$, we must have

$$l = \lim_{t \downarrow 0} \left\{ F'(G(h(t))) \right\}^{-1} h'(t) > 0. \tag{13.3.22}$$

This would follow if h' is continuous at $t = 0$ with $h'(0) < 0$ since $F'(G(f(x_0))) = F'(0) < 0$.

Collecting the requirements on f and h together, we obtain the result in the following theorem.

Theorem 13.3.1. *If f is nonnegative, twice continuously differentiable with bounded $f'' < 0$, $f(x_0) > 0$, and $f(0) = f(1) = 0$, and if h is continuously differentiable for $0 \le t \le \infty$, $h' < 0$, $h(0) = f(x_0)$, and $\lim_{t \uparrow \infty} h(t) = 0$, then $a(t)$ given by (13.3.21) and u given by (13.3.10) constitute the unique solution to problem (13.3.9).*

Proof. See the analysis of this problem preceding the theorem statement.   ☐

Problem 3. Boundary-Flux Measurement for a Finite Conductor with Zero Boundary Temperatures

We consider the problem of determining $a(t) > 0$ and $u = u(x, t)$ such that the pair (a, u) satisfies

$$
\begin{aligned}
u_t &= a(t)u_{xx}, & 0 < x < 1, & \quad 0 < t, \\
u(0, t) &= u(1, t) = 0, & 0 < t, & \\
u(x, 0) &= f(x), & 0 < x < 1, & \\
a(t)u_x(0, t) &= g(t), & 0 < t, &
\end{aligned}
\tag{13.3.23}
$$

where f and g are given functions.

From (13.1.19), we see that, for arbitrary $a(t) > 0$,

$$
u(x, t) = \sum_{k=1}^{\infty} A_k \exp\left\{ -k^2\pi^2 \int_0^t a(y)\, dy \right\} \sin k\pi x, \tag{13.3.24}
$$

where

$$
A_k = 2 \int_0^1 f(\xi) \sin k\pi\xi\, d\xi. \tag{13.3.25}
$$

Differentiating u with respect to x and setting $x = 0$, we arrive at the equation

$$
a(t) \sum_{k=1}^{\infty} k\pi A_k \exp\left\{ -k^2\pi^2 \int_0^t a(y)\, dy \right\} = g(t). \tag{13.3.26}
$$

Integrating with respect to t, it follows that

$$
\sum_{k=1}^{\infty} (k\pi)^{-1} A_k \left[1 - \exp\left\{ -k^2\pi^2 \int_0^t a(y)\, dy \right\} \right] = \int_0^t g(\tau)\, d\tau. \tag{13.3.27}
$$

Consider now the function

$$
F(\eta) = \sum_{k=1}^{\infty} (k\pi)^{-1} A_k \left[1 - \exp\{ -k^2\pi^2\eta \} \right], \qquad \eta > 0. \tag{13.3.28}
$$

It is easy to see that

$$
F(\eta) = \int_0^{\eta} v_x(0, \tau)\, d\tau, \tag{13.3.29}
$$

where v satisfies

$$
\begin{aligned}
v_t &= v_{xx}, & 0 < x < 1, & \quad 0 < t, \\
v(0, t) &= v(1, t) = 0, & 0 < t, & \\
v(x, 0) &= f(x), & 0 < x < 1. &
\end{aligned}
\tag{13.3.30}
$$

Assume now that f is positive for $0 < x < 1$, $f(0) = f(1) = 0$, $f'(0) > 0$, and f is sufficiently smooth so that $\sum_{k=1}^{\infty} (k\pi)^{-1} A_k < \infty$. From Theorem 15.4.1, it follows that $v_x(0, \eta) > 0$, $\eta > 0$. Also,

$$\lim_{\eta \uparrow \infty} F(\eta) = \lim_{\eta \uparrow \infty} \int_0^\eta v_x(0, \tau) \, d\tau = \sum_{k=1}^\infty (k\pi)^{-1} A_k > 0. \quad (13.3.31)$$

From $v_x(0,0) = f'(0)$ it follows that

$$\lim_{\eta \downarrow 0} F(\eta) = 0. \quad (13.3.32)$$

Consequently, $F(\eta)$ has an inverse function G and if $g > 0$ and

$$0 < \int_0^t g(\eta) \, d\eta < \sum_{k=1}^\infty (k\pi)^{-1} A_k < \infty, \quad (13.3.33)$$

then we can solve the equation

$$F\left(\int_0^t a(y) \, dy \right) = \int_0^t g(\tau) \, d\tau \quad (13.3.34)$$

to obtain

$$a(t) = \left\{ F'\left(G\left(\int_0^t g(\tau) \, d\tau \right) \right) \right\}^{-1} g(t), \quad (13.3.35)$$

which is positive and continuous at $t = 0$ with

$$a(0) = \{ f'(0) \}^{-1} g(0), \quad (13.3.36)$$

provided that g is continuous for $t \geq 0$.

    We collect all of these requirements into the following statement.

    Theorem 13.3.2. *If f is positive for $0 < x < 1$, $f(0) = f(1) = 0$, $f'(0) > 0$, and f is sufficiently smooth so that $\sum_{k=1}^\infty (k\pi)^{-1} A_k < \infty$ and if g is positive and continuous for $0 \leq t < \infty$ such that (13.3.33) is satisfied, then $a(t)$ given by (13.3.35) and u given by (13.3.24) constitute the unique solution to problem (13.3.23).*

    Proof. See the analysis of this problem preceding the theorem statement.   □

Problem 4.   Boundary-Flux Measurement for a Finite Conductor with Zero Initial Condition and Constant Boundary Temperatures

We consider the problem of the determination of $a(t) > 0$ and $u = u(x, t)$ such that (a, u) satisfies

$$
\begin{aligned}
u_t &= a(t) u_{xx}, & 0 < x < 1, \quad 0 < t, \\
u(0, t) &= \psi_0, & 0 < t, \\
u(1, t) &= \zeta_0, & 0 < t, \quad\quad (13.3.37) \\
u(x, 0) &= 0, & 0 < x < 1, \\
-a(t) u_x(0, t) &= g(t), & 0 < t,
\end{aligned}
$$

where ψ_0 and ζ_0 are given constants and g is a given function.

From (13.1.19) it follows for arbitrary $a(t) > 0$ that

$$u(x,t) = -\int_0^t \frac{\partial M\left(x, \int_\tau^t a(y)\, dy\right)}{\partial x} \psi_0 a(\tau)\, d\tau$$

$$+ \int_0^t \frac{\partial M\left(x-1, \int_\tau^t a(y)\, dy\right)}{\partial x} \varsigma_0 a(\tau)\, d\tau, \quad (13.3.38)$$

and from (13.1.27)

$$a(t)\left\{ M\left(0, \int_0^t a(y)\, dy\right)\psi_0 - M\left(-1, \int_0^t a(y)\, dy\right)\varsigma_0 \right\} = g(t),$$

$$(13.3.39)$$

where M is defined by (13.1.20). Integratng with respect to t, we obtain

$$F\left(\int_0^t a(y)\, dy\right) = \int_0^t g(\tau)\, d\tau, \quad (13.3.40)$$

where

$$F(\eta) = \int_0^\eta \{ M(0,s)\psi_0 - M(-1,s)\varsigma_0 \}\, ds, \quad (13.3.41)$$

and from (13.1.20)

$$M(0,s)\psi_0 - M(-1,s)\varsigma_0$$

$$= \pi^{-1/2}s^{-1/2}\left\{ \psi_0 + 2\psi_0 \sum_{n=1}^\infty \exp\left\{-\frac{n^2}{s}\right\} - 2\varsigma_0 \sum_{n=1}^\infty \exp\left\{-\frac{(2n-1)^2}{4s}\right\} \right\}.$$

$$(13.3.42)$$

For $\psi_0 > 0$ and $\varsigma_0 \leq 0$, it is clear that $F(\eta)$ is continuously differentiable for $\eta > 0$ and $F'(\eta) > 0$. Moreover, $F(0) = 0$, while $F(\eta) \geq 2\pi^{-1/2}\psi_0\eta^{1/2}$, which implies that $\lim_{\eta \uparrow \infty} F(\eta) = \infty$. Let G denote the inverse to F. If g is a positive continuous function for $0 < t$ such that $\int_0^t g(\tau)\, d\tau < \infty$, then we can invert (13.3.40) and differentiate with respect to t to obtain

$$a(t) = \left\{ F'\left(G\left(\int_0^t g(\tau)\, d\tau\right)\right) \right\}^{-1} g(t), \quad (13.3.43)$$

which is positive and continuous for $t > 0$ and integrable. Since $F'(s)$ is given by (13.3.42) and $G(0) = 0$ we see that for $a(0)$ to exist and be positive, it is sufficient that

$$\ell = \lim_{t \downarrow 0} \sqrt{G\left(\int_0^t g(\tau)\, d\tau\right)}\, g(t) \quad (13.3.44)$$

exist and be positive. Then, $a(t)$ would be positive and continuous for $t \geq 0$.

Collecting the above, we can summarize in the following statement.

Theorem 13.3.3. *For $\psi_0 > 0$ and $\zeta_0 \leq 0$, if g is a positive continuous function for $t > 0$ such that $\int_0^t g(\tau)\, d\tau < \infty$, then $a(t)$ given by (13.3.43), together with u given by (13.3.38), constitute the unique solution to problem (13.3.37). Moreover, if $\ell = \lim_{t \downarrow 0} g(t)\sqrt{G\left(\int_0^t g(\tau)\, d\tau\right)} > 0$, where G is the inverse function to $F(\eta) = \int_0^\eta \{ M(0, s)\psi_0 - M(-1, s)\zeta_0 \}\, ds$ and the integrand is defined by (13.3.42), then $a(t)$ is continuous and positive for $0 \leq t$.*

   Proof.   See the analysis preceding the statement of the theorem.   □

13.4.   A NONLINEAR INTEGRAL EQUATION FOR $a(t)$

We consider in this section the determination of a positive continuous function $a(t)$ defined on the interval $0 \leq t < T$ and a function $u = u(x, t)$ defined on $0 \leq x < \infty$, $0 \leq t < T$, such that the pair (a, u) satisfies

$$
\begin{aligned}
u_t &= a(t)u_{xx}, && 0 < x < \infty, && 0 < t < T, \\
u(x,0) &= 0, && 0 < x < \infty, \\
u(0,t) &= \psi(t), && 0 \leq t < T, && \psi(0) = 0, \\
-a(t)u_x(0,t) &= g(t), && 0 < t < T,
\end{aligned}
\tag{13.4.1}
$$

where ψ and g are given functions that are defined respectively on $0 \leq t < T$ and $0 < t < T$.

For this problem we shall make precise what we mean by a solution.

DEFINITION 13.4.1.   By a solution (a, u) of (13.4.1), we mean that

1. $a(t)$ is positive and continuous on $0 \leq t < T$;
2. $u(x, t)$ is continuous for $0 \leq t < T$, $0 \leq x < \infty$;
3. u_x, u_t, and u_{xx} exist and are continuous for $0 < x < \infty$, $0 < t < T$;
4. $\lim_{x \downarrow 0} u_x(x, t)$ exists for $0 < t < T$;
5. a and u satisfy (13.4.1), and
6. For any t, $0 < t < T$, u and u_x cannot grow faster than $C \exp\{x^\alpha\}$, $\alpha < 2$, as x tends to infinity.

Under this definition, with ψ continuously differentiable, the analysis of Sec. 13.1 via (13.1.28) yields the nonlinear integral equation

$$
a(t) = \frac{\pi^{1/2} g(t)}{\int_0^t \left[\psi'(\tau) / \left(\int_\tau^t a(y)\, dy \right)^{1/2} \right] d\tau}, \qquad 0 < t < T. \tag{13.4.2}
$$

We state without proof the following equivalence.

Lemma 13.4.1.   *Under the assumption that ψ is continuously differentiable and $\psi(0) = 0$, problem (13.4.1) possesses a unique solution if and only if* the nonlinear integral equation (13.4.2) *possesses a unique positive solution* $a(t)$ *that is continuous for* $0 \leq t < T$.

Proof. The proof is left as an exercise. ☐

Before turning to the analysis of the integral equation (13.4.2), it is convenient to list the following restrictions upon the data.

Assumption A

We shall assume that

1. ψ is continuously differentiable on every compact subset of $0 \leq t < T$;
2. $\psi(0) = 0$;
3. $\psi' > 0$, $0 < t < T$;
4. g is continuous for $0 \leq t < T$, and positive for $0 < t < T$; and
5. The function

$$h(t) = \frac{\pi^{1/2} g(t)}{\int_0^t \left(\psi'(\tau) / (t - \tau)^{1/2} \right) d\tau}, \qquad 0 < t < T,$$

satisfies

$$\lim_{t \downarrow 0} h(t) = h_0 > 0.$$

Next, it is convenient to define an operator from the integral equation.

DEFINITION 13.4.2. For any positive continuous function $a(t)$ defined on $0 \leq t < T$, let

$$\mathscr{F}a(t) = \frac{\pi^{1/2} g(t)}{\int_0^t \left[\psi'(\tau) d\tau \bigg/ \left(\int_\tau^t a(y) \, dy \right)^{1/2} \right]}, \qquad 0 < t < T. \quad (13.4.3)$$

The existence of a unique solution to the integral equation (13.4.2) is equivalent to the existence of a unique fixed point of the operator $\mathscr{F}$, as will be seen from the analysis of $\mathscr{F}$ given below.

Properties of the Operator $\mathscr{F}$
and an A Priori Estimate for $a(t)$

Since we shall make heavy use of the supremum and infinum of functions, we introduce the following notation.

DEFINITION 13.4.3. For any function $\varphi(t)$ defined for $0 \leq t < T$, let

$$s(\varphi, t) = \sup_{0 < \tau < t} \varphi(\tau), \qquad i(\varphi, t) = \inf_{0 < \tau < t} \varphi(\tau). \qquad (13.4.4)$$

We derive now our first estimate of $\mathscr{F}a(t)$ and the solution $a(t)$.

Lemma 13.4.2. *The function $\mathscr{F}a(t)$ of Definition 13.4.2 satisfies*

$$\sqrt{i(a, t)}\, i(h, t) \leq \mathscr{F}a(t) \leq \sqrt{s(a, t)}\, s(h, t), \qquad 0 < t < T, \quad (13.4.5)$$

where h is defined by Assumption A5.

Proof. Since $g > 0$ and $\psi' > 0$,

$$\mathscr{F}a(t) \leq \frac{\pi^{1/2} g(t)}{\int_0^t \psi'(\tau)\, d\tau \big/ \left(((t - \tau) s(a, t))^{1/2} \right)}$$

$$= \sqrt{s(a, t)}\, h(t) \leq \sqrt{s(a, t)}\, s(h, t). \qquad (13.4.6)$$

Likewise,

$$\mathscr{F}a(t) \geq \frac{\pi^{1/2} g(t)}{\int_0^t \psi'(\tau)\, d\tau \big/ \left(((t - \tau) i(a, t))^{1/2} \right)}$$

$$= \sqrt{i(a, t)}\, h(t) \geq \sqrt{i(a, t)}\, i(h, t). \qquad (13.4.7) \quad \square$$

As a corollary of this result, we obtain the following *a priori* estimate.

Lemma 13.4.3. *If $a(t)$ is a solution of equation (13.4.2), then*

$$i(h, t)^2 \leq a(t) \leq s(h, t)^2, \qquad 0 \leq t < T, \qquad (13.4.8)$$

where h is defined in Assumption A5.

Proof. By Lemma 13.4.2,

$$a(t) \leq \sqrt{s(a, t)}\, s(h, t), \qquad (13.4.9)$$

whence it follows that

$$s(a, t) \leq \sqrt{s(a, t)}\, s(h, t). \qquad (13.4.10)$$

Likewise,

$$a(t) \geq \sqrt{i(a, t)}\, i(h, t), \qquad (13.4.11)$$

whence it follows that

$$i(a, t) \geq \sqrt{i(a, t)}\, i(h, t). \qquad (13.4.12)$$

Combining (13.4.10) and (13.4.12), the result follows. $\square$

We now restrict our attention to the class of functions defined as follows.

DEFINITION 13.4.4. Let

$$\mathscr{G} = \left\{ a \in C([0, T)) | i(h, t)^2 \le a(t) \le s(h, t)^2 \right\}, \quad (13.4.13)$$

where h is defined by Assumption A5.

From Lemma 13.4.2 and (13.4.13), we obtain the following result.

Lemma 13.4.4. $\mathscr{F}$ maps $\mathscr{G}$ into $\mathscr{G}$.

Proof. From Lemma 13.4.2,

$$\mathscr{F}a(t) \le \sqrt{s(a, t)}\, s(h, t). \quad (13.4.14)$$

But, from (13.4.13),

$$\sqrt{s(a, t)} \le \sqrt{s(h, t)^2}. \quad (13.4.15)$$

Hence,

$$\mathscr{F}a(t) \le s(h, t)^2. \quad (13.4.16)$$

Likewise,

$$\mathscr{F}a(t) \ge i(h, t)^2. \quad (13.4.17) \quad \square$$

We can obtain one more easy property of the mapping $\mathscr{F}$.

Lemma 13.4.5. If a_1 and a_2 are in $\mathscr{G}$ and $a_1 \le a_2$, then $\mathscr{F}a_1 \le \mathscr{F}a_2$.

Proof. From the definition (13.4.3),

$$\frac{\pi^{1/2}g(t)}{\mathscr{F}a_2(t)} = \int_0^t \frac{\psi'(\tau)\, d\tau}{\left(\int_\tau^t a_2(y)\, dy \right)^{1/2}} \le \int_0^t \frac{\psi'(\tau)\, d\tau}{\left(\int_\tau^t a_1(y)\, dy \right)^{1/2}} = \frac{\pi^{1/2}g(t)}{\mathscr{F}a_1(t)},$$

$$(13.4.18)$$

whence it follows that

$$\mathscr{F}a_1(t) \le \mathscr{F}a_2(t). \quad (13.4.19) \quad \square$$

We study next the image of $\mathscr{G}$ under the mapping $\mathscr{F}$.

Lemma 13.4.6. The image $\mathscr{F}\mathscr{G}$ is an equicontinuous, uniformly bounded family of functions.

Proof. Since $i(h, t)^2 \le \mathscr{F}a(t) \le s(h, t)^2$ and $\lim_{t \to 0} i(h, t) = \lim_{t \to 0} s(h, t) = h_0$, via Assumption A5, it follows immediately that the family $\mathscr{F}\mathscr{G}$ is equicontinuous at $t = 0$. We turn now to the consideration of $\mathscr{F}\mathscr{G}$ at t, $0 < t < T$.

Let t_0 be fixed such that $t < t_0 < \min(2t, T)$ and let $\delta > 0$ satisfy $t < t + \delta < t_0$. Set

$$\Delta(a, \delta) = \int_0^{t+\delta} \frac{\psi'(\tau)\, d\tau}{\left(\int_\tau^{t+\delta} a(y)\, dy\right)^{1/2}} - \int_0^t \frac{\psi'(\tau)\, d\tau}{\left(\int_\tau^t a(y)\, dy\right)^{1/2}}.$$

(13.4.20)

Then,

$$|\Delta(a, \delta)| \le I_1 + I_2,$$

(13.4.21)

where

$$I_1 = \int_t^{t+\delta} \frac{\psi'(\tau)\, d\tau}{\left(\int_\tau^{t+\delta} a(y)\, dy\right)^{1/2}},$$

(13.4.22)

and

$$I_2 = \int_0^t \left\{ \left(\int_\tau^t a(y)\, dy\right)^{-1/2} - \left(\int_\tau^{t+\delta} a(y)\, dy\right)^{-1/2} \right\} \psi'(\tau)\, d\tau.$$

(13.4.23)

Now,

$$I_1 \le (i(h, t_0))^{-1} \int_t^{t+\delta} \frac{\psi'(\tau)}{(t+\delta - \tau)^{1/2}}\, d\tau \le C\delta^{1/2}, \qquad (13.4.24)$$

where C will denote a positive constant independent of $a(t)$ in the following discussion. Here,

$$C = 2(i(h, t_0))^{-1} \left\{ \sup_{t \le \tau \le t_0} \psi'(\tau) \right\},$$

(13.4.25)

which is finite by Assumption A1 and Assumption A3.

Next, it follows from $b_1^{-1/2} - b_2^{-1/2} = (b_2 - b_1) b_1^{-1/2} b_2^{-1/2} (b_1^{1/2} + b_2^{1/2})^{-1}$, $b_i > 0$, $i = 1, 2$, that

$$I_2 = \int_0^t \left(\int_t^{t+\delta} a(y)\, dy\right) \left(\int_\tau^{t+\delta} a(y)\, dy\right)^{-1/2} \left(\int_\tau^t a(y)\, dy\right)^{-1/2}$$

$$\cdot \left\{ \left(\int_\tau^{t+\delta} a(y)\, dy\right)^{1/2} + \left(\int_\tau^t a(y)\, dy\right)^{1/2} \right\}^{-1} \psi'(\tau)\, d\tau$$

$$\le s(h, t_0)^2 (i(h, t_0))^{-3} \int_0^t \delta(t+\delta - \tau)^{-1/2} \cdot (t - \tau)^{-1/2}$$

$$\cdot \left\{ (t+\delta - \tau)^{1/2} + (t - \tau)^{1/2} \right\}^{-1} \psi'(\tau)\, d\tau$$

$$= s(h, t_0)^2 (i(h, t_0))^{-3} \int_0^t \left\{ (t - \tau)^{-1/2} - (t+\delta - \tau)^{-1/2} \right\} \psi'(\tau)\, d\tau.$$

(13.4.26)

We now let η, $0 < \eta < \frac{1}{2}t$, denote a parameter that we shall select below. Then,

$$I_2 \leq C \int_0^\eta \frac{\psi'(\tau)}{(t-\tau)^{1/2}} d\tau$$

$$+ C \sup_{\eta \leq \tau \leq t_0} \psi'(\tau) \int_0^t \left\{ (t-\tau)^{-1/2} - (t+\delta-\tau)^{-1/2} \right\} d\tau$$

$$\leq C\sqrt{2} \, t^{-1/2}\psi(\eta) + 2C \sup_{\eta \leq \tau \leq t_0} \psi'(\tau) \left\{ t^{1/2} - (t+\delta)^{1/2} + \delta^{1/2} \right\}$$

$$\leq C\psi(\eta) + C\delta^{1/2} \sup_{\eta \leq \tau \leq t_0} \psi'(\tau). \tag{13.4.27}$$

Since ψ is continuous and $\psi(0) = 0$, we can select η sufficiently small so that $C\psi(\eta) < 2^{-1}\varepsilon$, $\varepsilon > 0$. Fixing η, we then can select δ sufficiently small so that $C\delta^{1/2}\sup_{\eta \leq \tau \leq t_0}\psi'(\tau) < 2^{-1}\varepsilon$. Consequently, for each $\varepsilon > 0$, there exists a $\delta_\varepsilon > 0$ independent of a such that

$$I_2 \leq \varepsilon$$

for all $0 < \delta < \delta_\varepsilon$. Combining this with (13.4.24), it follows that $\Delta(a, \delta)$ tends to zero uniformly with respect to $a \in \mathcal{G}$ as δ tends to zero from above. By a similar argument, $\Delta(a, \delta)$ tends to zero uniformly with respect to $a \in \mathcal{G}$ as δ tends to zero from below.

Consequently, the functions $\pi^{+1/2}g(t)\{\mathcal{F}a(t)\}^{-1}$ for $a \in \mathcal{G}$ are equicontinuous for $0 < t < T$. As g is continuous for $0 \leq t < T$, it follows that the functions $\mathcal{F}a(t)$ for $a \in \Gamma$ are equicontinuous. The uniform boundedness follows from Lemma 13.4.4. $\square$

Existence of a Fixed Point of the Operator $\mathcal{F}$

Let $a_0(t) = (i(h, t))^2$, $0 \leq t < T$. Then a_0 is in $\mathcal{G}$. By Lemma 13.4.4, $\mathcal{F}a_0 \in \mathcal{G}$, which implies that $\mathcal{F}a_0(t) \geq a_0(t)$. By Lemma 13.4.5, $\mathcal{F}^2a_0(t) \geq \mathcal{F}a_0(t)$, and by induction the sequence $\mathcal{F}^na_0(t)$ is a monotone increasing sequence of functions on $0 \leq t < T$. As $\mathcal{F}^na_0(t) \in \mathcal{G}$, $n = 1, 2, 3, \ldots$, they are bounded above by $s(h, t)^2$. Hence,

$$a(t) = \lim_{n \to \infty} \mathcal{F}^na_0(t)$$

exists for $0 \leq t < T$. But Lemma 13.4.6 implies that $\mathcal{F}^na_0(t)$, $n = 1, 2, 3, \ldots$, are equicontinuous and equibounded, which implies, via the Ascoli–Arzela theorem, that there exists a uniformly convergent subsequence on each compact subset of $0 \leq t < T$. This, together with the monotonicity of the sequence, implies that the entire sequence $\mathcal{F}^na_0$, $n = 1, 2, \ldots$, converges uniformly to $a(t)$ on each compact subset of $0 \leq t < T$.

The convergence of $\mathscr{F}"a_0(t)$ to $a(t)$ on $0 \le t \le T_0 < T$ implies that, for each τ, $0 < \tau < t \le T_0$,

$$\lim_{n \to \infty} \int_\tau^t \mathscr{F}"a_0(y)\, dy = \int_\tau^t a(y)\, dy,$$

whence it follows that

$$\lim_{n \to \infty} \frac{\psi'(\tau)}{\left(\int_\tau^t \mathscr{F}"a_0(y)\, dy \right)^{1/2}} = \frac{\psi'(\tau)}{\left(\int_\tau^t a(y)\, dy \right)^{1/2}}.$$

As

$$0 < \frac{\psi'(\tau)}{\left(\int_\tau^t a(y)\, dy \right)^{1/2}} \le \frac{\psi'(\tau)}{\left(\int_\tau^t \mathscr{F}"a_0(y)\, dy \right)^{1/2}} \le \frac{\psi'(\tau)}{\left(\int_\tau^t a_0(y)\, dy \right)^{1/2}},$$

which is integrable, it follows, from the Lebesgue dominated-convergence theorem, that

$$\lim_{n \to \infty} \int_0^t \frac{\psi'(\tau)\, d\tau}{\left(\int_\tau^t \mathscr{F}"a_0(y)\, dy \right)^{1/2}} = \int_0^t \frac{\psi'(\tau)\, d\tau}{\left(\int_\tau^t a(y)\, dy \right)^{1/2}},$$

while

$$\lim_{n \to \infty} \frac{\pi^{1/2} g(t)}{\mathscr{F}^{n+1} a_0(t)} = \frac{\pi^{1/2} g(t)}{a(t)},$$

whence it follows that $a(t) = \lim_{n \to \infty} \mathscr{F}"a_0(t)$ satisfies (13.4.2). We summarize the above analysis with the following statement.

Theorem 13.4.1.   *Under Assumption A, there exists a solution to the nonlinear integral equation* (13.4.2).

Proof.   See the analysis of this section preceding the statement of the theorem.   □

Uniqueness of the Fixed Point of the Operator $\mathscr{F}$

Suppose $a_i(t)$, $i = 1, 2$, are two solutions of the nonlinear integral equation (13.4.2). Then, we see that

$$\frac{\pi^{1/2} g(t)}{a_1(t)} - \frac{\pi^{1/2} g(t)}{a_2(t)}$$

$$= \int_0^t \left\{ \left(\int_\tau^t a_1(y)\, dy \right)^{-1/2} - \left(\int_\tau^t a_2(y)\, dy \right)^{-1/2} \right\} \psi'(\tau)\, d\tau.$$

$$(13.4.28)$$

Recalling $b_1^{-1/2} - b_2^{-1/2} = (b_2 - b_1)b_1^{-1/2}b_2^{-1/2}(b_1^{1/2} + b_2^{1/2})^{-1}$, $b_i > 0$, $i = 1, 2$, we can write (13.4.28) as

$$\frac{\pi^{1/2}g(t)(a_2(t) - a_1(t))}{a_1(t)a_2(t)}$$

$$= \int_0^t \left\{ \int_\tau^t (a_2(y) - a_1(y)) \, dy \right\} \left\{ \left(\int_\tau^t a_1(y) \, dy \right)^{-1/2} \left(\int_\tau^t a_2(y) \, dy \right)^{-1/2} \right\}$$

$$\cdot \left\{ \left(\int_\tau^t a_1(y) \, dy \right)^{1/2} + \left(\int_\tau^t a_2(y) \, dy \right)^{1/2} \right\}^{-1} \psi'(\tau) \, d\tau. \qquad (13.4.29)$$

Taking absolute values of both sides, and using the fact that $a_i(t) \in \mathscr{G}$, $i = 1, 2$, we obtain

$$|a_2(t) - a_1(t)| \le \frac{s(h, t)^4}{2\pi^{1/2}g(t)i(h, t)^3} \cdot \int_0^t \frac{\int_\tau^t |a_2(y) - a_1(y)| \, dy}{(t - \tau)^{3/2}} \psi'(\tau) \, d\tau$$

$$\le \frac{s(h, t)^4 s(|a_1 - a_2|, t)}{2i(h, t)^3 h(t)} \le \frac{s(h, t)^4}{2i(h, t)^4} s(|a_1 - a_2|, t).$$

$$(13.4.30)$$

Hence, it follows that

$$s(|a_1 - a_2|, t) \le \frac{s(h, t)^4}{2i(h, t)^4} s(|a_1 - a_2|, t) \qquad (13.4.31)$$

holds for $0 \le t < T$. But $h(t)$ is continuous, and as t tends to zero, $\lim_{t \downarrow 0} s(h, t) = \lim_{t \downarrow 0} i(h, t) = h_0$. Consequently, there exists a $t_0 > 0$ such that for all t, $0 \le t \le t_0$,

$$0 < \frac{s(h, t)^4}{2i(h, t)^4} < 1 - \varepsilon, \qquad 0 < \varepsilon < 1. \qquad (13.4.32)$$

Thus, for $0 \le t \le t_0$

$$s(|a_1 - a_2|, t) \le (1 - \varepsilon)s(|a_1 - a_2|, t), \qquad (13.4.33)$$

which implies that

$$a_1(t) \equiv a_2(t) \qquad (13.4.34)$$

for $0 \le t \le t_0$.

We return now to (13.4.30) and consider the inequality

$$|a_2(t)-a_1(t)| \le C\int_0^t (t-\tau)^{-3/2}\int_\tau^t |a_2(y)-a_1(y)|\, dy\, d\tau, \qquad t_0 \le t \le T_0,$$

(13.4.35)

where $0 < t_0 < T_0 < T$ and

$$C = \frac{s(h,T_0)^4 s(\psi',T_0)}{2\pi^{1/2}i(h,T_0)^3 \inf\limits_{t_0 \le t \le T_0} g(t)}.$$

(13.4.36)

Applying Fubini's Theorem,

$$\int_0^t (t-\tau)^{-3/2}\int_0^t |a_2(y)-a_1(y)|\, dy\, d\tau$$

$$= \int_0^t |a_2(y)-a_1(y)|\int_0^y (t-\tau)^{-3/2}\, d\tau\, dy$$

$$= 2\int_0^t |a_2(y)-a_1(y)|\left\{(t-y)^{-1/2}-t^{-1/2}\right\}\, dy$$

$$\le 2\int_0^t (t-y)^{-1/2}|a_2(y)-a_1(y)|\, dy.$$

(13.4.37)

Substituting (13.4.37) into (13.4.35), we obtain

$$|a_2(t)-a_1(t)| \le C\int_0^t (t-y)^{-1/2}|a_2(y)-a_1(y)|\, dy \quad (13.4.38)$$

for $t_0 \le t \le T_0$. As $a_2 \equiv a_1$ for $0 \le t \le t_0$, (13.4.38) is satisfied for all t, $0 \le t \le T_0$. From this inequality it follows from Lemma 8.4.2 (see Lemma 17.7.1) that $a_2(t) \equiv a_1(t)$ for $0 \le t \le T_0$. As T_0 is an arbitrary positive number less than T, we see that the solution to the nonlinear integral equation (13.4.2) is unique.

Theorem 13.4.2. *Under Assumption A, the solution to the nonlinear integral equation (13.4.2) is unique.*

Proof. See the analysis following Theorem 13.4.1 and preceding the statement of this theorem.   □

Continuous Dependence of the Solution upon the Data

Suppose now that $a_i(t)$, $i=1,2$, are the respective solutions of the nonlinear integral equation (13.4.2) corresponding to the data (ψ_i, g_i), $i=1,2$. Let h_i, $i=1,2$, denote the corresponding functions defined by Assumption A5.

Differencing the two equations (13.4.2), we obtain

$$a_2(t) - a_1(t) = \frac{(g_2(t) - g_1(t))a_1(t)}{g_1(t)}$$

$$+ \frac{a_1(t)a_2(t)}{\pi^{1/2}g_1(t)} \int_0^t \left(\int_\tau^t a_2(y)\,dy \right)^{-1/2} \{\psi_1'(\tau) - \psi_2'(\tau)\}\,d\tau$$

$$+ \frac{a_1(t)a_2(t)}{\pi^{1/2}g_1(t)} \int_0^t \left\{ \left(\int_\tau^t a_1(y)\,dy \right)^{-1/2} \right.$$

$$\left. - \left(\int_\tau^t a_2(y)\,dy \right)^{-1/2} \right\} \psi_1'(\tau)\,d\tau.$$

$$(13.4.39)$$

Taking absolute values, recalling that $a_i \in \mathscr{G}_i$, $i = 1, 2$, and utilizing (13.4.30), we obtain

$$|a_2(t) - a_1(t)| \le d(t) + \frac{s(h_1,t)^2 s(h_2,t)^2 s(|a_2 - a_1|,t)}{i(h_1,t)^2 i(h_2,t)(i(h_1,t) + i(h_2,t))},$$

$$(13.4.40)$$

where

$$d(t) = s(h_1,t)^2 s\left(\frac{|g_2 - g_1|}{g_1}, t \right) + \pi^{-1/2} s(h_1,t)^2 s(h_2,t)^2 i(h_2,t)^{-1} s(\gamma,t),$$

$$(13.4.41)$$

and

$$\gamma(t) = (g_1(t))^{-1} \int_0^t \frac{|\psi_1'(\tau) - \psi_2'(\tau)|}{\sqrt{t - \tau}}\,d\tau. \qquad (13.4.42)$$

First, we note for an $\varepsilon > 0$, $0 < \varepsilon < 1$, there exists t_0 positive yet sufficiently small such that

$$s(|a_2 - a_1|, t) \le d(t) + (1 - \varepsilon)s(|a_2 - a_1|, t), \qquad 0 < t \le t_0,$$

$$(13.4.43)$$

whence

$$s(|a_2 - a_1|, t) \le \varepsilon^{-1} d(t), \qquad 0 < t \le t_0. \qquad (13.4.44)$$

This is clear since $\lim_{t \downarrow 0} s(h_i, t) = \lim_{t \downarrow 0} i(h_i, t) = h_{0i}$, $i = 1, 2$, from whence the multiplier in front of $s(|a_1 - a_2|, t)$ on the right side of (13.4.40) becomes in the limit

$$0 < \frac{h_{02}}{(h_{01} + h_{02})} < 1. \qquad (13.4.45)$$

Combining (13.4.44) with the utilization of the first inequality on the right side of (13.4.30), we get, for $0 \le t \le T_0 < T$,

$$|a_2(t) - a_1(t)| \le \varepsilon^{-1} d(t) + C \int_0^t (t - y)^{-1/2} |a_2(y) - a_1(y)| \, dy,$$

(13.4.46)

where we have used (13.4.37) and

$$C = \frac{s(h_1, T_0)^2 s(h_2, T_0)^2 s(\psi_1', T_0)}{\pi^{1/2} i(h_1, T_0) i(h_2, T_0)(i(h_1, T_0) + i(h_2, T_0)) \inf_{t_0 \le t \le T_0} g_1(t)}$$

(13.4.47)

However, from (13.4.46) and Lemma 8.4.2 (see Lemma 17.7.1), it follows that

$$|a_2(t) - a_1(t)| \le C d(t),$$ (13.4.48)

where C depends critically upon the modulus of continuity of h_1 and h_2 at $t = 0$.

We collect the above analysis and present the following summary.

Theorem 13.4.3. *Let $a(t)$ denote the solution of the nonlinear integral equation (13.4.2) corresponding to the data (ψ, g) that satisfies Assumption A. Let $a_n(t)$, $n = 1, 2, \ldots$, denote the solutions of the nonlinear integral equation (13.4.2) corresponding respectively to the data (ψ_n, g_n) which also satisfy Assumption A. Then, if*

$$\lim_{n \to \infty} s\left(\frac{g_n - g}{g}, T_0\right) = 0,$$ (13.4.49)

$$\lim_{n \to \infty} s(\gamma_n, T_0) = 0,$$ (13.4.50)

where

$$\gamma_n(t) = (g(t))^{-1} \int_0^t \frac{|\psi'(\tau) - \psi_n'(\tau)|}{\sqrt{t - \tau}} \, d\tau,$$ (13.4.51)

and

$$\lim_{n \to \infty} s(|h - h_n|, T_0) = 0,$$ (13.4.52)

we conclude that

$$\lim_{n \to \infty} s(|a - a_n|, T_0) = 0.$$ (13.4.53)

Proof. In the analysis above, replace a_1, ψ, and g, with a, ψ, and g, and replace a_2, ψ_2, and g_2 with a_n, ψ_n, and g_n, The condition (13.4.52) allows us to replace the continuity of h_n at $t = 0$ with that of $h \pm \varepsilon$ for all n sufficiently large. As

$$0 < \frac{h_0 + \varepsilon}{2h_0 - \varepsilon} < 1$$ (13.4.54)

for ε sufficiently small but positive, the selection of t_0 can be made in a uniform way. Also, the condition (13.4.52) uniformizes in terms of h the constant C in (13.4.47) and the multipliers of $s(|g_n - g|/g, T_0)$ and $s(\gamma_n, T_0)$ in (13.4.40) for all n sufficiently large. From (13.4.48), (13.4.49), and (13.4.50) we obtain the result.   □

EXERCISES

13.1. Prove Lemma 13.4.1.
13.2. Give the details of the proof for Lemma 17.7.1.
13.3. Consider the problem of determining a pair $b(\cdot) > 0$ and $u = u(x, t)$ that satisfy

$$b(u)u_t = (b(u)u_x)_x, \qquad 0 < x < \infty, \quad 0 < t,$$
$$u(x, 0) = 0, \qquad 0 < x < \infty,$$
$$u(0, t) = f(t), \qquad 0 < t, \tag{A}$$
$$b(f(t))u_x(0, t) = g(t), \qquad 0 < t.$$

Give conditions on f and g that validate an explicit solution for $b(\cdot)$. *Hint*: Utilize the transformation

$$v(x, t) = \int_0^{u(x,t)} b(\xi)\, d\xi \tag{B}$$

to reduce the problem to that involving the heat equation.
13.4. Reconsider Exercise 13.3 for $b(u)u_t = a(t)(b(u)u_x)_x$, where $b(\cdot)$ and $a(t)$ are unknown. Try various combinations of data.

NOTES

The material here represents a blending of the work of Cannon [1, 2] and the work of Jones [6, 7, 8]. Section 20.4 on the nonlinear integral equation for the determination of the unknown diffusivity $a(t)$ is due to Jones [7].

REFERENCES

1. Cannon, J. R., Determination of an unknown coefficient in a parabolic differential equation, *Duke Math. J.*, **30** (1963), 313–323.
2. Cannon, J. R., Determination of certain parameters in heat conduction problems, *J. Math. Anal. Appl.*, **8** (1964), 188–201.
3. Cannon, J. R., and DuChateau, P., Determining unknown coefficients in a nonlinear heat conduction problem, *SIAM J. Appl. Math.*, **24** (1973), 298–314.

4. Cannon, J. R., and DuChateau, P., Determination of unknown physical proper-
    ties in heat conduction problems, *Int. J. Engng. Sci.*, **11** (1973), 783–794.
5. Douglas, J., Jr., and Jones, B. F., The determination of a coefficient in a
    parabolic differential equation; Part II. Numerical approximation, *J. Math.
    Mech.*, **11** (1962), 919–926.
6. Jones, B. F., "The Determination of a Coefficient in a Parabolic Differential
    Equation," Ph.D. Thesis, Rice University, 1961.
7. Jones, B. F., The determination of a coefficient in a parabolic differential
    equation; Part I: Existence and uniqueness, *J. Math. Mech.* **11** (1962), 907–918.
8. Jones, B. F., Various methods for finding unknown coefficients in parabolic
    differential equations, *Comm. Pure Appl. Math.*, **16** (1963), 33–44.

Chapter 14

Initial- and/or Boundary-Value Problems for General Regions with Hölder Continuous Boundaries

14.1. INTRODUCTION—FUNCTION SPACES

Until now our regions have been bounded by straight-line segments. Such problems have a wide application. However, there are numerous other problems for which the domain varies with time. For example, when a conductor melts and the liquid is drained away as it appears, the heat-conduction problem within the remaining solid involves the heat equation in a domain that is physically changing with time. This physical change must be represented somehow mathematically. For problems of one spatial dimension, this representation takes the form of a curve $x = s(t)$. For example, the problem

$$u_t = u_{xx}, \qquad 0 < x < s(t), \qquad 0 < t,$$
$$u(x,0) = f(x), \qquad 0 \leq x \leq b, \qquad s(0) = b,$$
$$u(0,t) = g(t), \qquad 0 < t, \tag{14.1.1}$$
$$u(s(t),t) = 0, \qquad 0 < t,$$

could very well represent the one spatial dimensional case of a melting ice slab for which the water is removed as it appears. (See Fig. 14.1.1.)

Our purpose here is to study the initial-boundary-value problems for the heat equation in the domain

$$D_T = \{(x,t) | s_1(t) < x < s_2(t), \qquad 0 < t \leq T\}, \tag{14.1.2}$$

213

where s_1 and s_2 are Hölder continuous functions of t over the interval $0 < t \le T$ and $a = s_1(0) \le s_2(0) = b$. For our study, it is convenient to introduce some function classes and norms. Until now we have been content to deal with piecewise-continuous functions and to use the uniform norms

$$\|\psi\|_T = \sup_{0 \le t \le T} |\psi(t)| \tag{14.1.3}$$

and

$$\|\psi\|_{[a,b]} = \sup_{a \le t \le b} |\psi(t)|. \tag{14.1.4}$$

We now undertake to generalize these norms and to introduce seminorms to measure the degree of Hölder continuity.

If I is an interval, we denote by $C^0(I)$, $C^1(I)$, and $C^\beta(I)$, $0 < \beta \le 1$, the continuous, the continuously differentiable, and the Hölder continuous (with exponent β) functions on I, respectively. Note that $C^\beta(I)$ with $\beta = 1$ denotes the Lipschitz continuous functions on I. We denote the Hölder constant (seminorm) by

$$|\psi|_\beta = \sup_{\substack{t,t+\delta \in I \\ \delta > 0}} \delta^{-\beta} |\psi(t+\delta) - \psi(t)| \tag{14.1.5}$$

for Hölder continuous functions ψ with exponent β on a closed interval I. For $I = (0, C]$, $C > 0$, we define $C^0_{(\nu)}(I)$, $0 < \nu \le 1$, as the subspace of $C(I)$ that consists of those functions ψ such that

$$\|\psi\|_C^{(\nu)} = \sup_{t \in I} t^{1-\nu} |\psi(t)| < \infty. \tag{14.1.6}$$

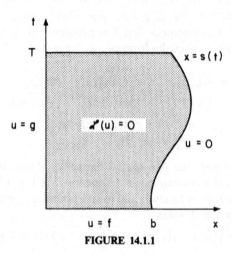

FIGURE 14.1.1

Also, we define $C^\varepsilon_{(\nu)}(I)$, $\varepsilon > 0$, as the subspace of $C^0_{(\nu)}(I)$ such that

$$|\psi|^{(\nu)}_\varepsilon = \sup_{\substack{t,\,t+\delta \in I \\ \delta > 0}} t^{1-\nu+\varepsilon}\delta^{-\varepsilon}|\psi(t+\delta)-\psi(t)| < \infty. \qquad (14.1.7)$$

Clearly, $\|\psi\|^{(\nu)}_C$ is a norm while $|\psi|^{(\nu)}_\varepsilon$ is a seminorm. Thus $C^0_{(\nu)}(I)$ is a Banach space under the norm $\|\psi\|^{(\nu)}_C$ and $C^\varepsilon_{(\nu)}(I)$ becomes one under the norm

$$\|\psi\|^{(\nu)}_{C,\varepsilon} = \|\psi\|^{(\nu)}_C + |\psi|^{(\nu)}_\varepsilon. \qquad (14.1.8)$$

In the remainder of this chapter we shall assume that $s_i \in C^\gamma([0,T])$, $i=1,2$, which means that $|s_i|_\gamma < \infty$, $i=1,2$. Moreover, we shall assume that

$$\tfrac{1}{2} < \gamma \le 1. \qquad (14.1.9)$$

14.2. THE SINGLE-LAYER POTENTIAL
$$w_\varphi(x,t,s) = \int_0^t K(x-s(\tau),t-\tau)\varphi(\tau)\,d\tau$$

From Sec. 1.4 and the definition (3.2.1) of the fundamental solution $K(x,t)$, it follows immediately that $w_\varphi(x,t,s)$ is a solution of the heat equation for $x < s(t)$, $0 < t \le T$, and $x > s(t)$, $0 < t \le T$, provided φ is integrable and $s \in C([0,T])$. Moreover, for $\varphi \in C^0_{(\nu)}((0,T])$,

$$|w_\varphi(x,t,s)| \le 2^{-1}\pi^{-1/2}\int_0^t (t-\tau)^{-1/2}|\varphi(\tau)|\,d\tau$$

$$\le 2^{-1}\pi^{-1/2}\|\varphi\|^{(\nu)}_t \int_0^t (t-\tau)^{-1/2}\tau^{\nu-1}\,d\tau$$

$$\le Ct^{\nu-(1/2)}\|\varphi\|^{(\nu)}_t, \qquad (14.2.1)$$

where C is a positive constant and where we have used the evaluation

$$\int_0^t \tau^{\alpha_1}(t-\tau)^{\alpha_2}\,d\tau = \frac{\Gamma(1+\alpha_1)\Gamma(1+\alpha_2)}{\Gamma(2+\alpha_1+\alpha_2)}t^{1+\alpha_1+\alpha_2}, \qquad (14.2.2)$$

$\alpha_1, \alpha_2 > -1$. We shall have occasion to use (14.2.2) several times in what follows.

Remark. We shall write $w_\varphi(x,t,s)$ as $w_\varphi(x,t)$ when no confusion can arise from the boundary s.

From (14.2.1) we see that $w_\varphi(x,t)$ is bounded for all x. We shall study now the derivatives of w_φ. Estimates for the derivatives of w_φ are most easily obtained by showing that w_φ is analytic in the spatial variable. We assume now that $s \in C^{\gamma}([0, T])$, $\frac{1}{2} < \gamma \leq 1$. Consider

$$K(z - s(\tau), t - \tau) = 2^{-1}\pi^{-1/2}(t - \tau)^{-1/2}\exp\left\{-\frac{(z - s(\tau))^2}{4(t - \tau)}\right\},$$

$$(14.2.3)$$

where $z = x + iy$ belongs to the domain

$$D_{\mu}^{+}(t) = \{z \mid x > s(t) \quad \text{and} \quad |y| < \mu|x - s(t)|\}, \qquad 0 < \mu < 1.$$

$$(14.2.4)$$

Now, for $z \in D_{\mu}(t)$,

$$\mathrm{Re}(z - s(\tau))^2 = (x - s(\tau))^2 - y^2$$

$$= (x - s(t))^2 - y^2 + 2(x - s(t))(s(t) - s(\tau)) + (s(t) - s(\tau))^2$$

$$\geq (1 - \mu^2)(x - s(t))^2 - \frac{(1 - \mu^2)}{2}(x - s(t))^2$$

$$+ \left[1 - \frac{2}{(1 - \mu^2)}\right](s(t) - s(\tau))^2$$

$$= \frac{(1 - \mu^2)}{2}(x - s(t))^2 - \left(\frac{1 + \mu^2}{1 - \mu^2}\right)(s(t) - s(\tau))^2, \quad (14.2.5)$$

since $2ab \leq \varepsilon a^2 + \varepsilon^{-1}b^2$ for any $\varepsilon > 0$. Consequently, for $z \in D_{\mu}(t)$,

$$\left|\exp\left\{-\frac{(z - s(\tau))^2}{4(t - \tau)}\right\}\right| \leq \exp\left\{\left(\frac{1 + \mu^2}{1 - \mu^2}\right)\frac{(s(t) - s(\tau))^2}{4(t - \tau)}\right\}$$

$$\cdot \exp\left\{-\frac{(1 - \mu^2)(x - s(t))^2}{8(t - \tau)}\right\}$$

$$\leq \exp\left\{\frac{(1 + \mu^2)}{(1 - \mu^2)}|s|_{\gamma}^2(t - \tau)^{2\gamma - 1}\right\}$$

$$\cdot \exp\left\{-\frac{(1 - \mu^2)(x - s(t))^2}{8(t - \tau)}\right\}. \quad (14.2.6)$$

Thus, we see that, for $z \in D_{\mu}^{+}(t)$,

$$|K(z - s(\tau), t - \tau)| \leq 2^{-1}\pi^{-1/2}(t - \tau)^{-1/2}\exp\left\{\frac{(1 + \mu^2)}{(1 - \mu^2)}|s|_{\gamma}^2 t^{2\gamma - 1}\right\}$$

$$(14.2.7)$$

and that for $\varphi \in C_{(\nu)}^0((0, T])$,

$$|w_\varphi(z, t)| \leq C \exp\left\{ \frac{(1+\mu^2)}{(1-\mu^2)} |s|_\gamma^2 t^{2\gamma-1} \right\} t^{\nu-(1/2)} \|\varphi\|_t^{(\nu)}. \quad (14.2.8)$$

Applying Theorem 10.1.1, we see that $w_\varphi(z, t)$ is analytic in the domain $D_\mu(t)$. We summarize this in the following statement.

Lemma 14.2.1. *For $\varphi \in C_{(\nu)}^0((0, T])$, the function $w_\varphi(z, t)$ is an analytic function of the complex variable $z = x + iy$ in the domains $D_\mu^+(t)$ and*

$$D_\mu^-(t) = \{ z \mid x < s(t) \quad and \quad |y| < \mu|x - s(t)| \}, \qquad 0 < \mu < 1.$$
$$(14.2.9)$$

Moreover, there exists a positive constant C such that for $z \in D_\mu^\pm(t)$,

$$|w_\varphi(z, t)| \leq C \exp\left\{ \frac{(1+\mu^2)}{(1-\mu^2)} |s|_\gamma^2 t^{2\gamma-1} \right\} t^{\nu-(1/2)} \|\varphi\|_t^{(\nu)}. \quad (14.2.10)$$

Proof. See the analysis preceding the statement of the Lemma. □

We can make an immediate application of this Lemma. Let $D_{\mu,\delta}^+(t)$ denote the translation of $D_\mu^+(t)$ a distance $\delta > 0$ in the direction of positive x while $D_{\mu,\delta}^-(t)$ denotes the translation of $D_\mu^-(t)$ a distance δ in the direction of negative x. For any $z \in D_{\mu,\delta}^\pm(t)$, the circle centered at z with radius $r = \delta \sin(\arctan \mu)$ is contained in $D_\mu^\pm(t)$. Applying the Cauchy–Riemann integral formula, we obtain the following result.

Lemma 14.2.2. *For $\delta > 0$, let*

$$D_{\mu,\delta}^+(t) = \{ z \mid x > s(t) + \delta \quad and \quad |y| < \mu|x - s(t) - \delta| \}, \quad (14.2.11)$$

and

$$D_{\mu,\delta}^-(t) = \{ z \mid x < s(t) - \delta \quad and \quad |y| < \mu|x - s(t) + \delta| \}, \quad (14.2.12)$$

where $0 < \mu < 1$. Then, for $z \in D_{\mu,\delta}^\pm(t)$,

$$\left| \frac{\partial^n w_\varphi}{\partial z^n}(z, t) \right| \leq C \exp\left\{ \frac{(1+\mu^2)}{(1-\mu^2)} |s|_\gamma^2 t^{2\gamma-1} \right\} t^{\nu-(1/2)} \|\varphi\|_t^{(\nu)}$$
$$\cdot n!(\delta \sin(\text{Arctan}\,\mu))^{-n}, \quad (14.2.13)$$

where C is a positive constant.

Proof. See the analysis preceding the statement of the Lemma. □

We turn now to the major interest of w_φ, namely, the existence of

$$\lim_{x \uparrow s(t)} \frac{\partial w_\varphi}{\partial x}(x,t) \quad \text{and} \quad \lim_{x \downarrow s(t)} \frac{\partial w_\varphi}{\partial x}(x,t).$$

Lemma 14.2.3. *For $\varphi \in C_{(\nu)}^0((0,T])$,*

$$\lim_{x \uparrow s(t)} \frac{\partial w_\varphi}{\partial x}(x,t) = 2^{-1}\varphi(t) + \int_0^t \frac{\partial K}{\partial x}(s(t) - s(\tau), t - \tau)\varphi(\tau)\, d\tau.$$

$$(14.2.14)$$

Proof. Since

$$\frac{\partial w_\varphi}{\partial x}(x,t) = \int_0^t \frac{\partial K}{\partial x}(x - s(\tau), t - \tau)\varphi(\tau)\, d\tau, \qquad (14.2.15)$$

we can write

$$\frac{\partial w_\varphi}{\partial x}(x,t) - 2^{-1}\varphi(t) - \int_0^t \frac{\partial K}{\partial x}(s(t) - s(\tau), t - \tau)\varphi(\tau)\, d\tau = I_1 + I_2 + I_3,$$

$$(14.2.16)$$

where

$$I_1 = \int_0^{t-\delta} \left\{ \frac{\partial K}{\partial x}(x - s(\tau), t - \tau) - \frac{\partial K}{\partial x}(s(t) - s(\tau), t - \tau) \right\} \varphi(\tau)\, d\tau,$$

$$(14.2.17)$$

$$I_2 = \int_{t-\delta}^t \frac{s(t) - s(\tau)}{2(t - \tau)} \{ K(s(t) - s(\tau), t - \tau) - K(x - s(\tau), t - \tau) \} \varphi(\tau)\, d\tau,$$

$$(14.2.18)$$

$$I_3 = -2^{-1}\varphi(t) + \int_{t-\delta}^t \frac{s(t) - x}{2(t - \tau)} K(x - s(\tau), t - \tau)\varphi(\tau)\, d\tau, \qquad (14.2.19)$$

and $0 < \delta < t$,

We begin our estimates with I_1. Set

$$F(\theta) = \frac{\partial K}{\partial x}(x\theta + s(t)(1 - \theta) - s(\tau), t - \tau). \qquad (14.2.20)$$

By the Mean-Value Theorem

$$|F(1) - F(0)| = |F'(\theta_0)|, \qquad 0 < \theta_0 < 1, \qquad (14.2.21)$$

where

$$F'(\theta_0) = \frac{\partial^2 K}{\partial x^2}(x\theta_0 + s(t)(1 - \theta_0) - s(\tau), t - \tau)(x - s(t)).$$

$$(14.2.22)$$

Since there exists a positive constant C_1 such that

$$\left|\frac{\partial^2 K}{\partial x^2}\right| \le C_1(t-\tau)^{-3/2}, \tag{14.2.23}$$

we see that

$$|F(1)-F(0)| \le C_1(t-\tau)^{-3/2}|x-s(t)| \tag{14.2.24}$$

and that

$$|I_1| \le C_1|x-s(t)|\int_0^{t-\delta}(t-\tau)^{-3/2}|\varphi(\tau)|\,d\tau$$

$$\le C_1\|\varphi\|_t^{(\nu)}|x-s(t)|\int_0^{t-\delta}(t-\tau)^{-3/2}\tau^{\nu-1}\,d\tau. \tag{14.2.25}$$

Now, for $0 < \delta \le t/2$, we can write

$$\int_0^{t-\delta}(t-\tau)^{-3/2}\tau^{\nu-1}\,d\tau \le \left(\frac{t}{2}\right)^{-3/2}\int_0^{t/2}\tau^{\nu-1}\,d\tau$$

$$+\left(\frac{t}{2}\right)^{\nu-1}\int_{t/2}^{t-\delta}(t-\tau)^{-3/2}\,d\tau$$

$$\le \nu^{-1}\left(\frac{t}{2}\right)^{\nu-(3/2)}+2\left(\frac{t}{2}\right)^{\nu-1}\delta^{-1/2}, \tag{14.2.26}$$

whence it follows that

$$|I_1| \le C_2\|\varphi\|_t^{(\nu)}|x-s(t)|\left\{\left(\frac{t}{2}\right)^{\nu-(3/2)}+\left(\frac{t}{2}\right)^{\nu-1}\delta^{-1/2}\right\}, \tag{14.2.27}$$

where C_2 is a positive constant. In what follows, C_i will denote positive constants that depend only upon ν and γ.

Consider now the integral I_2. Since

$$|K| \le 2^{-1}\pi^{-1/2}(t-\tau)^{-1/2}, \tag{14.2.28}$$

we see that for $s \in C^\gamma([0,T])$,

$$|I_2| \le C_3|s|_\gamma\int_{t-\delta}^t(t-\tau)^{\gamma-(3/2)}|\varphi(\tau)|\,d\tau$$

$$\le C_3|s|_\gamma\|\varphi\|_t^{(\nu)}\int_{t-\delta}^t(t-\tau)^{\gamma-(3/2)}\tau^{\nu-1}\,d\tau$$

$$\le C_4|s|_\gamma\|\varphi\|_t^{(\nu)}\delta^{\gamma-(1/2)}(t-\delta)^{\nu-1}. \tag{14.2.29}$$

We turn now to I_3. Since

$$\frac{1}{2} = \int_{-\infty}^t \frac{s(t)-x}{2(t-\tau)}K(x-s(t),t-\tau)\,d\tau \tag{14.2.30}$$

via the transformation $\rho = 2^{-1}(s(t) - x)(t - \tau)^{-1/2}$, we can write

$$I_3 = -\int_{-\infty}^{t-\delta} \frac{s(t) - x}{2(t - \tau)} K(x - s(t), t - \tau) \varphi(t) \, d\tau$$

$$+ \int_{t-\delta}^{t} \frac{s(t) - x}{2(t - \tau)} K(x - s(t), t - \tau)[\varphi(\tau) - \varphi(t)] \, d\tau$$

$$+ \int_{t-\delta}^{t} \frac{s(t) - x}{2(t - \tau)} \{ K(x - s(\tau), t - \tau) - K(x - s(t), t - \tau) \} \varphi(\tau) \, d\tau$$

$$= J_1 + J_2 + J_3. \tag{14.2.31}$$

For J_1, we see that

$$|J_1| \le \|\varphi\|_t^{(\nu)} t^{\nu - 1} \int_{-\infty}^{t-\delta} \frac{s(t) - x}{2(t - \tau)} K(x - s(t), t - \tau) \, d\tau$$

$$= \pi^{-1/2} \|\varphi\|_t^{(\nu)} t^{\nu - 1} \int_0^{(s(t) - x)/2\delta^{1/2}} \exp\{ -\rho^2 \} \, d\rho$$

$$\le 2^{-1} \pi^{-1/2} \|\varphi\|_t^{(\nu)} t^{\nu - 1} |s(t) - x| \delta^{-1/2}. \tag{14.2.32}$$

For J_2, we can obtain two estimates. The first follows trivially from (14.2.30). That is,

$$|J_2| \le \sup_{t-\delta \le \tau \le t} |\varphi(\tau) - \varphi(t)| \int_{t-\delta}^{t} \frac{s(t) - x}{2(t - \tau)} K(x - s(t), t - \tau) \, d\tau$$

$$\le \frac{1}{2} \sup_{t-\delta \le \tau \le t} |\varphi(\tau) - \varphi(t)|. \tag{14.2.33}$$

The second estimate can be obtained from the assumption that $\varphi \in C_{(\nu)}^{\varepsilon}((0, T])$. Recalling (14.1.6) and (14.1.7), we are led to the replacement of $[\varphi(\tau) - \varphi(t)]$ in (14.2.28) by

$$\{ |\varphi(\tau + (t - \tau)) - \varphi(\tau)| (t - \tau)^{-\varepsilon} \cdot \tau^{1 - \nu + \varepsilon} \} \tau^{\nu - 1 - \varepsilon} (t - \tau)^{+\varepsilon}.$$

As the term in the braces is dominated by $|\varphi|_\varepsilon^{(\nu)}$, we see that

$$|J_2| \le |\varphi|_\varepsilon^{\nu} (t - \delta)^{\nu - 1 - \varepsilon} \cdot \int_{t-\delta}^{t} \frac{s(t) - x}{2(t - \tau)} K(x - s(t), t - \tau)(t - \tau)^{+\varepsilon} \, d\tau.$$

$$\tag{14.2.34}$$

Since $\rho^{1 - (\gamma - (1/2))} \exp\{ -\rho^2 \}$ for $\rho > 0$ assumes its maximum value at $\rho = \{ 2^{-1}(1 - (\gamma - \frac{1}{2})) \}^{1/2}$, there exists a constant C_5 such that, after factoring out $|s(t) - x|^{\gamma - (1/2)}$, we are left with the estimate

$$|J_2| \le C_5 |\varphi|_\varepsilon^{\nu} (t - \delta)^{\nu - 1 - \varepsilon} |x - s(t)|^{\gamma - (1/2)}$$

$$\cdot \int_{t-\delta}^{t} (t - \tau)^{-(3/2) + (1/2)(1 - (\gamma - (1/2))) + \varepsilon} \, d\tau$$

$$\le C_5 |\varphi|_\varepsilon^{\nu} (t - \delta)^{\nu - 1 - \varepsilon} |x - s(t)|^{\gamma - (1/2)} t^{(1/4) - (\gamma/2) + \varepsilon}, \tag{14.2.35}$$

provided that

$$\varepsilon > \left(\frac{\gamma}{2} - \frac{1}{4} \right). \tag{14.2.36}$$

Note here that C_5 depends upon ε. Turning now to J_3, we consider the term

$$\{ K(x - s(\tau), t - \tau) - K(x - s(t), t - \tau) \} = F(1) - F(0), \tag{14.2.37}$$

where

$$F(\theta) = K(x - \theta s(\tau) - (1 - \theta)s(t), t - \tau), \qquad 0 \le \theta \le 1. \tag{14.2.38}$$

By the mean-value theorem,

$$|F(1) - F(0)| = |F'(\theta_0)|, \qquad 0 < \theta_0 < 1, \tag{14.2.39}$$

where

$$F'(\theta_0) = \frac{\partial K}{\partial x}(x - \theta_0 s(\tau) - (1 - \theta_0)s(t), t - \tau)(s(t) - s(\tau)). \tag{14.2.40}$$

Set

$$s_0(\tau) = \theta_0 s(\tau) + (1 - \theta_0)s(t). \tag{14.2.41}$$

Then, there exists a constant C_6 such that

$$|F'(\theta_0)| \le C_6 \frac{|x - s_0(\tau)|}{(t - \tau)^{3/2}} \exp \left\{ - \frac{(x - s_0(\tau))^2}{4(t - \tau)} \right\} \cdot |s|_\gamma (t - \tau)^\gamma. \tag{14.2.42}$$

Factoring out $|x - s(t)|^{\gamma - (1/2)}$ from $|x - s(t)|$ in the expression for J_3, replacing φ by $\|\varphi\|_t^\nu$, and using (14.2.42), we obtain

$$|J_3| \le C_6 \|\varphi\|_t^\nu (t - \delta)^{\nu - 1} |s|_\gamma |x - s(t)|^{(\gamma - (1/2))}$$

$$\cdot \int_{t - \delta}^t \frac{|x - s(t)|^{(3/2) - \gamma}}{(t - \tau)^{1 - \gamma}} \frac{|x - s_0(\tau)|}{(t - \tau)^{3/2}} \exp \left\{ - \frac{(x - s_0(\tau))^2}{4(t - \tau)} \right\} d\tau. \tag{14.2.43}$$

Observing that $|x - s(t)|^{(3/2) - \gamma} \le |s(t) - s_0(\tau)|^{(3/2) - \gamma} + |x - s_0(\tau)|^{(3/2) - \gamma}$, since $|a - b|^\alpha \le |a - c|^\alpha + |c - b|^\alpha$ holds for any a, b, c with $0 < \alpha < 1$, we obtain

$$|J_3| \le C_6 \|\varphi\|_t^{(\nu)} (t - \delta)^{\nu - 1} |s|_\gamma |x - s(t)|^{\gamma - (1/2)}$$

$$\cdot \int_{t - \delta}^t \left\{ \frac{|s(t) - s_0(\tau)|^{(3/2) - \gamma}}{(t - \tau)^{1 - \gamma}} + \frac{|x - s_0(\tau)|^{(3/2) - \gamma}}{(t - \tau)^{1 - \gamma}} \right\}$$

$$\cdot \frac{|x - s_0(\tau)|}{(t - \tau)^{(3/2)}} \exp \left\{ - \frac{(x - s_0(\tau))^2}{4(t - \tau)} \right\}. \tag{14.2.44}$$

Since $s(t) - s_0(\tau) = \theta_0(s(t) - s(\tau))$, it follows that

$$|s(t) - s_0(\tau)|^{(3/2)-\gamma} \leq |s|_\gamma^{(3/2)-\gamma} |t - \tau|^{\gamma((3/2)-\gamma)}. \qquad (14.2.45)$$

Consequently, we have

$$|J_3| \leq C_6 \|\varphi\|_t^{(\nu)} (t - \delta)^{\nu-1} |s|_\gamma |x - s(t)|^{\gamma-(1/2)}$$

$$\cdot \int_{t-\delta}^t \left\{ |s|_\gamma^{(3/2)-\gamma} \frac{|x - s_0(\tau)|}{(t - \tau)^{(5/2)-\gamma-\gamma((3/2)-\gamma)}} \right.$$

$$\left. + \frac{|x - s_0(\tau)|^{(5/2)-\gamma}}{(t - \tau)^{(5/2)-\gamma}} \right\} \exp\left\{ -\frac{(x - s_0(\tau))^2}{4(t - \tau)} \right\} d\tau. \qquad (14.2.46)$$

Recalling that $\rho^\beta \exp\{-\rho^2\} < C_7$, which depends only upon β, we employ $\rho = (|x - s_0(\tau)|)/(2(t - \tau)^{1/2})$ in the integral above to obtain

$$|J_3| \leq C_8 \|\varphi\|_t^{(\nu)} (t - \delta)^{\nu-1} |s|_\gamma |x - s(t)|^{(\gamma-(1/2))}$$

$$\cdot \left\{ |s|_\gamma^{(3/2)-\gamma} \int_{t-\delta}^t (t - \tau)^{-2+\gamma+\gamma((3/2)-\gamma)} d\tau + \int_{t-\delta}^t (t - \tau)^{-(5/4)+(\gamma/2)} d\tau \right\}$$

$$\leq C_9 \|\varphi\|_t^{(\nu)} |s|_\gamma |x - s(t)|^{\gamma-(1/2)} (t - \delta)^{\nu-1}$$

$$\cdot \left\{ |s|_\gamma^{(3/2)-\gamma} t^{\gamma+\gamma((3/2)-\gamma)-1} + t^{(\gamma/2)-(1/4)} \right\}, \qquad (14.2.47)$$

since $(\gamma/2) - (1/4)$ and $\gamma + \gamma((3/2) - \gamma) - 1$ are positive. The fact that $(\gamma/2) - (1/4) > 0$ follows easily from $\gamma > 1/2$. For $\gamma + \gamma((3/2) - \gamma) - 1$, we note that

$$\gamma + \gamma(\tfrac{3}{2} - \gamma) - 1 = (\gamma - \tfrac{1}{2})(2 - \gamma) > 0, \qquad (14.2.48)$$

for $\tfrac{1}{2} < \gamma \leq 1$.

In order to conclude the proof of Lemma 14.2.3, we collect the results (14.2.27), (14.2.29), (14.2.32), (14.2.33), and (14.2.47), to obtain

$$\left| \sum_{i=1}^3 I_i \right| < C_2 \|\varphi\|_t^\nu |x - s(t)| \left\{ \left(\frac{t}{2}\right)^{\nu-(3/2)} + \left(\frac{t}{2}\right)^{\nu-1} \delta^{-1/2} \right\}$$

$$+ C_4 |s|_\gamma \|\varphi\|_t^{(\nu)} \delta^{\gamma-(1/2)} (t - \delta)^{\nu-1}$$

$$+ 2^{-1} \|\varphi\|_t^{(\nu)} t^{\nu-1} |s(t) - x| \delta^{-1/2} + \frac{1}{2} \sup_{t-\delta \leq \tau \leq t} |\varphi(\tau) - \varphi(t)|$$

$$+ C_9 \|\varphi\|_t^{(\nu)} |s|_\gamma |x - s(t)|^{\gamma-(1/2)} (t - \delta)^{\nu-1}$$

$$\cdot \left\{ |s|_\gamma^{(3/2)-\nu} t^{\gamma+\gamma((3/2)-\gamma)-1} + t^{(\gamma/2)-(1/4)} \right\}. \qquad (14.2.49)$$

Let $\zeta > 0$ denote an arbitrary positive number. By selecting δ sufficiently small, we can force the second and fourth terms in (4.2.49) to be less than

$\zeta/5$. Once δ has been fixed, the remaining terms contain $|x - s(t)|^{\gamma-(1/2)}$ as a factor, and each can be made less than $\zeta/5$ by selecting x sufficiently close to $s(t)$. Hence,

$$\left|\sum_{i=1}^{3} I_i\right| < \zeta \tag{14.2.50}$$

for all x sufficiently close to $s(t)$ with $x < s(t)$. $\qquad\square$

Remark. For $0 < \mu \le t \le T$, φ is uniformly continuous, which permits a uniform choice of δ and consequently a uniform choice of x sufficiently close to $s(t)$. This will show that $(\partial w_\varphi / \partial x)(x, t)$ is two-dimensionally continuous at $(s(t), t)$ if $2^{-1}\varphi(t) + \int_0^t (\partial K/\partial x)(s(t) - s(\tau), t - \tau)\varphi(\tau)\, d\tau$ is continuous. Before turning to this question, we shall demonstrate the following result, which is a corollary of the previous argument.

Lemma 14.2.4. For $\varphi \in C_{(\nu)}^{\varepsilon}((0, T])$, there exists a positive constant $C = C(\varepsilon, \gamma, \nu, T, |s|_\gamma)$ such that, for $x < s(t)$,

$$\left|\frac{\partial w_\varphi}{\partial x}(x, t) - 2^{-1}\varphi(t) - \int_0^t \frac{\partial K}{\partial x}(s(t) - s(\tau), t - \tau)\varphi(\tau)\, d\tau\right|$$
$$\le C\|\varphi\|_{t, \varepsilon}^{(\nu)} t^{\nu-(3/2)} |x - s(t)|^{\gamma-(1/2)}, \tag{14.2.51}$$

provided that

$$\varepsilon > \left(\frac{\gamma}{2} - \frac{1}{4}\right). \tag{14.2.52}$$

Proof. In (14.2.49) we replace the fourth term on the right by the second estimate of $|J_2|$ given by (14.2.35). Since we required $0 < \delta \le t/2$, all factors of the form $(t - \delta)^{\nu-1}$ can be estimated above by $(t/2)^{\nu-1}$. Restricting now $|x - s(t)| < (t/2)$, we select $\delta = |x - s(t)|$, whence the result follows. $\qquad\square$

As corollaries of the arguments above, we can state without proof the following results.

Lemma 14.2.5. For $\varphi \in C_{(\nu)}^{0}((0, T])$,

$$\lim_{x \downarrow s(t)} \frac{\partial w_\varphi}{\partial x}(x, t) = -2^{-1}\varphi(t) + \int_0^t \frac{\partial K}{\partial x}(s(t) - s(\tau), t - \tau)\varphi(\tau)\, d\tau. \tag{14.2.53}$$

Lemma 14.2.6. For $\varphi \in C_{(\nu)}^{\varepsilon}((0, T])$, there exists a positive constant $C = C(\varepsilon, \gamma, \nu, T, |s|_\gamma)$ such that, for $x > s(t)$,

$$\left|\frac{\partial w_\varphi}{\partial x}(x, t) + 2^{-1}\varphi(t) - \int_0^t \frac{\partial K}{\partial x}(s(t) - s(\tau), t - \tau)\varphi(\tau)\, d\tau\right|$$
$$\le C\|\varphi\|_{t, \varepsilon}^{(\nu)} t^{\nu-(3/2)} |x - s(t)|^{\gamma-(1/2)}, \tag{14.2.54}$$

provided that

$$\varepsilon > \left(\frac{\gamma}{2} - \frac{1}{4} \right). \tag{14.2.55}$$

We consider now the question of the continuity of function

$$\Psi(t) = \int_0^t \frac{\partial K}{\partial x}(s(t) - s(\tau), t - \tau) \varphi(\tau) \, d\tau.$$

Lemma 14.2.7. *For* $\varphi \in C_{(\nu)}^0((0, T])$, $\Psi \in C_{(\nu)}^0((0, T])$, *and there exists a constant* $C = C(\nu, \gamma, |s|_\gamma)$ *such that*

$$\|\Psi\|_T^{(\nu)} \le C T^{\gamma - 1/2} \|\varphi\|_T^{(\nu)}.$$

Proof. We consider $\Psi(t + \delta) - \Psi(t)$. Since

$$\int_0^{t+\delta} \frac{\partial K}{\partial x}(s(t+\delta) - s(\tau), t + \delta - \tau) \varphi(\tau) \, d\tau$$

$$= \int_0^\delta \frac{\partial K}{\partial x}(s(t+\delta) - s(\tau), t + \delta - \tau) \varphi(\tau) \, d\tau$$

$$+ \int_0^t \frac{\partial K}{\partial x}(s(t+\delta) - s(\tau+\delta), t + \delta - (\tau+\delta)) \varphi(\tau+\delta) \, d\tau, \tag{14.2.56}$$

we can write $\Psi(t + \delta) - \Psi(t) = I_1 + I_2 + I_3$, where

$$I_1 = \int_0^\delta \frac{\partial K}{\partial x}(s(t+\delta) - s(\tau), t + \delta - \tau) \varphi(\tau) \, d\tau, \tag{14.2.57}$$

$$I_2 = \int_0^t \frac{\partial K}{\partial x}(s(t) - s(\tau), t - \tau)[\varphi(\tau+\delta) - \varphi(\tau)] \, d\tau, \tag{14.2.58}$$

and

$$I_3 = \int_0^t \left\{ \frac{\partial K}{\partial x}(s(t+\delta) - s(\tau+\delta), t + \delta - (\tau+\delta)) \right.$$

$$\left. - \frac{\partial K}{\partial x}(s(t) - s(\tau), t - \tau) \right\} \varphi(\tau+\delta) \, d\tau. \tag{14.2.59}$$

For I_1, we see that

$$|I_1| \le C_1 |s|_\gamma \|\varphi\|_t^{(\nu)} \int_0^\delta (t + \delta - \tau)^{-(3/2)+\gamma} \tau^{\nu-1} \, d\tau$$

$$\le C_1 |s|_\gamma \|\varphi\|_t^{(\nu)} \int_0^\delta (t + \delta - \tau)^{\nu-1-\varepsilon} \cdot (t + \delta - \tau)^{-(3/2)+\gamma-\nu+1+\varepsilon} \tau^{\nu-1} \, d\tau$$

$$\le C_1 |s|_\gamma \|\varphi\|_t^{(\nu)} t^{\nu-1-\varepsilon} (t + \delta)^{\gamma-(1/2)} \cdot \int_0^\delta (\delta - \tau)^{\varepsilon-\nu} \tau^{\nu-1} \, d\tau$$

$$\le C_2 |s|_\gamma \|\varphi\|_t^{(\nu)} (t + \delta)^{\gamma-(1/2)} t^{\nu-1-\varepsilon} \delta^\varepsilon, \tag{14.2.60}$$

where C_2 depends upon ν, γ, and ε.

For I_2, it follows immediately that

$$|I_2| \le C_3 |s|_\gamma \int_0^t (t-\tau)^{-(3/2)+\gamma} |\varphi(\tau+\delta) - \varphi(\tau)| \, d\tau$$

$$\le C_3 |s|_\gamma \bigg\{ \int_\eta^t (t-\tau)^{-(3/2)+\gamma} |\varphi(\tau+\delta) - \varphi(\tau)| \, d\tau$$

$$+ \int_0^\eta (t-\tau)^{-(3/2)+\gamma} |\varphi(\tau+\delta)| \, d\tau$$

$$+ \int_0^\eta (t-\tau)^{-(3/2)+\gamma} |\varphi(\tau)| \, d\tau \bigg\}$$

$$\le C_3 |s|_\gamma t^{\gamma-(1/2)} \sup_{\eta \le \tau \le t} |\varphi(\tau+\delta) - \varphi(\tau)|$$

$$+ 2C_3 |s|_\gamma \nu^{-1} \|\varphi\|_{t+\delta}^{(\nu)} \eta^\nu (t-\eta)^{-(3/2)+\gamma}. \tag{14.2.61}$$

It is convenient here to derive another estimate of I_2 for the case that $\varphi \in C_{(\nu)}^\varepsilon((0,T])$. Recalling (14.1.6), we see that

$$|I_2| \le C_3 |s|_\gamma |\varphi|_\varepsilon^{(\nu)} \int_0^t \delta^\varepsilon (t-\tau)^{-(3/2)+\gamma} \tau^{\nu-1-\varepsilon} \, d\tau$$

$$\le C_4 |s|_\gamma |\varphi|_\varepsilon^{(\nu)} t^{(\gamma-(1/2))+\nu-1-\varepsilon} \delta^\varepsilon$$

$$\le C_5 |s|_\gamma |\varphi|_\varepsilon^{(\nu)} t^{(\gamma-(1/2))} t^{\nu-1-\varepsilon} \delta^\varepsilon. \tag{14.2.62}$$

Turning now to I_3, we can apply the mean-value theorem to obtain

$$\left| \frac{\partial K}{\partial x}(s(t+\delta) - s(\tau+\delta), t-\tau) - \frac{\partial K}{\partial x}(s(t) - s(\tau), t-\tau) \right|$$

$$\le C_6 (t-\tau)^{-3/2} |s(t+\delta) - s(\tau+\delta) - s(t) + s(\tau)| \tag{14.2.63}$$

since $|\partial^2 K/\partial x^2| \le C_7 (t-\tau)^{-3/2}$. Next, we write

$$|s(t+\delta) - s(\tau+\delta) - s(t) + s(\tau)|$$

$$= |[s(t+\delta) - s(\tau+\delta)] - [s(t) - s(\tau)]|^{1-(\varepsilon/\gamma)}$$

$$\times |[s(t+\delta) - s(t)] - [s(\tau+\delta) - s(\tau)]|^{\varepsilon/\gamma}. \tag{14.2.64}$$

Recalling $s \in C^\gamma([0,T])$ it follows from (14.2.63) and (14.2.64) that

$$|I_3| \le C_6 |s|_\gamma \|\varphi\|_{t+\delta}^{(\nu)} \delta^\varepsilon \int_0^t (t-\tau)^{-(3/2)+\gamma-\varepsilon} \tau^{\nu-1} \, d\tau$$

$$\le C_8 |s|_\gamma \|\varphi\|_{t+\delta}^{(\nu)} t^{\gamma-(1/2)} t^{\nu-1-\varepsilon} \delta^\varepsilon. \tag{14.2.65}$$

Collecting the estimates (14.2.60), (14.2.61), and (14.2.65), it follows from the continuity of φ that Ψ is continuous at each t, $0 < t \le T$. This is clear since η can be selected arbitrarily small and then fixed and since φ is uniformly continuous on $0 < \eta \le t \le T$, which implies that the remaining terms can be made small via δ.

To see that $\Psi \in C^0_{(\nu)}((0, T])$, we consider the estimate

$$|\Psi(t)| \le C_9 |s|_\gamma \|\varphi\|_t^{(\nu)} \int_0^t (t - \tau)^{-(3/2)+\gamma} \tau^{\nu-1} d\tau$$

$$\le C_{10} |s|_\gamma \|\varphi\|_T^{(\nu)} t^{\nu-1} T^{\gamma-(1/2)} \tag{14.2.66}$$

whence it follows that

$$t^{1-\nu} |\Psi(t)| \le C_{10} |s|_\gamma \|\varphi\|_T^{(\nu)} T^{\gamma-(1/2)} \tag{14.2.67}$$

for all $0 < t \le T$.   □

Collecting now the estimates (14.2.60), (14.2.62) and (14.2.65), we obtain the following result as a corollary of the argument above.

Lemma 14.2.8. *For $\varphi \in C^\varepsilon_{(\nu)}((0, T])$, $\Psi \in C^\varepsilon_{(\nu)}((0, T])$ and there exists a constant $C = C(\nu, \gamma, \varepsilon, |s|_\gamma)$ such that*

$$\|\Psi\|_{T,\varepsilon}^{(\nu)} \le C T^{\gamma-1/2} \|\varphi\|_{T,\varepsilon}^{(\nu)}.$$

Remark. Lemmas 14.2.1 through 14.2.6 contain the information necessary to reduce initial-boundary-value problems to a system of integral equations. Lemmas 14.2.7 and 14.2.8 contain the information that is sufficient to demonstrate the unique solvability of the integral equations, namely, $\Psi: C^0_{(\nu)}((0, T]) \to C^0_{(\nu)}((0, T])$ is a contraction if T is sufficiently small.

14.3. THE INITIAL-BOUNDARY-VALUE PROBLEM WITH TEMPERATURE-BOUNDARY SPECIFICATION

We consider now the problem of finding the $u = u(x, t)$ that satisfies

$$u_t = u_{xx}, \qquad s_1(t) < x < s_2(t), \qquad 0 < t \le T,$$

$$u(x, 0) = f(x), \qquad a < x < b, \qquad s_1(0) = a, \qquad s_2(0) = b,$$

$$u(s_1(t), t) = g(t), \qquad 0 < t \le T,$$

$$u(s_2(t), t) = h(t), \qquad 0 < t \le T, \tag{14.3.1}$$

where the data f, g, and h are continuous functions whose degree of continuity we shall make precise later. (See Fig. 14.3.1.) Recall that we have assumed that the $s_i \in C^\gamma([0, T])$, $\gamma > \frac{1}{2}$, $i = 1, 2$. Here, we shall also assume that

$$\delta = \inf_{0 \le t \le T} |s_1(t) - s_2(t)| > 0, \tag{14.3.2}$$

and

$$\Delta = \sup_{0 \le t \le T} |s_1(t) - s_2(t)| < \infty. \tag{14.3.3}$$

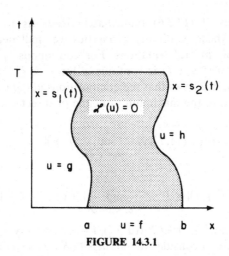

FIGURE 14.3.1

Now, the results of Sec. 14.2 as well as those of Chapters 6 and 7 motivate us to seek the solution u in the form

$$u(x,t) = v(x,t) + \frac{\partial w_{\varphi_1}}{\partial x}(x,t;s_1) + \frac{\partial w_{\varphi_2}}{\partial x}(x,t;s_2), \qquad (14.3.4)$$

where the $(\partial w_{\varphi_i}/\partial x)(x,t;s_i)$, $i=1,2$, are defined by the title of Sec. 14.2 while

$$v(x,t) = \int_{-\infty}^{\infty} K(x-\xi,t)f(\xi)\,d\xi, \qquad (14.3.5)$$

where the fundamental solution $K = K(x,t)$ is defined by (3.2.1) and f here denotes a continuous extension of f with compact support. From Lemmas 14.2.3 and 14.2.5, it is clear that, if (14.3.4) holds for $\varphi_i \in C_{(\nu)}^0((0,T])$, $i=1,2$, then φ_1 and φ_2 must satisfy

$$g(t) = v(s_1(t),t) - 2^{-1}\varphi_1(t) + \int_0^t \frac{\partial K}{\partial x}(s_1(t) - s_1(\tau), t-\tau)\varphi_1(\tau)\,d\tau$$

$$+ \int_0^t \frac{\partial K}{\partial x}(s_1(t) - s_2(\tau), t-\tau)\varphi_2(\tau)\,d\tau$$

$$h(t) = v(s_2(t),t) + 2^{-1}\varphi_2(t) + \int_0^t \frac{\partial K}{\partial x}(s_2(t) - s_2(\tau), t-\tau)\varphi_2(\tau)\,d\tau$$

$$+ \int_0^t \frac{\partial K}{\partial x}(s_2(t) - s_1(\tau), t-\tau)\varphi_1(\tau)\,d\tau, \qquad (14.3.6)$$

and conversely.

Consideration of (14.3.6) immediately leads us to the functions $v(s_i(t), t)$, $i = 1, 2$, whose continuity properties we shall need in order to study the system of integral equations. For continuous f, the results of Lemma 3.4.2 show that $v(x, t)$ is continuous for $-\infty < x < \infty$, $0 \le t \le T$. Since $s_i \in C^\gamma([0, T])$, it follows that the compositions $v(s_i(t), t)$ are continuous for $0 \le t \le T$. From the maximum principle or direct computations, we know that

$$|v(s_i(t), t)| \le \sup_{-\infty < x < \infty} |f(x)| \qquad (14.3.7)$$

and that

$$t^{1-\nu}|v(s_i(t), t)| \le T^{1-\nu} \sup_{-\infty < x < \infty} |f(x)|, \qquad i = 1, 2, \qquad (14.3.8)$$

which implies that $v(s_i(t), t) \in C_{(\nu)}^0((0, T])$, $i = 1, 2$, for any ν, $0 < \nu \le 1$.

Next, we need to consider the kernels $(\partial K/\partial x)(s_1(t) - s_2(\tau), t - \tau)$ and $(\partial K/\partial x)(s_2(t) - s_1(\tau), t - \tau)$. Clearly, for $t > \tau$, they are continuous in t and τ. Next, consider

$$(s_1(t) - s_2(\tau))^2 = ([s_1(t) - s_2(t)] + [s_2(t) - s_2(\tau)])^2$$

$$= (s_1(t) - s_2(t))^2 + 2(s_1(t) - s_2(t))$$

$$\cdot (s_2(t) - s_2(\tau)) + (s_2(t) - s_2(\tau))^2$$

$$\ge \delta^2 - \varepsilon\Delta^2 + (1 - \varepsilon^{-1})(s_2(t) - s_2(\tau))^2. \quad (14.3.9)$$

Selecting $\varepsilon = \delta^2/2\Delta^2$, we get

$$(s_1(t) - s_2(\tau))^2 \ge 2^{-1}\delta^2 - \left(\frac{2\Delta^2}{\delta^2} - 1\right)(s_2(t) - s_2(\tau))^2$$

$$\ge 2^{-1}\delta^2 - \left(\frac{2\Delta^2}{\delta^2} - 1\right)|s_2|_\gamma^2 |t - \tau|^{2\gamma}. \quad (14.3.10)$$

Consequently,

$$\exp\left\{-\frac{(s_1(t) - s_2(\tau))^2}{4(t - \tau)}\right\} \le \exp\left\{4^{-1}\left(\frac{2\Delta^2}{\delta^2} - 1\right)|s_2|_\gamma^2 T^{2\gamma - 1}\right\}$$

$$\cdot \exp\left\{-\frac{\delta^2}{8(t - \tau)}\right\}. \quad (14.3.11)$$

Since $\delta > 0$, it follows, from $\exp\{-x\} \le n! x^{-n}$ for $n \ge 1$, that there exist a constant $C_1 = C_1(\delta, n, T, \Delta, |s_2|_\gamma, \gamma)$ such that for $n \ge 1$,

$$\left|\frac{\partial K}{\partial x}(s_1(t) - s_2(\tau), t - \tau)\right| \le C_1(t - \tau)^{n - (3/2)} \quad (14.3.12)$$

and in a similar manner there exists a constant $C_2 = C_2(\delta, n, T, \Delta, |s_1|_\gamma, \gamma)$ such that for $n \geq 1$,

$$\left| \frac{\partial K}{\partial x} (s_2(t) - s_1(\tau), t - \tau) \right| \leq C_2 (t - \tau)^{n - (3/2)}. \qquad (14.2.13)$$

Consequently, from either the analysis of Sec. 14.2, or Sec. 8.2, Lemma 8.2.2, we see that the convolutions of these kernels with functions $\varphi \in C^0_{(\nu)}((0, T])$ are themselves $C^0_{(\nu)}((0, T])$ and the $\| \cdot \|^{(\nu)}_T$ of the convolutions are bounded by $C_3 \|\varphi\|^{(\nu)}_T T^{1/2}$.

As the other two convolutions in (14.3.6) possess the properties of Ψ in Lemma 14.2.7 for small T, the solution of the system (14.3.6) for $g \in C^0_{(\nu)}((0, T])$ and $h \in C^0_{(\nu)}((0, T])$ reduces to the fixed point of a contraction of $C^0_{(\nu)}((0, T])$ into itself. After the unique determination of φ_1 and φ_2 satisfying (14.3.6) for small t, the unique determination of φ_1 and φ_2 reduces to Corollary 8.2.1 since all data functions become continuous on compact intervals of $t > 0$.

We can summarize the above discussion in the following statement.

Lemma 14.3.1. *For $f \in C([a, b])$, $g \in C^0_{(\nu)}((0, T])$ and $h \in C^0_{(\nu)}((0, T])$, there exists unique functions $\varphi_i \in C^0_{(\nu)}((0, T])$, $i = 1, 2$, that satisfy (14.3.6) for $0 < t \leq T$. Moreover, the function*

$$u(x, t) = v(x, t) + \frac{\partial w_{\varphi_1}}{\partial x}(x, t; s_1) + \frac{\partial w_{\varphi_2}}{\partial x}(x, t; s_2) \qquad (14.3.14)$$

is a solution of (14.3.1), which may become unbounded at $x = a, b$ for $t = 0$ if g and h become unbounded.

Since we would like to apply the extended maximum principle to obtain unicity of the solution, we prove the following result.

Lemma 14.3.2. *For $\varphi \in C^0_{(1)}((0, T])$, there exists a positive constant $C = C(\gamma, |s|_\gamma, T)$ such that, for $0 < t \leq T$,*

$$\left| \frac{\partial w_\varphi}{\partial x}(x, t; s) \right| \leq C \|\varphi\|^{(1)}_t. \qquad (14.3.15)$$

Proof. We begin by noting that

$$(x - s(\tau))^2 = (x - s(t) + s(t) - s(\tau))^2$$

$$\geq (1 - \varepsilon)(x - s(t))^2 - (\varepsilon^{-1} - 1)(s(t) - s(\tau))^2 \qquad (14.3.16)$$

for $0 < \varepsilon < 1$ via the use of $2ab \leq \varepsilon a^2 + \varepsilon^{-1} b^2$. Hence, as we have seen before,

$$\exp\left\{-\frac{(x-s(\tau))^2}{4(t-\tau)}\right\} \le \exp\left\{(\varepsilon^{-1}-1)4^{-1}|s|_\gamma t^{2\gamma-1}\right\}$$

$$\cdot \exp\left\{-\frac{(1-\varepsilon)(x-s(t))^2}{4(t-\tau)}\right\}. \quad (14.3.17)$$

Thus, there exists a positive constant C_1 such that

$$\left|\frac{\partial w_\varphi}{\partial x}(x,t;s)\right| \le C_1\|\varphi\|_t^{(1)}\int_0^t\left\{\frac{|x-s(t)|}{(t-\tau)^{3/2}} + \frac{|s(t)-s(\tau)|}{(t-\tau)^{3/2}}\right\}$$

$$\cdot \exp\left\{-\frac{(1-\varepsilon)(x-s(t))^2}{4(t-\tau)}\right\} d\tau. \quad (14.3.18)$$

Now, using the transformation $\rho^2 = (1-\varepsilon)(x-s(t))^2/[4(t-\tau)]$, we can estimate the $|x-s(t)|/((t-\tau)^{3/2})$ term while the term $|s(t)-s(\tau)|$ $(t-\tau)^{-3/2}$ can be estimated using the fact that $s \in C^\gamma([0,T])$. □

For the case that $g \in C_{(1)}^0((0,T])$ and $h \in C_{(1)}^0((0,T])$, the equations (14.3.6) reduce to the case of those studied in Chapter 8, Corollary 8.4.1, whence it follows that the boundedness of f, g, and h implies the boundedness of φ_1 and φ_2. Consequently, u defined by (14.3.14) is in this case a bounded solution of (14.3.1). By the Extended Uniqueness Theorem 1.6.6, u is unique.

We summarize the analysis of this section in the following statement.

Theorem 14.3.1. *For the problem*

$$u_t = u_{xx}, \qquad s_1(t) < x < s_2(t), \qquad 0 < t \le T,$$

$$u(x,0) = f(x), \qquad a < x < b, \qquad s_1(0) = a, \qquad s_2(0) = b,$$

$$u(s_1(t),t) = g(t), \qquad 0 < t \le T,$$

$$u(s_2(t),t) = h(t), \qquad 0 < t \le T, \qquad (14.3.19)$$

where $f \in C([a,b])$, $g \in C_{(\nu)}^0((0,T])$, $h \in C_{(\nu)}^0((0,T])$, $0 < \nu \le 1$, $s_i \in C^\gamma([0,T])$, $i = 1,2, \frac{1}{2} < \gamma \le 1$,

$$\delta = \inf_{0 \le t \le T}|s_1(t) - s_2(t)| > 0, \qquad (14.3.20)$$

and

$$\Delta = \sup_{0 \le t \le T}|s_1(t) - s_2(t)| < \infty, \qquad (14.3.21)$$

there exists a solution u that has the representation

$$u(x,t) = \int_{-\infty}^{\infty} K(x-\xi,t)f(\xi)\,d\xi + \int_0^t \frac{\partial K}{\partial x}(x-s_1(\tau),t-\tau)\varphi_1(\tau)\,d\tau$$

$$+ \int_0^t \frac{\partial K}{\partial x}(x-s_2(\tau),t-\tau)\varphi_2(\tau)\,d\tau, \tag{14.3.22}$$

where $K = K(x,t)$ *is the fundamental solution defined by* (3.2.1), f *is an extension with compact support of the data* f, *and the pair of functions* $\varphi_i \in C^0_{(\nu)}((0,T])$, $i=1,2$, *is the unique solution of the system of integral equations* (14.3.6). *The solution u may become unbounded at* $x = a, b$, $t = 0$, *if g or h becomes unbounded at* $t = 0$. *For* $\nu = 1$, *g and h are bounded, and u is the unique bounded solution of* (14.3.19).

14.4. THE EXISTENCE OF u_x AT THE BOUNDARY FOR THE SOLUTION OF THE INITIAL-BOUNDARY-VALUE PROBLEM WITH TEMPERATURE-BOUNDARY SPECIFICATION

We shall study the conditions that must be specified upon the data to ensure the existence of $u_x(s_i(t),t)$, $i=1,2$, for the solution u of the problem

$$u_t = u_{xx}, \qquad s_1(t) < x < s_2(t), \qquad 0 < t \leq T,$$
$$u(x,0) = f(x), \qquad a < x < b, \qquad s_1(0) = a, \qquad s_2(0) = b, \tag{14.4.1}$$
$$u(s_1(t),t) = g(t), \qquad 0 < t \leq T,$$
$$u(s_2(t),t) = h(t), \qquad 0 < t \leq T,$$

where the data f, g, and h are continuous functions whose degree of continuity we shall specify later. The functions $s_i \in C^\gamma([0,T])$, $i=1,2$, $\frac{1}{2} < \gamma \leq 1$, and

$$\delta = \inf_{0 \leq t \leq T} |s_1(t) - s_2(t)| > 0. \tag{14.4.2}$$

Since the property of the existence of $u_x(s_i(t),t)$, $i=1,2$, is purely local in nature, we need only consider a neighborhood of $(s_1(t),t)$ and a neighborhood of $(s_2(t),t)$. Let us fix our attention upon a neighborhood of $(s_2(t),t)$. By a translation and a relabeling of variables, it is easy to see that the existence of $u_x(s_2(t),t)$ is equivalent to the existence of $w_x(s(t),t)$, where w satisfies

$$w_t = w_{xx}, \qquad 0 < x < s(t), \qquad 0 < t \leq T,$$
$$w(x,0) = f(x), \qquad 0 < x < b, \qquad s(0) = b,$$
$$w(0,t) = g(t), \qquad 0 < t \leq T, \tag{14.4.3}$$
$$w(s(t),t) = h(t), \qquad 0 < t \leq T,$$

where f, g, and h are continuous functions, $f(0) = g(0)$, $f(b) = h(0)$, and the degree of continuity of f, g, and h will be specified below. Also, $s \in C^\gamma([0, T])$ and

$$\delta_1 = \inf_{0 \leq t \leq T} s(t) > 0. \tag{14.4.4}$$

Now, the existence of $w_x(s(t), t)$ will follow from representation

$$w(x, t) = v(x, t) + z_\varphi(x, t) + w_\varphi(x, t; s), \tag{14.4.5}$$

where

$$w_\varphi(x, t; s) = \int_0^t K(x - s(\tau), t - \tau) \varphi(\tau) \, d\tau, \tag{14.4.6}$$

$$z_\varphi(x, t) = -2 \int_0^t \frac{\partial K}{\partial x}(x, t - \tau) \{ g(\tau) - w_\varphi(0, \tau; s) \} \, d\tau,$$

and

$$v(x, t) = \int_0^b G(x, \xi, t) f(\xi) \, d\xi, \tag{14.4.7}$$

where $G(x, \xi, t)$ is the Green's function for the quarter plane defined by (4.3.4), and $K = K(x, t)$ is the fundamental solution defined by (3.2.1). The representation (14.4.5) for w is equivalent to the existence of a continuous $\varphi = \varphi(t)$, which satisfies the integral equation

$$h(t) = v(s(t), t) + z_\varphi(s(t), t) + w_\varphi(s(t), t; s), \tag{14.4.8}$$

which is a Volterra integral equation of the first kind for φ.

When you consider

$$w_\varphi(s(t), t; s) = \int_0^t (4\pi)^{-1/2} (t - \tau)^{-1/2} \exp \left\{ -\frac{(s(t) - s(\tau))^2}{4(t - \tau)} \right\} \varphi(\tau) \, d\tau, \tag{14.4.9}$$

you note that

$$\lim_{\tau \uparrow t} \exp \left\{ -\frac{(s(t) - s(\tau))^2}{4(t - \tau)} \right\} = 1 \tag{14.4.10}$$

since $s \in C^\gamma([0, T])$ with $\gamma > \frac{1}{2}$. Consequently, $w_\varphi(s(t), t; s)$ for τ close to t behaves like

$$(4\pi)^{-1/2} \int_0^t (t - \tau)^{-1/2} \varphi(\tau) \, d\tau, \tag{14.4.11}$$

and we are led to apply the Abel inversion operator

$$(AF)(t) = \frac{1}{\pi} \frac{d}{dt} \int_0^t \frac{F(\eta)}{(t - \eta)^{1/2}} \, d\eta \tag{14.4.12}$$

to the equation (14.4.8). As we shall see below, this will reduce (14.4.8) to an equivalent Volterra integral equation of the second kind. Unfortunately, Theorem 8.1.2 is not applicable here. We shall have to study the effect of the operator A on each term in (14.4.8).

We shall begin with $w_\varphi(s(t), t; s)$. Now,

$$(Aw_\varphi)(t) = \frac{1}{\pi} \frac{d}{dt} \int_0^t (t-\eta)^{-1/2} \int_0^\eta (4\pi)^{-1/2}$$

$$\cdot (\eta-\tau)^{-1/2} \exp\left\{ -\frac{(s(\eta)-s(\tau))^2}{4(\eta-\tau)} \right\} \varphi(\tau) \, d\tau \, d\eta,$$

$$(14.4.13)$$

which, from Fubini's theorem, becomes

$$(Aw_\varphi)(t) = 2^{-1}\pi^{-3/2} \frac{d}{dt} \int_0^t \varphi(\tau) \left\{ \int_\tau^t (t-\eta)^{-1/2} \right.$$

$$\left. \cdot (\eta-\tau)^{-1/2} \exp\left\{ -\frac{(s(\eta)-s(\tau))^2}{4(\eta-\tau)} \right\} d\eta \right\} d\tau.$$

$$(14.4.14)$$

By adding and subtracting $\exp\{-(s(t)-s(\tau))^2/[4(t-\tau)]\}$ it follows from

$$\pi = \int_\tau^t (t-\eta)^{-1/2} (\eta-\tau)^{-1/2} \, d\eta \qquad (14.4.15)$$

that

$$\lim_{\tau \uparrow t} \int_\tau^t (t-\eta)^{-1/2} (\eta-\tau)^{-1/2} \exp\left\{ -\frac{(s(\eta)-s(\tau))^2}{4(\eta-\tau)} \right\} d\eta$$

$$= \pi + \lim_{\tau \uparrow t} \int_\tau^t (t-\eta)^{-1/2} (\eta-\tau)^{-1/2}$$

$$\cdot \left\{ \exp\left\{ -\frac{(s(\eta)-s(\tau))^2}{4(\eta-\tau)} \right\} - \exp\left\{ -\frac{(s(t)-s(\tau))^2}{4(t-\tau)} \right\} \right\} d\eta.$$

$$(14.4.16)$$

Applying the mean-value theorem and using the evaluations

$$\int_\tau^t (t-\eta)^{-1/2} (\eta-\tau)^{2\gamma-(3/2)} \, d\eta = \int_\tau^t (t-\eta)^{2\gamma-(3/2)} (\eta-\tau)^{-(1/2)} \, d\eta$$

$$= C_1 (t-\tau)^{2\gamma-1}, \qquad (14.4.17)$$

it follows that the second term on the righthand side of (14.4.16) is $\mathcal{O}(|t-\tau|^{2\gamma-1})$. Since $\gamma > \frac{1}{2}$, its limit is zero, and

$$\lim_{\tau \uparrow t} \int_\tau^t (t-\eta)^{-1/2} (\eta-\tau)^{-1/2} \exp\left\{ -\frac{(s(\eta)-s(\tau))^2}{4(\eta-\tau)} \right\} d\eta = \pi. \quad (14.4.18)$$

Consequently, Leibniz' rule implies

$$(Aw_\varphi)(t) = 2^{-1}\pi^{-1/2}\varphi(t) + 2^{-1}\pi^{-3/2}\int_0^t \varphi(\tau)\frac{d}{dt}$$

$$\cdot\left\{\int_\tau^t (t-\eta)^{-1/2}(\eta-\tau)^{-1/2}\exp\left\{-\frac{(s(\eta)-s(\tau))^2}{4(\eta-\tau)}\right\}d\eta\right\}d\tau,$$

$$(14.4.19)$$

which leads us to the study of

$$\frac{d}{dt}\left\{\int_\tau^t (t-\eta)^{-1/2}(\eta-\tau)^{-1/2}\exp\left\{-\frac{(s(\eta)-s(\tau))^2}{4(\eta-\tau)}\right\}d\eta\right\}.$$

$$(14.4.20)$$

For this purpose, the following result is useful.

Lemma 14.4.1. *Let* $h(t,\tau)$ *be defined and continuous in both arguments for* $\tau < t \leq T$ *and satisfy*

$$|h(t,\tau)-h(\eta,\tau)| \leq \sum_{i=1}^l C_i(\eta-\tau)^{-\alpha_i}(t-\eta)^{+\beta_i}, \qquad \tau < \eta \leq t \leq T,$$

$$(14.4.21)$$

for some constants C_i, $i=1,\ldots,l$, *and exponents* $\alpha_i < \frac{1}{2} < \beta_i$, $i=1,\ldots,l$. *Let*

$$z(t,\tau) = \int_\tau^t (t-\eta)^{-1/2}(\eta-\tau)^{-1/2}h(\eta,\tau)\,d\eta. \qquad (14.4.22)$$

Then, $(d/dt)z(t,\tau)$ *exists for* $\tau < t \leq T$ *and is given by*

$$\frac{d}{dt}z(t,\tau) = \frac{1}{2}\int_\tau^t (t-\eta)^{-3/2}(\eta-\tau)^{-1/2}[h(t,\tau)-h(\eta,\tau)]\,d\eta.$$

$$(14.4.23)$$

Moreover, there exists a constant $C = C(\alpha_i,\beta_i)$ *such that*

$$\left|\frac{d}{dt}z(t,\tau)\right| \leq C\sum_{i=1}^l (t-\tau)^{\beta_i-\alpha_i-1}, \qquad (14.4.24)$$

and there exists a constant C_ε *such that*

$$\left|\frac{d}{dt}z(t+\delta,\tau) - \frac{d}{dt}z(t,\tau)\right| \leq C_\varepsilon\delta^\varepsilon \sum_{i=1}^l (t-\tau)^{\beta_i-\alpha_i-1-c\varepsilon},$$

$$\tau < t \leq t+\delta \leq T, \quad (14.4.25)$$

where

$$c = \{\max\beta_i + (\max\alpha_i - \min\alpha_i)\}(\min\beta_i)^{-1} \qquad (14.4.26)$$

and ε *is sufficiently small that*

$$\min_i\left(\beta_i - \tfrac{1}{2}, \tfrac{1}{2} - \alpha_i\right) > c\varepsilon. \qquad (14.4.27)$$

Proof. We begin with the consideration of the difference quotient

$$\{z(t+\delta,\tau)-z(t,\tau)\}\delta^{-1} = \delta^{-1}\left[\int_\tau^{t+\delta}(t+\delta-\eta)^{-1/2}(\eta-\tau)h(\eta,\tau)\,d\eta\right.$$
$$\left.-\int_\tau^t(t-\eta)^{-1/2}(\eta-\tau)h(\eta,\tau)\,d\eta\right],$$

$$(14.4.28)$$

and use the fact that

$$\int_\tau^{t+\delta}(t+\delta-\eta)^{-1/2}(\eta-\tau)^{-1/2}h(t,\tau)\,d\eta$$
$$= \int_\tau^t(t-\eta)^{-1/2}(\eta-\tau)^{-1/2}h(t,\tau)\,d\eta = \pi h(t,\tau) \qquad (14.4.29)$$

to introduce $h(t,\tau)$ into the difference quotient to obtain

$$\{z(t+\delta,\tau)-z(t,\tau)\}\delta^{-1}$$
$$= \delta^{-1}\left[\int_\tau^{t+\delta}(t+\delta-\eta)^{-1/2}(\eta-\tau)^{-1/2}[h(\eta,\tau)-h(t,\tau)]\,d\eta\right.$$
$$\left.-\int_\tau^t(t-\eta)^{-1/2}(\eta-\tau)^{-1/2}[h(\eta,\tau)-h(t,\tau)]\,d\eta\right]$$
$$= J_1 + J_2, \qquad (14.4.30)$$

where

$$J_1 = \delta^{-1}\int_t^{t+\delta}(t+\delta-\eta)^{-1/2}[h(\eta,\tau)-h(t,\tau)]\,d\eta \qquad (14.4.31)$$

and

$$J_2 = \int_\tau^t\left[(t+\delta-\eta)^{-1/2}-(t-\eta)^{-1/2}\right]\delta^{-1}(\eta-\tau)^{-1/2}$$
$$\cdot[h(\eta,\tau)-h(t,\tau)]\,d\eta. \qquad (14.4.32)$$

First, we see from (14.4.21) that

$$|J_1| \le \delta^{-1}\sum_{i=1}^l C_i(t-\tau)^{-(1/2)-\alpha_i}\int_t^{t+\delta}(t+\delta-\eta)^{-1/2}(\eta-t)^{\beta_i}\,d\eta.$$

$$(14.4.33)$$

But, from (14.2.2),

$$\int_t^{t+\delta}(t+\delta-\eta)^{-1/2}(\eta-t)^{\beta_i}\,d\eta = \tilde{C}_i\delta^{\beta_i+1/2}. \qquad (14.4.34)$$

Hence,

$$J_1 = \mathcal{O}\left(\sum_{i=1}^{l} \delta^{\beta_i - (1/2)}\right) \tag{14.4.35}$$

and

$$\lim_{\delta \to 0} J_1 = 0. \tag{14.4.36}$$

Next, we see that for all $\delta > 0$,

$$|\{(t+\delta-\eta)^{-1/2} - (t-\eta)^{-1/2}\}\delta^{-1}| \le \tfrac{1}{2}(t-\eta)^{-3/2}. \tag{14.4.37}$$

Consequently, from (14.4.21) it follows that the integrand of J_2 is dominated by

$$\frac{1}{2}\sum_{i=1}^{l} C_i (t-\eta)^{-(3/2)+\beta_i}(\eta-\tau)^{-(1/2)-\alpha_i}. \tag{14.4.38}$$

Since $-(3/2)+\beta_i > -1$ and $-(1/2)-\alpha_i > -1$, it follows from (14.2.2) that

$$\int_{\tau}^{t}(t-\eta)^{-(3/2)+\beta_i}(\eta-\tau)^{-(1/2)-\alpha_i}\,d\eta = \tilde{C}_i(t-\tau)^{\beta_i-\alpha_i-1}. \tag{14.4.39}$$

Consequently, the Lebesgue dominated-convergence theorem implies that

$$\lim_{\delta \to 0} J_2 = \frac{1}{2}\int_{\tau}^{t}(t-\eta)^{-3/2}(\eta-\tau)^{-1/2}[h(t,\tau)-h(\eta,\tau)]\,d\eta. \tag{14.4.40}$$

Thus, we have demonstrated (14.4.23), and by collecting (14.4.38) and (14.4.39), we have shown (14.4.24).

It remains to be shown that (14.4.25) holds. We see that

$$\frac{d}{dt}z(t+\delta,\tau) - \frac{d}{dt}z(t,\tau) = I_1 + I_2 + I_3, \tag{14.4.41}$$

where

$$I_1 = \frac{1}{2}\int_{\tau}^{\tau+\delta}(t+\delta-\eta)^{-3/2}(\eta-\tau)^{-1/2}\{h(t+\delta,\tau)-h(\eta,\tau)\}\,d\eta, \tag{14.4.42}$$

$$I_2 = \frac{1}{2}\int_{\tau}^{t}(t-\eta)^{-3/2}\left[(\eta+\delta-\tau)^{-1/2}-(\eta-\tau)^{-1/2}\right]$$
$$\cdot\{h(t+\delta,\tau)-h(\eta+\delta,\tau)\}\,d\eta, \tag{14.4.43}$$

and

$$I_3 = \frac{1}{2}\int_{\tau}^{t}(t-\eta)^{-3/2}(\eta-\tau)^{-1/2}\{(h(t+\delta,\tau)-h(\eta+\delta,\tau))$$
$$-(h(t,\tau)-h(\eta,\tau))\}\,d\eta. \tag{14.4.44}$$

We begin with I_1 and note that, from (14.4.21), we have

$$|I_1| \le \frac{1}{2}\sum_{i=1}^{l} C_i\int_{\tau}^{\tau+\delta}(t+\delta-\eta)^{-(3/2)+\beta_i}(\eta-\tau)^{-(1/2)-\alpha_i}\,d\eta. \tag{14.4.45}$$

Then, since $(\frac{1}{2} - \alpha_i - \varepsilon) > 0$,

$$(t + \delta - \eta)^{-(3/2) + \beta_i} = (t + \delta - \eta)^{-1 + \beta_i - \alpha_i - \varepsilon}(t + \delta - \eta)^{\alpha_i - (1/2) + \varepsilon}$$
$$\leq (t - \tau)^{-1 + \beta_i - \alpha_i - \varepsilon}(\tau + \delta - \eta)^{\alpha_i - (1/2) + \varepsilon}.$$

$$(14.4.46)$$

Hence, from (14.2.2),

$$|I_1| \leq \frac{1}{2} \sum_{i=1}^{l} C_i (t - \tau)^{-1 + \beta_i - \alpha_i - \varepsilon}$$
$$\cdot \int_{\tau}^{\tau + \delta} (\tau + \delta - \eta)^{\alpha_i - (1/2) + \varepsilon}(\eta - \tau)^{-(1/2) - \alpha_i} d\eta$$

$$\leq C_{1,\varepsilon} \sum_{i=1}^{l} (t - \tau)^{-1 + \beta_i - \alpha_i - \varepsilon} \delta^{\varepsilon}. \qquad (14.4.47)$$

Next, we consider I_2 and note that

$$\left|\left[(\eta + \delta - \tau)^{-1/2} - (\eta - \tau)^{-1/2}\right]\right|$$
$$\leq (\tfrac{1}{2})^{\varepsilon}(\eta - \tau)^{-(3/2)\varepsilon} \delta^{\varepsilon} \cdot 2^{1 - \varepsilon}\left\{(\eta - \tau)^{-(1/2)(1 - \varepsilon)}\right\}. \qquad (14.4.48)$$

Consequently, from (14.4.21) we have

$$|I_2| \leq C_{2,\varepsilon} \delta^{\varepsilon} \sum_{i=1}^{l} C_i \int_{\tau}^{t} (t - \eta)^{-(3/2) + \beta_i}(\eta - \tau)^{-(1/2) - \alpha_i - \varepsilon} d\eta$$

$$\leq C_{3,\varepsilon} \delta^{\varepsilon} \sum_{i=1}^{l} (t - \tau)^{\beta_i - \alpha_i - 1 - \varepsilon}, \qquad (14.4.49)$$

via (14.2.2). For I_3, we see that

$$|\{(h(t + \delta, \tau) - h(\eta + \delta, \tau)) - (h(t, \tau) - h(\eta, \tau))\}|$$
$$\leq \{|h(t + \delta, \tau) - h(\eta + \delta, \tau)| + |h(t, \tau) - h(\eta, \tau)|\}^{1 - \beta^{-1}\varepsilon}$$
$$\cdot \{|h(t + \delta, \tau) - h(t, \tau)| + |h(\eta + \delta, \tau) - h(\eta, \tau)|\}^{\beta^{-1}\varepsilon}$$
$$\leq \left\{\sum_{i=1}^{l} C_i\left[(t - \eta)^{\beta_i}(\eta + \delta - \tau)^{-\alpha_i} + (t - \eta)^{\beta_i}(\eta - \tau)^{-\alpha_i}\right]\right\}^{1 - \beta^{-1}\varepsilon}$$
$$\cdot \left\{\sum_{i=1}^{l} C_i\left[\delta^{\beta_i}(t - \tau)^{-\alpha_i} + \delta^{\beta_i}(\eta - \tau)^{-\alpha_i}\right]\right\}^{\beta^{-1}\varepsilon}$$
$$\leq \left\{2\sum_{i=1}^{l} C_i(t - \eta)^{\beta_i}(\eta - \tau)^{-\alpha_i}\right\}^{1 - \beta^{-1}\varepsilon}\left\{2\sum_{i=1}^{l} C_i(\eta - \tau)^{-\alpha_i}\right\}^{\beta^{-1}\varepsilon} \delta^{\varepsilon}$$

$$(14.4.50)$$

if $\delta < 1$ and $\beta = \min_i \beta_i$. If $\alpha = \max_i \alpha_i$, then for $(t - \tau) < 1$,

$$(\eta - \tau)^{-\alpha_i} \le (\eta - \tau)^{-\alpha}. \tag{14.4.51}$$

From $(a + b)^\alpha \le a^\alpha + b^\alpha$, $a, b \ge 0, 0 < \alpha \le 1$, and (14.4.51), it follows that

$$|\{(h(t + \delta, \tau) - h(\eta + \delta, \tau)) - (h(t, \tau) - h(\eta, \tau))\}|$$

$$\le 2 \left(\sum_{i=1}^{l} C_i^{\beta^{-1}\varepsilon} \right) \left(\sum_{i=1}^{l} C_i^{1 - \beta^{-1}\varepsilon}(t - \eta)^{\beta_i - \beta_i\beta^{-1}\varepsilon} \right.$$

$$\left. \cdot (\eta - \tau)^{-\alpha_i + \alpha_i\beta^{-1}\varepsilon - \alpha\beta^{-1}\varepsilon} \right) \delta^\varepsilon$$

$$\le C_{4,\varepsilon} \delta^\varepsilon \sum_{i=1}^{l} (t - \eta)^{\beta_i - \beta_i\beta^{-1}\varepsilon}(\eta - \tau)^{-\alpha_i + \alpha_i\beta^{-1}\varepsilon - \alpha\beta^{-1}\varepsilon}.$$

$$\tag{14.4.52}$$

Consequently, substituting (14.4.52) into I_3 and integrating via (14.2.2), we obtain

$$|I_3| \le C_{5,\varepsilon} \delta^\varepsilon \sum_{i=1}^{l} (t - \tau)^{1 - (3/2) + \beta_i - \beta_i\beta^{-1}\varepsilon - \alpha_i - (1/2) + \alpha_i\beta^{-1}\varepsilon - \alpha\beta^{-1}\varepsilon},$$

$$\tag{14.4.53}$$

provided that ε is sufficiently small that

$$-\tfrac{3}{2} + \beta_i - \beta_i\beta^{-1}\varepsilon > -1, \qquad i = 1, \ldots, l,$$

and

$$-\alpha_i - \tfrac{1}{2} + \alpha_i\beta^{-1}\varepsilon - \alpha\beta^{-1}\varepsilon > -1, \qquad i = 1, \ldots, l.$$

Now,

$$1 + \alpha_i - \beta_i + \beta_i\beta^{-1}\varepsilon + \left(\alpha\beta^{-1}\varepsilon - \alpha_i\beta^{-1}\varepsilon \right)$$

$$\le 1 + \alpha_i - \beta_i + \left\{ \frac{\max \beta_i}{\min \beta_i} + \frac{\max \alpha_i - \min \alpha_i}{\min \beta_i} \right\} \varepsilon. \tag{14.4.54}$$

Thus, for $(t - \tau) < 1$, it follows that

$$|I_3| \le C_{6,\varepsilon} \delta^\varepsilon \sum_{i=1}^{l} (t - \tau)^{\beta_i - \alpha_i - 1 - c\varepsilon}, \tag{14.4.55}$$

where

$$c = \{ \max \beta_i + (\max \alpha_i - \min \alpha_i) \}(\min \beta_i)^{-1}. \tag{14.4.56}$$

Combining (14.4.47), (14.4.49), and (14.4.55), we obtain the result (14.4.25) provided that ε is sufficiently small that

$$\min_i \left(\beta_i - \tfrac{1}{2}, \tfrac{1}{2} - \alpha_i \right) > c\varepsilon. \tag{14.4.57} \qquad \square$$

We consider now the function

$$h_1(t, \tau) = \exp\left\{ -\frac{(s(t) - s(\tau))^2}{4(t - \tau)} \right\} \tag{14.4.58}$$

and write

$$h_1(t, \tau) - h_1(\eta, \tau) = \int_0^1 \frac{d}{d\theta} \exp\left\{ -\frac{(\theta s(t) + [1 - \theta] s(\eta) - s(\tau))^2}{4(\theta t + [1 - \theta] \eta - \tau)} \right\} d\theta. \tag{14.4.59}$$

As

$$\frac{d}{d\theta} \exp\left\{ -\frac{(\theta s(t) + (1 - \theta) s(\eta) - s(\tau))^2}{4(\theta t + (1 - \theta) \eta - \tau)} \right\}$$

$$= \left[-\frac{(\theta s(t) + (1 - \theta) s(\eta) - s(\tau))}{2(\theta t + (1 - \theta) \eta - \tau)} \cdot (s(t) - s(\eta)) \right.$$

$$\left. + \frac{(\theta s(t) + (1 - \theta) s(\eta) - s(\tau))^2}{4(\theta t + (1 - \theta) \eta - \tau)^2} \cdot (t - \tau) \right]$$

$$\cdot \exp\left\{ -\frac{(\theta s(t) + (1 - \theta) s(\eta) - s(\tau))^2}{4(\theta t + (1 - \theta) \eta - \tau)} \right\}, \tag{14.4.60}$$

we begin our estimate by considering the ratio in (14.4.60), which can be rewritten as

$$\rho = \frac{(\theta(s(t) - s(\tau)) + (1 - \theta)(s(\eta) - s(\tau)))}{(\theta(t - \tau) + (1 - \theta)(\eta - \tau))}. \tag{14.4.61}$$

But,

$$|\theta(s(t) - s(\tau)) + (1 - \theta)(s(\eta) - s(\tau)))|$$

$$\leq |s|_\gamma (\theta(t - \tau)^\gamma + (1 - \theta)(\eta - \tau)^\gamma)$$

$$\leq |s|_\gamma (\theta(t - \tau) + (1 - \theta)(\eta - \tau))^\gamma, \tag{14.4.62}$$

since $(\theta a + (1 - \theta) b)^\alpha \geq \theta a^\alpha + (1 - \theta) b^\alpha$ holds for $a, b > 0$, $0 < \theta$, $\alpha < 1$. Hence,

$$|\rho| \leq |s|_\gamma (\theta(t - \tau) + (1 - \theta)(\eta - \tau))^{\gamma - 1} = |s|_\gamma ((\eta - \tau) + \theta(t - \eta))^{\gamma - 1} \tag{14.4.63}$$

Since the exponential in (14.4.60) is bounded by 1, the substitution of

(14.4.63) into (14.4.60) yields

$$\int_0^1 |\rho| \cdot |s(t) - s(\eta)| \, d\theta \le |s|_\gamma^2 (\eta - \tau)^{\gamma-1} (t-\eta)^\gamma, \qquad (14.4.64)$$

while

$$\int_0^1 |\rho|^2 (t-\eta) \, d\theta \le |s|_\gamma^2 \int_0^1 ((\eta - \tau) + \theta(t-\eta))^{2\gamma-2} (t-\eta) \, d\theta$$

$$\le |s|_\gamma \int_0^1 (\eta - \tau)^{\gamma-1} (t-\eta)^{\gamma-1} \theta^{\gamma-1} (t-\eta) \, d\theta$$

$$\le \gamma^{-1} |s|_\gamma^2 (\eta - \tau)^{\gamma-1} (t-\eta)^\gamma \qquad (14.4.65)$$

since $(a+b)^{-2\beta} \le a^{-\beta} b^{-\beta}, 0 < \beta < 1, a > 0, b > 0$. Combining (14.4.64) and (14.6.65), we obtain

$$|h_1(t,\tau) - h_1(\eta,\tau)| \le C|s|_\gamma^2 (\eta - \tau)^{\gamma-1} (t-\eta)^\gamma. \qquad (14.4.66)$$

Since $\frac{1}{2} < \gamma \le 1$, Lemma 14.4.1 yields the existence of

$$\frac{d}{dt} z_1(t,\tau) = \frac{d}{dt} \left\{ \int_\tau^t (t-\eta)^{-1/2} (\eta - \tau)^{-1/2} \exp\left\{ -\frac{(s(\eta) - s(\tau))^2}{4(\eta - \tau)} \right\} d\eta \right\}, \qquad (14.4.67)$$

its boundedness by Const. $(t-\tau)^{2\gamma-2}$ and its Hölder continuity for some ε sufficiently small.

We consider now the function

$$z_\varphi(s(t),t) = z_1(s(t),t) + z_{2,\varphi}(s(t),t), \qquad (14.4.68)$$

where

$$z_1(s(t),t) = -2 \int_0^t \frac{\partial K}{\partial x}(s(t), t-\tau) g(\tau) \, d\tau, \qquad (14.4.69)$$

and

$$z_{2,\varphi}(s(t),t) = 2 \int_0^t \frac{\partial K}{\partial x}(s(t), t-\sigma) \int_0^\sigma K(s(\sigma), \sigma - \tau)\varphi(\tau) \, d\tau d\sigma$$

$$= \int_0^t \left\{ 2 \int_\tau^t \frac{\partial K}{\partial x}(s(t), t-\sigma) K(s(\sigma), \sigma - \tau) \, d\sigma \right\} \varphi(\tau) \, d\tau$$

$$= \int_0^t L(t,\tau)\varphi(\tau) \, d\tau, \qquad (14.4.70)$$

where

$$L(t,\tau) = 2 \int_\tau^t \frac{\partial K}{\partial x}(s(t), t-\sigma) K(s(\sigma), \sigma - \tau) \, d\sigma. \qquad (14.4.71)$$

By the linearity of A and z_φ,

$$Az_\varphi = Az_1 + Az_{2,\varphi}. \qquad (14.4.72)$$

We consider first

$$(Az_{2,\varphi})(t) = \pi^{-1} \frac{d}{dt} \int_0^t (t-\eta)^{-1/2} \int_0^\eta L(\eta,\tau)\varphi(\tau)\,d\tau\,d\eta$$

$$= \pi^{-1/2} \frac{d}{dt} \int_0^t \varphi(\tau)\left[\int_\tau^t (t-\eta)^{-1/2} L(\eta,\tau)\,d\eta \right] d\tau. \qquad (14.4.73)$$

Let

$$h_2(\eta,\tau) = (\eta-\tau)^{1/2} L(\eta,\tau). \qquad (14.4.74)$$

Then,

$$h_2(t,\tau) - h_2(\eta,\tau) = (t-\tau)^{1/2} L(t,\tau) - (\eta-\tau)^{1/2} L(\eta,\tau)$$

$$= (t-\tau)^{1/2} \{ L(t,\tau) - L(\eta,\tau) \}$$

$$+ L(\eta,\tau)\{ (t-\tau)^{1/2} - (\eta-\tau)^{1/2} \}. \qquad (14.4.75)$$

Thus,

$$L(t,\tau) - L(\eta,\tau) = 2 \int_\eta^t \frac{\partial K}{\partial x}(s(t), t-\sigma) K(s(\sigma), \sigma-\tau)\,d\sigma$$

$$+ 2 \int_\tau^\eta \left\{ \frac{\partial K}{\partial x}(s(t), t-\sigma) - \frac{\partial K}{\partial x}(s(\eta), \eta-\sigma) \right\}$$

$$\cdot K(s(\sigma), \sigma-\tau)\,d\sigma = I_1 + I_2. \qquad (14.4.76)$$

From (14.4.4), we see that there exists a positive constant $C_1 = C_1(p)$ such that

$$\left| \frac{\partial K}{\partial x}(s(t), t-\sigma) \right| \le C_1 (t-\sigma)^p \qquad (14.4.77)$$

for any exponent p. Hence,

$$|I_1| \le C_2 (\eta-\tau)^{1/2}(t-\eta)^p \qquad (14.4.78)$$

for any exponent p. Next,

$$\frac{\partial K}{\partial x}(s(t), t-\sigma) - \frac{\partial K}{\partial x}(s(\eta), \eta-\sigma)$$

$$= \int_0^1 \frac{d}{d\theta} \left\{ \frac{\partial K}{\partial x}(\theta s(t) + (1-\theta)s(\eta), \quad \theta t + (1-\theta)\eta - \sigma) \right\} d\theta$$

$$= \int_0^1 \left[[s(t) - s(\eta)] \frac{\partial^2 K}{\partial x^2} + \frac{\partial^2 K}{\partial x\,\partial t}(t-\eta) \right] d\theta. \qquad (14.4.79)$$

Since $\partial^2 K/\partial x^2$ and $\partial^2 K/(\partial x\,\partial t)$ are uniformly bounded for $x \geq \delta_1$, $0 \leq \sigma \leq \theta t + (1-\theta)\eta \leq T$,

$$\left|\frac{\partial K}{\partial x}(s(t), t-\sigma) - \frac{\partial K}{\partial x}(s(\eta), \eta-\sigma)\right| \leq C_3\{|t-\eta|^\gamma + |t-\eta|\}$$

$$\leq C_4|t-\eta|^\gamma. \tag{14.4.80}$$

Hence, we see that

$$I_2 \leq C_5(t-\eta)^\gamma(\eta-\tau)^{1/2}. \tag{14.4.81}$$

Combining (14.4.78) and (14.4.81) for I_1 and I_2, we see that

$$(t-\tau)^{1/2}|L(t,\tau) - L(\eta,\tau)|$$

$$\leq C_2(\eta-\tau)^{1/2}(t-\tau)^{1/2}(t-\eta)^p + C_5(t-\eta)^\gamma(t-\tau)^{1/2}(\eta-\tau)^{1/2}$$

$$\leq C_2(\eta-\tau)^{-1/4}(t-\eta)^p + C_5(\eta-\tau)^{(1/2)+(1/4)}(\eta-\tau)^{-1/4}$$

$$\cdot(t-\tau)^{1/2}(t-\eta)^\gamma \leq C_6(\eta-\tau)^{-1/4}(t-\eta)^\gamma. \tag{14.4.82}$$

Again, since $s(t) \geq \delta_1 > 0$, there exists for each positive p, $C_7 = C_7(p)$ such that

$$\left|\frac{\partial K}{\partial x}(s(t), t-\sigma)\right| \leq C_7(t-\sigma)^p \tag{14.4.83}$$

for any positive p. Hence,

$$|L(\eta,\tau)| \leq C_8(\eta-\tau)^p \tag{14.4.84}$$

for any positive p. Now,

$$\left\{(t-\tau)^{1/2} - (\eta-\tau)^{1/2}\right\} = \int_0^1 \frac{1}{2}\left\{(\theta t + (1-\theta)\eta - \tau)^{-1/2}\right\} \cdot (t-\eta)\,d\theta \tag{14.4.85}$$

and

$$(\theta t + (1-\theta)\eta - \tau) = (\theta(t-\tau) + (1-\theta)(\eta-\tau)). \tag{14.4.86}$$

Hence,

$$|\{(t-\tau)^{1/2} - (\eta-\tau)^{1/2}\}| \leq (t-\eta)(t-\tau)^{-1/2}\int_0^1 \frac{1}{2}\theta^{-1/2}\,d\theta$$

$$\leq (t-\eta)(t-\tau)^{-1/2} \leq (t-\eta)(\eta-\tau)^{-1/2}. \tag{14.4.87}$$

Thus, from (14.4.83) and (14.4.87), we see that

$$|L(\eta,\tau)|\,|\{(t-\tau)^{1/2} - (\eta-\tau)^{1/2}\}| \leq C_8(t-\eta)(\eta-\tau)^{p-(1/2)}$$

$$\leq C_9(t-\eta)^\gamma(\eta-\tau)^{-1/4}. \tag{14.4.88}$$

Combining (14.4.82) and (14.4.88), we see that

$$|h_2(t, \tau) - h_2(\eta, \tau)| \le C_{10}(t - \eta)^{\gamma}(\eta - \tau)^{-1/4}. \qquad (14.4.89)$$

Also, from (14.4.84) we see that

$$\lim_{\tau \uparrow t} \int_{\tau}^{t} (t - \tau)^{-1/2} L(\eta, \tau) \, d\eta = 0. \qquad (14.4.90)$$

Thus, by Lemma 14.4.1, it follows from (14.4.89) that

$$(Az_{2, \varphi})(t) = \pi^{-1} \int_0^t \varphi(\tau) \frac{d}{dt} \left\{ \int_{\tau}^t (t - \eta)^{-1/2}(\eta - \tau)^{-1/2} h_2(\eta, \tau) \, d\eta \right\} d\tau,$$

$$(14.4.91)$$

where

$$\frac{d}{dt} z_2(t, \tau) = \frac{d}{dt} \int_{\tau}^t (t - \eta)^{-1/2}(\eta - \tau)^{-1/2} h_2(\eta, \tau) \, d\eta \qquad (14.4.92)$$

exists, satisfies

$$\left| \frac{d}{dt} z_2(t, \tau) \right| \le C_{11}(t - \tau)^{\gamma - (1/4) - 1}, \qquad (14.4.93)$$

and is Hölder continuous for exponent ε sufficiently small.

It is of interest to summarize the results thus far in the following statement.

Lemma 14.4.2. *For continuous φ,*

$$A(z_{2, \varphi} + w_{\varphi})(t) = 2^{-1} \pi^{-1/2} \varphi(t) + \int_0^t H(t, \tau) \varphi(\tau) \, d\tau, \qquad (14.4.94)$$

where $H(t, \tau)$ is a continuous kernel such that

$$|H(t, \tau)| \le C(t - \tau)^{\alpha - 1}, \qquad (14.4.95)$$

for $C = C(T, \gamma, |s|_{\gamma})$ and $\alpha = \alpha(\gamma) > 0$ and such that

$$|H(t + \delta, \tau) - H(t, \tau)| \le C_{\varepsilon}(t - \tau)^{\alpha - 1 - c\varepsilon} \delta^{\varepsilon}, \qquad (14.4.96)$$

where $c = c(\gamma) > 0$, $C_{\varepsilon} = C_{\varepsilon}(c, \varepsilon)$, and $\varepsilon > 0$ with $\alpha - c\varepsilon > 0$.

Now, we turn to

$$(Az_1)(t) = -\frac{2}{\pi} \frac{d}{dt} \int_0^t (t - \eta)^{-1/2} \int_0^{\eta} \frac{\partial K}{\partial x}(s(\eta), \eta - \tau) g(\tau) \, d\tau \, d\eta$$

$$= -\frac{2}{\pi} \frac{d}{dt} \int_0^t g(\tau) \left\{ \int_{\tau}^t (t - \eta)^{-1/2} \frac{\partial K}{\partial x}(s(\eta), \eta - \tau) \, d\eta \right\} d\tau.$$

$$(14.4.97)$$

Let

$$h_3(\eta, \tau) = (\eta - \tau)^{1/2} \frac{\partial K}{\partial x}(s(\eta), \eta - \tau). \qquad (14.4.98)$$

The techniques for estimating $h_2(\eta, \tau)$ above apply directly here and yield a kernel $H_1(t, \tau)$ with properties similar to $H(t, \tau)$ such that

$$(A z_1)(t) = \int_0^t H_1(t, \tau) g(\tau)\, d\tau. \tag{14.4.99}$$

Now, it is no loss of generality to assume that $f(b) = h(0) = 0$ since the addition or subtraction of a constant cannot affect the differentiability of w. Also, we assume that $f \in C([0, b])$ and that

$$|f(\xi)| = |f(\xi) - f(b)| \le C_f |\xi - b|^\beta.$$

Applying the operator A to $v(s(t), t)$ yields

$$(A v)(t) = \frac{1}{\pi} \frac{d}{dt} \int_0^t (t - \eta)^{-1/2} v(s(\eta), \eta)\, d\eta. \tag{14.4.100}$$

We note that this integral is in the form of the other integrals above for $\tau = 0$. Consequently, we define

$$h_4(\eta, 0) = \eta^{1/2} v(s(\eta), \eta) \tag{14.4.101}$$

and turn to its analysis to demonstrate the applicability of Lemma 14.4.1. Consider

$$h_4(t, 0) - h_4(\eta, 0) = t^{1/2} v(s(t), t) - \eta^{1/2} v(s(\eta), \eta)$$

$$= \int_0^1 \frac{d}{d\theta} \left\{ \theta t + (1 - \theta)\eta \right)^{1/2} v(\theta s(t) + 1 - \theta) s(\eta), \theta t + (1 - \theta)\eta \right\}\, d\theta. \tag{14.4.102}$$

Differentiating with respect to θ, we obtain a three-termed expression

$$V = \tfrac{1}{2} (\theta t + (1 - \theta)\eta)^{-1/2} v \cdot (t - \eta) + (\theta t + (1 - \theta)\eta)^{1/2} v_x \cdot (s(t) - s(\eta))$$

$$+ (\theta t + (1 - \theta)\eta)^{1/2} v_t \cdot (t - \eta), \tag{14.4.103}$$

where v, v_x, and v_t are evaluated at the point $(\theta s(t) + (1 - \theta) s(\eta), \theta t + (1 - \theta)\eta)$.

It is convenient here to estimate v, v_x, and v_t before estimating V.

Lemma 14.4.3. *For $f \in C([0, b])$ with $f(b) = 0$ and*

$$|f(\xi)| = |f(\xi) - f(b)| < C_f |\xi - b|^\beta,$$

$0 < \beta \le 1$, $C_f > 0$, $|x - b| < |s|_\gamma t^\gamma$, *and* $x \ge \delta_1 > 0$, *there exist positive constants* C_1, C_2, *and* C_3 *such that*

$$|v(x, t)| \le C_1 \{ \|f\|_{[0, b]} + C_f \} t^{\beta/2}, \tag{14.4.104}$$

$$\left| \frac{\partial v}{\partial x}(x, t) \right| \le C_2 \{ \|f\|_{[0, b]} + C_f \} t^{(\beta/2) - (1/2)} \tag{14.4.105}$$

and

$$\left|\frac{\partial v}{\partial t}(x,t)\right| \le C_3\{\|f\|_{[0,b]} + C_f\}t^{(\beta/2)-1}, \tag{14.4.106}$$

where the constants depend upon $|s|_\gamma$.

   Proof. As

$$|f(\xi)| = |f(\xi) - f(b)| \le C_f|\xi - b|^\beta$$

$$\le C_f\{|x - \xi|^\beta + |s|_\gamma^\beta t^{\gamma\beta}\} \tag{14.4.107}$$

and as

$$\exp\left\{-\frac{(x+\xi)^2}{4t}\right\} \le \exp\left\{-\frac{\delta_1^2}{4t}\right\}, \tag{14.4.108}$$

we need only estimate

$$I_1 = \int_0^b K(x - \xi, t)|x - \xi|^\beta\,d\xi, \tag{14.4.109}$$

$$I_2 = \int_0^b \frac{\partial K}{\partial x}(x - \xi, t)|x - \xi|^\beta\,d\xi, \tag{14.4.110}$$

$$I_3 = \int_0^b \frac{\partial K}{\partial t}(x - \xi, t)|x - \xi|^\beta\,d\xi \tag{14.4.111}$$

and similar terms with $|x - \xi|^\beta$ replaced by $t^{\gamma\beta}$. Using the transformation $\rho = (x - \xi)/2t^{1/2}$, we see immediately that

$$|I_1| \le C_1 t^{\beta/2}, \tag{14.4.112}$$

and

$$|I_2| \le C_2 t^{(\beta/2)-1/2} \tag{14.4.113}$$

$$|I_3| \le C_3 t^{(\beta/2)-1}. \tag{14.4.114}$$

Since $\gamma > \frac{1}{2}$, the above estimates hold for the terms involving $t^{\gamma\beta}$. Employing the fact that

$$\exp\left\{-\frac{\delta_1^2}{4t}\right\} \le p!\left(\frac{4t}{\delta_1^2}\right)^p, \tag{14.4.115}$$

we easily estimate the second term in the Green's function and obtain the results. □

   As $x = \theta s(t) + (1 - \theta)s(\eta)$ satisfies the Lemma 14.4.3, we replace t by $(\theta t + (1 - \theta)\eta)$ and see that there exists a constant C_4 such that

$$|V| \le C_4\{(t - \eta)(\theta(t - \eta) + \eta)^{(\beta/2)-(1/2)} + (t - \eta)^\gamma(\theta(t - \eta) + \eta)^{\beta/2}\}$$

$$\le C_5\{(t - \eta)\eta^{-((1/2)-(\beta/2))} + (t - \eta)^\gamma\}, \tag{14.4.116}$$

where $C_5 = C_5(|s|_\gamma, T, \beta)$. Combining this with (14.4.102), we see that

$$|h_4(t,0) - h_4(\eta,0)| \leq C_5 \left\{ (t - \eta)\eta^{-((1/2)-(\beta/2))} + (t - \eta)^\gamma \right\}.$$

$$(14.4.117)$$

Consequently, $(Av)(t)$ exists and is Hölder continuous for exponent $\varepsilon > 0$ and ε sufficient small. Moreover, $(Av)(t) \in C^0_{(\nu)}((0, T])$ for $\nu = \min(\gamma, (\beta + 1)/2)$ and $(Av)(t) \in C^\varepsilon_{(\nu_\varepsilon)}((0, T])$ for $\nu_\varepsilon = \nu - (c - 1)\varepsilon$. Here, c depends upon γ and β.

Considering now $h(t)$ and defining

$$h_5(\eta, 0) = \eta^{1/2} h(\eta),$$

$$(14.4.118)$$

it follows that, if $h \in C^{\gamma_1}([0, T])$, $\gamma_1 > \frac{1}{2}$, then

$$|h_5(t,0) - h_5(\eta,0)| \leq C_1(t - \eta)^{\gamma_1} + C_2(t - \eta)\eta^{\gamma_1 - (1/2)}. \quad (14.4.119)$$

This implies, via Lemma 14.4.1, that $(Ah)(t)$ exists and is Hölder continuous for exponent $\varepsilon > 0$ and sufficiently small. Moreover, $(Ah)(t) \in C^0_{(\gamma_1)}((0, T])$ and $(Ah)(t) \in C^\varepsilon_{(\gamma_\varepsilon)}((0, T])$, where $\gamma_\varepsilon = \gamma_1 - (c - 1)\varepsilon$ and where here c depends upon γ_1.

Collecting all of the results of the above analysis, we can summarize it in the following statement.

Lemma 14.4.4. *If* $s \in C^\gamma([0, T])$, $\frac{1}{2} < \gamma \leq 1$, $g \in C^0_{(\nu)}((0, T])$, $0 < \nu \leq 1$, $h \in C^{\gamma_1}([0, T])$, $\gamma_1 > \frac{1}{2}$, $f \in C([0, b])$, $h(0) = f(b) = 0$, $|f(\xi)| < C_f |\xi - b|^\beta$, $0 < \beta \leq 1$, $C_f > 0$, and $\varphi \in C^0_{(\nu_1)}((0, T])$, $0 < \nu_1 \leq 1$, then the Abel operator*

$$(AF)(t) = \pi^{-1} \frac{d}{dt} \int_0^t (t - \eta)^{-1/2} F(\eta)\, d\eta \qquad (14.4.120)$$

can be applied to the integral equation (see Eqs. (14.4.5) through (14.4.8)):

$$h(t) = v(s(t), t) + z_\varphi(s(t), t) + w_\varphi(s(t), t; s), \quad (14.4.121)$$

and an equivalent Volterra integral equation of the second kind

$$\varphi(t) = G(t) + \int_0^t H(t, \tau)\varphi(\tau)\, d\tau \qquad (14.4.122)$$

is obtained. Moreover, the data $G \in C^0_{(\nu_2)}((0, T]) \cap C^\varepsilon_{(\nu_\varepsilon)}((0, T))$, *where* $\nu_2 = \nu_2(\nu, \gamma, \gamma_1, \beta)$, $\nu_\varepsilon = \nu - c(\nu, \gamma_1, \beta, \gamma)\varepsilon$, $\varepsilon > 0$, *and the kernel* $H(t, \tau)$ *satisfies the properties stated in Lemma 14.4.2.*

Proof. The equivalence of (14.4.121) and (14.4.122) follows from the reversibility of the application of A. For example, $AF \equiv 0$ implies that

$$B(t) \equiv \int_0^t (t - \eta)^{-1/2} F(\eta)\, d\eta = 0 \quad \text{if } \lim_{t \downarrow 0} \int_0^t (t - \eta)^{-1/2} F(\eta)\, d\eta = 0.$$

Then,

$$(AB)(t) = \frac{d}{dt} \int_0^t F(\eta) \, d\eta = 0, \qquad (14.4.123)$$

whence follows

$$F(t) = 0. \qquad (14.4.124)$$

Considering each term in (14.4.121), we see that all are $\mathcal{O}(t^\delta)$, $\delta > 0$. Hence

$$\int_0^t (t - \eta)^{-1/2} \eta^\delta \, d\eta = \mathcal{O}(t^{\delta + (1/2)}) \qquad (14.4.125)$$

and the equivalence is complete.

When we consider the Banach space $C_{(\nu_2)}^0((0, T])$, we see that

$$B_\varphi(t) = G(t) + \int_0^t H(t, \tau) \varphi(\tau) \, d\tau \qquad (14.4.126)$$

is a contraction of $C_{(\nu_2)}^0((0, T])$ into itself for T sufficiently small. Hence, there exists a unique solution $\varphi \in C_{(\nu_2)}^0((0, T])$ of (14.4.126) for T sufficiently small. However, the techniques of Chapter 8 apply, to yield a unique solution $\varphi \in C_{(\nu_2)}^0((0, T])$ for any $T > 0$.

We summarize the results above in the following statement.

Theorem 14.4.1. *For the solution w of*

$$\begin{aligned}
w_t &= w_{xx}, & 0 < x < s(t), & \quad 0 < t \leq T, \\
w(x, 0) &= f(x), & 0 < x < b, & \quad s(0) = b, \\
w(0, t) &= g(t), & 0 < t \leq T, & \\
w(s(t), t) &= h(t), & 0 < t \leq T, &
\end{aligned} \qquad (14.4.127)$$

where $s \in C^\gamma([0, T])$, $\gamma > \frac{1}{2}$, $\delta_1 = \inf_{0 \leq t \leq T} s(t) > 0$, $g \in C_{(\nu)}^0((0, T])$, $0 < \nu \leq 1$, $h \in C^{\gamma_1}([0, T])$, $\gamma_1 > \frac{1}{2}$, $f \in C([0, b])$, $h(0) = f(b) = 0$, and

$$|f(\xi)| = |f(\xi) - f(b)| \leq C_f |\xi - b|^\beta, \qquad 0 < \beta \leq 1,$$

where C_f is a positive constant,

$$\lim_{x \uparrow s(t)} w_x(x, t) \qquad (14.4.128)$$

exists. Moreover, $w_x(x, t)$ is two-dimensionally continuous at $x = s(t)$ for $0 < t \leq T$.

Corollary 14.4.1. *The condition of $h(0) = f(b) = 0$ can be replaced with $h(0) = f(b)$ and the Theorem 14.4.1 still holds.*

Corollary 14.4.2. *For T sufficiently small, the solution u of Problem (14.4.1) possesses a two-dimensionally continuous u_x at $x = s_i(t)$, $i = 1, 2$, $0 < t \leq T$ if $s_i \in C^\gamma([0, T])$, $i = 1, 2$, $\gamma > \frac{1}{2}$, $\delta = \inf_{0 \leq t \leq T} |s_1(t) - s_2(t)| > 0$, $f \in C([a, b])$, $|f(\xi) - f(a)| \leq C_f |\xi - a|^\beta$, $|f(\xi) - f(b)| \leq C_f |\xi - b|^\beta$, $0 < \beta \leq 1$, $C_f > 0$, $g \in C^{\gamma_1}([0, T])$, $h \in C^{\gamma_1}([0, T])$, $\gamma_1 > \frac{1}{2}$, $f(a) = g(0)$, and $f(b) = h(0)$.*

Proof. The interval $[a, b]$ can be split in two at $(a + b)/2$. Simple affine transformation reduces the case of each boundary to that of Theorem 14.4.1. Since u is continuous at $t = 0$, $u[(a + b)/2, t] \in C^0_{(\nu)}((0, T])$ for $\nu = 1$.

□

Corollary 14.4.3. *Under the hypotheses of Corollary 14.4.2, with the exception of the limitation of the size of T, the solution u of Problem (14.4.1) possesses a two-dimensionally continuous u_x at $x = s_i(t)$, $i = 1, 2$, $0 < t \leq T$.*

Proof. The existence and continuity of u_x at $x = s_i(t)$, $i = 1, 2$, is equivalent to the problem of representing u in the form

$$u(x, t) = v(x, t) + \frac{\partial w_{\varphi_1}}{\partial x}(x, t; s_1) + w_{\varphi_2}(x, t, s_2), \quad (14.4.129)$$

which reduces to the pair of integral equations

$$g(t) = v(s_1(t), t) + \frac{1}{2}\varphi_1(t) + \int_0^t \frac{\partial K}{\partial x}(s_1(t) - s_1(\tau), t - \tau)\varphi_1(\tau)\, d\tau$$

$$+ \int_0^t K(s_1(t) - s_2(\tau), t - \tau)\varphi_2(\tau)\, d\tau,$$

(14.4.130)

$$h(t) = v(s_2(t), t) + \int_0^t \frac{\partial K}{\partial x}(s_2(t) - s_1(\tau), t - \tau)\varphi_1(\tau)\, d\tau$$

$$+ \int_0^t K(s_2(t) - s_2(\tau), t - \tau)\varphi_2(\tau)\, d\tau,$$

where

$$v(x, t) = \int_a^b K(x - \xi, t) f(\xi)\, d\xi. \quad (14.4.131)$$

The analysis leading up to Theorem 14.4.1 can be applied with the result that the second equation can be converted into an equivalent form,

$$\phi_2(t) = G(t) + \int_0^t H_1(t, \tau)\phi_1(\tau)\, d\tau + \int_0^t H_2(t, \tau)\phi_2(\tau)\, d\tau,$$

(14.4.132)

via the Abel transformation operator.

□

Corollary 14.4.4. *Under the hypotheses on s_i, $i = 1, 2$, g, and h of Corollary 14.4.2, if, for some t_0, $0 < t_0 < T$,*

$$\int_{s_1(t_0)}^{s_2(t_0)} [u_x(x, t_0)]^2\, dx < \infty, \quad (14.4.133)$$

then $u(x, t_0) \in C^{1/2}([s_1(t_0), s_2(t_0)])$ and u_x exists and is two-dimensionally continuous at $x = s_i(t)$, $i = 1, 2$, for $t_0 < t \leq T$.

Proof. The result follows from

$$|u(x_1,t_0)-u(x_2,t_0)| \le \int_{x_1}^{x_2} |u_x(x,t_0)|\, dx \le \left(\int_{x_1}^{x_2} |u_x|^2\, dx \right)^{1/2} |x_1-x_2|^{1/2}$$

$$(14.4.134)$$

and Corollary 14.4.3.   □

EXERCISES

14.1. Specify conditions on the data s_1, s_2, f, g, and h that guarantee existence of a solution u of the problem

$$u_t = u_{xx}, \qquad s_1(t) < x < s_2(t), \qquad 0 < t \le T,$$
$$u(x,0) = f(x), \qquad a < x < b, \qquad s_1(0) = a, \qquad s_2(0) = b,$$
$$u_x(s_1(t),t) = g(t), \qquad 0 < t \le T, \qquad \text{(A)}$$
$$u_x(s_2(t),t) = h(t), \qquad 0 < t \le T,$$

via a reduction to a system of integral equations.

14.2. Demonstrate the uniqueness of the solution to Problem (A).

14.3. Repeat Exercises 14.1 and 14.2 for the problem

$$u_t = x_{xx}, \qquad s_1(t) < x < s_2(t), \qquad 0 < t \le T,$$
$$u(x,0) = f(x), \qquad a < x < b, \qquad s_1(0) = a, \qquad s_2(0) = b,$$
$$u_x(s_1(t),t) = G(t, u(s_1(t),t)), \qquad 0 < t \le T, \qquad \text{(B)}$$
$$u_x(s_2(t),t) = H(t, u(s_2(t),t)), \qquad 0 < t \le T.$$

14.4. Assume that $s_i \in C^\gamma((-\infty,\infty))$, $\gamma > 1/2$, $i=1,2$, with $|s_i|_\gamma < \infty$, $i=1,2$, $\delta = \inf_{-\infty < t < \infty} |s_1(t) - s_2(t)| > 0$ and $-B \le s_1(t) < s_2(t) \le B$, $-\infty < t < \infty$, where B is a positive constant. Show that the problem

$$u_t = u_{xx}, \qquad s_1(t) < x < s_2(t), \qquad -\infty < t < \infty,$$
$$u(x_1(t),t) = g(t), \qquad -\infty < t < \infty, \qquad \text{(C)}$$
$$u(s_2(t),t) = h(t), \qquad -\infty < t < \infty,$$

possesses a solution u provided that g and h are continuous functions that satisfy

$$\sup_{-\infty < t \le T} |g(t)| < C \qquad \text{(D)}$$

and

$$\sup_{-\infty < t \le T} |h(t)| < C, \qquad \text{(E)}$$

where C is a positive constant and T is some negative constant. Show that u satisfies

$$\sup_{\substack{s_1(t)<x<s_2(t)\\-\infty<t<T}} |u(x,t)| < C \tag{F}$$

and u is unique within the class of solutions v such that

$$\lim_{t\to-\infty}\sup\left[\left\{\sup_{s_1(t)<x<s_2(t)}|v(x,t)|\right\}\exp\left\{\frac{\pi^2}{4B^2}t\right\}\right]=0. \tag{G}$$

14.5. Show that if s_i, $i=1,2$, g, and h are periodic with period P, then the solution u of Problem (C) is periodic in t with period P.

14.6. Let $\{s_i^{(n)}\}$, $i=1,2$, denote two sequences of functions such that $s_i^{(n)}\in C^\gamma([0,T])$, $i=1,2$, $n=1,2,\dots$, $\frac12<\gamma\le1$, and that

$$|s_i^{(n)}|_\gamma\le m, \qquad i=1,2,\dots \tag{H}$$

We assume also that

$$\lim_{n\to\infty}s_i^{(n)}(t)=s_i(t), \qquad i=1,2, \tag{I}$$

uniformly for $0\le t\le T$. This necessarily implies that $s_i\in C^\gamma([0,T])$, $\frac12<\gamma\le1$, $i=1,2$, and that

$$|s_i|_\gamma\le m, \qquad i=1,2. \tag{J}$$

Let $u=u(x,t)$ denote the bounded solution of

$$u_t=u_{xx}, \qquad s_1(t)<x<s_2(t), \qquad 0<t\le T,$$
$$u(x,0)=f(x), \qquad a<x<b, \qquad s_1(0)=a, \qquad s_2(0)=b,$$
$$u(s_1(t),t)=g(t), \qquad 0<t\le T, \tag{K}$$
$$u(s_2(t),t)=h(t), \qquad 0<t\le T,$$

where $f\in C(-\infty,\infty)$, $g\in C_0^{(1)}((0,T])$ and $h\in C_0^{(1)}((0,T])$. Let $u^{(n)}=u^{(n)}(x,t)$ denote the bounded solution of Problem (K) with $s_i^{(n)}$ substituted for s_i.

Show that for $s_1(t)<x<s_2(t)$, $0<t\le T$,

$$\lim_{n\to\infty}u^{(n)}(x,t)=u(x,t), \tag{L}$$

where the $u^{(n)}(x,t)=g(t)$, $x<s_1^{(n)}(t)$, and $u^{(n)}(x,t)=h(t)$, $x>s_2^{(n)}(t)$, so that every $u^{(n)}$ is defined throughout the domain $D_T=\{(x,t)|s_1(t)<x<s_2(t),0<t\le T\}$. Here, it is assumed that

$$\delta=\inf_{0\le t\le T}|s_1(t)-s_2(t)|>0 \tag{M}$$

and

$$\Delta=\sup_{0\le t\le T}|s_1(t)-s_2(t)|<\infty. \tag{N}$$

14.7.   As a corollary of Exercise 14.6, show that, on each compact subset of

$$D_T = \{(x, t) | s_1(t) < x < s_2(t), \quad 0 < t \leq T\},$$

the $\lim_{n \to \infty} u^{(n)}(x, t) = u(x, t)$ is uniform.

NOTES

The solution of initial-boundary-value problems in domains with boundaries that are Hölder continuous with exponent greater than $1/2$ was presented by Gevrey [2]. The presentation in Secs. 14.1–14.4 was based upon the work of Cannon, Henry, and Kotlow [1]. The remaining material is a straightforward application of ideas from previous chapters.

REFERENCES

1.   Cannon, J. R., Henry, D. B., and Kotlow, D. B., Classical solutions of the one-dimensional two-phase Stefan problem, *Ann. di Mat. Pura ed. Appl.* (IV), **107** (1976), 311–341.
2.   Gevrey, M., Sur les équations aux dérivées partielles du type parabolique, *J. Math. Pures Appl.*, **9** (1913), 305–471.

Chapter 15

Some Properties of Solutions in General Domains

Recalling the discussion of Sec. 1.5 concerning parabolic domains D_T and their boundaries B_T, we shall study some properties of solutions of $u_t = u_{xx}$ in D_T that are continuous in $D_T \cup B_T$.

15.1. UNIFORM CONVERGENCE OF SOLUTIONS

We shall demonstrate the following result.

Theorem 15.1.1. *If u_n, $n = 1, 2, \ldots$, satisfy the heat equation in D_T and the u_n, $n = 1, 2, \ldots$, are uniformly convergent on B_T, then u_n, $n = 1, 2, 3, \ldots$, converge uniformly in $D_T \cup B_T$ to a function u that satisfies the heat equation in D_T and is continuous in $D_T \cup B_T$.*

Proof. Since u_n, $n = 1, 2, 3, \ldots$, are uniformly convergent on B_T, for each $\varepsilon > 0$, there exists N_ε such that for all $m, n > N_\varepsilon$,

$$\sup_{B_T} |u_n - u_m| < \varepsilon. \tag{15.1.1}$$

As u_n, $n = 1, 2, \ldots$, satisfy the heat equation in D_T, it follows from the Maximum Principle that

$$\sup_{D_T \cup B_T} |u_n - u_m| < \varepsilon. \tag{15.1.2}$$

Hence, u_n, $n = 1, 2, 3, \ldots$, form a uniform Cauchy sequence in $D_T \cup B_T$ and converge uniformly to a function u that is continuous in $D_T \cup B_T$.

In order to show that u satisfies the heat equation in D_T, we select $(x_0, t_0) \in D_T$. There exists a $\delta > 0$ such that the rectangle R_δ

$$R_\delta = \{(x, t) | x_0 - \delta \le x \le x_0 + \delta, t_0 - \delta \le t \le t_0\} \subset D_T.$$

Under the change of variables

$$\xi = \frac{x - (x_0 - \delta)}{2\delta} \quad \text{and} \quad \frac{t - (t_0 - \delta)}{4\delta^2},$$

R_δ is transformed into

$$R_* = \left\{(\xi, \tau) | 0 \le \xi \le 1, \quad 0 \le \tau \le \frac{1}{4\delta}\right\}.$$

Also, the solutions u_n are transformed into

$$U_n(\xi, \tau) = u_n(2\delta\xi + (x_0 - \delta), 4\delta^2\tau + (t_0 - \delta)),$$

which satisfy $\partial U_n / \partial \tau = \partial^2 U_n / \partial \xi^2$ in the parabolic interior of R_*, are continuous in R_*, and converge uniformly to

$$U(\xi, \tau) = u(2\delta\xi + (x_0 - \delta), 4\delta^2\tau + (t_0 - \delta))$$

in R_*. From Theorem 6.3.1,

$$U_n(\xi, \tau) = \int_0^1 \{\theta(\xi - \zeta, t) - \theta(\xi + \zeta, t)\} U_n(\zeta, 0) \, d\zeta$$

$$-2\int_0^\tau \frac{\partial \theta}{\partial \xi}(\xi, \tau - \eta) U_n(0, \eta) \, d\eta + 2\int_0^\tau \frac{\partial \theta}{\partial \xi}(\xi - 1, \tau - \eta) U_n(1, \eta) \, d\eta$$

holds for $0 < \xi < 1$, $0 < \tau \le 1/4\delta$. From the uniform convergence of $U_n(\xi, \tau)$ to $U(\xi, \tau)$, we see that U possesses the same representation for $0 < \xi < 1$, $0 < \tau \le 1/4\delta$, which implies that $\partial U / \partial \tau = \partial^2 U / \partial \xi^2$ in the parabolic interior of R_*. Carrying out the inverse transformation, we see that u satisfies the heat equation in the parabolic interior of R_δ and consequently at (x_0, t_0). As (x_0, t_0) is any point in D_T, we conclude that u satisfies the heat equation throughout D_T. □

We turn now to the consideration of when a sequence of solutions behaves as a normal family and possesses uniformly convergent subsequences on compact subsets of D_T. From Theorem 15.1.1, we see that such limits necessarily satisfy the heat equation on those subsets.

Theorem 15.1.2. *If u_n, $n = 1, 2, 3, \ldots$, satisfy the heat equation in D_T, are continuous in $D_T \cup B_T$, and there exists a positive constant C independent of n such that*

$$|u_n(x, t)| \le C, \qquad n = 1, 2, 3, \ldots, \tag{15.1.3}$$

for all $(x, t) \in D_T \cup B_T$, then there exists a subsequence u_{n_j}, $j = 1, 2, \ldots$, which converges pointwise to a function u defined over D_T. Moreover, the convergence is uniform on compact subsets of D_T, u satisfies the heat equation in D_T, and $|u(x, t)| < C$ for $(x, t) \in D_T$.

Proof. Let $D_T^{(m)} m = 1, 2, \ldots$, denote a sequence of subdomains of D_T each of which possesses a boundary $B_T^{(m)}$, which consists of two Hölder continuous boundary curves $s_i^{(m)}(t)$, $i = 1, 2$, and an interval on the characteristic $t = m^{-1}$ such that $D_T = \bigcup_{m=1}^{\infty} D_T^{(m)}$. Note that such a sequence of subdomains can be constructed so that $D_T^{(m)} \cup B_T^{(m)} \subset D_T^{(m+1)}$, $m = 1, 2, \ldots$.

Consider now the sequence of solutions u_n, $n = 1, 2, 3, \ldots$, on the subdomain $D_T^{(m+1)}$. Now, in $D_T^{(m+1)}$

$$u_n(x, t) = \int_{-\infty}^{\infty} K\left(x - \xi, t - (m+1)^{-1}\right) \tilde{u}_n\left(\xi, (m+1)^{-1}\right) d\xi$$

$$+ \int_{(m+1)^{-1}}^{t} \frac{\partial K}{\partial x}\left(x - s_1^{(m+1)}(\tau), t - \tau\right) \varphi_1^{(n)}(\tau) d\tau$$

$$+ \int_{(m+1)^{-1}}^{t} \frac{\partial K}{\partial x}\left(x - s_2^{(m+1)}(\tau), t - \tau\right) \varphi_2^{(n)}(\tau) d\tau,$$

$$(15.1.4)$$

where $\tilde{u}_n\left(\xi, (m+1)^{-1}\right)$ is a smooth extension of $u_n\left(\xi, (m+1)^{-1}\right)$ with compact support and the $\varphi_i^{(n)}(t)$, $i = 1, 2$, satisfy the integral equations

$$u_n\left(s_1^{(m+1)}(t), t\right)$$

$$= \int_{-\infty}^{\infty} K\left(s_1^{(m+1)}(t) - \xi, t - (m+1)^{-1}\right) \tilde{u}_n\left(\xi, (m+1)^{-1}\right) d\xi - 2^{-1} \varphi_1^{(n)}(t)$$

$$+ \int_{(m+1)^{-1}}^{t} \frac{\partial K}{\partial x}\left(s_1^{(m+1)}(t) - s_1^{(m+1)}(\tau), t - \tau\right) \varphi_1^{(n)}(\tau) d\tau$$

$$+ \int_{(m+1)^{-1}}^{t} \frac{\partial K}{\partial x}\left(s_1^{(m+1)}(t) - s_2^{(m+1)}(\tau), t - \tau\right) \varphi_2^{(n)}(\tau) d\tau,$$

$$u_n\left(s_2^{(m+1)}(t), t\right)$$

$$= \int_{-\infty}^{\infty} K\left(s_2^{(m+1)}(t) - \xi, t - (m+1)^{-1}\right) \tilde{u}_n\left(\xi, (m+1)^{-1}\right) d\xi + 2^{-1} \varphi_2^{(n)}(t)$$

$$+ \int_{(m+1)^{-1}}^{t} \frac{\partial K}{\partial x}\left(s_2^{(m+1)}(t) - s_2^{(m+1)}(\tau), t - \tau\right) \varphi_2^{(n)}(\tau) d\tau$$

$$+ \int_{(m+1)^{-1}}^{t} \frac{\partial K}{\partial x}\left(s_2^{(m+1)}(t) - s_1^{(m+1)}(\tau), t - \tau\right) \varphi_1^{(n)}(\tau) d\tau. \quad (15.1.5)$$

Now, recalling the estimates of Sec. 14.3, we see that the estimates for the kernels in the above equation depend only upon m. Moreover, from (15.1.3) and Corollary 8.4.1, we see that $\varphi_i^{(n)}$, $i = 1, 2$, are bounded in absolute value by a constant $C_1 = C_1(m, T, C)$. As $\text{dist}(D_T^{(m)} \cup B_T^{(m)}, B_T^{(m+1)}) > 0$, we obtain, from Lemma 14.2.2 and Theorem 10.2.1, that there exists a constant $C_2 = C_2(C_1, m, \text{dist}(D_T^{(m)} \cup B_T^{(m)}, B_T^{(m+1)}))$ such that

$$\left| \frac{\partial u_n}{\partial x}(x, t) \right| \le C_2, \qquad n = 1, 2, \dots, \tag{15.1.6}$$

and

$$\left| \frac{\partial u_n}{\partial t}(x, t) \right| \le C_2, \qquad n = 1, 2, \dots, \tag{15.1.7}$$

hold for all $(x, t) \in D_T^{(m)} \cup B_T^{(m)}$. From (15.1.6) and (15.1.7) we see that the sequence u_n, $n = 1, 2, 3, \dots$, is equi-Lipschitz-continuous on $D_T^{(m)} \cup B_T^{(m)}$ and uniformly bounded via (15.1.3). By the Ascoli–Arzela Theorem, there exists a uniformly convergent subsequence.

Now, let $u_{n,1}$, $n = 1, 2, \dots$, denote the uniformly convergent subsequence on $D_T^{(1)} \cup B_T^{(1)}$. Considering $u_{n,1}$, $n = 1, 2, \dots$, on $D_T^{(2)} \cup B_T^{(2)}$, we can select a uniformly convergent subsequence $u_{n,2}$, $n = 1, 2, \dots$, from it. Continuing in this manner we obtain an infinite set of sequences

$$u_{n,m}, \qquad n = 1, 2, \dots, \qquad m = 1, 2, 3, \dots,$$

such that $u_{n,m+1}$, $n = 1, 2, 3, \dots$, is a subsequence of $u_{n,m}$, $n = 1, 2, 3, \dots$, and for each $m = 1, 2, 3, \dots$, the $u_{n,m}$, $n = 1, 2, 3, \dots$, converge uniformly on $D_T^{(m)} \cup B_T^{(m)}$. Denoting the resulting limit function by u, it follows from Theorem 15.1.1 that u satisfies the heat equation in each $D_T^{(m)}$ and thus in D_T. Moreover, $|u(x, t)| < C$ since $|u_n(x, t)| < C$. $\qquad \square$

15.2. ANALYTICITY IN THE SPATIAL VARIABLE

In this section we show the following result.

Theorem 15.2.1. *If u is a solution of the heat equation in D_T, then $u(x, t)$ is an analytic function of x in each component of $\{t\} \cap D_T$.*

Proof. To demonstrate this result, it suffices to consider $(x_0, t_0) \in D_T$ and to show that

$$u(x, t_0) = \sum_{n=0}^{\infty} a_n (x - x_0)^n \qquad \text{for all } |x - x_0| < \alpha, \quad \alpha > 0,$$

where α denotes the radius of convergence of the power series.

Since $(x_0, t_0) \in D_T$, we observe, as in Theorem 15.1.1, that there exists a $\delta > 0$ such that the rectangle

$$R_\delta = \{(x, t) \mid x_0 - \delta \le x \le x_0 + \delta, \quad t_0 - \delta \le t \le t_0\} \subset D_T.$$

Under the transformation

$$\xi = \frac{x - (x_0 - \delta)}{2\delta} \quad \text{and} \quad \tau = \frac{t - (t_0 - \delta)}{4\delta^2},$$

R_δ is transformed into

$$R_* = \left\{ (\xi, \tau) \mid 0 \le \xi \le 1, \quad 0 \le \tau \le \frac{1}{4\delta} \right\}.$$

Also u is transformed into $U(\xi, \tau) = u(2\delta\xi + (x_0 - \delta), \ 4\delta^2\tau + (t_0 - \delta))$, which satisfies the heat equation in $0 < \xi < 1$, $0 < \tau \le 1/4\delta$ and is continuous in R_*. From Theorem 10.5.1, $U(\xi, 1/4\delta)$ is an analytic function of the complex variable ξ in the domain $D = \{ \xi \mid 0 < \text{Re}\,\xi < 1, \ |\text{Im}\,\xi| < \text{Re}\,\xi, \ |\text{Im}\,\xi| < 1 - \text{Re}\,\xi \}$. Consequently, for $|\xi - \tfrac{1}{2}| \le \rho < 1/(2\sqrt{2})$, $U(\xi, 1/4\delta)$ possesses a uniformly absolutely convergent power series about the point $\xi = \tfrac{1}{2}$. In other words,

$$U\left(\xi, \frac{1}{4\delta} \right) = \sum_{n=0}^{\infty} c_n \left(\xi - \frac{1}{2} \right)^n \tag{15.2.1}$$

for $\tfrac{1}{2} - \rho \le \xi \le \tfrac{1}{2} + \rho$. Transforming back to x and t, we see that

$$u(x, t_0) = \sum_{n=0}^{\infty} c_n \left(\frac{x - x_0}{2\delta} \right)^n \tag{15.2.2}$$

for $|x - x_0| \le 2\delta\rho$ and that the uniform absolute convergence of the power series in (15.2.2) follows from the uniform absolute convergence of the power series in (15.2.1). $\qquad \square$

15.3. THE STRONG MAXIMUM PRINCIPLE

This section is devoted to the proof of the following generalization of the Weak Maximum Principle, which was demonstrated in Theorem 1.6.1.

  Theorem 15.3.1.   *If u satisfies the heat equation in D_T and is continuous in $D_T \cup B_T$, then for each t, $0 < t \le T$, either*

$$\inf_{B_t} u < u(x, t) < \sup_{B_t} u, \quad (x, t) \in D_T, \tag{15.3.1}$$

or

$$u \equiv constant \ in \ D_t \cup B_t, \tag{15.3.2}$$

where here we assume for convenience that $\{\tau\} \cap D_T$ possesses only one component.

  Proof.   The proof of Theorem 15.3.1 requires a few preliminary results. Our first lemma concerns a solution to the heat equation that we shall use as a comparison function to demonstrate in the second lemma that u cannot assume its maximum value at isolated points. From this it will follow that if u assumes its maximum value at $(x_0, t) \in D_T$, then $u \equiv$ maximum on $\{t\} \cap D_T$. In our fourth lemma we shall study a solution to the heat equation, which we shall use as a comparison function to show that $u \equiv$ maximum throughout $D_t \cup B_t$. Since similar results hold if u assumes its minimum value, the theorem will follow from the weak maximum principle:

$$\inf_{B_t} u \le u(x, t) \le \sup_{B_t} u.$$

Consider first the following result.

Lemma 15.3.1. *If* $m < M$, *and if* v *is the bounded solution of*

$$v_t = v_{xx}, \qquad x_0 - \delta_1 < x < x_0 + \delta_1, \qquad t_0 - \delta_2 < t \le t_0,$$
$$v(x_0 - \delta_1, t) = m, \qquad t_0 - \delta_2 < t \le t_0, \qquad (15.3.3)$$
$$v(x_0 + \delta_1, t) = m, \qquad t_0 - \delta_2 < t \le t_0,$$
$$v(x, t_0 - \delta_2) = M, \qquad x_0 - \delta_1 < x < x_0 + \delta_1, \qquad \delta_i > 0, \qquad i = 1, 2,$$

then

$$v(x_0, t_0) < M. \qquad (15.3.4)$$

Proof. Let $v = m + w$, where w is the bounded solution of

$$w_t = w_{xx}, \qquad x_0 - \delta_1 < x < x_0 + \delta_1, \qquad t_0 - \delta_2 < t \le t_0,$$
$$w(x_0 - \delta_1, t) = w(x_0 + \delta_1, t) = 0, \qquad t_0 - \delta_2 < t \le t_0, \qquad (15.3.5)$$
$$w(x, t_0 - \delta_2) = M - m, \qquad x_0 - \delta_1 < x < x_0 + \delta_1.$$

Since the solution

$$z(x, t) = \int_{x_0 - \delta_1}^{x_0 + \delta_1} K(x - \xi, t - (t_0 - \delta_2))(M - m) \, d\xi \qquad (15.3.6)$$

is positive for $x = x_0 \pm \delta_1, t_0 - \delta_2 < t \le t_0$, it follows from the Weak Maximum Principle that $z \ge w$. Hence,

$$v(x_0, t_0) \le m + z(x_0, t_0) = (1 - \alpha)m + \alpha M < M, \qquad (15.3.7)$$

since

$$0 < \alpha = \int_{x_0 - \delta_1}^{x_0 + \delta_1} K(x_0 - \xi, \delta_2) \, d\xi < 1. \qquad (15.3.8) \quad \square$$

The following result is an easy application of Lemma 15.3.1.

Lemma 15.3.2. *The solution* u *cannot assume its maximum value at a spatially isolated point in* D_T.

Proof. Let $M = \sup_{B_T} u$. Suppose that $(x_0, t_0) \in D_T$, $u(x_0, t_0) = M$, and there exists a $\delta_1 > 0$ such that $u(x_0 - \delta_1, t_0) < M$ and $u(x_0 + \delta_1, t_0) < M$.

Note that this is considerably weaker than $u(x, t_0) < M$, $0 < |x - x_0| \le \delta_1$. From the continuity of u, there exists a constant $m < M$ and $\delta_2 > 0$ such that $u(x_0 \pm \delta, t) \le m$ for $t_0 - \delta_2 \le t \le t_0$. Since $u(x, t_0 - \delta_2) \le M$, $x_0 - \delta_1 < x < x_0 + \delta_1$, we see that with v as in Lemma 15.3.1,

$$u(x_0, t_0) \le v(x_0, t_0) < M \qquad (15.3.9)$$

follows from Lemma 15.3.1 via the Weak Maximum Principle. From the contradiction (15.3.9), the result follows. □

From this result we acquire the following one.

Lemma 15.3.3. *If for some t, $0 < t \le T$, $M = \sup_{B_t} u$, and if*

$$K_M = \{x : (x, t) \in D_t, \ u(x, t) = M\} \ne \phi,$$

then $K_M = (\mathbb{R} \times \{t\}) \cap (D_T \cup B_T)$.

Proof. From the continuity of u, $K_M = \{u \equiv M\} \cap (\mathbb{R} \times \{t\})$ is closed. By Lemma 15.3.2, K_M cannot contain isolated points. Hence, every point in K_M must be an accumulation point of K_M. As $K_M \ne \phi$, there exists at least one accumulation point. The result follows from this and the analyticity of u with respect to x. □

In order to show that $u \equiv M$ throughout $D_t \cup B_t$, we need the following result.

Lemma 15.3.4. *If $M > m$ and if v is the bounded solution of*

$$v_t = v_{xx}, \qquad x_0 - \delta_1 < x < x_0 + \delta_1, \qquad t_0 - \delta_2 < t \le t_0,$$

$$v(x_0 \pm \delta_1, t) = M, \qquad t_0 - \delta_2 < t \le t_0, \qquad (15.3.10)$$

$$v(x, t_0 - \delta_2) = m, \qquad x_0 - \delta_1 < x < x_0 + \delta_1, \qquad \delta_i > 0, \qquad i = 1, 2,$$

then

$$v(x_0, t_0) < M. \qquad (15.3.11)$$

Proof. Let $v = M + w$, where w is the bounded solution of

$$w_t = w_{xx}, \qquad x_0 - \delta_1 < x < x_0 + \delta_1, \qquad t_0 - \delta_2 < t \le t_0,$$

$$w(x_0 \pm \delta_1, t) = 0, \qquad t_0 - \delta_2 < t \le t_0, \qquad (15.3.12)$$

$$w(x, t_0 - \delta_2) = m - M, \qquad x_0 - \delta_1 < x < x_0 + \delta_1.$$

From the Weak Maximum Principle,

$$w \le z = (m - M)\exp\left\{-\frac{\pi^2}{4\delta_1^2}(t - (t_0 - \delta_2))\right\} \cdot \sin \pi \left(\frac{x - (x_0 + \delta_1)}{2\delta_1}\right).$$

$$(15.3.13)$$

Hence,

$$v(x_0, t_0) \le M + z(x_0, t_0) = M(1 - \alpha) + \alpha m < M, \qquad (15.3.14)$$

since

$$0 < \alpha = \exp\left\{-\frac{\pi^2\delta_2}{4\delta_1^2}\right\} < 1. \qquad (15.3.15) \quad \square$$

We can now demonstrate the following result.

Lemma 15.3.5. *If for* $(x_0, t_0) \in D_T$, $u(x_0, t_0) = M = \sup_{B_{T_0}} u$, *then* $u \equiv M$ *for all* $(x, t) \in D_{t_0} \cup B_{t_0}$.

Proof. From Lemma 15.3.3, it follows that $u(x, t_0) \equiv M$ for all $(x, t_0) \in D_{t_0} \cup B_{t_0}$. For all $(x, t) \in D_{t_0} \cup B_{t_0}$ such that $(x, t_0) \in D_{t_0} \cup B_{t_0}$, a direct application of the Weak Maximum Principle and Lemma 15.3.4 yields the fact that $u(x, t) \equiv M$. By analytic continuation $u(x, t) \equiv M$ for all $(x, t) \in D_{t_0} \cup B_{t_0}$ such that there exists (x_1, t) for which $(x_1, t_0) \in D_{t_0} \cup B_{t_0}$. If these points constitute all of $D_{t_0} \cup B_{t_0}$, then we are finished. If not, then we select a $t_1 < t_0$ for which there exists a point $(x_1, t_1) \in D_{t_0}$ such that $(x_1, t_0) \in D_{t_0}$ and repeat the argument. As D_{t_0} is connected and $\{t\} \cap D_{t_0}$ consists of only one component, every point in D_{t_0} can be reached by a descending stairstep curve with a finite number of steps. Hence, $u \equiv M$ in D_{t_0} and by continuity $u \equiv M$ in $D_{t_0} \cup B_{t_0}$. $\qquad \square$

Since a similar set of lemmas holds for the minimum value of u, we can complete the proof of the theorem by observing that, by the Weak Maximum Principle, either

$$\inf_{B_{t_0}} u < u(x, t_0) < \sup_{B_{t_0}} u \qquad \text{for all } (x, t_0) \in D_{t_0}, \qquad (15.3.16)$$

or there exists a point $(x_0, t_0) \in D_{t_0}$ such that

$$u(x_0, t_0) = \sup_{B_{t_0}} u, \qquad (15.3.17)$$

or

$$u(x_0, t_0) = \inf_{B_{t_0}} u. \qquad (15.3.18)$$

In the latter two cases, Lemma 15.3.5 implies that $u \equiv$ constant in $D_{t_0} \cup B_{t_0}$.
$\qquad \square$

15.4. BEHAVIOR OF u_x AT A BOUNDARY POINT WHERE u ASSUMES ITS MAXIMUM OR MINIMUM VALUE

Under certain conditions on the boundary in the neighborhood of a point at which u assumes its maximum or minimum value, we shall show that if $u \not\equiv$ const., and u_x exists at that point, then $u_x \neq 0$. We assume that

$$D_T = \{(x, t) | s_1(t) < x < s_2(t), \quad 0 < t \leq T\}$$

and that s_2 is lower Lipschitz continuous while s_1 is upper Lipschitz continuous. We shall let L denote the Lipschitz constant in either case.

Let us assume that $u \neq \text{const.}$, and that $u(s_2(t_0), t_0) = \sup_{B_{t_0}} u, t_0 > 0$. It is no loss of generality to assume that $\sup_{B_{t_0}} u = 0$ and, via a translation, that $s_2(t_0) = t_0 = 0$. Since s_2 is lower Lipschitz continuous, we have that

$$Lt \leq s_2(t), \qquad t \leq 0. \tag{15.4.1}$$

Since $u \neq \text{const.}$, it follows that, for $x_0 < 0$, $u(x_0, 0) = -\eta$ for some positive η. Consider the triangle with vertices $(0,0)$, $(x_0, 0)$ and $(x_0, L^{-1}x_0)$. By the continuity of u, there exists a τ, $L^{-1}x_0 \leq \tau < 0$, such that $u(x_0, t) \leq -\eta/2$ for $\tau \leq t \leq 0$. Let

$$L_1 = \max\left(L, \frac{x_0}{\tau}\right). \tag{15.4.2}$$

Then we can restrict our attention to the triangle $\mathcal{T}$ with vertices $(0,0)$, $(x_0, 0)$, and (x_0, τ). On the line segment connecting $(0,0)$ and (x_0, τ), $u \leq 0$, while on the line segment connecting $(x_0, 0)$ and (x_0, τ), $u \leq -\eta/2$. We are now in a position to construct a function that bounds u from above.

Let

$$w = \varepsilon(1 - \exp\{-\alpha(x - L_1 t)\}), \tag{15.4.3}$$

where ε and α are positive constants to be selected below. Note first of all that for $(x, t) \in \mathcal{T}$, $(x - L_1 t) \leq 0$. Consequently, $-\alpha(x - L_1 t) \geq 0$ and $w \leq 0$. Differentiating with respect to x and t, we see that

$$w_{xx} - w_t = \varepsilon \exp\{-\alpha(x - L_1 t)\}(L_1 \alpha - \alpha^2) < 0 \tag{15.4.4}$$

if we select

$$\alpha = L_1 + 1. \tag{15.4.5}$$

It is not difficult to show that w assumes its minimum value at $(x_0, 0)$. We consider

$$w(x_0, 0) = \varepsilon(1 - \exp\{-(L_1 + 1)x_0\}) \tag{15.4.6}$$

and note that, if

$$0 < \epsilon < 2^{-1}\eta(\exp\{-(L_1 + 1)x_0\} - 1)^{-1}, \tag{15.4.7}$$

then

$$w(x, t) \geq w(x_0, 0) > -2^{-1}\eta. \tag{15.4.8}$$

Now, set

$$z = w - u. \tag{15.4.9}$$

On the segment connecting $(0,0)$ and (x_0, τ), $z \geq 0$ since $w = 0$ and $u \leq 0$. On the segment connecting $(x_0, 0)$ and (x_0, τ), it follows, from

$u(x_0, t) \le - \eta/2$ and (15.4.8), that $z \ge 0$. As $\mathcal{L}(z) = z_{xx} - z_t < 0$ in $\mathcal{T}$, it follows that z cannot possess a negative minimum. Hence $z \ge 0$ throughout $\mathcal{T}$.

We turn now to the difference quotient for u at $(0,0)$ and note that, for $x < 0$, $z \ge 0$ implies that

$$\frac{u(x,0) - u(0,0)}{x} > \frac{w(x,0) - w(0,0)}{x} \tag{15.4.10}$$

since $u(0,0) = w(0,0) = 0$. Hence,

$$\liminf_{x \to 0} \frac{u(x,0) - u(0,0)}{x} > \liminf_{x \to 0} \frac{w(x,0) - w(0,0)}{x}$$

$$= \frac{\partial w}{\partial x}(0,0)$$

$$= (L_1 + 1)\varepsilon > 0. \tag{15.4.11}$$

Consequently, if u_x exists at $(0,0)$, then it must be positive.

The remaining cases can be reduced to this one. For a minimum at $(s_2(t_0), t_0)$, we can employ $-u$ and consider it a maximum. Hence, at a minimum at $(s_2(t_0), t_0)$, $u_x < 0$. For a maximum or minimum at $(s_1(t_0), t_0)$, we can employ the change of variable $\xi = -x$, which interchanges the role of s_1 and s_2. From the negative sign in the change of variable, we see that at a maximum at $(s_1(t_0), t_0)$, $u_x < 0$ while at a minimum, $u_x > 0$.

We can summarize all of this in the following statement.

Theorem 15.4.1 (Hopf). *For $D_T = \{(x, t) | s_1(t) < x < s_2(t),$ $0 < t \le T\}$, where $s_2(t)$ is lower Lipschitz continuous and $s_1(t)$ is upper Lipschitz continuous, let u denote a solution to the heat equation in D_T that is continuous in $D_T \cup B_T$ and that is not identically constant in each D_t, $0 < t \le T$. If u assumes its maximum value at $(s_2(t_0), t_0)$, $t_0 > 0$, then*

$$\lim_{x \uparrow s_2(t_0)} \inf \frac{u(x, t_0) - u(s_2(t_0), t_0)}{x - s_2(t_0)} > 0. \tag{15.4.12}$$

If in this case u_x exists at $(s_2(t_0), t_0)$, then

$$u_x(s_2(t_0), t_0) > 0. \tag{15.4.13}$$

If u assumes its minimum value at $(s_2(t_0), t_0)$, $t_0 > 0$, then

$$\lim_{x \uparrow s_2(t_0)} \sup \frac{u(x, t_0) - u(s_2(t_0), t_0)}{x - s_2(t_0)} < 0. \tag{15.4.14}$$

If in this case u_x exists at $(s_2(t_0), t_0)$, then

$$u_x(s_2(t_0), t_0) < 0. \tag{15.4.15}$$

If u assumes its maximum value at $(s_1(t_0), t_0)$, $t_0 > 0$, then

$$\lim_{x \downarrow s_1(t_0)} \sup \frac{u(x, t_0) - u(s_1(t_0), t_0)}{x - s_1(t_0)} < 0. \tag{15.4.16}$$

If in this case u_x exists at $(s_1(t_0), t_0)$, then

$$u_x(s_1(t_0), t_0) < 0. \tag{15.4.17}$$

If u assumes its minimum value at $(s_1(t_0), t_0)$, $t_0 > 0$, then

$$\lim_{x \downarrow s_1(t_0)} \inf \frac{u(x, t_0) - u(s_1(t_0), t_0)}{x - s_1(t_0)} > 0. \tag{15.4.18}$$

If in this case, u_x exists at $(s_1(t_0), t_0)$ then

$$u_x(s_1(t_0), t_0) > 0. \tag{15.4.19}$$

Remark. We shall find this theorem quite useful in the study of the one-phase Stefan problem in Chapters 17 and 18.

EXERCISES

15.1.   Use Theorem 15.4.1 and a Maximum Principle argument to show the unicity of solutions to problems involving the specification of u_x on a boundary.

We shall now demonstrate the Extended Comparison Theorem in the special case of one bounded discontinuity on the parabolic boundary B_T, since a complicated but straightforward extension of the argument can be made to handle the case of a finite number of bounded discontinuities on B_T. We assume that $T > 0$ and that B_T contains the line segment $-A \leq x \leq A$, $t = 0$, where A is a positive constant. Moreover, we assume that u and v are solutions of $u_t = u_{xx}$ in D_T that are continuous in $D_T \cup B_T$ except at $x = t = 0$. We assume that u and v are bounded in absolute value by the constant C throughout $D_T \cup B_T$ except at $x = t = 0$ and that on B_T, $u \leq v$ except at $x = t = 0$. Now, an immediate application of the Comparison Theorem for $t < 0$ enables us to assume, without loss of generality, that $D_T \subset \mathbf{R}_+^2 = \{(x, t) | t > 0\}$.

We consider now the domain

$$\tilde{D}_T^\delta = \mathbf{R}_+^2 - \{(x, t) | -\delta \leq x \leq \delta \quad \text{and} \quad 0 \leq t \leq \delta\}.$$

Clearly, the parabolic boundary $\tilde{B}_T^\delta$ consists of the union of the line segments $-\infty < x \leq -\delta$, $t = 0$; $\delta \leq x < \infty$, $t = 0$; $x = -\delta$, $0 \leq t \leq \delta$; $x = \delta$, $0 \leq t \leq \delta$; and $-\delta \leq x \leq \delta$, $t = \delta$. In $\tilde{D}_T^\delta$, we consider the problem

$$z_t^\delta = z_{xx}^\delta, \qquad (x,t) \in \tilde{D}_T^\delta,$$

$$z^\delta(x,0) = 0, \qquad -\infty < x < -2\delta,$$

$$z^\delta(x,0) = C\delta^{-1}(2\delta + x), \qquad -2\delta \le x \le -\delta,$$

$$z^\delta(-\delta,t) = C, \qquad 0 \le t \le \delta,$$

$$z^\delta(x,\delta) = C, \qquad -\delta \le x \le \delta, \tag{A}$$

$$z^\delta(\delta,t) = C, \qquad 0 \le t \le \delta,$$

$$z^\delta(x,0) = C\delta^{-1}(2\delta - x), \qquad \delta \le x \le 2\delta,$$

$$z^\delta(x,0) = 0, \qquad 2\delta < x < \infty.$$

A solution, which we denote by z^δ, can be constructed from the representations given in Theorem 4.3.1 and Theorem 3.5.1. We note that z^δ is continuous in $\tilde{D}_T^\delta \cup \tilde{B}_T^\delta$.

Consider now the domain $D_T \cap \tilde{D}_T^\delta$. The solutions u, v, and z^δ are continuous in this domain and on its parabolic boundary. Moreover, on the parabolic boundary, $u \le v + z^\delta$. By the Comparison Theorem 1.6.3, $u \le v + z^\delta$ in $D_T \cap \tilde{D}_T^\delta$. Given a particular point $(x_1, t_1) \in D_T$. It follows that for all δ sufficiently small, $(x_1, t_1) \in D_T \cap \tilde{D}_T^\delta$. Consequently,

$$u(x_1, t_1) \le v(x_1, t_1) + z^\delta(x_1, t_1) \tag{B}$$

for all δ sufficiently small. Hence the result will follow provided we can show that

$$\lim_{\delta \downarrow 0} z^\delta(x_1, t_1) = 0. \tag{C}$$

15.2. Show that the $\lim_{\delta \downarrow 0} z^\delta(x_1, t_1) = 0$.
15.3. Prove Theorem 1.6.5 (*Extended Comparison Theorem*).
15.4. Prove Theorem 1.6.6 (*Extended Uniqueness Theorem*).

NOTES

Uniform convergence of sequences of solutions and compactness properties of bounded sequences of solutions are analogous to those for harmonic functions found in Kellogg [4]. The strong maximum principle for general parabolic differential equations can be found in Protter and Weinberger [6], who present the work of Nirenberg [5]. The behavior of u_x at a point of maximum or minimum of u on the boundary was demonstrated by Hopf [3] for elliptic differential equations. Friedman [1,2] studied this behavior for parabolic differential equations. See also Protter and Weinberger [6].

REFERENCES

1. Friedman, A., *Partial Differential Equations of Parabolic Type*. Prentice-Hall, Inc., Englewood Cliffs, N.J., 1964.
2. Friedman, A., Remarks on the maximum principle for parabolic equations and its applications, *Pacific J. of Math.*, **8**, (1958), 201–211.
3. Hopf, E., A remark on linear elliptic differential equations of second order, *Proc. Amer. Math. Soc.*, **3** (1952), 791–793.
4. Kellogg, O. D., *Foundations of Potential Theory*. Dover Publications, Inc., New York, 1953.
5. Nirenberg, L. A., A strong maximum principle for parabolic equations, *Comm. on Pure and Appl. Math.*, **6** (1953), 167–177.
6. Protter, M. H., and Weinberger, H. F., *Maximum Principles in Differential Equations*. Prentice-Hall, Inc., Englewood Cliffs, N.J., 1967.

Chapter 16

The Solution in a General Region with Temperature-Boundary Specification: The Method of Perron–Poincaré

16.1. INTRODUCTION

We consider the problem

$$u_t = u_{xx}, \qquad s_1(t) < x < s_2(t), \qquad 0 < t \le T,$$

$$u(x,0) = f(x), \qquad a < x < b, \qquad s_1(0) = a, \qquad s_2(0) = b,$$

$$u(s_1(t), t) = g(t), \qquad 0 < t \le T,$$

$$u(s_2(t), t) = h(t), \qquad 0 < t \le T,$$

$$(16.1.1)$$

where f, g, h, s_1, and s_2 are continuous functions of their respective arguments, $f(a) = g(0)$, and $f(b) = h(0)$. (See Fig. 16.1.1.) Let

$$D_T = \{(x,t) | s_1(t) < x < s_2(t), \qquad 0 < t \le T\} \qquad (16.1.2)$$

and let B_T denote its boundary. For what follows it is convenient to define

$$\mathcal{L}(v) = v_{xx} - v_t. \qquad (16.1.3)$$

Denoting by F the triple (g, f, h) on B_T, Problem (16.1.1) condenses to the statement

$$\mathcal{L}(u) = 0 \qquad \text{in } D_T,$$

$$u = F \qquad \text{on } B_T. \qquad (16.1.4)$$

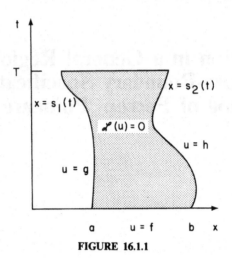

FIGURE 16.1.1

From the definition of D_T, we see that for, any point $(x, t) \in D_T$, there exists a closed polygonal region P with one side a line segment contained in $\{T\} \cap D_T$, another a line segment contained in $\{\tau\} \cap D_T$, $0 < \tau < t$, and the remaining boundary of P consisting of two nonintersecting polygonal curves whose line segments are not characteristics and which connect, respectively, the left endpoint of the line segment on $t = \tau$ to the left endpoint of the line segment on $t = T$ and the right endpoint to the right endpoint, such that $(x, t) \in P^o$, the interior of P.

Remark. We shall call all such polygonal regions P *admissible*. Moreover, we shall utilize only *admissible* P in the discussion below.

For all such admissible P, we define the following operator M_P on continuous functions in $D_T \cup B_T$.

DEFINITION 16.1.1. For $v \in C(D_T \cap B_T)$ and any admissible $P \subset D_T$,

$$M_P v = \begin{cases} v, & (x, t) \notin P, \\ v_*, & (x, t) \in P, \end{cases} \tag{16.1.5}$$

where

$$\begin{aligned} \mathscr{L}(v_*) &= 0 && \text{in } P^o, \\ v_* &= v && \text{on } \partial P, \end{aligned} \tag{16.1.6}$$

where ∂P denotes the parabolic boundary of P.

We collect in the following statement some properties of $M_P v$ that are easy consequences of Theorem 14.3.1, the linearity of $\mathscr{L}$, and the Weak Maximum Principle and the Extended Comparison Theorem 1.6.5.

Lemma 16.1.1.

Property A: $M_P v$ is well defined.

Property B: $M_P v \in C(D_T \cup B_T)$ if $v \in C(D_T \cup B_T)$.

Property C: If $v \equiv const.$, then $M_P v \equiv const.$

Property D: If $v_1 \geq v_2$, then $M_P v_1 \geq M_P v_2$.

Property E: If $a \equiv const.$, then

$$M_P a v = a M_P v.$$

Property F: $M_P(v_1 + v_2) = M_P v_1 + M_P v_2.$

Property G: $\mathscr{L}(v) = 0$ in D_T if and only if $M_P v \equiv v$ for all $P \subset D_T$.

Proof. The proof is left to the reader.   □

We shall use this notion of $M_P v$ and its properties to define superparabolic and subparabolic functions in the next section and to study their properties.

16.2. SUPERPARABOLIC AND SUBPARABOLIC FUNCTIONS

We begin this section with the following definition.

DEFINITION 16.2.1. For $\psi \in C(D_T \cup B_T)$, ψ is superparabolic in D_T if $\psi \geq M_P \psi$ for all admissible $P \subset D_T$, and ψ is subparabolic in D_T if $\psi \leq M_P \psi$ for all admissible $P \subset D_T$. With respect to the function F on B_T, ψ is an upper function if ψ is superparabolic in D_T and $\psi \geq F$ on B_T, and ψ is a lower function if ψ is subparabolic in D_T and $\psi \leq F$ on B_T.

In the following statement we collect some elementary properties of superparabolic and subparabolic functions.

Lemma 16.2.1.

Property 1:   If $\mathscr{L}(\psi) = 0$ in D_T, then ψ is both superparabolic and subparabolic in D_T.

Property 2:   If ψ is subparabolic (superparabolic) and $\mathscr{L}(v) = 0$ in D_T, then $\psi \pm v$ is subparabolic (superparabolic) in D_T.

Property 3:   If ψ_1 and ψ_2 are subparabolic (superparabolic) in D_T, then $\psi_1 + \psi_2$ is subparabolic (superparabolic) in D_T.

Property 4:   If ψ_1 is superparabolic in D_T and ψ_2 is subparabolic in D_T, then $\psi_1 - \psi_2$ is superparabolic in D_T.

Property 5:   The function ψ_1 is superparabolic in D_T if and only if $-\psi_1$ is subparabolic in D_T.

Proof. Property 1 follows from Property G of Lemma 16.1.1. For Property 2, we see that $M_P(\psi \pm v) = M_P \psi \pm M_P v = M_P \psi \pm v \geq \psi \pm v$. Property 3 follows immediately from Property F, which was used for Property 2. Property 4 and Property 5 follow easily from the definition.   □

We can demonstrate a strong minimum principle for superparabolic functions.

Lemma 16.2.2. *If ψ is superparabolic in D_T, then ψ assumes its minimum value on B_T.*

Proof. Let $m = \min_{D_T \cup B_T} \psi$ and assume that $\psi(x_0, t_0) = m$ for $(x_0, t_0) \in D_T$. Consider the point $(s_2(\tau), \tau) \in B_T$ for $0 < \tau < t_0$. Let $x_n = s_2(\tau) - n^{-1}$. Since $(x_n, \tau) \in D_T$, we can construct an admissible $P_n \subset D_T$ with rightmost lower corner (x_n, τ) such that $(x_0, t_0) \in P_n^o$. Let $\varphi = M_{P_n} \psi$. Then $\varphi \geq m$ on ∂P_n implies that $\varphi \geq m$ in P_n. But $\psi(x_0, t_0) = m \geq \varphi(x_0, t_0)$ implies that $\varphi(x_0, t_0) = m$. By the Strong Maximum Principle, Theorem 15.3.1, we see that $\varphi \equiv m$ for all $(x, t) \in P_n$ such that $\tau \leq t \leq t_0$. As $\psi(x_n, \tau) = \varphi(x_n, \tau) = m$, it follows that $\psi = m$ in every neighborhood of $(s_2(\tau), \tau)$. From the continuity of ψ in $D_T \cup B_T$, it follows that $\psi(s_2(\tau), \tau) = m$.   □

Now we show that every upper function is greater than or equal to every lower function.

Lemma 16.2.3. *If ψ_1 is an upper function and ψ_2 is a lower function, then $\psi_1 \geq \psi_2$.*

Proof. The function $\varphi = \psi_1 - \psi_2$ is superparabolic, by Property 4. Hence, by Lemma 16.2.2, φ has its minimum on B_T. But, on B_T, $\psi_1 \geq F \geq \psi_2$, which implies that $\varphi \geq 0$ on B_T and throughout D_T.   □

The next two results enable us to construct upper functions from upper functions.

Lemma 16.2.4. *If $\psi_1, \ldots, \psi_n$ are upper functions, then $\psi = \min(\psi_1, \ldots, \psi_n)$ is an upper function.*

Proof. Clearly, $\psi \geq F$ on B_T. Also, from the Weak Maximum Principle, $M_P \psi_i \geq M_P \psi$ for each i. Hence, at a point $Q \in D_T$, $\psi(Q) = \psi_{i_0}(Q) \geq M_P \psi_{i_0}(Q) \geq M_P \psi(Q)$.   □

Lemma 16.2.5. *If ψ is an upper function, then $M_{P_1} \psi$ is an upper function for each $P_1 \subset D_T$.*

Proof. First, $M_{P_1} \psi \geq F$ on B_T. Now, we must show that $M_P(M_{P_1} \psi) \leq M_{P_1} \psi$ for all $P \subset D_T$. There are four cases that an arbitrary P can have relative to P_1. (See Fig. 16.2.1.) First, suppose $P \cap P_1 = \phi$. Then over P, $M_{P_1} \psi = \psi \geq M_P \psi = M_P(M_{P_1} \psi)$. Second, suppose that $P \subset P_1$. Then as $\mathscr{L}(M_{P_1} \psi) = 0$ in P_1^o, $M_P(M_{P_1} \psi) = M_{P_1} \psi$ in P and elsewhere by definition. Third, suppose that $P_1 \subset P$. Now, $M_P \psi \leq \psi$. Consequently, $M_P \psi \leq \psi = M_{P_1} \psi$

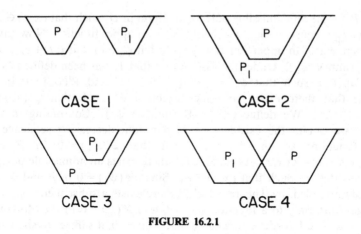

FIGURE 16.2.1

on ∂P_1. From the Maximum Principle, it follows that $M_P \psi \le M_{P_1} \psi$ in P_1. As $\psi = M_{P_1} \psi$ in $D_T - P_1^o$, we see that $M_P(M_{P_1} \psi) \le M_{P_1} \psi$ throughout D_T. Finally, we consider the case that $P \cap P_1 \ne \phi$ and neither inclusion holds. As $M_{P_1} \psi \le \psi$ we see that, on the boundary of P, $M_P(M_{P_1} \psi) = M_{P_1} \psi \le \psi$ and thus from the maximum principle $M_P(M_{P_1} \psi) \le M_P \psi \le \psi$ in P. Hence, in $P - (P \cap P_1)$, $\psi = M_{P_1} \psi \ge M_P \psi \ge M_P(M_{P_1} \psi)$. Next, we consider $P \cap P_1$. On $\partial(P \cap P_1) \cap \partial P$, $M_P(M_{P_1} \psi) = M_{P_1} \psi$, while on $\partial(P \cap P_1) \cap \partial P_1$, $M_{P_1} \psi = \psi \ge M_P(M_{P_1} \psi)$. Since $\mathscr{L}(M_P(M_{P_1} \psi)) = \mathscr{L}(M_{P_1} \psi) = 0$ in $P \cap P_1$, it follows from the Maximum Principle that $M_{P_1} \psi \ge M_P(M_{P_1} \psi)$ in $P \cap P_1$. As the same inequality holds in $P - (P \cap P_1)$, we see that $M_{P_1} \psi \ge M_P(M_{P_1} \psi)$ throughout $D_T \cup B_T$.

16.3. EXISTENCE OF u SATISFYING THE HEAT EQUATION IN D_T

Since F is continuous on B_T, it is bounded above by a constant C_1. As $\mathscr{L}(C_1) = 0$ in D_T, C_1 is an upper function. Consequently, the class $\mathscr{U}$ of upper functions is nonempty.

DEFINITION 16.3.1. Set

$$u(x, t) = \inf_{\psi \in \mathscr{U}} \psi(x, t) \tag{16.3.1}$$

for each $(x, t) \in (D_T \cup B_T)$.

We shall show that $\mathscr{L}(u) = 0$ in D_T and, under certain conditions on B_T, that $u = F$ on B_T and $u \in C(D_T \cup B_T)$. In order to show that $\mathscr{L}(u) = 0$ in D_T, it suffices to show that $\mathscr{L}(u) = 0$ at each point in D_T. Let $(x, t) \in D_T$

and let P denote an admissible polygonal region in D_T such that $(x, t) \in P^0$. Consider $\psi_0 \equiv$ const. $\geq \sup_{B_T} F$. Clearly, ψ_0 is an upper function. Now, given $\varepsilon > 0$, there exists an upper function $\tilde{\psi}$ such that $\tilde{\psi}(x, t) < u(x, t) + \varepsilon/2$. Let $\psi_1 = M_P(\min(\tilde{\psi}, \psi_0))$. Under the assumption that ψ_n has been defined as an upper function such that $\psi_n(x, t) < u(x, t) + \varepsilon/2^n$ and $\mathcal{L}(\psi_n) = 0$ in P, we note that there exists an upper function ψ^* such that $\psi^*(x, t) < u(x, t) + \varepsilon/2^{n+1}$. We define $\psi_{n+1} = M_P(\min(\psi^*, \psi_n))$. Continuing in this manner, we obtain the existence of a monotone-decreasing sequence of upper functions ψ_n, $n = 1, 2, \ldots$, such that $\mathcal{L}(\psi_n) = 0$ in P and $\lim_{n \to \infty} \psi_n(x, t) = u(x, t)$. As $(x, t) \in P^0$, there exists an admissible polygonal region $P_1 \subset P^0$ such that $(x, t) \in P_1^0$. Since $\mathcal{L}(\psi_n) = 0$ in P and ψ_n are uniformly bounded, by Theorem 15.1.2, there exists a subsequence ψ_{n_i} that converges uniformly to a function φ on P_1 and $\mathcal{L}(\varphi) = 0$ in P_1. Moreover, $\varphi(x, t) = u(x, t)$. In order to conclude that $\mathcal{L}(u) = 0$, it suffices to show that $\varphi \equiv u$ in $(P_1 \cap D_t)^0$. Now, by construction, $u \leq \varphi$ in $(P_1 \cap D_t)^0$. Suppose that there exists a point $(x_1, t_1) \in (P_1 \cap D_t)^0$ at which $u(x_1, t_1) < \varphi(x_1, t_1)$. Now we can construct a polygonal region $P_2 \subset P_1^0$ such that $(x_1, t_1) \in \partial P_2$ and $(x, t) \in P_2^0$. Consider the upper function $\tilde{\psi}^*$, which satisfies

$$u(x_1, t_1) < \tilde{\psi}^*(x_1, t_1) < \varphi(x_1, t_1),$$

and define $\zeta_i = M_{P_2}(\min(\tilde{\psi}^*, \psi_{n_i}))$, $i = 1, 2, \ldots$ Clearly, ζ_i is a sequence of upper functions. Moreover, as ψ_{n_i} converges uniformly to φ on ∂P_2, ζ_i converges uniformly to $M_{P_2}(\min(\tilde{\psi}^*, \varphi))$ on P_2. Now, $\min(\tilde{\psi}^*, \varphi) \leq \varphi$ on ∂P_2 and $\tilde{\psi}^*(x_1, t_1) < \varphi(x_1, t_1)$. Consequently, by the Strong Maximum Principle,

$$M_{P_2}(\min(\tilde{\psi}^*, \varphi))(x, t) < M_{P_2} \varphi(x, t) = u(x, t).$$

But ζ_i are upper functions and $\lim_{i \to \infty} \zeta_i(x, t) = M_{P_2}(\min(\tilde{\psi}^*, \varphi))(x, t)$. Consequently, for i sufficiently large there exists a ζ_i such that $\zeta_i(x, t) < u(x, t)$, which is a contradiction. Hence, $u \equiv \varphi$ in $(P_1 \cap D_t)^0$, which implies that $\mathcal{L}(u) = 0$ in D_T.

We can summarize the above results in the following statement.

Theorem 16.3.1. *For the class $\mathcal{U}$ of upper functions for F on B_T, the function u defined by*

$$u(x, t) = \inf_{\psi \in \mathcal{U}} \psi(x, t), \qquad (x, t) \in D_T \cup B_T, \qquad (16.3.2)$$

satisfies the heat equation in D_T.

Proof. See the analysis preceding the theorem statement. □

Remark. For $Q \in B_T$ and for $\varepsilon > 0$, there exists an upper function ψ such that $\psi(Q) < u(Q) + \varepsilon$. As ψ is continuous, there exists a neighborhood N of Q such that, for any point $Q' \in N \cap (D_T \cup B_T)$,

$$\psi(Q') < u(Q) + 2\varepsilon.$$

Hence, it follows, from $u(Q') \leq \psi(Q')$, that $u(Q') \leq u(Q)+2\varepsilon$ throughout $N \cap (D_T \cup B_T)$ and

$$u(Q) \leq \sup_{N \cap (D_T \cup B_T)} u \leq u(Q)+2\varepsilon.$$

Since such a neighborhood can be found for each $\varepsilon > 0$, we see that

$$u(Q) = \lim_{Q' \to Q} \sup u \qquad (16.3.3)$$

for any $Q \in B_T$.

16.4. BOUNDARY BEHAVIOR OF u

The remark following Theorem 16.3.1 demonstrated a rudimentary form of boundary behavior of u, which is independent of any properties of the boundary. We shall consider now various properties of the boundary B_T which imply that $u = F$ on B_T and that $u \in C(D_T \cup B_T)$.

We shall begin our consideration with the notion of a *barrier* at a point $Q \in B_T$.

DEFINITION 16.4.1. A function $w_Q(Q')$, which is defined and continuous in $D_T \cup B_T$, is a *barrier* at Q if

$$w_Q \text{ is superparabolic in } D_T \cup B_T, \qquad (16.4.1)$$

$$w_Q(Q) = 0, \qquad (16.4.2)$$

$$w_Q(Q') > 0, \qquad Q' \in D_T \cup B_T, \qquad Q' \neq Q. \qquad (16.4.3)$$

Theorem 16.4.1. If B_T possesses a barrier at $Q \in B_T$, then $u(Q) = F(Q)$ and u is two-dimensionally continuous at Q.

Proof. Since F is continuous on B_T, for $\varepsilon > 0$, there exists a neighborhood η_ε of Q such that, for $Q' \in B_T \cap \eta_\varepsilon$, $|F(Q')-F(Q)| < \varepsilon$. Now, on the closed bounded set $(D_T \cup B_T)- \eta_\varepsilon, w_Q$, the barrier at Q, is continuous and positive. Hence, there is a positive δ such that $w_Q > \delta$ for all $Q' \in (D_T \cup B_T)- \eta_\varepsilon$. We consider now the function

$$\psi(Q') = F(Q)+\varepsilon+\delta^{-1}\Big(\sup_{B_T}|F|+1\Big)w_Q(Q'). \qquad (16.4.4)$$

For $Q' \in B_T \cap \eta_\varepsilon$, $F(Q)+\varepsilon \geq F(Q')$, while, for $Q' \in B_T - \eta_\varepsilon$,

$$\delta^{-1}\Big(\sup_{B_T}|F|+1\Big)w_Q(Q') \geq F(Q').$$

Hence, $\psi(Q') \geq F(Q')$ for all $Q' \in B_T$. From Lemma 16.1.1 and Lemma 16.2.1, ψ is an upper function in $D_T \cup B_T$. Consequently, $F(Q) \leq u(Q) \leq$

$F(Q) + \varepsilon$. Also,

$$\tilde{\psi}(Q') = F(Q) - \varepsilon - \delta^{-1}\left(\sup_{B_T}|F| + 1\right)w_Q(Q') \qquad (16.4.5)$$

is a lower function. From Lemma 16.2.3, $\tilde{\psi}(Q') \le u(Q') \le \psi(Q')$. However, w_Q is continuous and $w_Q(Q) = 0$. Thus, there exists a neighborhood $\tilde{\eta}_\varepsilon$ of Q such that, for all $Q' \in (D_T \cup B_T) \cap \tilde{\eta}_\varepsilon$.

$$F(Q) - 2\varepsilon \le u(Q') \le F(Q) + 2\varepsilon. \qquad (16.4.6)$$

As Eq. (16.4.6) holds for each $\varepsilon > 0$, we conclude that $u(Q) = F(Q)$ and u is continuous at Q. $\square$

Now, we shall consider some specific properties of B_T that allow us to conclude the existence of a barrier.

Lemma 16.4.1.   *If* $(x_0, 0) \in B_T$, *then* $w = 2t + (x - x_0)^2$ *is a barrier.*

Proof. $\mathscr{L}(w) = 0$, $w(x_0, 0) = 0$, and $w > 0$ in $D_T \cup B_T - \{(x_0, 0)\}$. $\square$

Recall now that D_T is defined by (16.1.2), which involves the continuous functions s_1 and s_2 defined on $0 \le t \le T$. We concern ourselves now with the points $(s_i(t), t)$, $0 < t \le T$, $i = 1, 2$, of B_T.

Lemma 16.4.2.   *If at* t_0, $0 \le t_0 \le T$, s_2 *is left-upper-Hölder-continuous with Hölder exponent* $\gamma > \frac{1}{2}$, *then there exists a barrier at* $(s_2(t_0), t_0)$.

Proof.   The continuity condition on s_2 is simply the restriction that there exists an $\eta > 0$ and $C > 0$ such that

$$s_2(t) \le s_2(t_0) + C(t_0 - t)^\gamma, \qquad t_0 - \eta \le t \le t_0. \qquad (16.4.7)$$

Set

$$s(t) = s_2(t_0) + C(t_0 - t)^\gamma, \qquad t_0 - \eta \le t \le t_0, \qquad (16.4.8)$$

and consider the problem

$$w_t = w_{xx}, \qquad -\infty < x < s(t), \qquad t_0 - \eta < t \le t_0,$$
$$w(s(t), t) = \eta^{-1}(t_0 - t), \qquad (16.4.9)$$
$$w(x, t_0 - \eta) = 1, \qquad -\infty < x < s(t_0 - \eta).$$

Since $s(t) \in C^\gamma([t_0 - \eta, t_0])$, it follows, from the analysis of Chapter 14, that a solution $w(x, t)$ of Problem (16.4.9) exists and continuously assumes its initial and boundary values.

We extend w above t_0 by solving

$$w_t = w_{xx}, \qquad -\infty < x < \infty, \qquad t_0 < t \le T,$$
$$w(x, t_0) = \begin{cases} w(x, t_0), & -\infty < x \le s(t_0), \\ 1 - \exp\{-(x - s(t_0))\}, & s(t_0) < x < \infty. \end{cases}$$

$$(16.4.10)$$

For $t < (t_0 - \eta)$, we extend w by

$$w \equiv 1, \qquad t < (t_0 - \eta). \qquad (16.4.11)$$

Clearly, $\mathcal{L}(w) = 0$ in D_T, $w(s_2(t_0), t_0) = 0$, and $w(x, t) > 0$ in $(D_T \cup B_T) - \{(s_2(t_0), t_0)\}$. Hence w is a barrier at $(s_2(t_0), t_0)$. $\qquad \square$

As a corollary of the argument of Lemma 16.4.2, we obtain the corresponding result for s_1.

Lemma 16.4.3. *If at t_0, $0 < t_0 < T$, s_1 is left-lower-Hölder-continuous with Hölder exponent $\gamma > \frac{1}{2}$, then there exists a barrier at $(s_1(t_0), t_0)$.*

Proof. The proof is left to the reader. $\qquad \square$

Remark. Left-lower-Hölder-continuous means that there exists $\eta > 0$ and $C > 0$ such that

$$s_1(t) \geq s_1(t_0) - C(t_0 - t)^\gamma \qquad (16.4.12)$$

for $t_0 - \eta \leq t \leq t_0$.

As an example of curves for which barriers exist at every point we prove the following result.

Lemma 16.4.4. *If s_1 is decreasing and if s_2 is increasing, then B_T for such a D_T possesses a barrier at every point.*

Proof. Lemma 16.4.1 covers the points of $B_T \cap \{t = 0\}$. Consider the case of $(s_2(t_0), t_0)$, $0 < t_0 \leq T$. Since $s_2(t) \leq s_2(t_0)$ for $0 \leq t \leq t_0$, $s_2(t)$ is left-upper-Lipschitz-continuous at $t = t_0$ with Lipschitz constant $C = 0$. Hence, Lemma 16.4.2 applies immediately. Similarly, $s_1(t)$ is left-lower-Lipschitz-continuous with Lipschitz constant $C = 0$. Lemma 16.4.3 applies, and concludes the proof. $\qquad \square$

For continuous boundary curves s_1 and s_2, Lemma 16.4.4 is an example of the utilization of the translation of a single barrier function. If the boundary curves s_1 and s_2 are such that we can utilize the translation of a single barrier function, then we need only obtain a barrier for $D_t \cup B_t$ at $(s_i(t), t)$, $i = 1, 2$, since translations of it will work for $(s_i(\tau), \tau)$, $i = 1, 2$ with τ near t. From this we obtain

$$\lim_{\substack{x \to s_i(\tau) \\ x \in D_T}} u(x, \tau) = F(s_i(\tau), \tau)$$

uniformly for τ in a neighborhood of t.

From the uniform continuity of F in a neighborhood of t, it follows that u is two-dimensionally continuous at $(s_i(t), t)$, $i = 1, 2$, and $u(s_i(t), t)$ assumes the respective values of the data F.

In order to find a large class of barrier functions that can be used as translates, the following lemma is useful.

Lemma 16.4.5. *Let $a < 0$ and let φ be a positive-valued, continuously differentiable function defined on $a \le t < 0$ with $\lim_{t \uparrow 0} \varphi(t) = 0$. Let*

$$D_1 = \{(x, t) \mid a < t < 0, \quad -\varphi(t) < x < \varphi(t)\}$$

and

$$D_2 = \{(x, t) \mid a < t < 0, \quad -\varphi(t) < x < L\},$$

where $L \ge \sup_{a \le t < 0} \varphi(t)$. Then there exists a barrier at $(0,0)$ for D_1 if and only if there exists a barrier at $(0,0)$ for D_2.

Proof. Clearly, any barrier at $(0,0)$ for $D_2 \supset D_1$ is a barrier at $(0,0)$ for D_1. So, we need only prove that the existence of a barrier at $(0,0)$ for D_1 implies the existence of a barrier at $(0,0)$ for D_2. For this, we let F denote a continuous, nonnegative, not identically zero, function defined on the parabolic boundary of D_2 such that $F = 0$ at $(0,0)$. Since φ is continuously differentiable for $a \le t < 0$ and thus locally Lipschitz-continuous, the solution u exists and assumes continuously all of its initial and boundary values for all t such that $a \le t < 0$. We shall show that

$$\alpha \equiv \lim_{\substack{(x,t) \to 0 \\ (x,t) \in D_2}} \sup u \tag{16.4.13}$$

is zero. Let g be defined on the parabolic boundary of D_1 as

$$g = \begin{cases} F & \text{on } \partial D_1 \cap \partial D_2, \\ u & \text{on } \partial D_1 \cap D_2. \end{cases}$$

From the continuous differentiability of φ it follows from the Maximum Principle that the solution is again u in D_1. Reflecting across the t-axis we see that $\tilde{u}(x, t) = u(-x, t)$ is the solution of the problem in D_1 with data $\tilde{g}(x, t) = g(-x, t)$. The existence of a barrier at $(0,0)$ for D_1 enables us to conclude that

$$\lim_{(x,t) \to (0,0)} \sup(u(x,t) + \tilde{u}(x,t)) \le \lim_{\substack{(x,t) \to (0,0) \\ (x,t) \in \partial D_1}} \sup(g(x,t) + \tilde{g}(x,t)) \le \alpha$$

$$\tag{16.4.14}$$

As $\tilde{u}(0, t) = u(0, t)$, we see that

$$\limsup_{t \uparrow 0} 2u(0, t) \le \alpha \tag{16.4.15}$$

and

$$\limsup_{t \uparrow 0} u(0, t) \le \frac{\alpha}{2}. \tag{16.4.16}$$

Since $D_3 = D_1 \cap \{x < 0\} \subset D_1$, the barrier at $(0,0)$ for D_1 is also a barrier for this smaller region D_3. Also, we note that, trivially, a barrier at $(0,0)$ exists for the region

$$D_4 = \{(x, t) \mid a \le t < 0, 0 < x < L\}.$$

Consequently, from (16.4.15) and $F(0) = 0$, it follows that

$$\lim_{\substack{(x,t) \to 0 \\ (x,t) \in D_i}} \sup u(x,t) \le \frac{\alpha}{2}, \qquad i = 3,4. \tag{16.4.17}$$

But this implies that

$$\alpha \equiv \lim_{\substack{(x,t) \to 0 \\ (x,t) \in D_2}} \sup u(x,t) \le \frac{\alpha}{2}. \tag{16.4.18}$$

Thus, $\alpha = 0$ and the solution u constitutes a barrier for D_2. □

As a corollary of the argument of Lemma 16.4.5, we can state the following result.

Lemma 16.4.6. *Let φ and D_1 be as defined in Lemma 16.4.5. Let*

$$D_2 = \{(x,t) | a < t < 0, \quad -L < x < \varphi(t)\},$$

where $L \ge \sup_{a \le t \le 0} \varphi(t)$. Then there exists a barrier at $(0,0)$ for D_1 if and only if there exists a barrier at $(0,0)$ for D_2.

As we shall see, the symmetry of D_1 is a great asset in the determination of the existence of a barrier at $(0,0)$ for a large class of functions φ.

Lemma 16.4.7. *Let D be defined by the inequality $x^2 < 4t \log \rho(t)$, $t < 0$, where ρ is a positive, continuously differentiable function that decreases monotonically to zero as t increases to zero, such that*

$$\lim_{t \uparrow 0} t \log \rho(t) = 0 \tag{16.4.19}$$

and

$$\lim_{\varepsilon \uparrow 0} \int_{t_0}^{\varepsilon} \frac{\rho(\eta) \, d\eta}{\eta} = -\infty. \tag{16.4.20}$$

Then there exists a barrier at $(0,0)$ for D.

Proof. We shall construct the barrier at $(0,0)$ from the function ρ. We begin with the definition of the function φ via

$$6 \log \varphi(t) = \int_{t_0}^{t} \frac{\rho(\eta) \, d\eta}{\eta}, \tag{16.4.21}$$

where t_0 is a chosen negative constant and t is restricted to be above t_0. Clearly, from (16.4.20) it follows that φ tends to zero when t increases to zero. Moreover,

$$6 \frac{\varphi'}{\varphi} = t^{-1} \rho(t) < 0. \tag{16.4.22}$$

Consequently, φ decreases monotonically to zero.

Next, we define

$$f(t) = -\tfrac{1}{2}\rho(t)\varphi(t). \tag{16.4.23}$$

As $f < 0$ and $f(t)$ increases monotonically to zero as t increases to zero, we conclude that $f'(t) \geq 0$. Now, we also see that

$$\varphi' = 6^{-1}t^{-1}\rho(t)\varphi(t) = -\tfrac{1}{3}t^{-1}f(t). \tag{16.4.24}$$

Consequently, for

$$v = f(t)e^{-x^2/(4t)} + \varphi(t), \qquad t < 0, \tag{16.4.25}$$

we see that

$$\mathscr{L}(v) = -e^{-x^2/(4t)}\left\{ f'(t) + \frac{f(t)}{2t} + \varphi'(t)e^{x^2/(4t)}\right\} \tag{16.4.26}$$

and, from (16.4.22) and (16.4.23), it follows that in D

$$\mathscr{L}(v) < 0, \qquad t < 0, \qquad t_0 < t < 0, \tag{16.4.27}$$

where t_0 is sufficiently close to zero.

We shall leave to the reader the elementary application of the Maximum-Principle argument to show that if $\mathscr{L}(v) < 0$, then v is superparabolic. Next, we see that

$$v(0, t) = f(t) + \varphi(t) = \left(1 - \tfrac{1}{2}\rho(t)\right)\varphi(t) > 0, \tag{16.4.28}$$

when t is sufficiently near zero.

Remark. We could restrict ρ to be less than one without loss of generality, since the local behavior of ρ near $t = 0$ is the major point in the existence or nonexistence of a barrier at a point.

The equation of the level curve $v(x, t) = 0$ is given by the equation

$$-\tfrac{1}{2}\rho(t)e^{-x^2/(4t)} + 1 = 0, \tag{16.4.29}$$

since $\varphi > 0$. Solving (16.4.29) for x, we obtain

$$\begin{cases} x^2 = 4t(\log\rho(t) - \log 2), \text{ or} \\ x = \pm(4t(\log\rho(t) - \log 2))^{1/2}. \end{cases} \tag{16.4.30}$$

But the region determined by these curves contains D. Hence, $v > 0$ and superparabolic in D. From the symmetry of the function v, the maximum value for any $t < 0$ occurs at $x = 0$. From (16.4.28) and the fact that $\varphi(t)$ tends to zero as $t \uparrow 0$, it follows that

$$\lim_{\substack{(x,t) \to (0,0) \\ (x,t) \in D}} v(x, t) = 0. \tag{16.4.31}$$

Consequently, v is a barrier for D.

Remark. Some examples of functions ρ that satisfy the conditions of Lemma 16.4.7 are

$$\rho(t) = \{|\log|t|\,|\}^{-1},$$

$$\rho(t) = \{|\log|t|\,|\cdot\log|\log|t|\,|\}^{-1}, \qquad\qquad (16.4.32)$$

$$\rho(t) = \{|\log|t|\,|\cdot\log|\log|t|\,|\cdot\log\log|\log|t|\,|\}^{-1}. \quad \text{etc.}$$

Clearly, Lemmas 16.4.5, 16.4.6, and 16.4.7 provide a generalization of Lemmas 16.4.2 and 16.4.3. Further extensions can be made, but the methods employed are more advanced than those used thus far in this presentation.

We shall conclude our discussion on barriers here by demonstrating that Lemma 16.4.7, with $\rho(t) = \{|\log|t|\,|\}^{-1}$, is nearly best possible.

> **Lemma 16.4.8.** *Let D be the domain that is bounded by the curve*
>
> $$x^2 = -4(1+\varepsilon)t\log|\log|t|\,|$$
>
> *and the line*
>
> $$t = t_0 < 0.$$
>
> *Then, for any positive ε and $|t_0|$ sufficiently small, there cannot exist a barrier for D at $(0,0)$.*

Proof. We demonstrate first that, in a certain subdomain D_1 of D with common boundary point $(0,0)$ there exists a subparabolic function that possesses a discontinuity at $(0,0)$.

Consider the function

$$v(x,t) = -\{|\log|t|\,|\}^{-(1+\varepsilon_1)}\cdot\exp\left\{-\frac{x^2}{4t}k\right\} + \{\log|\log|t|\,|\}^{-1}, \qquad t < 0,$$
$$(16.4.33)$$

where ε_1 and k, $\frac{1}{2} < k < 1$, are certain positive constants that will be chosen below. Differentiating with respect to x and t, we obtain

$$\mathscr{L}(v) = +\exp\left\{-\frac{x^2 k}{4t}\right\}\left[\frac{x^2(k-k^2)}{4t^2|\log|t|\,|^{1+\varepsilon_1}} + \frac{1+\varepsilon_1}{t|\log|t|\,|^{2+\varepsilon_1}} + \frac{k}{2t|\log|t|\,|^{1+\varepsilon_1}}\right]$$

$$+\frac{1}{\log^2|\log|t|\,|\cdot\log|t|\cdot t}. \qquad (16.4.34)$$

Now, factoring out $1/(t|\log|t|\,|^{1+\varepsilon_1})$ from $\mathscr{L}(v)$, it follows that the sign of $\mathscr{L}(v)$ is determined by the sign of the expression

$$\frac{x^2(k-k^2)}{4t} + \frac{1+\varepsilon_1}{|\log|t|\,|} + \frac{k}{2} - \exp\left\{\frac{x^2}{4t}k\right\}\frac{|\log|t|\,|^{\varepsilon_1}}{\log^2|\log|t|\,|}. \quad (16.4.35)$$

In fact, $\mathscr{L}(v) > 0$ if the expression (16.4.35) is negative. We set $|t|$ sufficiently small so that

$$\frac{1+\varepsilon_1}{|\log|t||} < \frac{k}{2}. \tag{16.4.36}$$

Now, the second and third terms are estimated above by k. Hence the first and fourth terms must be employed to dominate k for any choice of x. Consider $|x|$ small. Clearly we must employ the fourth term. In this case the expression will be negative if

$$k < \exp\left\{\frac{x^2 k}{4t}\right\} \frac{|\log|t||^{\varepsilon_1}}{\log^2|\log|t||}, \tag{16.4.37}$$

which is true if

$$\log k < \frac{x^2 k}{4t} + \varepsilon_1 \log|\log|t|| - 2\log\log|\log|t||. \tag{16.4.38}$$

Assuming $|t|$ sufficiently small so that

$$\tfrac{1}{2}\varepsilon_1 \log|\log|t|| > 2\log\log|\log|t||, \tag{16.4.39}$$

we see that (16.4.38) is true if

$$\frac{x^2}{4|t|} < \frac{\varepsilon_1}{2k}\log|\log|t||. \tag{16.4.40}$$

On the other hand, for $|x|$ large, the expression (16.4.35) will be negative if

$$k < \frac{x^2(k-k^2)}{4|t|}, \tag{16.4.41}$$

which is true if

$$\frac{1}{1-k} < \frac{x^2}{4|t|}. \tag{16.4.42}$$

We select $|t|$ now sufficiently small so that

$$\frac{\varepsilon_1}{2k}\log|\log|t|| > \frac{1}{1-k}. \tag{16.4.43}$$

Then, for arbitrary x, we have that at least one of (16.4.37) and (16.4.41) is satisfied. Thus, for $|t|$ sufficiently small, expression (16.4.35) is negative, which implies that $\mathscr{L}(v) > 0$. By an easy application of the Maximum Principle, it follows that v is subparabolic for $|t|$ sufficiently small.

We consider now the level curve determined by the equation

$$v(x,t) = c < 0. \tag{16.4.44}$$

This can be solved to obtain the equation

$$x^2 = -4t\left[\frac{1+\varepsilon_1}{k}\log|\log|t|| + \frac{1}{k}\log\left(-c + \frac{1}{\log|\log|t||}\right)\right]. \tag{16.4.45}$$

We shall denote by D_1 the domain that is bounded by this curve and the line $t = t_0 < 0$, where $|t_0|$ is sufficiently small so that $\mathcal{L}(v) > 0$ in D_1. *Note that along the curve given in (16.4.45), $v \equiv c < 0$, while for $x = 0$ and $t \uparrow 0$, v tends to zero.*

Now if k is near one and ε_1 is sufficiently small, then $D_1 \subset D_2$, provided that $|t_0|$ is sufficiently small. Looking back through the proof, we see that only a finite number of selections of $|t_0|$ was needed, to accomplish this construction.

We can now show that there cannot exist a barrier for D at $(0,0)$. If there were a barrier for D at $(0,0)$, then it would also be a barrier for D_1 at $(0,0)$. Consider the solution to the heat equation in D_1 with boundary values c on the curve (16.4.42) and initial data $v(x, t_0)$. Then, u is continuous at $(0,0)$ and

$$\lim_{\substack{(x,t) \to (0,0) \\ (x,t) \in D_1}} u = c < 0. \tag{16.4.46}$$

However, as $v(x, t)$ is subparabolic in D_1, $u(x, t) \geq v(x, t)$ at all points of D_1. Thus,

$$\liminf_{t \uparrow 0} u(0, t) \geq \lim_{t \uparrow 0} v(0, t) \geq 0, \tag{16.4.47}$$

which is a contradiction. Consequently, there cannot exist a barrier for D at $(0,0)$. $\square$

Remark. For any domain that contains a translate of D such that the translate of $(0,0)$ coincides with a boundary point of that domain, it follows that there cannot exist a barrier for the domain at that boundary point. Hence, no guarantee can be given that a solution to the heat equation will assume its boundary values continuously at that boundary point.

EXERCISES

16.1. Prove Lemma 16.1.1.
16.2. Show that $u(x, t) = \sup_{\psi \in L} \psi(x, t)$, where L is the class of lower functions, satisfies $\mathcal{L}(u) = 0$. *Hint.* Redo Secs. 16.2 and 16.3.
16.3. For the $u(x, t)$ defined in Exercise 16.2, show that $u(Q) = \lim_{Q' \to Q} \inf u$ for any $Q \in B_T$.
16.4. Demonstrate that the $\rho(t)$ given by (16.4.32) satisfy the conditions of Lemma 16.4.7.

NOTES

Extending the ideas of Perron [3] for the Dirichlet problem, Sternberg [6] demonstrated existence using the ideas of superparabolic and subparabolic functions. Petrovsky [4] studied the boundary behavior of solutions with respect to the behavior of the boundary itself. He derived almost necessary and sufficient conditions on the behavior of the boundary in order that a boundary point be regular. Landis [2] has announced necessary and sufficient conditions for the regularity of a boundary point. Recently, Evans and Gariepy [1] have demonstrated necessary and sufficient conditions for the regularity of a boundary point. The treatment of existence given here follows that given by Petrovsky [5] for the Dirichlet problem, as modified and presented by Jim Douglas, Jr., in a series of lectures at Rice University in 1959. The treatment of the boundary behavior of u is taken from Petrovsky [4], employing some modifications of proofs communicated to the author by B. Frank Jones.

REFERENCES

1. Evans, L. C., and R. Gariepy, A. Wiener criterion for the heat equation, *Arch. Rat. Mech. Anal.*, **78** (1982), 293–314.
2. Landis, E. M., Necessary and sufficient conditions for regularity of a boundary point in the Dirichlet problem for the heat-conduction equation, *Soviet Math. Dokl.*, **10** (1969), 380–384.
3. Perron, O., Eine neue Behändlung der Randwertaufgabe der $\Delta u = 0$, *Math. Zeit.*, **18** (1923), 42–54.
4. Petrovsky, I. G., Zur ersten Randwertaufgabe der Wärmeleitungsgleichung, *Composito Math.*, **1** (1935), 383–419.
5. Petrovsky, I. G., *Lectures on Partial Differential Equations.* Interscience Pub., Inc., New York and London, 1954.
6. Sternberg, W., Über die Gleichung der Wärmeleitung, *Math. Annalen.*, **101** (1929), 394–398.

Chapter 17

The One-Phase Stefan Problem with Temperature-Boundary Specification

17.1. INTRODUCTION

The one-phase Stefan problem is one of the simplest examples of a free-boundary-value problem for the heat equation. Physically, we consider the situation of a half-plane region of ice at 0°C in contact with an infinite strip of water. We ask the question "If circulation in the water is ignored, can we determine the location of the ice–water interface from the initial and boundary temperature of the water and the initial location of the interface?" A moment's reflection leads us to the realization that the domain of the temperature u is now one of the unknowns of the problem. Consequently, we must utilize the physics of the situation to derive another relation between the solution and its domain of definition.

Setting $x = 0$ on the known face of the infinite strip of water and setting the positive x-axis perpendicular to it and pointing in the direction of the ice (see Fig. 17.1.1), we can describe the domain of the temperature of the water by

$$D_T = \{(x, t) | 0 < x < s(t), \quad 0 < t \leq T\},$$

where T is a fixed positive constant and $x = s(t)$ denotes the location of the water–ice interface at time t relative to the known water interface at $x = 0$. Ignoring circulation in the water, specifying initial and boundary temperatures, and specifying the initial position of the interface, we have the

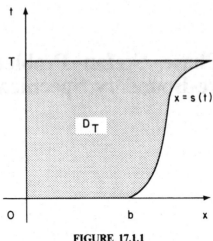

FIGURE 17.1.1

problem

$$u_t = \kappa u_{xx}, \qquad 0 < x < s(t), \qquad 0 < t \le T,$$

$$u(0, t) = f(t) \ge 0, \qquad 0 < t \le T,$$

$$u(x, 0) = \varphi(x) \ge 0, \qquad 0 \le x \le b = s(0),$$

$$u(s(t), t) = 0, \qquad 0 < t \le T,$$

for the determination of u and D_T. (See Fig. 17.1.2.) We have seen that we must determine an additional condition relating u and D_T in order to uniquely determine a solution. As it stands now, the problem stated above has a solution for any reasonable $s(t)$ such that $s(0) = b$. Our task now is to employ the physics of the situation to determine a mathematical relation between u and s which, when added to the problem stated above, will uniquely determine u and s. Since we have assumed that the ice is at 0°C, then all of the heat energy arriving at the ice–water interface must be utilized in the melting process. The rate of change of s with respect to time is proportional to an energy rate. This energy rate can only be supplied via the flow rate of energy from the water at the interface. Utilizing Fourier's law (1.1.1), we obtain

$$\alpha \dot{s}(t) = - k u_x(s(t), t),$$

where α is the product of the density of the ice and the latent heat of fusion, $k = c\rho\kappa$ is the thermal conductivity of the water, and $\dot{s}$ denotes the derivative of s with respect to t. In our study of the one-phase Stefan problem, we shall assume that the various physical constants α, k, and κ are one. This can be achieved by changes of scale for x, t, and u, and is left as an exercise

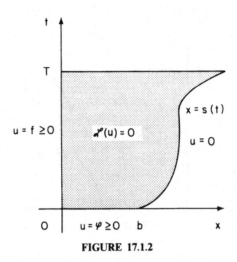

FIGURE 17.1.2

for the reader. Thus, all of the arguments given below are valid for arbitrary positive α, k, and κ.

In this chapter we shall consider the Stefan problem of determining two functions $u = u(x, t)$ and $s = s(t)$, such that the pair satisfies

$$
\begin{aligned}
\mathscr{L}(u) \equiv u_{xx} - u_t = 0, \quad & 0 < x < s(t), \quad 0 < t \le T, \\
u(0, t) = f(t) \ge 0, \quad & 0 < t \le T, \\
u(x, 0) = \varphi(x) \ge 0, \quad & 0 \le x \le b \equiv s(0), \\
u(s(t), t) = 0, \quad & 0 < t \le T,
\end{aligned}
\tag{17.1.1}
$$

and

$$
\dot{s}(t) = -u_x(s(t), t), \qquad 0 < t \le T. \tag{17.1.2}
$$

(See Fig. 17.1.3.) It is convenient here to define what we mean by a solution of (17.1.1) for a given continuous $s(t)$ and the Stefan problem (17.1.1)–(17.1.2).

DEFINITION 17.1.1. A solution of (17.1.1) for a known continuous $s(t)$ is a function $u = u(x, t)$ defined in $D_T \cup B_T$ such that u_{xx}, $u_t \in C(D_T)$, u satisfies the conditions of (17.1.1) and $u \in C(D_T \cup B_T)$ except at points of discontinuity of f and φ. Also, u is bounded in D_T, which implies that at a point of discontinuity of f or φ, $0 \le \liminf u \le \limsup u < \infty$.

DEFINITION 17.1.2. A solution (s, u) of the Stefan problem (17.1.1)–(17.1.2) is a pair of functions $s = s(t)$ and $u = u(x, t)$ such that $s(0) = b$, $s(t) > 0$, $0 < t \le T$, $s \in C^1((0, T]) \cap C^0([0, T])$, u satisfies (17.1.1)

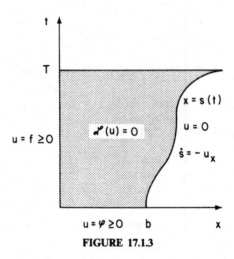

FIGURE 17.1.3

for this s in the sense of Definition 17.1.1, $u_x(s(t), t)$ exists and is continuous, and s and u satisfy (17.1.2).

Our first result is an easy consequence of Green's Theorem.

Lemma 17.1.1. *The condition*

$$\dot{s}(t) = -u_x(s(t), t) \tag{17.1.2}$$

can be replaced by the equivalent condition

$$s(t)^2 = b^2 + 2\int_0^t f(\tau)\, d\tau + 2\int_0^b \xi \varphi(\xi)\, d\xi - 2\int_0^{s(t)} \xi u(\xi, t)\, d\xi. \tag{17.1.3}$$

Proof. Setting $\mathscr{L}(u) = 0$ and $v = -x$ in the Green's Identity

$$\int_D \{v\mathscr{L}(u) - uMv\}\, dx\, dt = \oint_{\partial D} \{(vu_x - uv_x)\, dt + uv\, dx\},$$

where $Mv \equiv v_{xx} + v_t$, we obtain the identity

$$\oint_{\partial D} \{(u - xu_x)\, dt - xu\, dx\} = 0. \tag{17.1.4}$$

Let (s, u) be a solution of (17.1.1)–(17.1.2). Applying the identity to

$$D = D_T \cap \{(\xi, \tau) | 0 < \varepsilon \le \tau \le t\}$$

with $t \le T$ and letting ε tend to zero, we obtain (17.1.3).

Conversely, if (s, u) satisfies (17.1.1) and (17.1.3) and if u_x exists and is continuous at the boundary $x = s(t)$, then it follows, from the identity

(17.1.3) and (17.1.4), that

$$s(t)^2 = b^2 - 2\int_0^t s(\tau)u_x(s(\tau),\tau)\,d\tau. \qquad (17.1.5)$$

Differentiating (17.1.5), we see that (17.1.2) is satisfied.   □

Remark. The equivalence in Lemma 17.1.1 was defined in the proof to be valid when u_x exists and is continuous at the boundary $x = s(t)$. From Theorem 14.4.1, we see that, for a solution u of (17.1.1), u_x will exist and be continuous at the boundary $x = s(t)$ for s that is Hölder continuous with Hölder exponent greater than $\frac{1}{2}$. In what follows below, we shall obtain solutions of (17.1.1)–(17.1.2) with Lipschitz-continuous s, which, on the basis of Theorem 14.4.1, will validate the equivalence in Lemma 17.1.1.

17.2. MONOTONE DEPENDENCE UPON THE DATA AND UNIQUENESS OF THE SOLUTION

From the Lemma 17.1.1 we see that we can use either (17.1.2) or (17.1.3) interchangeably in discussing properties of solutions of the Stefan problem (17.1.1)–(17.1.2). The property of monotone dependence upon the data is a property that seems to require the use of each condition in order to complete the demonstration. Throughout our discussions in this section we shall assume the existence of solutions to (17.1.1)–(17.1.2).

We shall begin with the following result.

Lemma 17.2.1. *Let (s_1, u_1) and (s_2, u_2) be solutions to the Stefan problem (17.1.1)–(17.1.2) corresponding, respectively, to the data (b_1, φ_1, f_1) and (b_2, φ_2, f_2). If $b_1 < b_2$, $\varphi_1 \le \varphi_2$, and $f_1 \le f_2$, then $s_1 < s_2$.*

Proof. Assume the contrary to the assertion and let $t_0 > 0$ be the first t for which $s_1(t) = s_2(t)$. Then, $\dot{s}_1(t_0) \ge \dot{s}_2(t_0)$. By the Strong Maximum Principle, Theorem 15.3.1, $u_2(s_1(t), t) > 0$ for $0 < t < t_0$, where we have assumed, without loss of generality, that either $\varphi_2 \ne 0$ or $f_2 \ne 0$. Hence, $u_2 - u_1 > 0$ in the region $0 < x < s_1(t)$, $0 < t < t_0$. Since both u_1 and u_2 vanish at the boundary point $(s_1(t_0), t_0)$, it follows, from Theorem 15.4.1 that

$$\dot{s}_1(t_0) - \dot{s}_2(t_0) = u_{2x}(s_1(t_0), t_0) - u_{1x}(s_1(t_0), t_0) < 0,$$

which is a contradiction.   □

We see here the use of (17.1.2) in order to produce a partial result on monotonicity, which now is employed with (17.1.3) to produce the following general result.

Theorem 17.2.1 (Monotonicity). *Let (s_1, u_1) and (s_2, u_2) be solutions to the Stefan problem* (17.1.1)–(17.1.2) *corresponding, respectively, to the data (b_1, φ_1, f_1) and (b_2, φ_2, f_2). If $b_1 \leq b_2$, $\varphi_1 \leq \varphi_2$, and $f_1 \leq f_2$, then $s_1 \leq s_2$.*

Proof. For $\delta > 0$ let (s_2^δ, u_2^δ) be a solution of Problem (17.1.1)–(17.1.2), which corresponds to the data $(b_2 + \delta, \varphi_2, f_2)$, where we extend φ_2 as zero outside of $0 \leq x \leq b_2$. From Lemma 17.2.1 we see that $s_1 < s_2^\delta$. Now, (s_2, u_2) and (s_2^δ, u_2^δ), which are solutions of (17.1.1)–(17.1.2), must each satisfy their respective versions of (17.1.3). We subtract those versions to obtain

$$s_2^\delta(t)^2 - s_2(t)^2 = (b_2 + \delta)^2 - b_2^2 - 2\int_0^{s_2^\delta(t)} \xi \left\{ u_2^\delta(\xi, t) - u_2(\xi, t) \right\} d\xi$$

$$- 2\int_{s_2(t)}^{s_2^\delta(t)} \xi u_2^\delta(\xi, t) \, d\xi, \tag{17.2.1}$$

where we have used Lemma 17.2.1 again to assert $s_2(t) < s_2^\delta(t)$. By the Maximum Principle $u_2^\delta \geq 0$ and $u_2^\delta - u_2 \geq 0$. Hence,

$$s_2^\delta(t)^2 \leq s_2(t)^2 + (b_2 + \delta)^2 - b_2^2. \tag{17.2.2}$$

Therefore,

$$s_1(t)^2 < s_2(t)^2 + (b_2 + \delta)^2 - b_2^2. \tag{17.2.3}$$

As $\delta > 0$ can be selected as small as desired, it follows that $s_1(t) \leq s_2(t)$. □

As a corollary of Theorem 17.2.1, we obtain the following result.

Theorem 17.2.2 (Uniqueness). *There can exist at most one solution to the Stefan problem* (17.1.1)–(17.1.2).

Proof. If (s_1, u_1) and (s_2, u_2) both correspond to the data (b, φ, f), then, from Theorem 17.2.1, $s_1 \leq s_2$ and $s_2 \leq s_1$, which implies that $s_1 \equiv s_2$. From this, the Maximum Principle, applied via the Extended Uniqueness Theorem 1.6.6, yields $u_1 \equiv u_2$. □

We conclude this section with a very important observation on the behavior of the free boundary s.

Lemma 17.2.2. *The free boundary s is an increasing function.*

Proof. From the general assumptions $\varphi \geq 0$ and $f \geq 0$, it follows from the Maximum Principle that $u \geq 0$. If $u \equiv 0$, then $u_x \equiv 0$, which implies that $\dot{s} \equiv 0$. On the other hand, if $u > 0$, then Theorem 15.4.1 implies that $u_x(s(t), t) < 0$, which means that $\dot{s}(t) > 0$. In either case, $\dot{s}(t) \geq 0$, which means that s must be an increasing function. □

Remark. If $\varphi \not\equiv 0$ or $f \not\equiv 0$ in every neighborhood of $t = 0$, then the Strong Maximum Principle, Theorem 15.3.1, implies that $\dot{s}(t) > 0$, which means that s must be a strictly increasing function of t.

Remark.   These results are in line with physical reality. Monotonicity is just the mathematical statement that ice melts faster when the initial and/or boundary temperature of the water is raised. For example a cube of ice thrown into boiling water will disappear faster than a similar cube thrown into cold water. The free boundary $s(t)$ being an increasing function says that ice at $0°C$ cannot freeze water. It can only melt.

17.3. EXISTENCE

We begin our study of existence with the important special case of $b = 0$, $\varphi \equiv 0$, and $f \equiv \lambda > 0$.

It is easy to see that the function

$$v(x,t) = \int_0^{x/(2t^{1/2})} \exp\{-\rho^2\} \, d\rho \tag{17.3.1}$$

satisfies the heat equation. By differentiation, we obtain

$$v_x(x,t) = 2^{-1}t^{-1/2}\exp\left\{-\frac{x^2}{4t}\right\}. \tag{17.3.2}$$

Now, given a positive constant c, we shall assume that

$$s(t) = ct^{1/2}. \tag{17.3.3}$$

Consequently,

$$\dot{s}(t) = 2^{-1}ct^{-1/2}. \tag{17.3.4}$$

Next, we note that

$$v_x(s(t),t) = 2^{-1}t^{-1/2}\exp\left\{-\frac{c^2}{4}\right\}. \tag{17.3.5}$$

Consequently,

$$c\exp\left\{\frac{c^2}{4}\right\}v_x(s(t),t) = \dot{s}(t). \tag{17.3.6}$$

Consider now the function

$$u(x,t) = \Lambda - c\exp\left\{\frac{c^2}{4}\right\}v(x,t), \tag{17.3.7}$$

where Λ is a positive constant that is defined as

$$\Lambda = c\exp\left\{\frac{c^2}{4}\right\}\int_0^{c/2}\exp\{-\rho^2\} \, d\rho. \tag{17.3.8}$$

Then u satisfies the heat equation in $0 < x < ct^{1/2}$, $0 < t$, $u(0,t) \equiv \Lambda$ for $0 < t$, $u(ct^{1/2},t) = 0$, and $-u_x(ct^{1/2},t) = 2^{-1}ct^{-1/2}$. Moreover, u is bounded.

Thus, proceeding from a knowledge of the free boundary $s(t) = ct^{1/2}$, $c > 0$, we see that there exists a function u such that the pair $(ct^{1/2}, u)$ satisfies (17.1.1)–(17.1.2) for the data $b = 0$, $\varphi \equiv 0$, and $f \equiv \Lambda(c)$. This is not quite what we had in mind. However, the function $\Lambda(c)$ is continuous, $\Lambda(0) = 0$, $\lim_{c \to +\infty} \Lambda(c) + \infty$, and $\Lambda'(c) > 0$, $0 < c < \infty$, since Λ is the product of three positive increasing functions. Thus, for each $\lambda > 0$, the equation

$$\lambda = \Lambda(c) \tag{17.3.9}$$

possesses a unique solution $c = c(\lambda)$. Defining

$$u^\lambda(x, t) = \lambda - c(\lambda) \exp\left\{\frac{c(\lambda)^2}{4}\right\} v(x, t), \tag{17.3.10}$$

we conclude that $(c(\lambda)t^{1/2}, u^\lambda)$ is the unique solution for the Stefan problem (17.1.1)–(17.1.2) for the data $b = 0$, $\varphi \equiv 0$, and $f \equiv \lambda > 0$.

We summarize the preceding discussion in the following statement.

Lemma 17.3.1. *For the data $b = 0$, $\varphi \equiv 0$ (φ is irrelevant in this case), and $f \equiv \lambda$, where λ is a positive constant, there exists a unique positive constant $c = c(\lambda)$ such that the pair*

$$s(t) = c(\lambda)t^{1/2}, \qquad 0 < t, \tag{17.3.11}$$

and

$$u^\lambda(x, t) = \lambda - c(\lambda) \exp\left\{\frac{c(\lambda)^2}{4}\right\} v(x, t), \tag{17.3.12}$$

where

$$v(x, t) = \int_0^{x/(2t^{1/2})} \exp\{-\rho^2\} \, d\rho, \tag{17.3.13}$$

is the unique solution to the Stefan problem (17.1.1)–(17.1.2).

We continue our study of existence with the important case of:

Assumption A. $b > 0, 0 \le \varphi(x) \le N(b - x), 0 \le x \le b$, *N is a positive constant, and φ and f are nonnegative piecewise-continuous functions.*

Under this assumption we can derive an important *a priori* estimate on $u_x(s(t), t)$ for any solution u of (17.1.1) for an increasing boundary s.

Lemma 17.3.2. *Under Assumption A, let u denote the solution of (17.1.1) for the nondecreasing boundary s. Then there exists a positive constant C depending only on b, $M = \max(\max_{0 \le t \le T} f(t), \max_{0 \le x \le b} \varphi(x))$, and the Lipschitz constant N in Assumption A, such that*

$$0 \le \rho^{-1} u(s(t) - \rho, t) \le C \tag{17.3.14}$$

for all $0 \le t \le T$ and $0 < \rho < b$.

Proof. Since $f, \varphi \geq 0$, it follows from the Maximum Principle that $u \geq 0$. Set

$$C = \max\{Mb^{-1}, N\} \tag{17.3.15}$$

and define for each t_0, $0 \leq t_0 \leq T$,

$$w(x) = C\{s(t_0) - x\}. \tag{17.3.16}$$

Since $\mathscr{L}(w) = 0$, $w(0) = Cs(t_0) \geq Cb \geq M \geq f$, $w(x) \geq C(b - x) \geq N(b - x) \geq \varphi$ and, for $0 \leq t \leq t_0$, $w(s(t)) = C\{s(t_0) - s(t)\} \geq 0$, it follows from the Maximum Principle that $u(x, t) \leq w(x)$ in the region $0 \leq x \leq s(t)$, $0 \leq t \leq t_0$. Thus,

$$0 \leq \rho^{-1} u(s(t_0) - \rho, t_0) \leq w(s(t_0) - \rho)\rho^{-1} = \rho^{-1} C\rho = C \tag{17.3.17}$$

for any $0 \leq t_0 \leq T$. $\qquad\qquad\square$

Remark. In the event that $u_x(s(t_0), t_0)$ exists, the lemma shows that

$$-C \leq u_x(s(t_0), t_0) \leq 0. \tag{17.3.18}$$

For each τ, $0 < \tau < b$, we now construct a family of approximations (s^τ, u^τ) to the solution of (17.1.1)–(17.1.2) by utilizing equation

$$\dot{s}^\tau(t) = -u_x^\tau(s^\tau(t - \tau), t - \tau), \tag{17.3.19}$$

which amounts to a retardation of the argument in the free-boundary equation (17.1.2). Let

$$\chi^\tau = \begin{cases} 1, & 0 \leq x \leq b - \tau, \\ 0, & b - \tau < x \leq b, \end{cases} \tag{17.3.20}$$

and $\varphi^\tau = \chi^\tau \varphi$. In the first interval $0 \leq t \leq \tau$, we set $s^\tau(t) \equiv b$ and define u^τ to be the unique solution of (17.1.1) in the region $0 \leq x \leq s^\tau(t)$, $0 \leq t \leq \tau$, where u and s have been, respectively, replaced by s^τ and φ^τ. Clearly, $u_x^\tau(b, t)$ exists and is continuous for $0 \leq t \leq \tau$. Moreover, by Lemma 17.3.2, $-C \leq u_x^\tau(b, t) \leq 0$ for $0 \leq t \leq \tau$. Proceeding by induction, we assume that (s^τ, u^τ) has been constructed for $0 \leq t \leq n\tau$, that $s^\tau \in C^1$, that $u_x^\tau(s^\tau(t), t)$ exists and is continuous, that $-C \leq u_x^\tau(s^\tau(t), t) \leq 0$, and that, if $t \geq \tau$,

$$s^\tau(t) = b - \int_\tau^t u_x^\tau(s^\tau(\eta - \tau), \eta - \tau)\, d\eta. \tag{17.3.21}$$

(See Fig. 17.3.1.) In the next interval $n\tau \leq t \leq (n + 1)\tau$ we define $s^\tau(t)$ by (17.3.21) and solve (17.1.1) for $u^\tau(x, t)$ in the region $0 \leq x \leq s^\tau(t)$, $n\tau \leq t \leq (n + 1)\tau$. By the inductive hypothesis on u_x^τ, s^τ is a C^1 function satisfying $0 \leq \dot{s}^\tau(t) \leq C$. Hence, from Eq. (17.3.14) of Lemma 17.3.2, Theorem 14.4.1, $u_x^\tau(s^\tau(t), t)$ exists and is continuous. By Lemma 17.3.2, $-C \leq u_x^\tau(s^\tau(t), t) \leq 0$, which completes the induction step. Hence, the approximation (s^τ, u^τ) can be constructed for $0 < t \leq T$.

Lemma 17.3.3. *Under Assumption A, there exists a unique solution (s, u) to the Stefan problem (17.1.1)–(17.1.2). Moreover, the free boundary s,* *which is C^1 and nondecreasing, satisfies*

$$0 \le \dot{s}(t) \le C, \tag{17.3.22}$$

where C is defined by (17.3.15) and the constants in Lemma 17.3.2.

    Proof. Since $0 \le \dot{s}^\tau(t) \le C$ and $b \le s^\tau(t) \le b + CT$ holds for every τ, $0 < \tau < b$, the functions s^τ are equicontinuous and uniformly bounded. By the Theorem of Ascoli–Arzela, we can choose a subsequence $\{s^{\tau_i}\}$ such that as τ_i tends to zero the s^{τ_i} converge uniformly on $0 \le t \le T$ to a nondecreasing, Lipschitz-continuous function s. Let u denote the solution of (17.1.1) for this particular s. Now, given any $\varepsilon > 0$, we can take τ_i sufficiently small so that

$$|s^{\tau_i}(t) - s(t)| \le C^{-1}\varepsilon, \qquad 0 \le t \le T, \tag{17.3.23}$$

and that

$$|\varphi^{\tau_i}(x) - \varphi(x)| \le N\tau^i \le \varepsilon, \tag{17.3.24}$$

$0 \le x \le b$. Extending the functions u and u^{τ_i} by setting them equal to zero outside of their natural domains of definition, we can apply the Maximum Principle and Lemma 17.3.2 to the difference $u^{\tau_i} - u$ to obtain, from (17.3.23) and (17.3.24), for all τ_i sufficiently small, the bound

$$|u^{\tau_i}(x,t) - u(x,t)| \le \varepsilon \tag{17.3.25}$$

for $0 \le x \le \max(s^{\tau_i}(t), s(t))$ and $0 \le t \le T$. Consequently, the sequence u^{τ_i} converges uniformly to u.

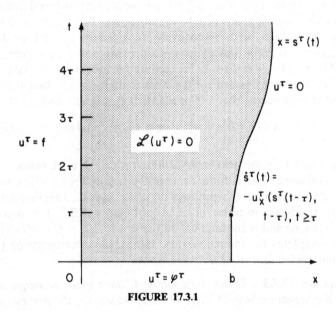

FIGURE 17.3.1

We now apply the Green's identity (17.1.4) to the approximation (s^{τ_i}, u^{τ_i}) for $0 \le \xi \le s^{\tau_i}(\eta)$, $0 \le \eta \le t \le T$. Utilizing (17.3.19) and (17.3.21), we obtain

$$2\int_0^t s^{\tau_i}(\eta)\dot{s}^{\tau_i}(\eta + \tau_i)\,d\eta = 2\int_0^t f(\eta)\,d\eta + 2\int_0^b \xi\varphi^{\tau_i}(\xi)\,d\xi$$
$$- 2\int_0^{s^{\tau_i}(t)} \xi u^{\tau_i}(\xi, t)\,d\xi. \qquad (17.3.26)$$

Now, the left side of (17.3.26) can be written as

$$s^{\tau_i}(t + \tau_i)^2 - s^{\tau_i}(\tau_i)^2 - 2\int_0^t \{s^{\tau_i}(\eta + \tau_i) - s^{\tau_i}(\eta)\}\dot{s}^{\tau_i}(\eta + \tau_i)\,d\eta$$
$$= s^{\tau_i}(t + \tau_i)^2 - b^2 + \mathcal{O}(\tau_i), \qquad (17.3.27)$$

since $0 \le \dot{s}^{\tau}(t) \le C$ and $|s^{\tau_i}(\eta + \tau_i) - s^{\tau_i}(\eta)| = |\dot{s}^{\tau_i}(\tilde{\eta})|\tau_i$ via the Mean-Value Theorem. Thus, taking the limit in (17.3.26) as modified by (17.3.27) as τ_i tends to zero, it follows, from the uniform convergence of s^{τ_i} to s, φ^{τ_i} to φ, and u^{τ_i} to u, that (s, u) satisfies (17.1.3). But as s is Lipschitz-continuous, u_x is continuous at $x = s(t)$, and Lemma 17.1.1 applies. Hence, (s, u) is the solution to the Stefan problem (17.1.1)–(17.1.2) under Assumption A. Uniqueness has been demonstrated in Theorem 17.2.2.   □

We continue our discussion of the existence of the solution to the Stefan problem (17.1.1)–(17.1.2) with the

Assumption B. $b = 0$, there is no φ, and $f(0) \ne 0$, $f(t) \ge 0$, and f is piecewise-continuous (and continuous at 0).

Under Assumption B, we see that there exists an $\eta > 0$ such that f is continuous on $[0, \eta]$ and there exist two positive constants λ, Λ, such that

$$0 < \lambda \le f(t) \le \Lambda \qquad (17.3.28)$$

for $0 \le t \le \eta$. We shall restrict our attention to $0 \le t \le \eta$. Afterwards we shall show that Lemma 17.3.3 applies, to extend the solution into $\eta \le t \le T$.

For each $b > 0$, let (s^b, u^b) denote the solution of the Stefan problem (17.1.1)–(17.1.2) that corresponds to the data $(b, \varphi \equiv 0, f)$. From the monotonicity Lemma 17.2.1 and the existence Lemma 17.3.1, we see that there exists a positive constant $c(\lambda)$ such that

$$c(\lambda)t^{1/2} < s^b(t) \qquad (17.3.29)$$

for $0 \le t \le \eta$. From the Maximum Principle, $0 \le u^b \le \Lambda$. Considering the Stefan problem (see Fig. 17.3.2)

$$\begin{aligned}
\mathcal{L}(v) &= 0, & b < x < s(t), & \quad 0 < t \le \eta, \\
v(b, t) &= u^b(b, t), & 0 < t \le \eta, \\
v(s(t), t) &= 0, & 0 < t \le \eta, \\
-v_x(s(t), t) &= \dot{s}(t), & 0 < t \le \eta,
\end{aligned} \qquad (17.3.30)$$

it follows from the Uniqueness Theorem 17.2.2 that $s = s^b$ and $v = u^b$

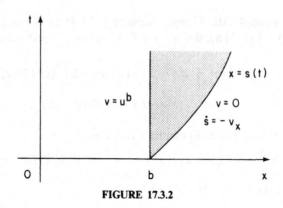

FIGURE 17.3.2

restricted to $b \leq x \leq s^b(t)$, $0 \leq t \leq \eta$. Replacing now $u^b(b,t)$ by Λ in (17.3.30), we obtain a translation of the solution given by Lemma 17.3.1. Utilizing the Monotonicity Theorem 17.2.1, we see that there exists a positive constant $c(\Lambda)$ such that

$$s^b(t) - b \leq c(\Lambda)t^{1/2} \tag{17.3.31}$$

for $0 \leq t \leq \eta$. Combining (17.3.29) and (17.3.31), we see that

$$c(\lambda)t^{1/2} < s^b(t) \leq b + c(\Lambda)t^{1/2}. \tag{17.3.32}$$

We can use (17.3.32) to derive a uniform estimate of $u_x^b(s^b(t), t)$ and thus $\dot{s}^b(t)$. For each $t_0 > 0$, set

$$w(x) = \Lambda c(\lambda)^{-1}t_0^{-1/2}(s^b(t_0) - x). \tag{17.3.33}$$

Clearly, $\mathcal{L}(w) = 0$, $w(s^b(t)) = \Lambda c(\lambda)^{-1}t_0^{-1/2}(s^b(t_0) - s^b(t)) \geq 0$ for $0 \leq t \leq t_0$, since s^b is nondecreasing, and

$$w(0) = \Lambda c(\lambda)^{-1}t_0^{-1/2}s^b(t_0) \geq \Lambda \geq f(t), \qquad 0 \leq t \leq t_0.$$

By the Maximum Principle, $w(x) \geq u^b(x,t)$ for $0 \leq x \leq s^b(t)$, $0 \leq t \leq t_0$. Now, at $(s^b(t_0), t_0)$, $w(s^b(t_0)) = u^b(s^b(t_0), t_0) = 0$. Hence,

$$0 \leq \rho^{-1}u^b(s^b(t_0) - \rho, t_0) \leq \Lambda c(\lambda)^{-1}t_0^{-1/2}, \tag{17.3.34}$$

which implies that

$$-\Lambda c(\lambda)^{-1}t_0^{-1/2} \leq u_x^b(s^b(t_0), t_0) \leq 0. \tag{17.3.35}$$

From this we derive the result that

$$0 \leq \dot{s}^b(t) \leq \Lambda c(\lambda)^{-1}t^{-1/2} \tag{17.3.36}$$

holds for $0 < t \leq \eta$.

Now, let $0 < b \leq 1$ and set $\sigma^b(t) = ts^b(t)$. Then,

$$0 \leq \sigma^b(t) \leq \eta(1 + c(\Lambda)\eta^{1/2}) \qquad \text{for } 0 \leq t \leq \eta.$$

Also, $\dot{\sigma}^b(t) = s^b(t) + t \dot{s}^b(t)$, so that $0 \leq \dot{\sigma}^b(t) \leq 1 + (c(\Lambda) + \Lambda c(\lambda)^{-1}) \eta^{1/2}$ for $0 \leq t \leq \eta$. Consequently, the σ^b are equicontinuous and uniformly bounded. By the Theorem of Ascoli–Arzela, there exists a subsequence that we shall denote by σ_n, which converges uniformly to a Lipschitz nondecreasing function σ as the corresponding sequence of b_n's tends to zero. Let $s = t^{-1}\sigma(t)$ from (17.3.32), we see that

$$c(\lambda) t^{1/2} \leq s(t) \leq c(\Lambda) t^{1/2}. \tag{17.3.37}$$

The fact that $s(t)$ is continuous and nondecreasing allows us to use the existence Theorem 16.3.1, coupled with the Lemma 16.4.4, to solve for the bounded solution u of (17.1.1) for this particular $s(t)$.

It remains to be shown that (s, u) satisfies (17.1.3). Let $s_n = t^{-1}\sigma_n(t)$ and u_n denote the corresponding u^{b_n}. Then (s_n, u_n) satisfy (17.1.1)–(17.1.2) for some $b_n > 0$, $\varphi \equiv 0$, and f. We need only show that, for each $t > 0$, $u_n(x, t)$ converges uniformly to $u(x, t)$ with respect to x as b_n tends to zero. First, we note from the Monotonicity Lemma 17.2.1 that $s(t) < s_{n+1}(t) \leq s_n(t)$. Consequently, from (17.3.34) with $\rho = s_n(t) - s(t)$, we see that

$$0 \leq u_n(s(t), t) \leq \Lambda c(\lambda)^{-1} t^{-1/2}(s_n(t) - s(t))$$
$$= \Lambda c(\lambda)^{-1} t^{-3/2}(\sigma_n(t) - \sigma(t)), \tag{17.3.38}$$

for $0 < t \leq \eta$. Consider now the difference $v_n(x, t) = u_n(x, t) - u(x, t)$ in the region $0 \leq x \leq s(t)$, $0 < \delta \leq t \leq t_0$. Clearly, $\mathscr{L}(v_n) = 0$, $v_n(0, t) = 0$, $\delta \leq t \leq t_0$, $0 \leq v_n(x, \delta) \leq 2\Lambda$, $0 \leq x \leq s(\delta)$, and $0 \leq v_n(s(t), t) \leq \Lambda c(\lambda)^{-1} \delta^{-3/2}(\sigma_n(t) - \sigma(t))$, $\delta \leq t \leq t_0$. We can show that $|v_n(x, t_0)| < \varepsilon$ for any positive ε. First, recall (17.3.37), which states that $s(\delta) \leq c(\Lambda) \delta^{1/2}$. The estimate for $|v_n(x, t_0)|$ can be broken into two pieces. The function v_n can be dominated via the Maximum Principle, by a solution to an initial-value problem with data 2Λ over an interval of length $c(\Lambda) \delta^{1/2}$ plus the contribution from the boundary s. For δ sufficiently small, the contribution from the initial-value problem can be made uniformly smaller than $\varepsilon/2$. Fixing δ, $\Lambda c(\lambda)^{-1} t^{-3/2}(\sigma_n(t) - \sigma(t))$ tends uniformly to zero as the b_n tend to zero. Hence, for b_n sufficiently small, the contribution from the boundary s can be made less than $\varepsilon/2$ uniformly via the Maximum Principle. Thus $u_n(x, t_0)$ tends uniformly to $u(x, t_0)$ for each $t_0 > 0$. Now, (17.1.3) for (s_n, u_n) is

$$s_n(t_0)^2 = b_n^2 + 2 \int_0^{t_0} f(\tau) \, d\tau - 2 \int_0^{s_n(t_0)} \xi u_n(\xi, t_0) \, d\xi, \tag{17.3.39}$$

and, taking limits as b_n tends to zero, we get

$$s(t_0)^2 = 2 \int_0^{t_0} f(\tau) \, d\tau - 2 \int_0^{s(t_0)} \xi u(\xi, t_0) \, d\xi \tag{17.3.40}$$

for each $t_0 > 0$. Since $s(t)$ is Lipschitz continuous on each compact interval $0 < \delta \leq t \leq \eta$, and $0 \leq \rho^{-1} u(s(t) - \rho, t) \leq \Lambda c(\lambda)^{-1} \delta^{-1/2}$ for $0 < \delta \leq t \leq \eta$, it follows, from Theorem 14.4.1 that $u_x(s(t), t)$ exists and is continuous. From the Equivalence Lemma 17.1.1, we see that (s, u) is the solution for the Stefan problem (17.1.1)–(17.1.2) in the case of Assumption B on the data. We summarize this analysis in the following statement.

Lemma 17.3.4.   *Under Assumption B, there exists a unique solution* (s, u) *to the Stefan problem* (17.1.1)–(17.1.2).

Proof.   The analysis preceding the statement of the Lemma demonstrates the existence for $0 \leq t \leq \eta$ for some $\eta > 0$. Considering the problem (17.1.1)–(17.1.2) for $2^{-1}\eta \leq t \leq T$ with data $b = s(2^{-1}\eta)$, $\varphi = u(x, 2^{-1}\eta)$, and f, Lemma 17.3.3 applies immediately, to yield a solution for $2^{-1}\eta \leq t \leq T$. The Uniqueness Theorem 17.2.2 implies that the two solutions agree for $2^{-1}\eta \leq t \leq \eta$, and that the application of Lemma 17.3.3 produces an extension of the solution to the interval $0 \leq t \leq T$.   □

We are now in a position to state the main existence result.

Theorem 17.3.1.   *If f and φ are nonnegative, piecewise-continuous functions such that, when $b > 0$, either $\varphi \not\equiv 0$ or $f \not\equiv 0$ in $0 \leq t < \varepsilon$ for each positive ε or, when $b = 0$ and φ is irrelevant, $f \not\equiv 0$ in $0 \leq t < \varepsilon$ for each positive ε, then there exists a unique solution (s, u) to the Stefan problem* (17.1.1)–(17.1.2).

Proof.   *Case $b > 0$.*   We begin the proof by defining

$$\varphi^\mu(x) = \begin{cases} \varphi(x), & 0 \leq x < b - \mu, \\ 0, & b - \mu \leq x \leq b \end{cases} \qquad (17.3.41)$$

for each μ satisfying $0 < \mu < b$. Since φ and f are bounded, it follows that, for each μ satisfying $0 < \mu < b$, there exists an $N(\mu) = \Lambda\mu^{-1}$, where Λ is a constant that bounds both f and φ, such that Assumption A of Lemma 17.3.3 is satisfied. In other words $0 \leq \varphi^\mu(x) \leq N(\mu)(b - x)$. Hence, for each μ satisfying $0 < \mu < b$, there exists a unique solution (s^μ, u^μ) of the Stefan problem (17.1.1)–(17.1.2) corresponding to the data (b, φ^μ, f). From the Monotonicity Theorem 17.2.1, we see that, for $\mu_2 < \mu_1$, $s^{\mu_1} \leq s^{\mu_2}$. Set $\mu_n = 2^{-n}b$, $n = 1, 2, 3, \ldots$, and consider the sequence $(s^{\mu_n}, u^{\mu_n})\}$. The sequence $\{s^{\mu_n}\}$ is a monotone-increasing sequence of functions that is bounded above via the Monotonicity Theorem 17.2.1 by the boundary of the solution corresponding to the data $(b + 1, \varphi, f)$ where φ has been extended as 0 over $b < x \leq b + 1$. Consequently, there exists a function $s(t) = \lim_{n \to \infty} s^{\mu_n}(t)$. From the remark following Lemma 17.2.2, it follows that the s^{μ_n} are strictly increasing functions of t. Hence for $\sigma > 0$, $s^{\mu_1}(\sigma) - b > 0$. Setting $M = (s^{\mu_1}(\sigma) - b)^{-1}\Lambda$, we can apply the Maximum Principle again to show that, for $\sigma \leq t_0 \leq T$,

$$0 \leq u^{\mu_n}(x, t) \leq M(s^{\mu_n}(t_0) - x) \qquad (17.3.42)$$

for $0 \le x \le s^{\mu_n}(t)$, $0 \le t \le t_0$. As $u^{\mu_n}(s^{\mu_n}(t_0), t_0) = 0$, we see that

$$- M \le u_x^{\mu_n}(s^{\mu_n}(t_0), t_0) \le 0, \qquad (17.3.43)$$

provided that $0 < \sigma \le t_0 \le T$. Hence,

$$0 < \dot{s}^{\mu_n}(t) \le M \qquad (17.3.44)$$

for $\sigma \le t \le T$. Hence, s is nondecreasing and Lipschitz-continuous with Lipschitz constant M for $\sigma \le t \le T$. In order to demonstrate the continuity of s at $t = 0$, we consider the solution (ρ, w) of the Stefan problem

$$\mathscr{L}(w) = 0, \qquad b < x < \rho(t), \qquad 0 < t \le T,$$
$$w(b, t) = \Lambda, \qquad 0 < t \le T,$$
$$w(\rho(t), t) = 0, \qquad 0 < t \le T, \qquad (17.3.45.)$$
$$\dot{\rho}(t) = - w_x(\rho(t), t), \qquad 0 < t \le T,$$
$$\rho(0) = b.$$

From Lemma 17.3.1, (ρ, w) exists and, from the Monotonicity Theorem 17.2.1, $s^{\mu_1}(t) \le s(t) \le \rho(t)$ for $0 \le t \le T$. As $s^{\mu_1}(0) = \rho(0) = b$ and ρ is Hölder continuous with exponent $\frac{1}{2}$ at $t = 0$, it follows that $s(t)$ is continuous at $t = 0$, and its continuity is no worse than Hölder with exponent $\frac{1}{2}$. Since s is nondecreasing, the results of Chapter 16 apply again, to allow us to obtain the solution u of (17.1.1) for this particular s. An argument analogous to that contained between (17.3.38) and (17.3.40) allows us to conclude that, for any $t_0 > 0$, $u^{\mu_n}(x, t_0)$ tends uniformly to $u(x, t_0)$ as n tends to infinity. Hence, the expression (17.1.3) for (s^{μ_n}, u^{μ_n}) converges to the one for (s, u). As (17.3.42) holds for u and s, it follows from Theorem 14.4.1 that $u_x(s(t), t)$ exists and is continuous for $0 < t \le T$. Thus, Lemma 17.1.1 implies that (s, u) is the solution to the Stefan problem (17.1.1)–(17.1.2) for this case.

   Case $b = 0$:   Consider the sequence $\{(s_n, u_n)\}$ of solutions to the Stefan problem (17.1.1)–(17.1.2) for the data $[(1/n), 0, f]$. Now, $\{s_n\}$ is a monotonically decreasing sequence of strictly monotone-increasing functions of t. The proof of this case follows the methods of the case of $b > 0$ or that of Lemma 17.3.4, provided that a lower bound for the sequence $\{s_n\}$ can be found. For each $\sigma > 0$, there exists a τ_σ satisfying $0 < \tau_\sigma < \sigma$, at which $f(\tau_\sigma) > 0$ and f is continuous. Consider the solution (ρ, v) of the Stefan problem

$$\mathscr{L}(v) = 0, \qquad 0 < x < \rho(t), \qquad \tau_\sigma < t \le T,$$
$$v(0, t) = f(t), \qquad \tau_\sigma < t \le T,$$
$$v(\rho(t), t) = 0, \qquad \tau_\sigma < t \le T, \qquad (17.3.46)$$
$$\dot{\rho}(t) = - v_x(\rho(t), t), \qquad \tau_\sigma < t \le T,$$
$$\rho(\tau_\sigma) = 0,$$

which exists via Lemma 17.3.4. The function $\rho(t)$ is a strictly monotone-increasing function of t for $\tau_o < t \leq T$. Applying Theorem 17.2.1, we see that $\rho(t) < s_n(t)$ for $\tau_o \leq t \leq T$. The value $\rho(\sigma) > 0$ provides us with the lower estimate necessary to repeat the arguments above.   □

17.4. SOME SPECIAL SOLUTIONS

Recalling Theorem 2.6.1 (with $f = 0$ and $g = -\dot{s}(t)$), we obtain, under the assumption of analyticity,

$$u(x, t) = \sum_{n=1}^{\infty} \frac{1}{(2n)!} \frac{\partial^n}{\partial t^n} [x - s(t)]^{2n}. \qquad (17.4.1)$$

Now, it is easy to see that $\mathscr{L}(u) = 0$, $u(s(t), t) = 0$, and $u_x(s(t), t) = -\dot{s}(t)$, provided that all the various series converge. We shall leave the question of convergence to the reader.

It is of interest to consider boundaries of the form $s(t) = \beta t^\alpha$, where

$$\alpha = \frac{1 + \gamma}{2}, \qquad 0 < \gamma < \infty. \qquad (17.4.2)$$

This yields the solution

$$v_\beta^\alpha(x, t) = \sum_{n=1}^{\infty} \frac{1}{(2n)!} \frac{\partial^n}{\partial t^n} [x - \beta t^\alpha]^{2n}. \qquad (17.4.3)$$

In order to apply the Monotonicity Theorem to obtain some asymptotic estimates, we must consider the function

$$v_\beta^\alpha(0, t) = \sum_{n=1}^{\infty} \frac{\Gamma(2\alpha n + 1)}{\Gamma(2n + 1)\Gamma(\gamma n + 1)} \beta^{2n} t^{\gamma n}. \qquad (17.4.4)$$

Now,

$$\frac{\Gamma(2\alpha n + 1)}{\Gamma(2n + 1)\Gamma(\gamma n + 1)} = \frac{1}{n!} \prod_{k=0}^{n-1} \frac{(n + \gamma n - k)}{(2n - k)}. \qquad (17.4.5)$$

In order to estimate the product in (17.4.5), it suffices to consider the function

$$f(x) = \frac{c - x}{d - x}, \qquad 0 \leq x \leq n - 1, \qquad (17.4.6)$$

where c and d are positive constants. Differentiating with respect to x, we see that

$$f'(x) = \frac{c - d}{(d - x)^2}. \qquad (17.4.7)$$

Consequently, if $c > d$, then $f' > 0$ and f is increasing, while if $c < d$, then

$f' < 0$ and f is decreasing. When $c = d$, $f' = 0$ and $f \equiv 1$. Here, $d = 2n$ and $c = n(1 + \gamma)$. Thus, if $\gamma > 1$, $c > d$, and

$$\alpha = \frac{1 + \gamma}{2} \leq f(x) \leq \frac{\gamma n + 1}{n + 1} \leq \gamma. \qquad (17.4.8)$$

On the other hand, if $\gamma < 1$, $c < d$, and

$$\gamma \leq f(x) \leq \frac{1 + \gamma}{2} = \alpha. \qquad (17.4.9)$$

Applying these results to (17.4.5), we obtain the estimates

$$\frac{1}{n!} \alpha^n \leq \frac{\Gamma(2\alpha n + 1)}{\Gamma(2n + 1)\Gamma(\gamma n + 1)} \leq \frac{1}{n!} \gamma^n, \qquad \gamma > 1, \qquad (17.4.10)$$

$$\frac{\Gamma(2\alpha n + 1)}{\Gamma(2n + 1)\Gamma(\gamma n + 1)} = \frac{1}{n!}, \qquad \gamma = 1, \qquad (17.4.11)$$

and

$$\frac{1}{n!} \gamma^n \leq \frac{\Gamma(2\alpha n + 1)}{\Gamma(2n + 1)\Gamma(\gamma n + 1)} \leq \frac{1}{n!} \alpha^n, \qquad 0 < \gamma < 1. \qquad (17.4.12)$$

From these estimates, we can estimate $v_\beta^\alpha(0, t)$ in (17.4.4). We obtain

$$\exp\{\alpha\beta^2 t^\gamma\} - 1 \leq v_\beta^\alpha(0, t) \leq \exp\{\gamma\beta^2 t^\gamma\} - 1, \qquad \gamma > 1, \qquad (17.4.13)$$

$$v_\beta^\alpha(0, t) = \exp\{\beta^2 t\} - 1, \qquad \gamma = 1, \qquad (17.4.14)$$

and

$$\exp\{\gamma\beta^2 t^\gamma\} - 1 \leq v_\beta^\alpha(0, t) \leq \exp\{\alpha\beta^2 t^\gamma\} - 1, \qquad 0 < \gamma < 1. \qquad (17.4.15)$$

The estimates of $v_\beta^\alpha(0, t)$, corresponding to the boundary βt^α, $\alpha = (1 + \gamma)/2$, along with the elementary solution $c(\lambda)t^{1/2}$, corresponding to the boundary data λ, will enable us to give good asymptotic estimates of the behavior of the boundary as t tends to zero. These estimates depend upon the asymptotic behavior of the boundary data as t tends to zero.

The elementary solution $c(\lambda)t^{1/2}$, corresponding to the boundary data λ, will also enable us to study the asymptotic behavior at $t = \infty$. For other behaviors of the boundary at $t = \infty$, we first demonstrate the following result

Lemma 17.4.1. *The series $\sum_{n=1}^\infty a_n t^n$ is asymptotic to $\exp\{\rho t\}$, $\rho > 0$, as t tends to infinity if*

$$\lim_{n \to \infty} n! a_n \rho^{-n} = 1. \qquad (17.4.16)$$

Proof. From (17.4.16), there exists, for each $\varepsilon > 0$, an N_ε such that, for all $n > N_\varepsilon$,

$$1 - \varepsilon \leq n! a_n \rho^{-n} \leq 1 + \varepsilon. \qquad (17.4.17)$$

Then, we can write

$$\sum_{n=1}^{\infty} a_n t^n = \sum_{n=1}^{N_\varepsilon} a_n t^n + \sum_{n=N_\varepsilon}^{\infty} n! a_n \rho^{-n} \frac{(\rho t)^n}{n!}. \qquad (17.4.18)$$

From this and (17.4.17), it follows that

$$\sum_{n=1}^{N_\varepsilon} a_n t^n + (1-\varepsilon) \sum_{N=N_\varepsilon}^{\infty} \frac{(\rho t)^n}{n!} \le \sum_{n=1}^{\infty} a_n t^n \le \sum_{n=1}^{N_\varepsilon} a_n t^n + (1+\varepsilon) \sum_{n=N_\varepsilon}^{\infty} \frac{(\rho t)^n}{n!},$$

$$(17.4.19)$$

which can be rewritten as

$$P_\varepsilon(t) + (1-\varepsilon)\exp\{\rho t\} \le \sum_{n=1}^{\infty} a_n t^n \le Q_\varepsilon(t) + (1+\varepsilon)\exp\{\rho t\},$$

$$(17.4.20)$$

where P_ε and Q_ε are polynomials in t. Dividing by $\exp\{\rho t\}$ and taking limits, we see that

$$1 - \varepsilon \le \lim_{t \to \infty} \inf \exp\{-\rho t\} \sum_{n=1}^{\infty} a_n t^n \le \lim_{t \to \infty} \sup \exp\{-\rho t\} \sum_{n=1}^{\infty} a_n t^n \le 1 + \varepsilon.$$

$$(17.4.21)$$

Since ε is arbitrary, it follows that

$$\lim_{t \to \infty} \exp\{-\rho t\} \sum_{n=1}^{\infty} a_n t^n = 1. \qquad (17.4.22) \quad \square$$

Returning now to $v_\beta^\alpha(0, t)$, we consider the coefficients

$$a_n = \frac{\Gamma(2\alpha n + 1)\beta^{2n}}{\Gamma(2n + 1)\Gamma(\gamma n + 1)}. \qquad (17.4.23)$$

We consider $n! a_n \rho^{-n}$ for a given ρ, and ask for the choice of β so that $\lim_{n \to \infty} n! a_n \rho^{-n} = \text{const.}$ For this, we recall the asymptotic behavior of

$$\Gamma(az + b) \sim \sqrt{2\pi}\, e^{-az}(az)^{az+b-1/2} \qquad (17.4.24)$$

as real z tends to infinity. Substituting this into the expression for $n! a_n \rho^{-n}$, we get

$$n! a_n \rho^{-n} = \frac{\sqrt{2\pi}\, e^{-n}(n)^{n+1/2}\sqrt{2\pi}\, e^{-2\alpha n}(2\alpha n)^{2\alpha n+1/2}\beta^{2n}}{\sqrt{2\pi}\, e^{-2n}(2n)^{2n+1/2}\sqrt{2\pi}\, e^{-\gamma n}(\gamma n)^{\gamma n+1/2}\rho^n}$$

$$= \left(\frac{1+\gamma}{2\gamma}\right)^{1/2}\left[\frac{(1+\gamma)^{1+\gamma}\beta^2}{2^2\gamma^\gamma\rho}\right]^n. \qquad (17.4.25)$$

From this, we see that

$$\lim_{n \to \infty} n! a_n \rho^{-n} = \left(\frac{(1+\gamma)}{2\gamma} \right)^{1/2} \tag{17.4.26}$$

if

$$\beta = 2\gamma^{\gamma/2} \rho (1+\gamma)^{-(1+\gamma)/2}. \tag{17.4.27}$$

Now, an easy application of Lemma 17.4.1 yields the following result.

Lemma 17.4.2. For $\beta = 2\gamma^{\gamma/2}\rho(1+\gamma)^{-(1+\gamma)/2}$,

$$\lim_{t \to \infty} \frac{v_\beta^\alpha(0, t)}{((1+\gamma)/2\gamma)^{1/2} \exp\{\rho t^\gamma\}} = 1. \tag{17.4.28}$$

Proof. The proof is left to the reader. □

As a corollary of Lemma 17.4.2 we have the following result.

Lemma 17.4.3. For $\beta = 2\gamma^{\gamma/2}\rho(1+\gamma)^{-(1+\gamma)/2}$,

$$\lim_{t \to \infty} \frac{\log v_\beta^\alpha(0, t)}{\rho t^\gamma} = 1. \tag{17.4.29}$$

Proof. The proof is left to the reader. □

17.5. ASYMPTOTIC BEHAVIOR OF THE FREE BOUNDARY AT $t = +\infty$

Under the assumptions stated in Theorem 17.3.1, we are assured that there exists a unique solution of the Stefan problem (17.1.1)–(17.1.2) for each $T > 0$. As T tends to infinity, the solution extends with it. Consequently, it is of interest to study the behavior of the free boundary $s(t)$ as t tends to infinity. The equivalent condition (17.1.3) is quite useful here.

We begin our study with the following general result.

Theorem 17.5.1. If

$$\int_0^\infty f(t)\, dt = +\infty, \tag{17.5.1}$$

then

$$\lim_{t \to \infty} s(t) = +\infty. \tag{17.5.2}$$

If

$$\int_0^\infty f(t)\, dt < \infty, \tag{17.5.3}$$

then

$$\lim_{t \to \infty} s(t) = l^{1/2}, \tag{17.5.4}$$

where

$$l = b^2 + 2 \int_0^b \xi \varphi(\xi) \, d\xi + 2 \int_0^\infty f(\tau) \, d\tau. \tag{17.5.5}$$

Proof. We consider first the case that f has compact support, that is, that f vanishes for all t sufficiently large. Now, condition (17.1.3) states that

$$s(t)^2 = b^2 + 2 \int_0^b \xi \varphi(\xi) \, d\xi + 2 \int_0^t f(\tau) \, d\tau - 2 \int_0^{s(t)} \xi u(\xi, t) \, d\xi. \tag{17.5.6}$$

As $u \geq 0$, then

$$s(t)^2 \leq b^2 + 2 \int_0^b \xi \varphi(\xi) \, d\xi + 2 \int_0^t f(\tau) \, d\tau \leq l \tag{17.5.7}$$

for all t. Thus, $s(t) \leq l^{1/2}$. By the Maximum Principle, $0 \leq u \leq w$, where w satisfies

$$\mathcal{L}(w) = 0, \qquad 0 < x < l^{1/2}, \qquad 0 < t,$$

$$w(0, t) = f(t), \qquad 0 < t,$$

$$w(x, 0) = \begin{cases} \varphi(x), & 0 \leq x \leq b, \\ 0, & b < x \leq l^{1/2}, \end{cases} \tag{17.5.8}$$

$$w(l^{1/2}, t) = 0, \qquad 0 < t.$$

However, since $f \equiv 0$ for $t > t_0$, where t_0 is sufficiently large and fixed, it follows from Chapter 6 that $w = \mathcal{O}(\exp\{-\alpha t\})$, where $\alpha = \alpha(l) > 0$. Hence, w tends uniformly to zero as t tends to infinity. Consequently, u must tend uniformly to zero as t tends to infinity. We see then that

$$\lim_{t \to \infty} 2 \int_0^{s(t)} \xi u(\xi, t) \, d\xi = 0, \tag{17.5.9}$$

from which follows the result (17.5.4) in this case.

For the remaining case of a general f, we set

$$f_n(t) = \begin{cases} f(t), & 0 < t \leq n, \\ 0, & n < t < \infty, \end{cases} \tag{17.5.10}$$

and we consider the corresponding l_n and s_n. By the Monotonicity Theorem 17.2.1, $s_n(t) \leq s(t) \leq l^{1/2}$. However, we have just shown that

$$l_n^{1/2} = \lim_{t \to \infty} s_n(t) \leq \lim_{t \to \infty} \inf s(t) \leq \lim_{t \to \infty} \sup s(t) \leq l^{1/2}. \tag{17.5.11}$$

Letting n tend now to infinity, we see that l_n tends to l.   □

The next result shows that when $\int_0^\infty f(t) \, dt = \infty$, the behavior of the boundary at $t = \infty$ depends only upon f.

Theorem 17.5.2. *For $\int_0^\infty f(t)\,dt = \infty$, let (s, u) be the solution of the Stefan problem (17.1.1)–(17.1.2) and let (σ, v) be the solution of*

$$\mathscr{L}(v) = 0, \qquad 0 < x < \sigma(t), \qquad t_0 < t < \infty,$$
$$v(0, t) = f(t), \qquad t_0 < t < \infty,$$
$$v(\sigma(t), t) = 0, \qquad t_0 < t < \infty, \qquad (17.5.12)$$
$$\dot{\sigma}(t) = -v_x(\sigma(t), t), \qquad t_0 < t < \infty,$$
$$\sigma(t_0) = 0$$

for any $t_0 > 0$. Then as t tends to infinity,

$$\frac{s(t)}{\sigma(t)} = 1 + \mathscr{O}\!\left(\frac{1}{\sigma(t)}\right) \qquad (17.5.13)$$

and

$$\lim_{t \to \infty} \frac{s(t)}{\sigma(t)} = 1. \qquad (17.5.14)$$

Proof. From the Monotonicity Theorem 17.2.1, we have $\sigma(t) \le s(t)$. By the Maximum Principle, it follows that $0 \le v \le u$ for $0 \le x \le \sigma(t)$, $t_0 \le t$. Applying (17.1.3) to (s, u) for the initial time t_0, we get

$$s(t)^2 - s(t_0)^2 - 2\int_0^{s(t_0)} \xi u(\xi, t_0)\,d\xi = 2\int_{t_0}^t f(\tau)\,d\tau - 2\int_0^{s(t)} \xi u(\xi, t)\,d\xi$$

$$\le 2\int_{t_0}^t f(\tau)\,d\tau - 2\int_0^{\sigma(t)} \xi v(\xi, t)\,d\xi = \sigma(t)^2.$$

$$(17.5.15)$$

Thus,

$$\sigma(t)^2 \le s(t)^2 \le \sigma(t)^2 + s(t_0)^2 + 2\int_0^{s(t_0)} \xi u(\xi, t_0)\,d\xi. \qquad (17.5.16)$$

Dividing by $\sigma(t)^2$, we get

$$1 \le \frac{s(t)^2}{\sigma(t)^2} \le 1 + \left\{ s(t_0)^2 + 2\int_0^{s(t_0)} \xi u(\xi, t_0)\,d\xi \right\} \sigma(t)^{-2}. \qquad (17.5.17)$$

Hence,

$$1 \le \frac{s(t)^2}{\sigma(t)^2} \le 1 + \mathscr{O}\!\left(\frac{1}{\sigma(t)^2}\right), \qquad (17.5.18)$$

and the result follows from Theorem 17.5.1. $\qquad\qquad\qquad\qquad\qquad\square$

Recalling that $\|f\|_{[a, b]} = \sup_{a \le \tau \le b} |f(\tau)|$, we consider those f's for which $\|f\|_{[t_0, \infty]}$ is finite, and obtain a sharp relation between the integral of f and s at $t = \infty$.

Theorem 17.5.3. *If* $\int_0^\infty f(\tau)\, d\tau = +\infty$ *and* $\lim_{t_0 \to \infty} \|f\|_{[t_0, \infty]} = 0$, *then, for the solution (s, u) of the Stefan problem* (17.1.1)–(17.1.2),

$$\lim_{t \to \infty} s(t) \left(2 \int_0^t f(\tau)\, d\tau \right)^{-1/2} = 1. \qquad (17.5.19)$$

Proof. For any $t_0 > 0$, let (σ, v) denote the solution of the problem (17.5.12). By the Maximum Principle, $0 \leq v(x, t) \leq \|f\|_{[t_0, t]}$. As

$$\sigma(t)^2 = 2 \int_{t_0}^t f(\tau)\, d\tau - 2 \int_0^{\sigma(t)} \xi v(\xi, t)\, d\xi$$

$$\geq 2 \int_{t_0}^t f(\tau)\, d\tau - \|f\|_{[t_0, t]} \sigma(t)^2, \qquad (17.5.20)$$

it follows that

$$\sigma(t)^2 \geq \left\{ 1 + \|f\|_{[t_0, t]} \right\}^{-1} 2 \int_{t_0}^t f(\tau)\, d\tau. \qquad (17.5.21)$$

From the Maximum Principle, we have $u \geq 0$, and from the Monotonicity Theorem 17.2.1 and (17.1.3) for (s, u), we get

$$\sigma(t)^2 \leq s(t)^2 = b^2 + 2 \int_0^b \xi \varphi(\xi)\, d\xi + 2 \int_0^t f(\tau)\, d\tau - 2 \int_0^{s(t)} \xi u(\xi, t)\, d\xi$$

$$\leq b^2 + 2 \int_0^b \xi \varphi(\xi)\, d\xi + 2 \int_0^t f(\tau)\, d\tau. \qquad (17.5.22)$$

Using (17.5.21), dividing by $2 \int_0^t f(\tau)\, d\tau$, and taking the limit as t tends to infinity, we see that

$$\left(\frac{1}{1 + \|f\|_{[t_0, \infty]}} \right) \leq \lim_{t \to \infty} \inf \left[\frac{s(t)^2}{2 \int_0^t f(\tau)\, d\tau} \right]$$

$$\leq \lim_{t \to \infty} \sup \left[\frac{s(t)^2}{2 \int_0^t f(\tau)\, d\tau} \right] \leq 1. \qquad (17.5.23)$$

As (17.5.23) holds for each $t_0 > 0$, we can let t_0 tend to infinity, demonstrating the result (17.5.19). $\qquad \square$

The following corollaries to Theorem 17.5.3 are most easily stated using the symbol $\sim$ for asymptotic limits; i.e., the ratio of the two functions tends to one in the limit as t tends to infinity.

Corollary 17.5.1. *If* $f(t) \sim \rho t^{-\delta}$, $\rho > 0$, $0 < \delta < 1$, *then*

$$s(t) \sim \left(\frac{2\rho}{1 - \delta} \right)^{1/2} t^{(1-\delta)/2}. \qquad (17.5.24)$$

Proof. The proof is left to the reader. $\qquad \square$

Corollary 17.5.2. *If* $f(t) \sim \rho t^{-1}$, $\rho > 0$, *then* $s(t) \sim (2\rho \log t)^{1/2}$.

Proof. The proof is left to the reader. $\qquad \square$

For some results for some particular functions f that do not go to zero as t tends to infinity, we state the following result.

Theorem 17.5.4. *If* $f(t) \sim \rho$, $\rho > 0$, *then* $s(t) \sim \beta t^{1/2}$, *where* β *satisfies*

$$\rho = \sum_{n=1}^{\infty} \frac{n!}{(2n)!} \beta^{2n}. \tag{17.5.25}$$

If $\log f(t) \sim \rho t^{\gamma}$, $\rho > 0$, $0 < \gamma < \infty$, *then* $s(t) \sim \beta t^{(1+\gamma)/2}$, *where* β *satisfies*

$$\beta = 2\rho^{1/2}\gamma^{\gamma/2}(1+\gamma)^{-(1+\gamma)/2}. \tag{17.5.26}$$

Proof. For the first part of the theorem, we can use either the special form of the solution given by Lemma 17.3.1 or its form given by v_{β}^{α} (see Eq. (17.4.3)) for $\alpha = \frac{1}{2}$. For the second part, we use v_{β}^{α} of (17.4.3) and Lemmas 17.4.1, 17.4.2, and 17.4.3. In either case the proof utilizes the Monotonicity Theorem 17.2.1.

Let $\varepsilon, 0 < \varepsilon < \beta$, be given. From the hypothesis of the asymptotic behavior on $f(t)$, it follows, from the asymptotic behavior of $v_{\beta}^{\alpha}(0, t)$, that there exists a $t_0 > 0$ and sufficiently large so that $v_{\beta-\varepsilon}^{\alpha}(0, t) \leq f(t) \leq v_{\beta+\varepsilon}^{\alpha}(0, t)$ for $t \geq t_0$. Denoting by $\sigma_{-\varepsilon}(t)$, $\sigma(t)$, and $\sigma_{+\varepsilon}(t)$ the free boundaries in the respective solutions of (17.5.12), the Monotonicity Theorem 17.2.1 implies that $\sigma_{-\varepsilon}(t) \leq \sigma(t) \leq \sigma_{+\varepsilon}(t)$. But Theorem 17.5.2 states that $\sigma_{-\varepsilon}(t) \sim (\beta - \varepsilon)t^{\alpha}$, $\sigma_{+\varepsilon}(t) \sim (\beta + \varepsilon)t^{\alpha}$, and $\sigma(t) \sim s(t)$. Hence,

$$\beta - \varepsilon = \lim_{t \to \infty} \frac{\sigma_{-\varepsilon}(t)}{t^{\alpha}} \leq \lim_{t \to \infty} \inf \frac{s(t)}{t^{\alpha}}$$

$$\leq \lim_{t \to \infty} \sup \frac{s(t)}{t^{\alpha}}$$

$$\leq \lim_{t \to \infty} \frac{\sigma_{+\varepsilon}(t)}{t^{\alpha}} = \beta + \varepsilon. \tag{17.5.27}$$

Since ε is arbitrary, $s(t) \sim \beta t^{\alpha}$ for $\alpha = (1+\gamma)/2$, $0 \leq \gamma < \infty$. $\square$

17.6. ASYMPTOTIC BEHAVIOR OF THE FREE BOUNDARY AT $t = 0$

We have seen in Sec. 17.3 that, if $b = 0$ and the boundary data $f(0) \neq 0$, then there exist two positive constants λ, Λ, such that $0 < \lambda \leq f(t) \leq \Lambda$ for $0 \leq t \leq \eta$, $\eta > 0$. From this it followed, from the Monotonicity Theorem 17.2.1 and Lemma 17.3.1, that there exist positive constants $c_1 = c_1(\lambda)$ and $c_2 = c_2(\Lambda)$, such that

$$c_1 t^{1/2} \leq s(t) \leq c_2 t^{1/2} \tag{17.6.1}$$

for $0 \leq t \leq \eta$. This gives the asymptotic behavior of the free boundary s at $t = 0$ for the case that $f(0) > 0$.

In order to supplement the existence theorem 17.3.1 in its application to continuous dependence of the solution upon the data, we shall apply the Monotonicity Theorem 17.2.1 to the special solutions v_β^α of Sec. 17.4, in order to obtain the asymptotic behavior at $t = 0$ for classes of functions f.

Theorem 17.6.1. *For $b = 0$, suppose that the boundary datum f satisfies*

$$lt^\gamma \le f(t) \le Lt^\gamma, \qquad 0 \le t \le T, \tag{17.6.2}$$

where l, L, and γ are positive constants. Then, there exist positive constants $\beta_1 = \beta_1(\gamma, T)$ and $\beta_2 = \beta_2(\gamma, T)$ such that the free boundary s of the solution of the Stefan problem (17.1.1)–(17.1.2) satisfies

$$\beta_1 t^\alpha \le s(t) \le \beta_2 t^\alpha, \qquad 0 \le t \le T, \tag{17.6.3}$$

where $\alpha = (1 + \gamma)/2$.

Proof. From Eqs. (17.4.13) through (17.4.15), we see that $v_\beta^\alpha(0, t)$ satisfies inequalities of the form

$$\exp\{c_1 \beta^2 t^\gamma\} - 1 \le v_\beta^\alpha(0, t) \le \exp\{c_2 \beta^2 t^\gamma\} - 1, \tag{17.6.4}$$

where c_1 and c_2 are positive constants depending upon γ. In order to apply the Monotonicity Theorem 17.2.1, we need only select β_1 satisfying

$$\exp\{c_2 \beta_1^2 T^\gamma\} - 1 = lT^\gamma; \tag{17.6.5}$$

that is,

$$\beta_1 = \{c_2^{-1} T^{-\gamma} \log(lT^\gamma + 1)\}^{1/2}. \tag{17.6.6}$$

This implies that $v_{\beta_1}^\alpha(0, t) \le lt^\gamma$ for $0 \le t \le T$. It is easy to see that

$$\beta_2 = c_1^{-1/2} L^{1/2} \tag{17.6.7}$$

allows us to conclude that $v_{\beta_2}^\alpha(0, t) \ge Lt^\gamma, 0 \le t \le T$. □

17.7. CONTINUOUS DEPENDENCE UPON THE DATA

We shall study the dependence of the free boundary s of the solution of the Stefan problem (17.1.1)–(17.1.2) upon the data b, φ, and f.

Throughout most of this section we shall use the following result.

Lemma 17.7.1. *Let $\varphi(t)$ satisfy*

$$0 \le \varphi(t) \le \psi(t) + C_1 \int_0^t \frac{\varphi(\tau)}{(t - \tau)^{1/2}} d\tau, \qquad 0 \le t \le T, \tag{17.7.1}$$

where $C_1 \ge 0$ and $\psi(t)$ is nonnegative and nondecreasing. Then

$$0 \le \varphi(t) \le [1 + 2C_1 t^{1/2}] \psi(t) \exp\{\pi C_1^2 t\} \le C_2 \psi(t) \tag{17.7.2}$$

with

$$C_2 = \left[1 + 2C_1 T^{1/2}\right] \exp\left\{\pi C_1^2 T\right\}. \tag{17.7.3}$$

Proof. Apply the Abel integral operator to (17.7.1) to derive

$$0 \le \varphi(t) \le \left[1 + 2C_1 t^{1/2}\right] \psi(t) + \pi C_1^2 \int_0^t \varphi(\tau) \, d\tau. \tag{17.7.4}$$

An application of Gronwall's Lemma 8.4.2 yields the result. □

We consider first the case $b > 0$ when the data φ and f satisfy Assumption A of Sec. 17.3. Recalling Lemma 17.3.2, the boundary s has Lipschitz constant $C = \max\{Mb^{-1}, N\}$, where $M = \max(\max_{0 \le t \le T} f(t), \max_{0 \le x \le b} \varphi(x))$ and N is the Lipschitz constant such that $0 \le \varphi(x) \le N(b - x)$. We shall demonstrate the following result.

Lemma 17.7.2. *For $i = 1, 2$, let (s_i, u_i) be solutions of the Stefan problem* (17.1.1)–(17.1.2) *corresponding to the data (b_i, φ_i, f_i) satisfying Assumption A and $0 < b_1 \le b_2$. Then there exists a positive constant $C_3 = C_3(b_1, C, T)$ such that*

$$|s_2(t) - s_1(t)| \le C_3 \Big\{ (b_2 - b_1) + \int_0^t |f_2(\tau) - f_1(\tau)| \, d\tau$$

$$+ \int_0^{b_1} \xi |\varphi_2(\xi) - \varphi_1(\xi)| \, d\xi + \int_{b_1}^{b_2} \xi \varphi_2(\xi) \, d\xi \Big\} \tag{17.7.5}$$

for $0 \le t \le T$.

Proof. Set $\alpha(t) \equiv \min(s_1(t), s_2(t))$, $\beta(t) \equiv \max(s_1(t), s_2(t))$, $\delta(t) \equiv \beta(t) - \alpha(t)$, and

$$j(t) \equiv \begin{cases} 2, & s_1(t) \le s_2(t), \\ 1, & s_1(t) > s_2(t). \end{cases} \tag{17.7.6}$$

Now, s_i, $i = 1, 2$, are Lipschitz-continuous, whence it follows that α and β are Lipschitz-continuous with Lipschitz constant C. As $s_1(t) + s_2(t) \ge b_1 + b_2 \ge 2b_1$, we subtract the corresponding relations (17.1.3), and obtain

$$\delta(t) \le (b_2 - b_1) + \frac{1}{b_1} \int_0^t |f_2(\tau) - f_1(\tau)| \, d\tau$$

$$+ \frac{1}{b_1} \int_0^{b_1} \xi |\varphi_2(\xi) - \varphi_1(\xi)| \, d\xi + \frac{1}{b_1} \int_{b_1}^{b_2} \xi \varphi_2(\xi) \, d\xi$$

$$+ \frac{1}{b_1} \int_0^{\alpha(t)} \xi |u_2(\xi, t) - u_1(\xi, t)| \, d\xi + \frac{1}{b_1} \int_{\alpha(t)}^{\beta(t)} \xi u_{j(t)}(\xi, t) \, d\xi$$

$$= (b_2 - b_1) + I_1 + I_2 + I_3 + I_4 + I_5. \tag{17.7.7}$$

We begin the estimate with I_4. From Lemma 17.3.2,

$$|u_2(\alpha(t), t) - u_1(\alpha(t), t)| \le C\delta(t).$$

From the Maximum Principle, $|u_2 - u_1|$ is dominated in $0 \leq x \leq \alpha(t)$, $0 \leq t \leq T$, by $w_1 + w_2$, where w_1 satisfies

$$\mathcal{L}(w_1) = 0, \qquad 0 < x < \infty, \qquad 0 < t \leq T,$$
$$w_1(0, t) = |f_2(t) - f_1(t)|, \qquad 0 \leq t \leq T, \qquad (17.7.8)$$
$$w_1(x, 0) = \begin{cases} |\varphi_2(x) - \varphi_1(x)|, & 0 \leq x \leq b_1, \\ 0, & b_1 < x < \infty, \end{cases}$$

and w_2 satisfies

$$\mathcal{L}(w_2) = 0, \qquad 0 < x < \alpha(t), \qquad 0 < t \leq T,$$
$$w_2(0, t) = 0, \qquad 0 \leq t \leq T,$$
$$w_2(x, 0) = 0, \qquad 0 \leq x \leq b_1, \qquad (17.7.9)$$
$$w_2(\alpha(t), t) = C\delta(t), \qquad 0 \leq t \leq T.$$

Applying the identity (17.1.4) to (17.7.8) yields

$$\frac{1}{b_1} \int_0^{\alpha(t)} \xi w_1(\xi, t) \, d\xi < \frac{1}{b_1} \int_0^{\infty} \xi w_1(\xi, t) \, d\xi = \frac{1}{b_1} \int_0^t |f_2(\tau) - f_1(\tau)| \, d\tau$$
$$+ \frac{1}{b_1} \int_0^{b_1} \xi |\varphi_2(\xi) - \varphi_1(\xi)| \, d\xi = I_1 + I_2. \quad (17.7.10)$$

Next, we see that w_2 can be written in the form

$$w_2(x, t) = \int_0^t \mu(\tau) \left\{ \frac{\partial K}{\partial x} (x - \alpha(\tau), t - \tau) + \frac{\partial K}{\partial x} (x + \alpha(\tau), t - \tau) \right\} d\tau,$$
$$(17.7.11)$$

where $K = K(x, t)$ is the fundamental solution defined by (3.2.1), and $\mu(t)$ is the solution of the integral equation

$$C\delta(t) = \frac{1}{2}\mu(t) + \int_0^t \mu(\tau) \left\{ \frac{\partial K}{\partial x} (\alpha(t) - \alpha(\tau), t - \tau) \right.$$
$$\left. + \frac{\partial K}{\partial x} (\alpha(t) + \alpha(\tau), t - \tau) \right\} d\tau. \quad (17.7.12)$$

As α is Lipschitz-continuous with Lipschitz constant C, there exists a constant C_4, where C_4 and all that follow depend only upon C, b_1, and T, such that

$$0 \leq |\mu(t)| \leq 2C\|\delta\|_{[0, t]} + C_4 \int_0^t \frac{|\mu(\tau)|}{(t - \tau)^{1/2}} d\tau, \qquad (17.7.13)$$

where

$$\|\delta\|_{[0, t]} = \sup_{0 \leq \tau \leq t} |\delta(\tau)|. \qquad (17.7.14)$$

An application of Lemma 17.7.1 yields

$$|\mu(t)| \le C_5 \|\delta\|_{[0,t]}.$$ (17.7.15)

Now,

$$\frac{1}{b_1} \int_0^{\alpha(t)} \xi w_2(\xi, t)\, d\xi = \frac{1}{b_1} \int_0^t \mu(\tau) \left\{ \int_0^{\alpha(t)} \xi \left\{ \frac{\partial K}{\partial x}(\xi - \alpha(\tau), t - \tau) \right. \right.$$

$$\left. \left. + \frac{\partial K}{\partial x}(\xi + \alpha(\tau), t - \tau) \right\} d\xi \right\} d\tau$$

$$= \frac{1}{b_1} \int_0^t \mu(\tau) \left\{ \alpha(t)[K(\alpha(t) - \alpha(\tau), t - \tau) \right.$$

$$+ K(\alpha(t) + \alpha(\tau), t - \tau)]$$

$$- \int_0^{\alpha(t)} [K(\xi - \alpha(\tau), t - \tau)$$

$$\left. + K(\xi + \alpha(\tau), t - \tau)]\, d\xi \right\} d\tau$$

$$\le C_6 \int_0^t \frac{|\mu(\tau)|}{(t - \tau)^{1/2}}\, d\tau \le C_7 \int_0^t \frac{\|\delta\|_{[0,\tau]}}{(t - \tau)^{1/2}},$$ (17.7.16)

since $0 \le \alpha(t) \le b_2 + CT$. Hence,

$$I_4 \le I_1 + I_2 + C_7 \int_0^t \frac{\|\delta\|_{[0,\tau]}}{(t - \tau)^{1/2}}\, d\tau.$$ (17.7.17)

We consider now the term I_5. For whatever value $j(t)$ assumes, the Maximum Principle yields the fact that $u_{j(t)}$ is dominated by the sum $z_1 + z_2$, where z_1 satisfies

$$\mathcal{L}(z_1) = 0, \qquad \alpha(t) < x < \beta(t), \qquad 0 < t \le T,$$

$$z_1(\alpha(t), t) = z_1(\beta(t), t) = 0, \qquad 0 < t \le T,$$ (17.7.18)

$$z_1(x, 0) = \varphi_2(x), \qquad b_1 \le x \le b_2,$$

and z_2 satisfies

$$\mathcal{L}(z_2) = 0, \qquad \alpha(t) < x < \infty, \qquad 0 < t \le T,$$

$$z_2(\alpha(t), t) = C\delta(t), \qquad 0 < t \le T,$$ (17.7.19)

$$z_2(x, 0) = 0, \qquad b_1 \le x < \infty.$$

Since $\varphi_2(x) \ge 0$ for $b_1 \le x \le b_2$, it follows, from Theorem 15.4.1, that $(\partial z_1/\partial x)(\alpha(t), t) \ge 0$ and $(\partial z_1/\partial x)(\beta(t), t) \le 0$. Applying the identity

(17.1.3) to (17.7.18), we obtain

$$\int_{\alpha(t)}^{\beta(t)} \xi z_1(\xi, t)\, d\xi = \int_{b_1}^{b_2} \xi \varphi_2(\xi)\, d\xi + \int_0^t \beta(\tau)\frac{\partial z}{\partial x}(\beta(\tau), \tau)\, d\tau$$

$$- \int_0^t \alpha(\tau)\frac{\partial z}{\partial x}(\alpha(\tau), \tau)\, d\tau$$

$$\leq \int_{b_1}^{b_2} \xi \varphi_2(\xi)\, d\xi. \tag{17.7.20}$$

Hence,

$$\frac{1}{b_1}\int_{\alpha(t)}^{\beta(t)} \xi z_1(\xi, t)\, d\xi \leq I_3. \tag{17.7.21}$$

Now,

$$z_2(x, t) = \int_0^t \sigma(\tau)\frac{\partial K}{\partial x}(x - \alpha(\tau), t - \tau)\, d\tau, \tag{17.7.22}$$

where $\sigma(t)$ is the solution of

$$C\delta(t) = -\frac{1}{2}\sigma(t) + \int_0^t \sigma(\tau)\frac{\partial K}{\partial x}(\alpha(t) - \alpha(\tau), t - \tau)\, d\tau. \tag{17.7.23}$$

following the derivation given in (17.7.16), it follows in a similar manner that

$$\int_{\alpha(t)}^{\beta(t)} \xi z_2(\xi, t)\, d\xi \leq \int_{\alpha(t)}^{\infty} \xi z_2(\xi, t)\, d\xi \leq C_8 \int_0^t \frac{\|\delta\|_{[0,\tau]}}{(t - \tau)^{1/2}}\, d\tau. \tag{17.7.24}$$

Combining the estimates (17.7.7), (17.7.17), (17.7.21), and (17.7.24), it follows that

$$\|\delta\|_{[0, t]} \leq (b_2 - b_1) + 2I_1 + 2I_2 + 2I_3 + C_9 \int_0^t \frac{\|\delta\|_{[0,\tau]}}{(t - \tau)^{1/2}}\, d\tau,$$

$$\tag{17.7.25}$$

whence the result follows via an application of Lemma 17.7.1.   □

Next, we turn to the case that $b = 0$ and the datum f satisfies $lt^\gamma \leq f(t) \leq Lt^\gamma$, $0 \leq t \leq T$, where $1 \leq \gamma < \infty$. Now, from Theorem 17.6.1, it follows that there exist constants $\beta_i = \beta_i(l, L, \gamma, T)$, $i = 1, 2$, such that $\beta_1 t^{(1+\gamma)/2} \leq s(t) \leq \beta_2 t^{(1+\gamma)/2}$. Employing the lower bound $\beta_1 t^{(1+\gamma)/2}$, we can apply the Maximum Principle to obtain

$$0 \leq u(x, t) \leq Lt_0^\gamma \beta_1^{-1} t_0^{-(1+\gamma)/2}(s(t_0) - x) \quad \text{for } 0 \leq x \leq s(t), \quad 0 \leq t \leq t_0.$$

From this it follows that

$$0 \leq \dot{s}(t) \leq L\beta_1^{-1} t^{(\gamma/2) - (1/2)} \leq L\beta_1^{-1} T^{(\gamma - 1)/2},$$

since $\gamma \geq 1$. Consequently, s is Lipschitz-continuous with Lipschitz constant

$L\beta_1^{-1}T^{(\gamma-1)/2}$. We shall apply these facts in demonstrating the following result.

Lemma 17.7.3. *For* $i=1,2$, *let* (s_i, u_i) *be solutions of the Stefan problem* (17.1.1)–(17.1.2) *corresponding to the data* $(0,0,f_i)$, *where* $lt^\gamma \le f_i(t) \le Lt^\gamma$, $0 \le t \le T$ *for the positive constants* l, L, *and* $\gamma \ge 1$. *Then, there exists a positive constant* C *that depends only upon* l, L, γ, *and* T, *such that*

$$|s_1(t) - s_2(t)| \le C\|f_1 - f_2\|_{[0,t],\gamma}, \qquad (17.7.26)$$

where

$$\|f_1 - f_2\|_{[0,t],\gamma} = \sup_{0 \le \tau \le t} \tau^{(1-\gamma)/2}|f_1(\tau) - f_2(\tau)|. \qquad (17.7.27)$$

Proof. As in the proof of Lemma 17.7.2, we set

$$\alpha(t) = \min(s_1(t), s_2(t)), \quad \beta(t) = \max(s_1(t), s_2(t)),$$

and $\delta(t) = \beta(t) - \alpha(t)$. We note, from the discussion preceding this lemma, that there exist constants $\beta_i = \beta_i(l, L, \gamma, T)$, $i=1,2$, such that $\beta_1 t^{(1+\gamma)/2} \le s_i(t) \le \beta_2 t^{(1+\gamma)/2}$, $i=1,2$. Moreover, the s_i are Lipschitz-continuous with Lipschitz constant $L\beta_1^{-1}T^{(\gamma-1)/2}$, whence it follows that $\alpha(t)$ and $\beta(t)$ are Lipschitz-continuous with Lipschitz constant $L\beta_1^{-1}T^{(\gamma-1)/2}$. Noting that $s_1(t) + s_2(t) \ge 2\beta_1 t^{(1+\gamma)/2}$, and differencing the relations (17.1.3), we obtain

$$\delta(t) \le 2^{-1}\beta_1^{-1}t^{-(1+\gamma)/2}\int_0^t|f_1(\tau) - f_2(\tau)|\,d\tau$$

$$+ 2^{-1}\beta_1^{-1}t^{-(1+\gamma)/2}\int_0^{\alpha(t)}\xi|u_2(\xi,t) - u_1(\xi,t)|\,d\xi$$

$$+ 2^{-1}\beta_1^{-1}t^{-(1+\gamma)/2}\int_{\alpha(t)}^{\beta(t)}\xi u_{j(t)}(\xi,t)\,d\xi = I_1 + I_2 + I_3,$$

$$(17.7.28)$$

where $j(t)$ is defined by (17.7.6).

We consider I_1 first. Now,

$$I_1 = 2^{-1}\beta_1^{-1}t^{-(1+\gamma)/2}\int_0^t|f_1(\tau) - f_2(\tau)|\,d\tau$$

$$\le 2^{-1}\beta_1^{-1}t^{-1}\int_0^t\tau^{[(1-\gamma)/2]}|f_1(\tau) - f_2(\tau)|\,d\tau$$

$$\le 2^{-1}\beta_1^{-1}\|f_1 - f_2\|_{[0,t],\gamma}, \qquad (17.7.29)$$

where $\|f_1 - f_2\|_{[0,t],\gamma}$ is defined by (17.7.27).

Next, we can consider I_2, and follow the estimation procedure of I_4 of Lemma 17.7.2 (see Eqs. (17.7.8) through (17.7.17)). The estimation of w_1 is exactly the same with $\varphi_1 = \varphi_2 = 0$, and yields an estimate like (17.7.29). The estimation of w_2 in (17.7.16) is the same, and from $\alpha(t) \le \beta_2 t^{(1+\gamma)/2}$ we obtain

$$I_2 \le I_1 + C_1 \int_0^t \frac{|\mu(\tau)| \, d\tau}{(t-\tau)^{1/2}}, \qquad (17.7.30)$$

where $\mu(t)$ is defined by (17.7.12) and C_1 is a positive constant depending only on l, L, γ, and T. Recalling that $0 \le u_i(x, t_0) \le Lt_0^{\gamma} \beta_1^{-1} t_0^{-(1+\gamma)/2}$ $(s(t_0) - x)$, it follows that the constant C multiplying $\delta(t)$ in (17.7.12) depends only on l, L, γ, and T. Consequently, the Lipschitz continuity of $\alpha(t)$ allows us to deduce, via Eqs. (17.7.13) through (17.7.15), that there exists a positive constant $C_2 = C_2(l, L, \gamma, T)$ such that

$$|\mu(\tau)| \le C_2 \|\delta\|_{[0, \tau]}. \qquad (17.7.31)$$

Consequently,

$$I_2 \le I_1 + C_3 \int_0^t \frac{\|\delta\|_{[0, \tau]}}{(t-\tau)^{1/2}} \, d\tau, \qquad (17.7.32)$$

where C_3 has the same dependence as C_2.

For the term I_3 we need only consider z_2 of (17.7.22). Using the technique of (17.7.16) and $\beta_1 t^{(1+\gamma)/2} \le \alpha(t) \le \beta(t) \le \beta_2 t^{(1+\gamma)/2}$, it follows again that there exists a constant C_4 such that

$$I_3 \le 2^{-1} \beta_1^{-1} t^{-(1+\gamma)/2} \int_{\alpha(t)}^{\beta(t)} \xi z_2(\xi, t) \, d\xi \le C_4 \int_0^t \frac{\|\delta\|_{[0, \tau]}}{(t-\tau)^{1/2}} \, d\tau. \qquad (17.7.33)$$

Combining (17.7.29), (17.7.32), and (17.7.33), we obtain

$$0 \le \|\delta\|_{[0, t]} \le C_5 \|f_1 - f_2\|_{[0, t], \gamma} + C_6 \int_0^t \frac{\|\delta\|_{[0, \tau]}}{(t-\tau)^{1/2}} \, d\tau, \qquad (17.7.34)$$

where C_5 and C_6 are positive constants that depend upon l, L, γ, and T. An application of Lemma 17.7.1 to (17.7.34) yields the result. $\qquad \square$

We conclude this discussion with the following general result.

Theorem 17.7.1. *If (s_n, u_n) are the solutions of the Stefan problem (17.1.1)–(17.1.2) corresponding to the data (b_n, φ_n, f_n), if b_n converges to b as n tends to infinity, and if φ_n tends to φ and f_n tends to f uniformly as n tends to infinity, then s_n tends to s and u_n tends to u uniformly as n tends to infinity, where (s, u) is the solution of the Stefan problem (17.1.1)–(17.1.2) corresponding to the data (b, φ, f).*

Proof. First, we can apply the Monotonicity Theorem 17.2.1 to generate a monotone-increasing sequence of free boundaries converging to s and a monotone-decreasing sequence of free boundaries converging to s. Given a curve from each sequence, it follows again, from the Monotonicity

Theorem 17.2.1, that for all n sufficiently large, all s_n are between these two curves. Hence, we obtain the uniform convergence of the s_n to s.

The uniform convergence of the u_n to u can be accomplished via a combination of the Maximum Principle and potential theoretic estimates similar to those used in the proof of the existence Theorem 17.3.1.   ☐

17.8. REGULARITY OF THE FREE BOUNDARY

It is not difficult to demonstrate using the methods above that there exists a solution (s_δ, u_δ), $\delta > 0$, of the problem

$$\mathcal{L}(u_\delta) = 0, \qquad 0 < x < s_\delta(t), \qquad 0 < t \leq T,$$
$$u_\delta(0, t) = f(t), \qquad 0 < t \leq T,$$
$$u_\delta(x, 0) = \varphi(x), \qquad 0 \leq x \leq b = s_\delta(0), \qquad (17.8.1)$$
$$u_\delta(s_\delta(t), t) = 0, \qquad 0 < t \leq T,$$
$$\dot{s}_\delta(t) = \delta^{-1} u(s_\delta(t) - \delta, t), \qquad 0 < t \leq T.$$

From the interior regularity of u_δ, it follows immediately that $s_\delta \in C^\infty((0, T])$. As (s_δ, u_δ) look like approximations to (s, u) for δ small, it is natural to conjecture that $s \in C^\infty((0, T])$ and to proceed to demonstrate this fact. Using (17.8.1) with $\varphi \equiv 0$ and $b > 0$, the author and C. Denson Hill employed a Maximum Principle argument that is beyond the scope of this presentation to conclude that $s \in C^\infty((0, T])$.

Another approach that is more general is to note that, when $s \in C^\gamma((0, T])$, $\gamma > \frac{1}{2}$, we have $u_x(s(t), t) \in C^\nu((0, T])$, $\nu > 0$. This means that $\dot{s} \in C^\nu((0, T])$. Utilizing Schauder estimates for parabolic partial differential equations, you can show that the higher derivatives of u at $s(t)$ exist and are Hölder continuous. This allows you to differentiate $\dot{s}(t) = -u_x(s(t), t)$ to obtain $\ddot{s} \in C^{\nu_1}((0, T])$, $\nu_1 > 0$. Continuing in this fashion you obtain the following result.

Theorem 17.8.1.   *The free boundary $s \in C^\infty((0, T])$.*

Proof.   See references to the original papers.   ☐

Still further results can be obtained. We have seen in Chapter 10 that the contribution to the solution from the initial data is analytic.

Likewise, if $f(t)$ is analytic, its contribution is analytic. Since 0 is analytic on $s(t)$, a more detailed set of estimates yields:

Theorem 17.8.2.   *If f is analytic, then the free boundary s is analytic for $0 < t \leq T$.*

Proof.   See references to the original papers.   ☐

Remark. The proofs of both of these theorems are beyond the scope of this presentation, and the interested reader is referred to the notes for this chapter, which appear at the end of Chapter 18, for the references to the original papers.

EXERCISES

17.1. In the discussion preceding Eq. (17.1.1), show that the constants α, k, and κ can be set equal to one without loss of generality.

17.2. Prove Lemma 17.4.2.

17.3. Prove Lemma 17.4.3.

Chapter 18

The One-Phase Stefan Problem with Flux-Boundary Specification: Some Exercises

18.1. INTRODUCTION

The one-phase Stefan problem with flux-boundary specification is the problem of determining $s = s(t)$ and $u = u(x, t)$ such that the pair (s, u) satisfies

$$\mathcal{L}(u) = 0, \qquad 0 < x < s(t), \qquad 0 < t \le T,$$
$$-u_x(0, t) = h(t) > 0, \qquad 0 < t \le T,$$
$$u(x, 0) = \varphi(x) \ge 0, \qquad 0 \le x \le b \equiv s(0), \qquad (18.1.1)$$
$$u(s(t), t) = 0, \qquad 0 < t \le T,$$

and

$$\dot{s}(t) = -u_x(s(t), t), \qquad 0 \le t \le T, \qquad (18.1.2)$$

where T is a positive constant and $\mathcal{L}(u) \equiv u_{xx} - u_t$.

For a given $s(t)$, we can regard (18.1.1) as a problem for the determination of u.

DEFINITION 18.1.1. The real-valued function $u = u(x, t)$ is a solution of (18.1.1) for a given function $s = s(t)$ if u_x, u_t, and u_{xx} are continuous and $\mathcal{L}(u) = 0$ for $0 < x < s(t), 0 < t \le T, u(x, 0) = \varphi(x)$ at points of continuity of φ, and u is bounded at points of discontinuity of φ,

$u(s(t), t) = 0$, and $- u_x(0, t) = h(t)$ at points of continuity of h and u_x is bounded at points of discontinuity of h.

The definition of a solution to the Stefan problem can now be easily given.

DEFINITION 18.1.2. The pair (s, u) is a solution of the Stefan problem (18.1.1)–(18.1.2) if $s \in C^1((0, T]) \cap C([0, T])$, $s(0) = b$, u is a solution of (18.1.1) for this s, and s and u together satisfy (18.1.2).

Following the outline of Chapter 17, we demonstrate the equivalence of condition (18.1.2) with an integral condition.

Lemma 18.1.1.   *The condition*

$$\dot{s}(t) = - u_x(s(t), t), \qquad 0 < t \le T, \tag{18.1.3}$$

can be replaced by the equivalent condition

$$s(t) = b + \int_0^b \varphi(\xi) \, d\xi + \int_0^t h(\tau) \, d\tau - \int_0^{s(t)} u(\xi, t) \, d\xi, \qquad 0 < t \le T. \tag{18.1.4}$$

Proof.   Exercise 18.1. Hint: Set $\mathscr{L}(u) = 0$ and $v = -1$ in the Green's Identity.

Remark.   As in Chapter 17, we see that equivalence here is defined in the proof to be valid when u_x exists and is continuous at the boundary $x = s(t)$. We shall obtain boundaries here that are Hölder continuous with Hölder exponent greater than $\frac{1}{2}$, whence it follows, via Theorem 14.4.1, that u_x exists and is continuous.

Before proceeding with the discussion of monotone dependence and uniqueness, it is convenient to demonstrate a variation of the Weak Maximum Principle that is useful here.

Lemma 18.1.2 (Weak Maximum Principle).   *For $D_T = \{(x, t) | 0 < x < s(t), 0 < t \le T\}$, let v satisfy*

$$\mathscr{L}(v) = 0 \text{ in } D_T,$$
$$- v_x(0, t) \ge 0, 0 < t \le T,$$
$$v(x, 0) \ge 0, 0 \le x \le s(0), \tag{18.1.5}$$
$$v(s(t), t) \ge 0, 0 < t \le T,$$

in the sense of Definition 18.1.1. Then,

$$v(x, t) \ge 0 \text{ in } D_T. \tag{18.1.6}$$

Proof.   Let l denote a number exceeding $\sup_{0 \le t \le T} s(t)$ and consider for each $\varepsilon > 0$,

$$w^\varepsilon(x, t) = v(x, t) + \varepsilon(l - x). \tag{18.1.7}$$

Clearly,

$$\mathscr{L}(w^{\varepsilon}) = 0 \quad \text{in } D_T,$$

$$-w_x^{\varepsilon}(0, t) = -v_x(0, t) + \varepsilon > 0, \qquad 0 < t \leq T,$$

$$w^{\varepsilon}(x, 0) > 0, \qquad 0 \leq x < s(0), \tag{18.1.8}$$

$$w^{\varepsilon}(s(t), t) > 0, \qquad 0 < t \leq T.$$

We conclude that $w^{\varepsilon} > 0$ in D_T. Otherwise, from the Weak Minimum Principle, Theorem 1.6.1, the minimum of $w^{\varepsilon} \leq 0$ must occur on the boundary segment $x = 0$ for some $t_0 > 0$. But, $-w_x^{\varepsilon}(0, t_0) \geq \varepsilon > 0$ implies that $w^{\varepsilon} < 0$ in D_T and assumes its minimum value there. The Strong Maximum Principle, Theorem 15.3.1, implies that $w^{\varepsilon} \equiv c < 0$, which is a contradiction. Hence $w^{\varepsilon} > 0$ in D_T. Letting ε tend to zero, we see that $v(x, t) \geq 0$ in D_T. $\qquad\square$

As a corollary to lemma 18.1.2 we have the following result.

Lemma 18.1.3. *Any solution u of Problem* (18.1.1) *is nonnegative. Moreover, for any solution of* (18.1.1)–(18.1.2), *$\dot{s}(t) = u_x(s(t), t) \geq 0$. In the case that φ is positive and/or $h(t)$ is positive in each neighborhood of $t = 0$, $\dot{s}(t) = u_x(s(t), t) > 0, 0 < t \leq T$.*

Proof. Exercise 18.2. *Hint*: Use Lemma 18.1.2 and the Strong Maximum Principle of Theorem 15.3.1.

18.2. MONOTONE DEPENDENCE UPON THE DATA AND UNIQUENESS OF THE SOLUTION

The Weak Maximum Principle of Lemma 18.1.2 allows us to employ the same arguments as those in Sec. 17.2. We begin with the first result.

Lemma 18.2.1. *Let (s_1, u_1) and (s_2, u_2) be solutions to the Stefan problem* (18.1.1)–(18.1.2), *corresponding, respectively, to the data (b_1, φ_1, h_1) and (b_2, φ_2, h_2). If $b_1 < b_2$, $\varphi_1 \leq \varphi_2$, and $h_1 \leq h_2$, then $s_1 < s_2$.*

Proof. Exercise 18.3. *Hint*: See Lemma 17.2.1. Use Lemma 18.1.2 and the Strong Maximum Principle of Theorem 15.3.1. $\qquad\square$

As in Sec. 17.2, we now employ Lemma 18.2.1 and (18.1.4) to produce the following general result.

Theorem 18.2.1 (Monotonicity). *Let (s_1, u_1) and (s_2, u_2) be solutions to the Stefan problem* (18.1.1)–(18.1.2), *corresponding, respectively, to the data (b_1, φ_1, h_1) and (b_2, φ_2, h_2). If $b_1 \leq b_2$, $\varphi_1 \leq \varphi_2$, and $h_1 \leq h_2$, then $s_1 \leq s_2$.*

Proof. For $\delta > 0$, let $(s_2^{\delta}, u_2^{\delta})$ be a solution of (18.1.1)–(18.1.2) that corresponds to the data $(b_2 + \delta, \varphi_2, f_2)$, where φ_2 has been extended as zero outside of $0 \leq x \leq b_2$. From Lemma 18.2.1, we have $s_1 < s_2^\delta$. Now, (s_2, u_2) and (s_2^δ, u_2^δ) must each satisfy their respective versions of (18.1.4). Subtracting, we obtain

$$s_2^\delta(t) - s_2(t) = \delta - \int_0^{s_2(t)} \left\{ u_2^\delta(\xi, t) - u_2(\xi, t) \right\} d\xi - \int_{s_2(t)}^{s_2^\delta(t)} u_2^\delta(\xi, t) \, d\xi,$$

$$(18.2.1)$$

where we have used Lemma 18.2.1 to assert that $s_2(t) < s_2^\delta(t)$. An application of the Weak Maximum Principle, Lemma 18.1.2, yields $u_2^\delta \geq 0$ and $u_2^\delta - u_2 \geq 0$. Hence,

$$s_2^\delta(t) \leq s_2(t) + \delta \qquad (18.2.2)$$

and

$$s_1(t) \leq s_2(t) + \delta. \qquad (18.2.3)$$

As δ can be taken as small as desired, it follows that $s_1(t) \leq s_2(t)$.   □

As a corollary of Theorem 18.2.1 we obtain the following result.

Theorem 18.2.2 (Uniqueness). *There can exist at most one solution to the Stefan problem* (18.1.1)–(18.1.2).

     Proof. If (s_1, u_1) and (s_2, u_2) both correspond to the same data (b, φ, h), then $s_1 \leq s_2$ and $s_2 \leq s_1$ imply that $s_1 \equiv s_2$. From this, Lemma 18.1.2 implies that $u_1 \leq u_2$ and $u_2 \leq u_1$, whence $u_1 \equiv u_2$.   □

18.3. EXISTENCE

We begin our study of the existence of the solution to the Stefan problem (18.1.1)–(18.1.2) with the following assumption on the data φ and h.

     Assumption A. *h is a bounded, piecewise-continuous nonnegative function and when applicable, φ is a piecewise-continuous nonnegative function that satisfies $0 \leq \varphi(x) \leq N(b - x)$, where N is a positive constant.*

     As in Sec. 17.3, we can derive an *a priori* estimate upon $u_x(s(t), t)$ for any solution u of (18.1.1), with an increasing boundary s.

     Lemma 18.3.1. *Let u denote the solution of (18.1.1) with nondecreasing boundary s. Under Assumption A, there exists a positive constant $C = \max(N, M)$, where $M = \sup_{0 \leq t \leq T} h(t)$, such that*

$$0 \leq \rho^{-1} u(s(t) - \rho, t) \leq C \qquad (18.3.1)$$

for all $0 \leq t \leq T$ and $0 < \rho < s(t)$.

Proof. We can use the comparison function $w(x, t_0) = C(s(t_0) - x)$ and apply the Weak Maximum Principle, Lemma 18.1.2, to show that

$$0 \le u(x, t) \le C(s(t_0) - x) \qquad \text{for } 0 \le x \le s(t), \quad 0 \le t \le t_0.$$

As $u(s(t_0), t_0) = 0$ and t_0 is arbitrary, the result (18.3.1) follows. □

Remark. In the event that $u_x(s(t), t)$ exists, then (18.3.1) implies that

$$-C \le u_x(s(t), t) \le 0. \tag{18.3.2}$$

From this result we can derive our first existence result.

Lemma 18.3.2. *Under Assumption A with $b > 0$, there exists a unique solution (s, u) to the Stefan problem* (18.1.1)–(18.1.2). *Moreover, the free boundary s satisfies*

$$0 \le \dot{s}(t) \le C, \tag{18.3.3}$$

where C is defined in Lemma 18.3.1.

Proof. All of the construction using the retardation of the argument in Eq. (18.1.2) can be repeated here yielding a family (s^τ, u^τ), $0 < \tau < \tau_0$, where u^τ is the solution of (18.1.1) with boundary s^τ defined by

$$\dot{s}^\tau(t) = -u_x^\tau(s^\tau(t - \tau), t - \tau), \qquad t > \tau,$$
$$s(t) \equiv b, \qquad 0 \le t \le \tau. \tag{18.3.4}$$

From the construction,

$$0 \le \dot{s}^\tau(t) \le C, \qquad 0 < t \le T, \tag{18.3.5}$$

where $C = \max(N, M)$. The Ascoli–Arzela Theorem applies again to yield a limit curve s. We let u denote the solution of (18.1.1) for this s.

Let $\alpha^\tau(t) = \min(s^\tau(t), s(t))$ and consider $w = u^\tau - u$. On α^τ,

$$|w(\alpha^\tau(t), t)| \le C|s^\tau(t) - s(t)|$$

via Lemma 18.3.1, while $w(x, 0) = 0$, $0 \le x \le b - \tau$ and $|w(x, 0)| \le C\tau$, $b - \tau \le x \le b$. On $x = 0$, $w_x(0, t) = 0$. Consequently, a direct application of the Weak Maximum Principle Lemma 18.1.2 to $z^\tau(x, t) = C\|s^\tau - s\|_{[0, T]} + C\tau \pm w(x, t)$ yields the fact that, for $0 \le x \le \alpha(t)$, $0 \le t \le T$,

$$|u^\tau(x, t) - u(x, t)| \le C\{\|s^\tau - s\|_{[0, T]} + \tau\}, \tag{18.3.6}$$

where

$$\|s^\tau - s\|_{[0, T]} = \sup_{0 \le t \le T} |s^\tau(t) - s(t)|. \tag{18.3.7}$$

A similar estimate holds for $u^\tau - u$ in $\alpha^\tau(t) \le x \le \beta^\tau(t)$, $0 \le t \le T$, where $\beta^\tau(t) = \max(s^\tau(t), s(t))$. Hence, the uniform convergence of the subsequence s^τ to s implies the uniform convergence of the corresponding u^τ to u as τ tends to zero.

Now applying the identity (18.1.3) to u^τ yields

$$-\int_0^t u_x^\tau(s^\tau(\eta),\eta)\,d\eta = \int_0^{b-\tau}\varphi(\xi)\,d\xi + \int_0^t h(\eta)\,d\eta - \int_0^{s^\tau(t)} u^\tau(\xi,t)\,d\xi,$$

$$(18.3.8)$$

or

$$s^\tau(t+\tau) = b + \int_0^{b-\tau}\varphi(\xi)\,d\xi + \int_0^t h(\eta)\,d\eta - \int_0^{s^\tau(t)} u^\tau(\xi,t)\,d\xi,$$

$$(18.3.9)$$

whence it follows that (s,u) satisfies (18.1.4). As s is Lipschitz-continuous with Lipschitz constant C and $0 \le \varphi(x) \le N(b-x)$, it follows from Theorem 14.5.1 that $u_x(s(t),t)$ exists and is continuous. Thus, via Lemma 18.1.1, (s,u) is a solution of the Stefan problem (18.1.1)–(18.1.2). Uniqueness is assured via Theorem 18.2.2.   □

Using this result we can demonstrate the following result.

Lemma 18.3.3.   *Under Assumption A with $b = 0$ (and φ irrelevant), there exists a unique solution (s,u) to the Stefan problem (18.1.1)–(18.1.2). Moreover, the free boundary s satisfies*

$$0 \le \dot{s}(t) \le C, \qquad (18.3.10)$$

where here $C = \sup_{0 \le t \le T} h(t)$.

Proof.   Exercise 18.4 *Hint*: Let (s^b, u^b) denote the solutions corresponding to the data $(b,0,h)$. Use Lemma 18.3.1 and the argument of Lemma 18.3.2, beginning with the Ascoli–Arzela Theorem.   □

Applying these results and the methods of Sec. 17.3, we can remove the condition $0 \le \varphi(x) \le N(b-x)$. We conclude this section with the statement of the result.

Theorem 18.3.1 (Existence).   *If $b \ge 0$, h is a bounded, piecewise-continuous nonnegative function, and, when applicable, φ is a bounded piecewise-continuous nonnegative function, then there exists a unique solution (s,u) to the Stefan problem (18.1.1)–(18.1.2).*

Proof.   Exercise 18.5.   □

18.4.   ASYMPTOTIC BEHAVIOR OF THE FREE BOUNDARY AT $t = +\infty$

We begin our study with the following general result.

Theorem 18.4.1.   *If*

$$\lim_{t \to \infty} \int_0^t h(\tau)\,d\tau = +\infty, \qquad (18.4.1)$$

then

$$\lim_{t \to \infty} s(t) = +\infty. \tag{18.4.2}$$

If

$$\lim_{t \to \infty} \int_0^t h(\tau) \, d\tau = H, \qquad 0 \le H < \infty, \tag{18.4.3}$$

then

$$\lim_{t \to \infty} s(t) = l, \tag{18.4.4}$$

where

$$l = b + H + \int_0^b \varphi(x) \, dx. \tag{18.4.5}$$

Proof. Consider first the case of $h(t)$ with compact support; that is, $h(t) \equiv 0$ for $t > t_0$. Now, conditions (18.1.4) and (18.4.5) imply that

$$s(t) = l - \int_0^{s(t)} u(x, t) \, dx \tag{18.4.6}$$

for $t > t_0$.

By the Weak Maximum Principle of Lemma 18.1.2, $u(x, t)$ is dominated by $w_1(x, t) + w_2(x, t)$, where for $i = 1, 2$, $\mathscr{L}(w_i) = 0$ for $0 < x$, $0 < t$, and

$$-w_{1_x}(0, t) = 0, \qquad 0 < t,$$

$$w_1(x, 0) = \begin{cases} \varphi(x), & 0 \le x \le b, \\ 0, & b < x < \infty, \end{cases} \tag{18.4.7}$$

and

$$-w_{2_x}(0, t) = h(t), \qquad 0 < t,$$

$$w_2(x, 0) = 0. \tag{18.4.8}$$

Utilizing the solution formulas from Chapter 5 and making some elementary estimates, we see that, for $t > t_0$,

$$0 \le w_1(x, t) \le C_1 \|\varphi\|_{[0, b]} t^{-1/2} \tag{18.4.9}$$

and

$$0 \le w_2(x, t) \le C_2 \|h\|_{[0, t_0]} \left\{ \sqrt{t} - \sqrt{t - t_0} \right\}. \tag{18.4.10}$$

Clearly, w_1 tends uniformly to zero as t tends to infinity. By the Mean-Value Theorem,

$$\left\{ \sqrt{t} - \sqrt{t - t_0} \right\} \le \frac{1}{2} \left(\frac{t_0}{(t - t_0)^{1/2}} \right), \tag{18.4.11}$$

whence it follows that w_2 tends uniformly to zero as t tends to infinity. Thus, from $0 \le s(t) \le l$ and (18.4.6), we see that u tends uniformly to zero as t tends to infinity, and

$$\lim_{t \to \infty} s(t) = l. \qquad (18.4.12)$$

For the general $h(t)$, set

$$h_n(t) = \begin{cases} h(t), & 0 \le t \le n, \\ 0, & n < t, \end{cases} \qquad (18.4.13)$$

and define the corresponding $l^{(n)}$ and $s^{(n)}$. By the first argument, we have $\lim_{t \to \infty} s^{(n)}(t) = l^{(n)}$. However, $h_n \le h$. Theorem 18.2.1 implies that $s^{(n)}(t) \le s(t)$ and, since $u(x, t) \ge 0$, $s(t) \le l$. Hence,

$$l^{(n)} = \lim_{t \to \infty} s^{(n)}(t) \le \liminf_{t \to \infty} s(t) \le \limsup_{t \to \infty} s(t) \le l. \quad (18.4.14)$$

Letting n tend to infinity, we see that $l^{(n)}$ tends to l whether l is finite or not.
$\square$

The next result shows that when $\int_0^\infty h(t)\, dt = \infty$, the behavior of the boundary at $t = \infty$ depends only upon h.

Theorem 18.4.2. For $\int_0^\infty h(t)\, dt = \infty$, let (s, u) be the solution of the Stefan problem (18.1.1)–(18.1.2) and let (σ, v) be the solution of:

$$\mathcal{L}(v) = 0, \qquad 0 < x < \sigma(t), \qquad t_0 < t < \infty,$$

$$-v_x(0, t) = h(t), \qquad t_0 < t,$$

$$v(\sigma(t), t) = 0, \qquad t_0 < t, \qquad (18.4.15)$$

$$\dot{\sigma}(t) = -v_x(\sigma(t), t), \qquad t_0 < t,$$

$$\sigma(t_0) = 0.$$

Then, as t tends to infinity,

$$\frac{s(t)}{\sigma(t)} = 1 + \mathcal{O}\left(\frac{1}{\sigma(t)}\right), \qquad (18.4.16)$$

which implies that $s(t) \sim \sigma(t)$, since $\lim_{t \to \infty} \sigma(t) = \infty$.

Proof. From the Monotone Dependence Theorem 18.2.1, $\sigma(t) \le s(t)$. From the Weak Maximum Principle of Lemma 18.1.2, $v(x, t) \le u(x, t)$. Using (18.1.4) for both s and σ, we obtain

$$\sigma(t) \le s(t) = s(t_0) + \int_0^{s(t_0)} u(x, t_0)\, dx + \int_{t_0}^t h(\tau)\, d\tau - \int_0^{s(t)} u(x, t)\, dx$$

$$\le s(t_0) + \int_0^{s(t_0)} u(x, t_0)\, dx + \sigma(t). \qquad (18.4.17)$$

Hence

$$1 \le \frac{s(t)}{\sigma(t)} \le 1 + (\sigma(t))^{-1} \left\{ s(t_0) + \int_0^{s(t_0)} u(x, t_0) \, dx \right\}. \qquad (18.4.18)$$

□

We utilize Theorem 18.4.2 to link the behavior of s to h.

Theorem 18.4.3. *If (s, u) is the solution of the Stefan problem* (18.1.1)–(18.1.2), *if*

$$\lim_{t \to \infty} \int_0^t h(\tau) \, d\tau = \infty, \qquad (18.4.19)$$

and if

$$\lim_{t \to \infty} \int_0^t \frac{h(\tau)}{\sqrt{t - \tau}} \, d\tau = 0, \qquad (18.4.20)$$

then

$$s(t) \sim \int_0^t h(\tau) \, d\tau \qquad (18.4.21)$$

as t tends to infinity.

Proof. Consider $\sigma(t)$ defined by (18.4.15) with $t_0 = 0$. We have

$$\int_0^t h(\tau) \, d\tau - \int_0^{\sigma(t)} v(x, t) \, dx = \sigma(t) \le \int_0^t h(\tau) \, d\tau. \qquad (18.4.22)$$

Now, $0 \le v(x, t) \le w_2(x, t)$, where w_2 satisfies $\mathscr{L}(w_2) = 0$, $0 < x$, $0 < t$, and (18.4.8). Hence

$$\sigma(t) \ge \int_0^t h(\tau) \, d\tau - \int_0^{\sigma(t)} w_2(x, t) \, dx \ge \int_0^t h(\tau) \, d\tau - \sigma(t) C \int_0^t \frac{h(\tau)}{\sqrt{t - \tau}} \, d\tau,$$

$$(18.4.23)$$

where C is a positive constant resulting from the estimation of the formula for w_2. Thus, we obtain

$$\left\{ 1 + C \int_0^t \frac{h(\tau) \, d\tau}{\sqrt{t - \tau}} \right\} \sigma(t) \ge \int_0^t h(\tau) \, d\tau. \qquad (18.4.24)$$

Combining (18.4.24) with (18.4.22), we get

$$\left\{ 1 + C \int_0^t \frac{h(\tau) \, d\tau}{\sqrt{t - \tau}} \right\}^{-1} \le \frac{\sigma(t)}{\int_0^t h(\tau) \, d\tau} \le 1, \qquad (18.4.25)$$

whence the result follows. □

Remark. The relations (18.4.19) and (18.4.20) express the delicate area of where the total energy input is infinite but the boundary temperature tends to zero. For example, consider

$$h(t) = \begin{cases} 1, & 0 \le t \le 1, \\ t^{-\gamma}, & 1 \le t < \infty, \quad \tfrac{1}{2} < \gamma \le 1. \end{cases}$$

Recalling (14.2.2),

$$\int_0^t \frac{h(\tau)\,d\tau}{(t-\tau)^{1/2}} \le \int_0^t \frac{d\tau}{\tau^{\gamma}(t-\tau)^{1/2}} = \frac{\Gamma(1-\gamma)\Gamma(\tfrac{1}{2})}{\Gamma(\tfrac{3}{2}-\gamma)} t^{1-\gamma-1/2},$$

which tends to zero as t tends to infinity.

Using the special solution of Sec. 17.3 and 17.4, we can obtain some particular results for the case in which

$$\lim_{t \to \infty} \int_0^t \frac{h(\tau)\,d\tau}{\sqrt{t-\tau}} \ne 0.$$

Theorem 18.4.4. *Let (s,u) be the solution of the Stefan problem* (18.1.1)–(18.1.2). *If*

$$h(t) \sim t^{-\alpha} \sum_{n=1}^{\infty} \frac{\beta^{2n-1}\Gamma(2n\alpha - \alpha + 1)t^{2n\alpha - n}}{\Gamma(2n)\Gamma(2n\alpha - \alpha - n + 1)}, \qquad (18.4.26)$$

then

$$s(t) \sim \beta t^{\alpha}, \qquad \alpha \ge \tfrac{1}{2}. \qquad (18.4.27)$$

Proof. The proof is analogous to that of Theorem 17.5.4 and is left to the reader. □

Remark. For the cases $\alpha = \tfrac{1}{2}$ and $\alpha = 1$, the boundary behavior of the special solutions is easy to identify. Namely, for $\alpha = \tfrac{1}{2}$,

$$h(t) \sim t^{-1/2} \sum_{n=1}^{\infty} \frac{\Gamma(n+\tfrac{1}{2})\beta^{2n-1}}{\Gamma(2n)\Gamma(\tfrac{1}{2})}$$

and

$$s(t) \sim \beta t^{1/2}$$

while, for $\alpha = 1$,

$$h(t) \sim \beta \exp\{\beta^2 t\}$$

and

$$s(t) \sim \beta t.$$

18.5.   CONTINUOUS DEPENDENCE OF THE FREE BOUNDARY UPON THE DATA

We demonstrate here the following theorem.

Theorem 18.5.1. *Let (s_i, u_i), $i = 1,2$, be solutions of the Stefan problem* (18.1.1)–(18.1.2), *which correspond, respectively, to the data (b_i, φ_i, h_i),*

$i = 1, 2$. *Let the data* (b_i, φ_i, h_i), $i = 1, 2$, *satisfy the Assumption A for the same* N *and* $M \geq \sup_{0 \leq t \leq T} h_i(t)$, $i = 1, 2$. *Then there exists a constant* $C = C(N, M, T)$ *such that for* $b_2 \geq b_1$,

$$|s_1(t) - s_2(t)| \leq C \left\{ b_2 - b_1 + \int_0^{b_1} |\varphi_1(x) - \varphi_2(x)| \, dx + \int_{b_1}^{b_2} \varphi_2(x) \, dx \right.$$

$$\left. + \int_0^t |h_1(\tau) - h_2(\tau)| \, d\tau \right\}, \qquad 0 \leq t \leq T. \quad (18.5.1)$$

Proof. Let $\alpha(t) = \min(s_1(t), s_2(t))$, $\beta(t) = \max(s_1(t), s_2(t))$ and $\delta(t) = \beta(t) - \alpha(t)$. Differencing (18.1.4) for each i, we obtain

$$0 \leq \delta(t) \leq b_2 - b_1 + \int_0^t |h_1(\tau) - h_2(\tau)| \, d\tau + \int_0^{b_1} |\varphi_1(x) - \varphi_2(x)| \, dx$$

$$+ \int_{b_1}^{b_2} \varphi_2(x) \, dx + \int_0^{\alpha(t)} |u_1(x, t) - u_2(x, t)| \, dx$$

$$+ \int_{\alpha(t)}^{\beta(t)} u_{j(t)}(x, t) \, dt, \qquad (18.5.2)$$

where

$$j(t) = \begin{cases} 2, & s_2(t) \geq s_1(t), \\ 1, & s_2(t) < s_1(t). \end{cases} \quad (18.5.3)$$

As in the proof of Lemma 17.7.2, we shall estimate the last two terms in terms of the first four terms and δ.

From Lemma 18.3.1 and Lemma 18.3.2, we observe that $0 \leq u_i(x, t) \leq C_1(s_i(t) - x)$, where $C_1 = \max(M, N)$. Hence,

$$u_{j(t)}(\alpha(t), t) = |u_1(\alpha(t), t) - u_2(\alpha(t), t)| \leq C_1 \delta(t). \quad (18.5.4)$$

First, we shall estimate the integral of $|u_1 - u_2|$ on $0 \leq x \leq \alpha(t)$. Set

$$v(x, t) = u_1(x, t) - u_2(x, t) = v_1(x, t) + v_2(x, t) + v_3(x, t),$$

$$0 < x < \alpha(t), \qquad 0 < t \leq T, \quad (18.5.5)$$

where $\mathcal{L}(v_i) = 0$, $i = 1, 2, 3$, in $0 < x < \alpha(t)$, $0 < t \leq T$, and

$$-\frac{\partial v_1}{\partial x}(0, t) = 0, \qquad 0 < t \leq T,$$

$$v_1(\alpha(t), t) = 0, \qquad 0 < t \leq T,$$

$$v_1(x, 0) = \varphi_1(x) - \varphi_2(x), \qquad 0 \leq x \leq b_1,$$

$$(18.5.6)$$

$$-\frac{\partial v_2}{\partial x}(0, t) = h_1(t) - h_2(t), \qquad 0 < t \leq T,$$

$$v_2(\alpha(t), t) = 0, \qquad 0 < t \leq T, \quad (18.5.7)$$

$$v_2(x, 0) = 0, \qquad 0 \leq x \leq b_1,$$

and

$$-\frac{\partial v_3}{\partial x}(0,t) = 0, \qquad 0 < t \le T,$$

$$v_3(\alpha(t),t) = u_1(\alpha(t),t) - u_2(\alpha(t),t), \qquad 0 < t \le T, \quad (18.5.8)$$

$$v_3(x,0) = 0, \qquad 0 \le x \le b_1.$$

All the solutions v_i, $i = 1, 2, 3$, exist, since $\alpha(t)$ is Lipschitz-continuous.

Now, $|v_1(x,t)|$ is dominated via the Weak Maximum Principle, Lemma 18.1.2, by the solution w_1 of the heat equation in the quarter-plane $x > 0$, $t > 0$, such that

$$-\frac{\partial w_1}{\partial x}(0,t) = 0, \qquad 0 < t \le T,$$

$$w_1(x,0) = \begin{cases} |\varphi_1(x) - \varphi_2(x)|, & 0 \le x \le b_1, \\ 0, & b_1 < x < \infty. \end{cases} \qquad (18.5.9)$$

Thus,

$$\int_0^{\alpha(t)} |v_1(x,t)| \, dx \le \int_0^\infty w_1(x,t) \, dx = \int_0^{b_1} |\varphi_1(x) - \varphi_2(x)| \, dx.$$

$$(18.5.10)$$

An application of the Weak Maximum Principle, Lemma 18.1.2, again shows that $|v_2|$ is dominated by the solution w_2 of the heat equation in the quarter-plane $x > 0$, $t > 0$, such that

$$-\frac{\partial w_2}{\partial x}(0,t) = |h_1(t) - h_2(t)|, \qquad 0 < t \le T, \qquad (18.5.11)$$

$$w_2(x,0) = 0, \qquad 0 < x < \infty.$$

Hence,

$$\int_0^{\alpha(t)} |v_2(x,t)| \, dx \le \int_0^\infty w_2(x,t) \, dx = \int_0^t |h_1(\tau) - h_2(\tau)| \, d\tau.$$

$$(18.5.12)$$

Next, $|v_3|$ is dominated by w_3, where

$$\mathscr{L}(w_3) = 0, \qquad 0 < x < \alpha(t), \qquad 0 < t \le T,$$

$$-\frac{\partial w_3}{\partial x}(0,t) = 0, \qquad 0 < t \le T,$$

$$w_3(x,0) = 0, \qquad 0 \le x \le b_1, \qquad (18.5.13)$$

$$w_3(\alpha(t),t) = C\delta(t), \qquad 0 < t \le T.$$

By reflection about $x = 0$, w_3 can be extended as an even function into the domain $-\alpha(t) < x < \alpha(t)$, $0 < t \le T$, where it still satisfies the heat equation, and the condition at $x = 0$ is replaced by the condition $w_3(-\alpha(t),t) =$

$C\delta(t)$. By the Maximum Principle, $w_3(x, t) \le z(x, t) + z(-x, t)$, where

$$\mathcal{L}(z) = 0, \qquad -\alpha(t) < x < \infty, \qquad 0 < t \le T,$$

$$z(-\alpha(t), t) = C\delta(t), \qquad 0 < t \le T, \qquad (18.5.14)$$

$$z(x, 0) = 0, \qquad -b_1 < x < \infty.$$

As z has the representation

$$z(x, t) = \int_0^t \sigma(\tau) \frac{\partial K}{\partial x}(x + \alpha(\tau), t - \tau)\, d\tau, \qquad x > -\alpha(t), \tag{18.5.15}$$

where

$$C_1\delta(\tau) = -\tfrac{1}{2}\sigma(t) + \int_0^t \frac{\partial K}{\partial x}(\alpha(\tau) - \alpha(t), t - \tau)\sigma(\tau)\, d\tau, \quad (18.5.16)$$

it follows via Lemma 17.7.1 that

$$|\sigma(t)| \le C_2 \|\delta\|_{[0, t]}, \tag{18.5.17}$$

where $\|\delta\|_{[0, t]} = \sup_{0 \le \tau \le t} |\delta(\tau)|$ and $C_2 = C_2(T, C_1)$. Since

$$\int_0^{\alpha(t)} |v_3(x, t)|\, dx \le \tfrac{1}{2}\int_{-\alpha(t)}^{\alpha(t)} w_3(x, t)\, dx \le \int_{-\alpha(t)}^{\infty} z(x, t)\, dx \le C_3 \int_0^t \frac{|\sigma(\tau)|\, d\tau}{(t - \tau)^{1/2}}$$

$$\le C_4 \int_0^t \frac{\|\delta\|_{[0, \tau]}\, d\tau}{(t - \tau)^{1/2}}, \tag{18.5.18}$$

we can combine these estimates of v_i, $i = 1, 2, 3$, to obtain

$$\int_0^{\alpha(t)} |u_1(x, t) - u_2(x, t)|\, dx \le \int_0^{b_1} |\varphi_1(x) - \varphi_2(x)|\, dx$$

$$+ \int_0^t |h_1(\tau) - h_2(\tau)|\, d\tau + C_4 \int_0^t \frac{\|\delta\|_{[0, \tau]}}{(t - \tau)^{1/2}}\, d\tau, \tag{18.5.19}$$

where $C_4 = C_4(T, C_1)$.

By similar methods, we obtain

$$\int_{\alpha(t)}^{\beta(t)} u_{j(t)}(x, t)\, dx \le \int_{b_1}^{b_2} \varphi_2(x)\, dx + C_5 \int_0^t \frac{\|\delta\|_{[0, \tau]}}{(t - \tau)^{1/2}}, \tag{18.5.20}$$

where $C_5 = C_5(T, C_1)$. Combining (18.5.19) and (18.5.20) with (18.5.2), we obtain

$$0 \le \delta(t) \le b_2 - b_1 + 2\int_0^t |h_1(\tau) - h_2(\tau)| \, d\tau + 2\int_0^{b_1} |\varphi_1(x) - \varphi_2(x)| \, dx$$

$$+ 2\int_{b_1}^{b_2} \varphi_2(x) \, dx + C_6 \int_0^t \frac{\|\delta\|_{[0,\tau]}}{(t-\tau)^{1/2}} \, d\tau, \tag{18.5.21}$$

where $C_6 = C_6(C_1, T)$. From this, the result follows from Lemma 17.7.1.   □

NOTES ON CHAPTERS 17 AND 18.

The material presented here is taken from the work of Cannon, Douglas, Hill, and Primicerio [1,...,6]. For the infinite differentiability and analyticity of the free boundary, see Friedman [7,...,9], Li-Shang [10], and Schaefer [11]. For additional references on Stefan's problem, the reader is referred to Rubenstein [12] and Wilson, Solomon, and Trent [13].

REFERENCES FOR CHAPTERS 17 AND 18

1. Cannon, J. R., and Douglas, J., Jr., The stability of the boundary in a Stefan problem, *Ann. Scuol. Norm. Sup. di Pisa*, **21** (1967), 83–91.
2. Cannon, J. R., and Hill, C. D., Existence, uniqueness, stability, and monotone dependence in a Stefan problem for the heat equation, *J. Math. Mech.*, **17** (1967), 1–20.
3. Cannon, J. R., and Hill, C. D., Remarks on a Stefan problem, *J. Math. Mech.*, **17** (1967), 433–442.
4. Cannon, J. R., and Hill, C. D., On the infinite differentiability of the free boundary in a Stefan problem, *J. Math. Anal. and Appl.*, **22** (1968), 385–387.
5. Cannon, J. R., Hill, C. D., and Primicerio, M., The one-phase Stefan problem with temperature-boundary conditions, *Arch. Rat. Mech. and Anal.*, **39** (1970), 270–274.
6. Cannon, J. R., and Primicerio, M., Remarks on the one-phase Stefan problem for the heat equation with flux prescribed on the fixed boundary, *J. Math. Anal. and Appl.*, **35** (1971), 361–373.
7. Friedman, A., The Stefan problem in several space variables, *Trans. Amer. Math. Soc.*, **133** (1968), 51–87.
8. Friedman, A., One-dimensional Stefan problems with nonmonotone free boundary, *Trans. Amer. Math. Soc.*, **133** (1968), 89–114.
9. Friedman, A., Analyticity of the free boundary for the Stefan problem, *Arch. Rat. Mech. Anal.*, **61** (1976), 97–125.
10. Li-Shang, C., Existence and differentiability of the solution of the two-phase Stefan problem for quasilinear parabolic equations, *Chinese Math.-Acta*, **7** (1965), 481–496.
11. Schaefer, D. G., A new proof of the infinite differentiability of the free boundary in the Stefan problem, *J. Diff. Eq.*, **20** (1976), 226–269.
12. Rubenstein, L. I., The Stefan Problem, Vol. 27, *Translations of Math.* Monographs, American Math. Soc., Providence, R. I., 1967.
13. Wilson, D. G., Solomon, A. D., and Trent, J. S., "A Bibliography on moving-boundary problems, with key-word index." Oak Ridge National Laboratory Report No. CSD–44, U.S. Government Printing Office, 1979.

Chapter 19

The Inhomogeneous Heat Equation $u_t = u_{xx} + f(x, t)$

19.1. INTRODUCTION

We begin by noting that $f(x, t)$ in the equation

$$u_t = u_{xx} + f(x, t) \tag{19.1.1}$$

is a heat source or sink in terms of units of heat energy per unit time per unit length.

We begin our consideration of initial-boundary-value problems for Eq. (19.1.1) by solving the problem

$$
\begin{aligned}
u_t &= u_{xx} + f(x, t), & 0 < x < 1, & \quad 0 < t, \\
u(x, 0) &= \varphi(x), & 0 < x < 1, & \\
u(0, t) &= g(t), & 0 < t, & \\
u(1, t) &= h(t), & 0 < t.
\end{aligned}
\tag{19.1.2}
$$

By linearity of the heat operator, we can set $u = v + w$, where

$$
\begin{aligned}
v_t &= v_{xx}, & 0 < x < 1, & \quad 0 < t, \\
v(x, 0) &= \varphi(x), & 0 < x < 1, & \\
v(0, t) &= g(t), & 0 < t, & \\
v(1, t) &= h(t), & 0 < t,
\end{aligned}
\tag{19.1.3}
$$

and

$$w_t = w_{xx} + f(x,t), \qquad 0 < x < 1, \qquad 0 < t,$$
$$w(x,0) = 0, \qquad 0 < x < 1, \qquad (19.1.4)$$
$$w(0,t) = w(1,t) = 0, \qquad 0 < t.$$

From Chapter 6 we can write down the solution $v(x,t)$.

We have said nothing about conditions upon the data, and we shall ignore them throughout this formal discussion. Consider the functions $w(x,t)$ and $f(x,t)$. Expanding them in Fourier sine series, we obtain:

$$w(x,t) = \sum_{n=1}^{\infty} w_n(t)\sin n\pi x, \qquad (19.1.5)$$

and

$$f(x,t) = \sum_{n=1}^{\infty} f_n(t)\sin n\pi x, \qquad (19.1.6)$$

where

$$w_n(t) = 2\int_0^1 w(x,t)\sin n\pi x \, dx, \qquad (19.1.7)$$

and

$$f_n(t) = 2\int_0^1 f(x,t)\sin n\pi x \, dx, \qquad n = 1,2,3,\ldots \qquad (19.1.8)$$

Substituting these expressions into the differential equation and using the orthogonality of the $\sin m\pi x$, $m = 1,2,3,\ldots$, we obtain the equations

$$\dot{w}_n + n^2\pi^2 w_n = f_n, \qquad 0 < t,$$
$$w_n(0) = 0, \qquad n = 1,2,3,\ldots, \qquad (19.1.9)$$

for the determination of the w_n. Solving (19.1.9), we see that

$$w_n(t) = \int_0^t \exp\{-n^2\pi^2(t-\tau)\}f_n(\tau)\,d\tau, \qquad (19.1.10)$$

whence it follows that

$$w(x,t) = \int_0^t \left[\sum_{n=1}^{\infty} \exp\{-n^2\pi^2(t-\tau)\}f_n(\tau)\sin n\pi x\right] d\tau$$

$$= \int_0^t \int_0^1 \left\{2\sum_{n=1}^{\infty} \exp\{-n^2\pi^2(t-\tau)\}\sin n\pi x \sin n\pi\xi\right\} f(\xi,\tau)\,d\xi\,d\tau.$$

$$(19.1.11)$$

From Eq. (K) of Ex. 3.7 and (6.1.6), we can replace

$$2\sum_{n=1}^{\infty} \exp\{-n^2\pi^2(t-\tau)\}\sin n\pi x \sin n\pi\xi$$

by $\{\theta(x-\xi,t-\tau)-\theta(x+\xi,t-\tau)\}$ to obtain

$$w(x,t)=\int_0^t\int_0^1\{\theta(x-\xi,t-\tau)-\theta(x+\xi,t-\tau)\}f(\xi,\tau)\,d\xi\,d\tau,$$

$$(19.1.12)$$

where $\theta(x,t)$ is defined by (6.1.6) in terms of the fundamental solution $K(x,t)$. As $K(x,t)$ is the leading term singularity-wise in the expression for $\theta(x,t)$, we do not dwell on (19.1.12) but move directly to the study of the volume potential

$$z(x,t)=\int_0^t\int_{-\infty}^\infty K(x-\xi,t-\tau)f(\xi,\tau)\,d\xi\,d\tau. \qquad (19.1.13)$$

19.2. THE VOLUME POTENTIAL
$$z(x,t)=\int_0^t\int_{-\infty}^\infty K(x-\xi,t-\tau)f(\xi,\tau)\,d\xi\,d\tau$$

In this section we shall discuss conditions on f that guarantee that z, z_x, z_t, and z_{xx} exist as continuous functions for $t>0$. We begin our study of the potential $z(x,t)$ with the assumption that f is a bounded continuous function.

Consider the family of functions

$$z_h(x,t)=\int_0^{t-h}\int_{-\infty}^\infty K(x-\xi,t-\tau)f(\xi,\tau)\,d\xi\,d\tau, \qquad (19.2.1)$$

where $0<h<2^{-1}t$. The singularity of the kernel $K(x-\xi,t-\tau)$ occurs at $x=\xi$, $t=\tau$. Since $t-\tau\geq h>0$, it follows that $z_h(x,t)$ is continuous, and, from Leibniz's rule,

$$\frac{\partial z_h}{\partial x}(x,t)=\int_0^{t-h}\int_{-\infty}^\infty \frac{\partial K}{\partial x}(x-\xi,t-\tau)f(\xi,\tau)\,d\xi\,d\tau \qquad (19.2.2)$$

$$\frac{\partial^2 z_h}{\partial x^2}(x,t)=\int_0^{t-h}\int_{-\infty}^\infty \frac{\partial^2 K}{\partial x^2}(x-\xi,t-\tau)f(\xi,\tau)\,d\xi\,d\tau, \qquad (19.2.3)$$

and

$$\frac{\partial z_h}{\partial t}(x,t)=\int_{-\infty}^\infty K(x-\xi,h)f(\xi,t-h)\,d\xi$$

$$+\int_0^{t-h}\int_{-\infty}^\infty \frac{\partial K}{\partial t}(x-\xi,t-\tau)f(\xi,\tau)\,d\xi\,d\tau \qquad (19.2.4)$$

are also continuous.

We consider now $z(x,t)$ and $z_h(x,t)$. For $T>0$,

$$|z(x,t)-z_h(x,t)|=\int_{t-h}^t\int_{-\infty}^\infty K(x-\xi,t-\tau)f(\xi,\tau)\,d\xi\,d\tau\leq h\|f\|_T,$$

$$(19.2.5)$$

where

$$\|f\|_T = \sup_{\substack{-\infty < x < \infty \\ 0 < t < T}} |f(x, t)|. \qquad (19.2.6)$$

Hence, as h tends to zero, we see that z is the uniform limit of the continuous functions z_h on each compact subset of $-\infty < x < \infty$, $0 < t$. Hence, z is continuous for $-\infty < x < \infty$ and $0 < t$. As

$$|z(x, t)| \le t \|f\|_T, \qquad 0 < t < T, \qquad (19.2.7)$$

it follows that $z(x, t)$ is continuous at $t = 0$ and is equal to zero.

Next, we consider the function

$$g(x, t) = \int_0^t \int_{-\infty}^\infty \frac{-(x - \xi)}{2(t - \tau)} K(x - \xi, t - \tau) f(\xi, \tau) \, d\xi, \, d\tau, \quad (19.2.8)$$

where the kernel is $(\partial K / \partial x)(x - \xi, t - \tau)$. Then

$$\left| g(x, t) - \frac{\partial z_h}{\partial x}(x, t) \right|$$

$$= \int_{t-h}^t \int_{-\infty}^\infty \frac{\partial K}{\partial x}(x - \xi, t - \tau) f(\xi, \tau) \, d\xi \, d\tau$$

$$\le \|f\|_T \int_{t-h}^t (t - \tau)^{-1/2} \int_{-\infty}^\infty \frac{|x - \xi|}{2(t - \tau)^{1/2}} K(x - \xi, t - \tau) \, d\xi \, d\tau$$

$$\le 2\pi^{-1/2} \|f\|_T h^{1/2}. \qquad (19.2.9)$$

Thus, as h tends to zero, we see that g is the uniform limit of the continuous functions $\partial z_h / \partial x$ on each compact subset of $-\infty < x < \infty$, $0 < t$. Thus, g is continuous for $-\infty < x < \infty$ and $0 < t$. As

$$|g(x, t)| \le 2\pi^{-1/2} \|f\|_T t^{1/2}, \qquad 0 < t < T, \qquad (19.2.10)$$

it follows that $g(x, t)$ is continuous at $t = 0$ and is equal to zero. From

$$z_h(x, t) = \int_{x_0}^x \frac{\partial z_h}{\partial x}(\xi, t) \, d\xi + z_h(x_0, t) \qquad (19.2.11)$$

and the uniform convergence of z_h to z and $\partial z_h / \partial x$ to g on compact subsets of $-\infty < x < \infty$, $t > 0$, it follows that

$$z(x, t) = \int_{x_0}^x g(\xi, t) \, d\xi + z(x_0, t), \qquad (19.2.12)$$

whence

$$\frac{\partial z}{\partial x}(x, t) = g(x, t) = \int_0^t \int_{-\infty}^\infty \frac{\partial K}{\partial x}(x - \xi, t - \tau) f(\xi, \tau) \, d\xi \, d\tau$$

$$(19.2.13)$$

exists and is continuous.

In order to demonstrate the existence of $\partial z/\partial t$ and $\partial^2 z/\partial x^2$, we utilize the identity

$$\int_{-\infty}^\infty \frac{\partial K}{\partial t}(x - \xi, t - \tau) f(x, \tau) \, d\xi = \int_{-\infty}^\infty \frac{\partial^2 K}{\partial x^2}(x - \xi, t - \tau) f(x, \tau) \, d\xi$$

$$= f(x, \tau) \int_{-\infty}^\infty \frac{\partial^2 K}{\partial x^2}(x - \xi, t - \tau) \, d\xi$$

$$= f(x, \tau) \frac{\partial^2}{\partial x^2} \int_{-\infty}^\infty K(x - \xi, t - \tau) \, d\xi$$

$$= f(x, \tau) \frac{\partial^2}{\partial x^2} \{1\} \equiv 0, \qquad t > \tau,$$

$$(19.2.14)$$

and we assume that f is continuous and uniformly Hölder continuous with respect to x. From the identity (19.2.14) and (19.2.3), we can write

$$\frac{\partial^2 z_h}{\partial x^2}(x, t) = \int_0^{t-h} \int_{-\infty}^\infty \frac{\partial^2 K}{\partial x^2}(x - \xi, t - \tau) \{f(\xi, \tau) - f(x, \tau)\} \, d\xi \, d\tau.$$

$$(19.2.15)$$

Consider

$$\gamma(x, t) = \int_0^t \int_{-\infty}^\infty \frac{\partial^2 K}{\partial x^2}(x - \xi, t - \tau) \{f(\xi, \tau) - f(x, \tau)\} \, d\xi \, d\tau$$

$$(19.2.16)$$

and

$$\left| \gamma(x, t) - \frac{\partial^2 z_h}{\partial x^2}(x, t) \right| \leq \int_{t-h}^t \int_{-\infty}^\infty \left| \frac{\partial^2 K}{\partial x^2}(x - \xi, t - \tau) \right| |f|_\alpha |\xi - x|^\alpha \, d\xi \, d\tau$$

$$\leq \int_{t-h}^t 2^{-1} \pi^{-1/2} |f|_\alpha \int_{-\infty}^\infty \left\{ \frac{|x - \xi|^\alpha}{(t - \tau)} + \frac{|x - \xi|^{2+\alpha}}{2(t - \tau)^2} \right\}$$

$$\cdot \exp\left\{ -\frac{(x - \xi)^2}{4(t - \tau)} \right\} \cdot \frac{d\xi}{2(t - \tau)^{1/2}} \, d\tau$$

$$\leq C|f|_\alpha \int_{t-h}^t (t - \tau)^{(\alpha/2) - 1} \, d\tau \leq C|f|_\alpha h^{\alpha/2}, \quad (19.2.17)$$

where C is a positive constant obtained via the transformation $\rho = (x - \xi)/(2(t - \tau)^{1/2})$ and $|f|_\alpha$ is the Hölder seminorm for f with respect to x. From (19.2.17), it follows that $\gamma(x, t)$ is the uniform limit of the continuous functions $(\partial^2 z_h/\partial x^2)(x, t)$ on compact subsets of $-\infty < x < \infty$,

$0 < t$. Consequently, $\gamma(x, t)$ is continuous for $-\infty < x < \infty$, $0 < t$. Moreover,

$$|\gamma(x, t)| \le C|f|_\alpha t^{\alpha/2}, \tag{19.2.18}$$

which implies that $\gamma(x, t)$ is continuous at $t = 0$ with value $\gamma(x, 0) = 0$. As

$$\frac{\partial z_h}{\partial x}(x, t) = \int_{x_0}^x \frac{\partial^2 z_h}{\partial x^2}(\xi, t) \, d\xi + \frac{\partial z_h}{\partial x}(x_0, t), \tag{19.2.19}$$

it follows that

$$\frac{\partial z}{\partial x}(x, t) = \int_{x_0}^x \gamma(\xi, t) \, d\xi + \frac{\partial z}{\partial x}(x_0, t), \tag{19.2.20}$$

whence

$$\frac{\partial^2 z}{\partial x^2}(x, t) = \gamma(x, t) = \int_0^t \int_{-\infty}^\infty \frac{\partial^2 K}{\partial x^2}(x - \xi, t - \tau)\{f(\xi, \tau) - f(x, \tau)\} \, d\xi \, d\tau \tag{19.2.21}$$

exists and is continuous.
    Since

$$\lim_{h \downarrow 0} \int_{-\infty}^\infty K(x - \xi, h) f(\xi, t - h) \, d\xi = f(x, t), \tag{19.2.22}$$

we can employ the argument for $\partial^2 z / \partial x^2$ to show that for $-\infty < x < \infty$, $0 < t$,

$$\frac{\partial z}{\partial t}(x, t) = f(x, t) + \frac{\partial^2 z}{\partial x^2}(x, t) \tag{19.2.23}$$

exists and is continuous. Moreover,

$$\left| \frac{\partial z}{\partial t}(x, t) \right| \le \|f\|_t + C|f|_\alpha t^{\alpha/2}, \tag{19.2.24}$$

and

$$\frac{\partial z}{\partial t}(x, 0) = f(x, 0). \tag{19.2.25}$$

We summarize the analysis thus far in the following statement.

Lemma 19.2.1.   *For bounded continuous f in $-\infty < x < \infty$, $0 \le t$, which is uniformly Hölder continuous with exponent α, $0 < \alpha < 1$, with respect to x, the potential*

$$z(x, t) = \int_0^t \int_{-\infty}^\infty K(x - \xi, t - \tau) f(\xi, \tau) \, d\xi \, d\tau$$

possesses the following properties:

1) z, z_x, z_t, and z_{xx} are continuous;
2) $z_t = z_{xx} + f(x, t)$, $-\infty < x < \infty$, $0 < t$;
3) $|z(x, t)| \leq t \|f\|_t$, $-\infty < x < \infty$, $0 \leq t$,   where $\|f\|_t$

$$= \sup_{\substack{-\infty < x < \infty \\ 0 \leq \tau \leq t}} |f(x, \tau)|;$$

4) $|z_x(x, t)| \leq 2\pi^{-1/2} \|f\|_t t^{1/2}$, $-\infty < x < \infty$, $0 \leq t$;
5) $|z_{xx}(x, t)| \leq C |f|_\alpha t^{\alpha/2}$, $-\infty < x < \infty$, $0 \leq t$,   where C
   is a positive constant and

$$|f|_\alpha = \sup_{\substack{-\infty < x < \infty \\ \delta > 0, 0 \leq t}} \{|f(x + \delta, t) - f(x, t)| \delta^{-\alpha}\};$$

6) $|z_t(x, t)| \leq \|f\|_t + C |f|_\alpha t^{\alpha/2}$, $-\infty < x < \infty$, $0 \leq t$.

Proof.   See the analysis preceding the statement of the lemma. $\square$

Let us suppose now that f is bounded and locally Hölder continuous such that at (x, t) there exist positive $C(x, t)$ and $\delta(x, t)$ such that when $|\xi - x| < \delta(x, t)$ and $|t - \tau| < \delta(x, t)$,

$$|f(x, t) - f(\xi, \tau)| \leq C(x, t) \{|\xi - x|^\alpha + |t - \tau|^{\alpha/2}\}, \quad (19.2.26)$$

where $\alpha = \alpha(x, t)$ satisfies $0 < \alpha \leq 1$. Considering the identity (19.2.14), we see that we can insert $f(x, t)$ in the place of $f(x, \tau)$, whence follows

$$\gamma(x, t) = \int_0^t \int_{-\infty}^{\infty} \frac{\partial^2 K}{\partial x^2} (x - \xi, t - \tau) \{f(\xi, \tau) - f(x, t)\} \, d\xi \, d\tau.$$

$$(19.2.27)$$

Considering the difference $\gamma(x, t) - (\partial^2 z_h / \partial x^2)(x, t)$, where $0 < h < \delta(x, t)$, we obtain for a positive constant C_0

$$\left| \gamma(x, t) - \frac{\partial^2 z_h}{\partial x^2}(x, t) \right|$$

$$\leq \|f\|_t \int_{t-h}^{t} \int_{|x - \xi| > \delta(x, t)} \left| \frac{\partial^2 K}{\partial x^2}(x - \xi, t - \tau) \right| d\xi \, d\tau + C_0 C(x, t)$$

$$\cdot \int_{t-h}^{t} \int_{|x - \xi| < \delta(x, t)} \left\{ \frac{|x - \xi|^\alpha}{(t - \tau)} + \frac{|x - \xi|^{2+\alpha}}{(t - \tau)^2} \right.$$

$$\left. + \frac{1}{(t - \tau)^{1-(\alpha/2)}} + \frac{|x - \xi|^2}{(t - \tau)^{2-(\alpha/2)}} \right\}$$

$$\cdot \exp\left\{ \frac{(x - \xi)^2}{4(t - \tau)} \right\} \cdot \frac{d\xi}{2\sqrt{t - \tau}} \, d\tau = J_1 + J_2. \quad (19.2.28)$$

Consider J_2 first. We can bound it above by extending the ξ integration over all real ξ. This reduces the estimate of J_2 to exactly the estimate in (19.2.9). Hence,

$$J_2 \le C_1 C(x, t) h^{\alpha/2}, \tag{19.2.29}$$

where C_1 is a positive constant. Turning now to J_1, we see that there exists a positive constant C_2 such that

$$\left| \frac{\partial^2 K}{\partial x^2}(x - \xi, t - \tau) \right| \le C_2 \left\{ \frac{1}{(t - \tau)^{3/2}} + \frac{(x - \xi)^2}{(t - \tau)^{5/2}} \right\} \exp\left\{ -\frac{(x - \xi)^2}{4(t - \tau)} \right\}. \tag{19.2.30}$$

Employing the transformation $\rho = (x - \xi)/(2\sqrt{t - \tau})$, we obtain

$$J_1 \le C_3 \|f\|_t \int_{t-h}^t (t - \tau)^{-1} \int_{\delta(x, t)/(2\sqrt{t - \tau})}^\infty \{1 + \rho^2\} \cdot \exp\{-\rho^2\} \, d\rho \, d\tau$$

$$\le C_4 \|f\|_t \int_{t-h}^t (t - \tau)^{-1} \exp\left\{ -\frac{\delta(x, t)^2}{8(t - \tau)} \right\} d\tau$$

$$\le C_5 \|f\|_t (\delta(x, t))^{-2} h, \tag{19.2.31}$$

where C_i, $i = 3, 4, 5$, are positive constants and where we have used

$$(1 + \rho^2) \exp\{-\rho^2\} < C \exp\left\{ -\frac{\rho^2}{2} \right\} \quad \text{and} \quad \exp\{-x\} < x^{-1}, x > 0.$$

Combining (19.2.29) and (19.2.31), we see that

$$\left| \gamma(x, t) - \frac{\partial^2 z_h}{\partial x^2}(x, t) \right| \le C_1 C(x, t) h^{\alpha/2} + C_5 \|f\|_t (\delta(x, t))^{-2} h. \tag{19.2.32}$$

Consequently, if for each compact subset D of $-\infty < x < \infty$, $0 < t$, positive constants C, δ, and α can be found so that for each $(x, t) \in D$, $C(x, t) < C$, $\delta(x, t) > \delta$, and $1 \ge \alpha(x, t) \ge \alpha > 0$, then for each $(x, t) \in D$,

$$\left| \gamma(x, t) - \frac{\partial^2 z_h}{\partial x^2}(x, t) \right| \le C_1 C h^{\alpha/2} + C_5 \|f\|_t \delta^{-2} h, \tag{19.2.33}$$

which implies that the $\partial^2 z_h / \partial x^2$ converge uniformly to γ on each compact D as h tends to zero. Hence, $\gamma(x, t)$ is continuous and

$$\frac{\partial^2 z}{\partial x^2} = \gamma(x, t). \tag{19.2.34}$$

Likewise,

$$\frac{\partial z}{\partial t} = \frac{\partial^2 z}{\partial x^2} + f(x, t) \qquad (19.2.35)$$

exists and is continuous.

We summarize this in the following statement.

Lemma 19.2.2. *For bounded continuous f in $-\infty < x < \infty$, $0 < t$, which is uniformly Hölder continuous on each subset D of $-\infty < x < \infty$, $0 < t$, the potential*

$$z(x, t) = \int_0^t \int_{-\infty}^\infty K(x - \xi, t - \tau) f(\xi, \tau) \, d\xi \, d\tau$$

possesses the following properties:

1) z *and* z_x *are continuous for* $-\infty < x < \infty$, $0 \leq t$;
2) z_t *and* z_{xx} *are continuous for* $-\infty < x < \infty$, $0 < t$;
3) $z_t = z_{xx} + f(x, t)$, $-\infty < x < \infty$, $0 < t$;
4) $|z(x, t)| \leq \|f\|_t t$, $-\infty < x < \infty$, $0 \leq t$, where $\|f\|_t$
    $= \sup\limits_{\substack{-\infty < x < \infty \\ 0 \leq \tau \leq t}} |f(x, \tau)|$;
5) $|z_x(x, t)| \leq 2\pi^{-1/2} \|f\|_t t^{1/2}$, $-\infty < x < \infty$, $0 < t$.

Proof. See the analysis preceding Lemma 19.2.1 and the statement of this lemma.   □

It is of interest for applications of the potential z to estimate the Hölder continuity of the derivative z_x in the variable x. Since

$$\left| \frac{\partial z}{\partial x}(x + \delta, t) - \frac{\partial z}{\partial x}(x, t) \right| \leq 4\pi^{-1/2} \|f\|_t t^{1/2}, \qquad (19.2.36)$$

we see that, for $\delta \geq 1$,

$$\left| \frac{\partial z}{\partial x}(x + \delta, t) - \frac{\partial z}{\partial x}(x, t) \right| \leq 4\pi^{-1/2} \|f\|_t t^{1/2} \delta^\alpha \qquad (19.2.37)$$

for any positive α. Thus we can restrict our considerations to $0 < \delta < 1$. Let $\eta > 0$ be a positive parameter. Then,

$$\left| \frac{\partial z}{\partial x}(x + \delta, t) - \frac{\partial z}{\partial x}(x, t) \right| \leq \int_{t - \eta}^t \int_{-\infty}^\infty \left| \frac{\partial K}{\partial x}(x + \delta - \xi, t - \tau) \right| |f(\xi, \tau)| \, d\xi \, d\tau$$

$$+ \int_{t - \eta}^t \int_{-\infty}^\infty \left| \frac{\partial K}{\partial x}(x - \xi, t - \tau) \right| |f(\xi, \tau)| \, d\xi \, d\tau$$

$$+ \int_0^{t - \eta} \int_{-\infty}^\infty \left| \frac{\partial K}{\partial x}(x + \delta - \xi, t - \tau) \right.$$

$$\left. - \frac{\partial K}{\partial x}(x - \xi, t - \tau) \right| |f(\xi, \tau)| \, d\xi \, d\tau$$

$$= I_1 + I_2 + I_3. \qquad (19.2.38)$$

Recalling the estimate (19.2.9), we obtain

$$I_1 + I_2 \leq 4\pi^{-1/2}\|f\|_t \eta^{1/2}. \tag{19.2.39}$$

Thus, we can turn now to I_3. We can write it as

$$I_3 = \delta \int_0^{t-\eta} \int_{-\infty}^{\infty} \left| \left\{ \int_0^1 \frac{\partial^2 K}{\partial x^2}(x + \theta\delta - \xi, t - \tau)\, d\theta \right\} \right| \cdot |f(\xi, \tau)|\, d\xi\, d\tau. \tag{19.2.40}$$

Using Fubini's Theorem, we exchange the order of integration to obtain

$$I_3 \leq \delta \int_0^{t-\eta} \int_0^1 \int_{-\infty}^{\infty} \left| \frac{\partial^2 K}{\partial x^2}(x + \theta\delta - \xi, t - \tau) \right| \cdot |f(\xi, \tau)|\, d\xi\, d\theta\, d\tau$$

$$\leq C\|f\|_t \delta \int_0^{t-\eta} (t - \tau)^{-1}\, d\tau$$

$$\leq Ct\|f\|_t \delta \eta^{-1}, \tag{19.2.41}$$

where C is a positive constant obtained via the transformation $\rho = (x - \xi)/(2(t - \tau)^{1/2})$. Combining (19.2.39) with (19.2.41), we obtain

$$\left| \frac{\partial z}{\partial x}(x + \delta, t) - \frac{\partial z}{\partial x}(x, t) \right| \leq Ct\|f\|_t \delta \eta^{-1} + 4\pi^{-1/2}\|f\|_t \eta^{1/2}. \tag{19.2.42}$$

Since the left side of (19.2.42) is independent of η, we can select $\eta = \delta^{2/3}$, to obtain

$$\left| \frac{\partial z}{\partial x}(x + \delta, t) - \frac{\partial z}{\partial x}(x, t) \right| \leq (Ct + 4\pi^{-1/2})\|f\|_t \delta^{1/3}, \tag{19.2.43}$$

provided that $\delta^{2/3} < t$. On the other hand, if $\delta^{2/3} \geq t$, (19.2.36) implies that

$$\left| \frac{\partial z}{\partial x}(x + \delta, t) - \frac{\partial z}{\partial x}(x, t) \right| \leq 4\pi^{-1/2}\|f\|_t t^{1/2} \leq 4\pi^{-1/2}\|f\|_t \delta^{1/3}. \tag{19.2.44}$$

Selecting $\alpha = 1/3$ in (19.2.37) and combining it with (19.2.43) and (19.2.44), we obtain the following result.

Lemma 19.2.3. *For $0 \leq t$ and $0 < \delta$,*

$$\left| \frac{\partial z}{\partial x}(x + \delta, t) - \frac{\partial z}{\partial x}(x, t) \right| \leq (4\pi^{-1/2}[1 + t^{1/2}] + Ct)\|f\|_t \delta^{1/3}, \tag{19.2.45}$$

where

$$C = \pi^{-1/2} \int_0^{\infty} \{1 + 2\rho^2\}\exp\{-\rho^2\}\, d\rho. \tag{19.2.46}$$

Proof. See the analysis preceding the statement of the lemma. □

Remark.   The result of Lemma 19.2.3 shows that, for $0 \leq t \leq T$, $(\partial z/\partial x)$ is uniformly Hölder continuous with respect to x with Hölder exponent $\alpha = 1/3$, provided that f is bounded.

We shall leave to the reader the demonstration of similar results for each of the potentials

$$z_1(x,t) = \int_0^t \int_0^\infty G(x,\xi,t-\tau)f(\xi,\tau)\,d\xi\,d\tau, \qquad (19.2.47)$$

$$z_2(x,t) = \int_0^t \int_0^\infty N(x,\xi,t-\tau)f(\xi,\tau)\,d\xi\,d\tau, \qquad (19.2.48)$$

$$z_3(x,t) = \int_0^t \int_0^1 \{\theta(x-\xi,t-\tau) - \theta(x+\xi,t-\tau)\}f(\xi,\tau)\,d\xi\,d\tau,$$

$$(19.2.49)$$

and

$$z_4(x,t) = \int_0^t \int_0^1 \{\theta(x-\xi,t-\tau) + \theta(x+\xi,t-\tau)\}f(\xi,\tau)\,d\xi,\,d\tau,$$

$$(19.2.50)$$

where $G(x,\xi,t-\tau)$ is defined by (4.3.4), $N(x,\xi,t-\tau)$ is defined by Eq. (F) of the Exercise section of Chapter 3, and $\theta(x,t)$ is defined by (6.1.6).

19.3.   SOLUTIONS FOR SOME INITIAL-BOUNDARY-VALUE PROBLEMS FOR $u_t = u_{xx} + f(x,t)$

Recalling the analysis of Sec. 19.1, the solution to any initial-boundary-value problem for $u_t = u_{xx} + f(x,t)$ can be obtained from setting $u = v + w$, where $v_t = v_{xx}$ and $w_t = w_{xx} + f(x,t)$. Consequently, the list of potentials, $z, z_1, \ldots, z_4$, gives us the particular solutions w needed to reduce most of the problems back to the considerations for the heat equation given in Chapters 3–7. Before listing several problems and their solutions, we need to comment on unicity. Let $v = u_1 - u_2$, where u_i, $i = 1,2$, denote two solutions to an initial-boundary-value problem for $u_t = u_{xx} + f(x,t)$. By linearity of the equation, v satisfies $v_t = v_{xx}$ with zero initial-boundary values. Consequently, the unicity of solution reduces to that of unicity of the solution of the particular initial-boundary-value problem for the heat equation.

It is convenient here to list the basic assumptions on the data in the problems below.

Assumption A.   We shall assume that the source function $f(x,t)$ is bounded over each domain considered and that $f(x,t)$ is uniformly Hölder continuous on each compact subset of the domain under consideration. We shall also assume that the initial-boundary data is piecewise-continuous.

An Infinite Conductor

Theorem 19.3.1.   *For the problem*

$$u_t = u_{xx} + f(x,t), \qquad -\infty < x < \infty, \qquad 0 < t,$$
$$u(x,0) = \varphi(x), \qquad -\infty < x < \infty, \tag{19.3.1}$$

with f and φ satisfying Assumption A above, the function

$$u(x,t) = \int_{-\infty}^{\infty} K(x-\xi,t)\varphi(\xi)\,d\xi + \int_0^t \int_{-\infty}^{\infty} K(x-\xi,t-\tau)f(\xi,\tau)\,d\xi\,d\tau$$

$$\tag{19.3.2}$$

is the unique solution among the class of solutions that satisfy a growth condition of the form

$$|u(x,t)| \le C_1 \exp\{C_2|x|^{1+\alpha}\}, \tag{19.3.3}$$

where $\alpha < 1$, C_i, $i = 1, 2$, are positive constants, and $K = K(x,t)$ is the fundamental solution defined by (3.2.1).

Proof.   See Sec. 19.2 and Theorem 3.5.1.   □

A Semi-Infinite Conductor with Temperature Boundary Condition

Theorem 19.3.2.   *For the problem*

$$u_t = u_{xx} + f(x,t), \qquad 0 < x < \infty, \qquad 0 < t,$$
$$u(x,0) = \varphi(x), \qquad 0 < x < \infty, \tag{19.3.4}$$
$$u(0,t) = g(t), \qquad 0 < t,$$

with the data f, φ, and g satisfying Assumption A above, the function

$$u(x,t) = -2\int_0^t \frac{\partial K}{\partial x}(x,t-\tau)g(\tau)\,d\tau + \int_0^\infty G(x,\xi,t)\varphi(\xi)\,d\xi$$

$$+ \int_0^t \int_0^\infty G(x,\xi,t-\tau)f(\xi,\tau)\,d\xi\,d\tau, \tag{19.3.5}$$

is the unique solution among the class of solutions that satisfy a growth condition of the form (19.3.3). Here $K(x,t)$ is the fundamental solution defined by (3.2.1), and $G(x,\xi,t)$ is the Green's function defined by (4.3.4).

Proof.   See Sec. 19.2 and Theorem 4.3.1.   □

A Semi-Infinite Conductor with Flux-Boundary Condition

Theorem 19.3.3.   *For the problem*

$$u_t = u_{xx} + f(x,t), \qquad 0 < x < \infty, \qquad 0 < t,$$
$$u(x,0) = \varphi(x), \qquad 0 < x < \infty, \tag{19.3.6}$$
$$u_x(0,t) = g(t), \qquad 0 < t,$$

with the data f, φ, and g satisfying Assumption A above, the function

$$u(x, t) = -2 \int_0^t K(x, t - \tau) g(\tau) \, d\tau + \int_0^\infty N(x, \xi, t) \varphi(\xi) \, d\xi$$

$$+ \int_0^t \int_0^\infty N(x, \xi, t - \tau) f(\xi, \tau) \, d\xi \, d\tau \qquad (19.3.7)$$

is the unique solution among the class of solutions that satisfy a growth condition of the form (19.3.3). Here $K(x, t)$ is the fundamental solution defined by (3.2.1), and $N(x, \xi, t)$ is the Neumann's function defined by Eq. (F) of the Exercise Section of Chapter 3.

Proof.   See Sec. 19.2 and Theorem 5.1.1.   □

A Finite Conductor with Temperature-Boundary Conditions

Theorem 19.3.4.   *For the problem*

$$u_t = u_{xx} + f(x, t), \qquad 0 < x < 1, \qquad 0 < t,$$
$$u(x, 0) = \varphi(x), \qquad 0 < x < 1,$$
$$u(0, t) = g(t), \qquad 0 < t, \qquad (19.3.8)$$
$$u(1, t) = h(t), \qquad 0 < t,$$

with the data f, φ, g, and h satisfying Assumption A above, the function

$$u(x, t) = \int_0^1 \{\theta(x - \xi, t) - \theta(x + \xi, t)\} \varphi(\xi) \, d\xi - 2 \int_0^t \frac{\partial \theta}{\partial x}(x, t - \tau) g(\tau) \, d\tau$$

$$+ 2 \int_0^t \frac{\partial \theta}{\partial x}(x - 1, t - \tau) h(\tau) \, d\tau$$

$$+ \int_0^t \int_0^1 \{\theta(x - \xi, t - \tau) - \theta(x + \xi, t - \tau)\} f(\xi, \tau) \, d\xi \, d\tau \quad (19.3.9)$$

is the unique bounded solution. Here, $\theta(x, t)$ is defined by (6.1.6).

Proof.   See Sec. 19.2 and Theorem 6.3.1.   □

A Finite Conductor with Flux-Boundary Condition

Theorem 19.3.5.   *For the problem*

$$u_t = u_{xx} + f(x, t), \qquad 0 < x < 1, \qquad 0 < t,$$
$$u(x, 0) = \varphi(x), \qquad 0 < x < 1,$$
$$u_x(0, t) = g(t), \qquad 0 < t, \qquad (19.3.10)$$
$$u_x(1, t) = h(t), \qquad 0 < t,$$

with the data f, φ, g, and h satisfying Assumption A above, the function

$$u(x, t) = \int_0^1 \{\theta(x - \xi, t) + \theta(x + \xi, t)\} \varphi(\xi) \, d\xi - 2 \int_0^t \theta(x, t - \tau) g(\tau) \, d\tau$$

$$+ 2 \int_0^t \theta(x - 1, t - \tau) h(\tau) \, d\tau$$

$$+ \int_0^t \int_0^1 \{\theta(x - \xi, t - \tau) + \theta(x + \xi, t - \tau)\} f(\xi, \tau) \, d\xi \, d\tau$$

(19.3.11)

is the unique bounded solution. Here $\theta(x, t)$ is defined by (6.1.6).

   Proof.   See Sec. 19.2 and Theorem 6.4.1.   □

19.4.   INITIAL-BOUNDARY-VALUE PROBLEMS FOR GENERAL DOMAINS

We consider first the problem

$$u_t = u_{xx} + f(x, t), \qquad s_1(t) < x < s_2(t), \qquad 0 < t \le T,$$
$$u(x, 0) = \varphi(x), \qquad s_1(0) < x < s_2(0),$$
$$u(s_1(t), t) = g(t), \qquad 0 < t \le T,$$
$$u(s_2(t), t) = h(t), \qquad 0 < t \le T,$$

(19.4.1)

where $s_i(t) \in C([0, T])$, $i = 1, 2$, and the data f, φ, g, and h satisfy Assumption A of Sec. 19.3. First, we extend f by setting

$$f(x, t) = \begin{cases} f(s_1(t), t), & -\infty < x < s_1(t), \quad 0 < t < T, \\ f(s_2(t), t), & s_2(t) < x < \infty, \quad 0 < t \le T. \end{cases}$$

(19.4.2)

Then, the extended f is bounded continuous and uniformly Hölder continuous for each compact subset of

$$D_T = \{(x, t) | s_1(t) < x < s_2(t), \quad 0 < t \le T\}.$$

From Lemma 19.2.2 the potential $z(x, t)$ and its derivative $z_x(x, t)$ are continuous for $-\infty < x < \infty$, $0 \le t$. Consequently, $z(s_i(t), t)$ and $z_x(s_i(t), t)$, $i = 1, 2$, are continuous functions for $0 \le t \le T$. Let

$$v(x, t) = u(x, t) - z(x, t).$$

(19.4.3)

Then, v must satisfy

$$v_t = v_{xx}, \qquad s_1(t) < x < s_2(t), \qquad 0 < t \le T,$$
$$v(x, 0) = \varphi(x), \qquad s_1(0) < x < s_2(0),$$
$$v(s_1(t), t) = g(t) - z(s_1(t), t), \qquad 0 < t \le T,$$
$$v(s_2(t), t) = h(t) - z(s_2(t), t), \qquad 0 < t \le T.$$

(19.4.4)

Applying the results of Chapters 14 and 16, we obtain the following result.

Theorem 19.4.1. If the domain $D_T = \{(x,t) | s_1(t) < x < s_2(t), 0 < t \leq T\}$, where $s_i(t) \in C([0,T])$, $i = 1, 2$, possesses a barrier at every point of its parabolic boundary and if f, φ, g, and h satisfy Assumption A of Sec. 19.3, then there exists a unique solution $u = u(x,t)$ to Problem (19.4.1).

Proof. See Sec. 19.2 and Sec. 16.4. □

For the problem

$$
\begin{aligned}
u_t &= u_{xx} + f(x,t), & s_1(t) < x < s_2(t), & \quad 0 < t \leq T, \\
u(x,0) &= \varphi(x), & s_1(0) < x < s_2(0), & \\
u_x(s_1(t),t) &= g(t), & 0 < t \leq T, & \\
u_x(s_2(t),t) &= h(t), & 0 < t \leq T, &
\end{aligned}
\tag{19.4.5}
$$

we assume that $s_i(t) \in C^\gamma([0,T])$, $\gamma > \frac{1}{2}$, $i = 1, 2$. Setting

$$v(x,t) = u(x,t) - z(x,t), \tag{19.4.6}$$

we see that v must satisfy

$$
\begin{aligned}
v_t &= v_{xx}, & s_1(t) < x < s_2(t), & \quad 0 < t \leq T, \\
v(x,0) &= \varphi(x), & s_1(0) < x < s_2(0), & \\
v_x(s_1(t),t) &= g(t) - z_x(s_1(t),t), & 0 < t \leq T, & \\
v_x(s_2(t),t) &= h(t) - z_x(s_2(t),t), & 0 < t \leq T. &
\end{aligned}
\tag{19.4.7}
$$

From the analysis in Chapter 14 we obtain the result.

Theorem 19.4.2. If the domain $D_T = \{(x,t) | s_1(t) < x < s_2(t), 0 < t \leq T\}$ possesses bounding curves $s_i(t) \in C^\gamma([0,T])$, $\gamma > \frac{1}{2}$, $i = 1, 2$, and if f, φ, g, and h satisfy Assumption A of Sec. 19.3, then there exists a bounded solution u of Problem (19.4.5). Moreover, if $\varphi \in C^1([s_1(0), s_2(0)])$, then u is unique among the class of solutions for which u_x is bounded.

Proof. See Sec. 19.2 and Exercise 14.1 of the exercise section of Chapter 14. □

EXERCISES

19.1. Show that the problem of determining the unique solution u that satisfies

$$
\begin{aligned}
u_t &= u_{xx} + f(x,t), & 0 < x < \infty, & \quad 0 < t, \\
u(x,0) &= \varphi(x), & 0 < x < \infty, & \\
u_x(0,t) + \alpha(t)u(0,t) &= g(t), & 0 < t, &
\end{aligned}
\tag{A}
$$

and

$$|u(x,t)| \leq C_1 \exp\{C_2 |x|^{1+\gamma}\}, \qquad \gamma < 1, \tag{B}$$

where C_i, $i = 1, 2$, are positive constants and the data f, φ, α, and g satisfy Assumption A of Sec. 19.3, is equivalent to the problem of determining a unique, piecewise-continuous solution ψ to the integral equation

$$\psi(t) = d(t) + 2\alpha(t) \int_0^t K(0, t - \tau)\psi(\tau)\, d\tau, \qquad 0 < t, \qquad \text{(C)}$$

where

$$d(t) = g(t) - \alpha(t) \int_0^\infty N(0, \xi, t)\varphi(\xi)\, d\xi$$

$$- \alpha(t) \int_0^t \int_0^\infty N(0, \xi, t - \tau)f(\xi, \tau)\, d\xi\, d\tau \qquad \text{(D)}$$

and $K(x, t)$ is the fundamental solution defined by (3.2.1), and $N(x, \xi, t)$ is the Neumann's function defined by Equation (F) of the Exercise section of Chapter 3.

19.2.   Show that the problem of determining the unique solution u that satisfies

$$u_t = u_{xx} + f(x, t), \qquad 0 < x < \infty, \qquad 0 < t,$$

$$u(x, 0) = \varphi(x), \qquad 0 < x < \infty, \qquad \text{(E)}$$

$$u_t(0, t) + \alpha(t)u_x(0, t) + \beta(t)u(0, t) = g(t), \qquad 0 < t,$$

and

$$|u(x, t)| \le C_1 \exp\{C_2 |x|^{1+\gamma}\}, \qquad \gamma < 1, \qquad \text{(F)}$$

where C_i, $i = 1, 2$, are positive constants, the data f, α, β, and g satisfy Assumption A of Sec. 19.3, and φ satisfies Eq. (F) and is continuously differentiable, is equivalent to the problem of determining a unique solution ψ to the integral equation

$$\psi(t) = d(t) + \int_0^t \{2\alpha(t)K(0, t - \tau) - \beta(t)\}\psi(\tau)\, d\tau, \quad 0 < t, \quad \text{(G)}$$

where

$$d(t) = g(t) - \beta(t)\varphi(0) - \alpha(t) \int_0^\infty N(0, \xi, t)\varphi'(\xi)\, d\xi$$

$$- \alpha(t) \int_0^t \int_0^\infty \frac{\partial G}{\partial x}(0, \xi, t - \tau)f(\xi, \tau)\, d\xi\, d\tau, \qquad \text{(H)}$$

and where $K(x, t)$ is the fundamental solution defined by (3.2.1), $G(x, \xi, t)$ is the Green's function defined by (4.3.4), and $N(x, \xi, t)$ is the Neumann's function defined by equation (F) of the Exercise Section of Chapter 3.

19.3. Show that, for f and φ satisfying Assumption A of Sec. 19.3 and for continuous F and G, the bounded solution u of the problem

$$u_t = u_{xx} + f(x,t), \qquad 0 < x < 1, \qquad 0 < t,$$

$$u(x,0) = \varphi(x), \qquad 0 < x < 1,$$

$$u_x(0,t) = F(t, u(0,t)), \qquad 0 < t, \qquad \text{(I)}$$

$$u_x(1,t) = G(t, u(1,t)), \qquad 0 < t,$$

has the form

$$u(x,t) = \int_0^1 \{\theta(x - \xi, t) + \theta(x + \xi, t)\} \varphi(\xi) \, d\xi$$

$$-2\int_0^t \theta(x, t - \tau) F(\tau, \psi_1(\tau)) \, d\tau$$

$$+2\int_0^t \theta(x - 1, t - \tau) G(\tau, \psi_2(\tau)) \, d\tau$$

$$+\int_0^t \int_0^1 \{\theta(x - \xi, t - \tau) + \theta(x + \xi, t - \tau)\}$$

$$\times f(\xi, \tau) \, d\xi \, d\tau, \qquad \text{(J)}$$

where $\theta(x,t)$ is defined by (6.1.6), if and only if ψ_1 and ψ_2 are piecewise-continuous solutions of a system of integral equations. Moreover, the uniqueness of u is assured under the additional assumption that F and G are Lipschitz-continuous with respect to u.

NOTES

Levi [3] studied the inhomogeneous heat equation and Gevrey [1] presented some modifications and extensions. The treatment here follows that of Ladyzenskaja, Solonnikov, and Ural'ceva [2].

REFERENCES

1. Gevrey, M., Sur les équations aux dérivées partielles du type parabolique, *J. Math. Pures Appl.*, **9** (1913), 305–471.
2. Ladyzenskaja, O. A., Solonnikov, V. A., and Ural'ceva, N. N., *Linear and Quasilinear Equations of Parabolic Type*, Vol. 23; Translations of Math. Monographs. American Math. Soc., Providence, Rhode Island, 1968.
3. Levi, E. E., Sull'equazione del calore, *Annali di Mat. Pura ed. Appl.*, (III) **14** (1908), pp. 187–264.

Chapter 20

An Application of the Inhomogeneous Heat Equation:
The Equation $u_t = u_{xx} + F(x,t,u,u_x)$

20.1. INTRODUCTION

Our purpose here is not to study the most general form of F, but rather to demonstrate the applicability of the results of Chapter 19 to a wide range of problems.

When we consider the special case $F = u^{1+\alpha}$, where $u = u(t)$ satisfies

$$u_t = u^{1+\alpha}, \qquad 0 < t, \qquad \alpha > 0,$$
$$u(0) = u_0, \qquad u_0 > 0, \tag{20.1.1}$$

we obtain, via elementary considerations, the solution

$$u(t) = \left(u_0^{-\alpha} - \alpha t\right)^{-\alpha^{-1}}, \tag{20.1.2}$$

which tends to infinity as t tends to the finite time

$$t_\infty = \alpha^{-1} u_0^{-\alpha} \tag{20.1.3}$$

Hence, for general nonlinear F, we cannot expect to obtain a solution that is defined for arbitrary large time. Hence, we shall study

$$u_t = u_{xx} + F(x,t,u,u_x) \tag{20.1.4}$$

for the case that F has an asymptotic linear behavior in its last two arguments.

344

It is convenient here to state the assumption on F that we shall use throughout the remainder of this chapter.

Assumption A.   *For the function $F = F(x, t, u, p)$, we shall assume the following*:

1) *The function $F(x, t, u, p)$ is defined and continuous on the set*
$$\mathcal{S} = \{(x, t, u, p) \mid (x, t) \in D_T \cup B_T,$$
$$-\infty < u < \infty, \quad -\infty < p < \infty\},$$
   where D_T is the parabolic domain under consideration, with parabolic boundary B_T and $T > 0$.

2) *For each $C > 0$ and for $|u|, |p| < C$, the function $F(x, t, u, p)$ is uniformly Hölder continuous in x and t for each compact subset of D_T.*

3) *There exists a constant C_F such that*
$$|F(x, t, u_1, p_1) - F(x, t, u_2, p_2)| \leq C_F \{|u_1 - u_2| + |p_1 - p_2|\}$$
$$(20.1.5)$$
   holds for all (u_i, p_i), $i = 1, 2$.

4) *For unbounded D_T, we add the assumption that F is bounded for bounded u and p.*

Remark.   It is easy to write down examples of such functions; for example,
$$F(x, t, u, p) = \sin xt \cos u + \cos xt \sin p.$$
In fact, for most applications the nonlinear source F can be extended linearly for u and p beyond the range of physical reality.

Also, it is convenient here to define what we shall mean as a solution of $u_t = u_{xx} + F(x, t, u, u_x)$ in D_T.

DEFINITION 20.1.1.   By a solution to an initial-boundary-value problem for $u_t = u_{xx} + F(x, t, u, u_x)$ in D_T, we mean a function $u = u(x, t)$ defined in $D_T \cup B_T$ such that

1) $u \in C(D_T \cup B_T)$;
2) u_x, u_t, and $u_{xx} \in C(D_T)$;
3) u and u_x are bounded in D_T;
4) $u_t(x, t) \equiv u_{xx}(x, t) + F(x, t, u(x, t), u_x(x, t))$ for all $(x, t) \in D_T$; and
5) The initial and boundary conditions are satisfied by u, which means that, if u_x is involved, then we shall require $u_x \in C(D \cup B_T)$.

Remark.   If u_x is involved in a boundary condition, the last condition on the continuity of u_x on B_T is a bit restrictive. We shall leave to the reader the removal of such restrictions that appear below.

We turn now to the consideration of various initial-boundary-value problems for $u_t = u_{xx} + F(x, t, u, u_x)$. In the studies below, we shall find the following result useful.

Lemma 20.1.1. *For a solution of an initial-boundary-value problem for $u_t = u_{xx} + F(x, t, u, u_x)$ in D_T, the function*

$$g(x, t) \equiv F(x, t, u(x, t), u_x(x, t)) \qquad (20.1.6)$$

is bounded in D_T and is uniformly Hölder continuous in x and continuous in t on each compact subset of D_T.

Proof. We shall leave this and the minor modification of Lemma 19.2.1 and Lemma 19.2.2 as an exercise for the reader.   □

20.2. THE INITIAL-VALUE PROBLEM

We consider here the initial-value problem

$$u_t = u_{xx} + F(x, t, u, u_x), \qquad -\infty < x < \infty, \qquad 0 < t \leq T,$$
$$u(x, 0) = \varphi(x), \qquad -\infty < x < \infty, \qquad (20.2.1)$$

where we shall assume that φ is continuously differentiable and that φ and φ' are bounded.

Suppose now that (20.2.1) possesses a solution. Then, integrating the identity

$$(u(\xi, \tau) K(x - \xi, t - \tau))_\tau - (u_\xi(\xi, \tau) K(x - \xi, t - \tau))_\xi$$
$$+ (u(\xi, \tau) K_\xi(x - \xi, t - \tau))_\xi$$
$$= F(\xi, \tau, u(\xi, \tau), u_\xi(\xi, \tau)) K(x - \xi, t - \tau) \qquad (20.2.2)$$

over the domain $-\infty < \xi < \infty$, $0 < \tau < t$, we obtain the representation

$$u(x, t) = \int_{-\infty}^{\infty} K(x - \xi, t) \varphi(\xi) \, d\xi$$
$$+ \int_0^t \int_{-\infty}^{\infty} K(x - \xi, t - \tau) F(\xi, \tau, u(\xi, \tau), u_\xi(\xi, \tau)) \, d\xi \, d\tau$$

$$(20.2.3)$$

for $-\infty < x < \infty$, $0 < t \leq T$. On the other hand, suppose that the integral-differential equation (20.2.3) has a solution $u = u(x, t)$ such that for $0 \leq t \leq T$, u and u_x are continuous and bounded. Then, the first term on the right side of (20.2.3) is infinitely differentiable for $t > 0$ and its derivatives are bounded for t bounded away from zero. Next, we see that $F(\xi, \tau, u(\xi, \tau), u_\xi(\xi, \tau))$ is bounded and continuous, by Assumptions A1 and A4. Consequently, by Property 5 of Lemma 19.2.2 and Lemma 19.2.3, the second function on the right side of (20.2.3) and its x-derivative are uniformly Hölder continuous with respect to x for $-\infty < x < \infty$ and $0 \le t \le T$. Hence, for each compact subset of $-\infty < x < \infty$, $0 < t \le T$, u and u_x are uniformly Hölder continuous with respect to x for each (x, t) in that compact subset. Consequently, by Assumptions A2 and A3, $F(x, t, u(x, t), u_x(x, t))$ is uniformly Hölder continuous with respect to x on compact subsets of $-\infty < x < \infty$, $0 < t \le T$. Minor modifications of the proof of Lemma 19.2.1 enable us to conclude that u satisfies

$$u_t = u_{xx} + F(x, t, u, u_x)$$

for $-\infty < x < \infty$, $0 < t \le T$. Obviously, u satisfies the initial condition.

We summarize the above in the following statement.

Lemma 20.2.1. *The initial-value problem* (20.2.1) *possesses a unique solution if and only if the integral-differential equation* (20.2.3) *possesses a unique solution u such that u and u_x are continuous and bounded for* $-\infty < x < \infty$, $0 < t \le T$.

Proof. Recalling the analysis preceding the statement of the lemma, it suffices to discuss the equivalence of uniqueness; but this easily follows from the prior discussion, by contradiction. For example, if we assume the uniqueness of the solution of the integral-differential equation, then if (20.2.1) possesses two solutions, the integral-differential equation must possess two solutions, which is a contradiction.   □

We turn now to the treatment of the integral-differential equation (20.2.3). We first show that the solution exists for small time, and then extend this to existence for any finite T. For $\eta > 0$, let

$$\mathcal{B}_\eta = \left\{ v(x, t) : v, v_x \in C((-\infty, \infty) \times [0, \eta]) \quad \text{and} \quad \|v\|_\eta < \infty \right\},$$

$$(20.2.4)$$

where

$$\|v\|_\eta = \sup_{\substack{-\infty < x < \infty \\ 0 \le t \le \eta}} |v(x, t)| + \sup_{\substack{-\infty < x < \infty \\ 0 \le t \le \eta}} |v_x(x, t)|. \qquad (20.2.5)$$

The set $\mathcal{B}_\eta$ is a Banach space. From the discussion for Lemma 20.2.1, we see that the mapping

$$\mathcal{F}v(x, t) = \int_{-\infty}^{\infty} K(x - \xi, t)\varphi(\xi) \, d\xi$$

$$+ \int_0^t \int_{-\infty}^{\infty} K(x - \xi, t - \tau) F(\xi, \tau, v(\xi, \tau), v_\xi(\xi, \tau)) \, d\xi \, d\tau$$

$$(20.2.6)$$

for $-\infty < x < \infty$, $0 \le t \le \eta$, maps $\mathcal{B}_\eta$ into $\mathcal{B}_\eta$. Next, we estimate the continuity of the map $\mathcal{F}$. From Assumption A3, (20.1.5), we see that, for

$0 \le t \le \eta,$

$$|\mathscr{F}v_1(x,t) - \mathscr{F}v_2(x,t)| \le C_F t \|v_1 - v_2\|_\eta \le C_F \eta \|v_1 - v_2\|_\eta, \quad (20.2.7)$$

while

$$\left|\frac{\partial \mathscr{F}v_1}{\partial x}(x,t) - \frac{\partial \mathscr{F}v_2}{\partial x}(x,t)\right| \le C_F \|v_1 - v_2\|_\eta \int_0^t \int_{-\infty}^\infty \left|\frac{\partial K}{\partial x}(x-\xi, t-\tau)\right| d\xi \, d\tau$$

$$\le 2\pi^{-1/2} C_F \eta^{1/2} \|v_1 - v_2\|_\eta. \quad (20.2.8)$$

Combining (20.2.7) with (20.2.8), it follows that

$$\|\mathscr{F}v_1 - \mathscr{F}v_2\|_\eta \le \left(\eta + 2\pi^{-1/2}\eta^{1/2}\right) C_F \|v_1 - v_2\|_\eta. \quad (20.2.9)$$

We select η such that

$$C_F\left(\eta + 2\pi^{-1/2}\eta^{1/2}\right) < 1. \quad (20.2.10)$$

Hence, $\mathscr{F}$ is a contraction of $\mathscr{B}_\eta$ into $\mathscr{B}_\eta$. Thus, $\mathscr{F}$ has a unique fixed point $u = u(x,t) \in \mathscr{B}_\eta$.

In order to demonstrate that (20.2.3) has a unique solution for $0 < t \le T$, we proceed by induction and assume that (20.2.3) possesses a unique solution u such that u and u_x are bounded for $0 \le t \le m\eta$. For $m\eta < t \le (m+1)\eta$, we consider the mapping

$$\mathscr{F}_1 v(x,t) = \int_{-\infty}^\infty K(x-\xi, t)\varphi(\xi) \, d\xi$$

$$+ \int_0^{m\eta} \int_{-\infty}^\infty K(x-\xi, t-\tau) F\big(\xi, \tau, u(\xi,\tau), u_\xi(\xi,\tau)\big) \, d\xi \, d\tau$$

$$+ \int_{m\eta}^t \int_{-\infty}^\infty K(x-\xi, t-\tau) F\big(\xi, \tau, v(\xi,\tau), v_\xi(\xi,\tau)\big) \, d\xi \, d\tau.$$

$$(20.2.11)$$

Clearly, $\mathscr{F}_1$ maps $\tilde{\mathscr{B}}_\eta$ into $\tilde{\mathscr{B}}_\eta$, where

$$\tilde{\mathscr{B}}_\eta = \big\{ v(x,t) | v(x, t-m\eta) \in \mathscr{B}_\eta \big\}. \quad (20.2.12)$$

Moreover, the estimate (20.2.9) holds for $\mathscr{F}_1$. From (20.2.10), $\mathscr{F}_1$ is a contraction and hence possesses a unique fixed point $u = u(x,t) \in \tilde{\mathscr{B}}_\eta$. Since the fixed point of $\mathscr{F}_1$ matches continuously with $u(x, m\eta)$ and $u_x(x, m\eta)$, we see that (20.2.3) possesses a unique solution for $0 \le t \le (m+1)\eta$ such that the solution and its partial derivative with respect to x are bounded and continuous.

We can summarize all of the preceding discussion in the following statement.

Theorem 20.2.1. *If the initial data φ is continuously differentiable such that φ and φ' are bounded, and if F satisfies Assumption A of Sec. 20.1, then the initial-value problem (20.2.1) possesses a unique solution.*

Proof. See the analysis preceding the statement of the theorem. □

We can demonstrate now that the solution u and its derivative u_x depend continuously upon the data.

Theorem 20.2.2. *Let u_i, $i = 1, 2$, be solutions of (20.2.1) that correspond, respectively, to the data (φ_i, F_i), $i = 1, 2$, which satisfy the hypotheses of Theorem 20.2.1. Then there exists a positive constant $C = C(C_F, T)$ such that*

$$\|u_1 - u_2\|_T \le C \left\{ \sup_{-\infty < x < \infty} |\varphi_1(x) - \varphi_2(x)| + \sup_{-\infty < x < \infty} |\varphi_1'(x) - \varphi_2'(x)| \right.$$

$$\left. + \sup_{\substack{-\infty < x < \infty \\ 0 \le t \le T, \\ -\infty < u < \infty \\ -\infty < p < \infty}} |F_1(x, t, u, p) - F_2(x, t, u, p)| \right\}, \quad (20.2.13)$$

where the norm $\|u_1 - u_2\|_T$ is defined by (20.2.5), with η replaced by T.

Proof. The respective integral equations for u_1 and u_2 and their derivatives are differenced. Straightforward estimates reduce the estimate of $\|u_1 - u_2\|_t$ to the application of Lemma 8.4.2, and Lemma 8.4.1 (Gronwall). □

20.3. SOME INITIAL-BOUNDARY-VALUE PROBLEMS FOR ELEMENTARY DOMAINS

We shall utilize the solutions in Sec. 19.3, coupled with the analysis given in Sec. 20.2, in order to state some existence theorems.

A Semi-Infinite Conductor with Temperature Boundary Conditions

Theorem 20.3.1. *For the problem*

$$u_t = u_{xx} + F(x, t, u, u_x), \quad 0 < x < \infty, \quad 0 < t \le T,$$
$$u(x, 0) = \varphi(x), \quad 0 < x < \infty, \quad (20.3.1)$$
$$u(0, t) = g(t), \quad 0 < t \le T,$$

where F satisfies Assumption A of Sec. 20.1, φ is continuously differentiable such that φ and φ' are bounded, g is continuously differentiable, and $\varphi(0) = g(0)$, there exists a unique bounded solution $u = u(x, t)$ that is the unique *solution with bounded continuous derivative with respect to x of*

$$u(x,t) = -2\int_0^t \frac{\partial K}{\partial x}(x, t-\tau) g(\tau) \, d\tau + \int_0^\infty G(x, \xi, t) \varphi(\xi) \, d\xi$$
$$+ \int_0^t \int_0^\infty G(x, \xi, t-\tau) F\big(\xi, \tau, u(\xi, \tau), u_\xi(\xi, \tau)\big) \, d\xi \, d\tau.$$

$$(20.3.2)$$

where $K(x,t)$ is defined by (3.2.1) and $G(x, \xi, t)$ is defined by (4.3.4). Moreover, there exists a constant $C = C(T, C_F)$, where C_F is the Lipschitz constant of F, such that

$$\|u_1 - u_2\|_T \leq C\left\{ \sup_{0 < x < \infty} |\varphi_1(x) - \varphi_2(x)| + \sup_{0 < x < \infty} |\varphi_1'(x) - \varphi_2'(x)| \right.$$

$$+ \sup_{0 \leq t \leq T} |g_1(t) - g_2(t)| + \sup_{0 \leq t \leq T} |g_1'(t) - g_2'(t)|$$

$$\left. + \sup_{\substack{0 \leq x \leq \infty, \\ 0 \leq t \leq T, \\ -\infty < u < \infty, \\ \infty < p < \infty,}} |F_1(x, t, u, p) - F_2(x, t, u, p)| \right\}, \quad (20.3.3)$$

where u_i, $i = 1, 2$, are the solutions of (20.3.1) corresponding, respectively, to the data (φ_i, g_i, F_i), $i = 1, 2$, and $\|u_1 - u_2\|_T$ is defined by (20.2.5) with η replaced by T and $-\infty$ by 0.

Proof. The proof follows the pattern of that given in Sec. 20.2 with only minor modifications. □

A Semi-Infinite Conductor with Flux Boundary Conditions

Theorem 20.3.2. *For the problem*

$$u_t = u_{xx} + F(x, t, u, u_x), \qquad 0 < x < \infty, \qquad 0 < t \leq T,$$
$$u(x, 0) = \varphi(x), \qquad 0 < x < \infty, \qquad (20.3.4)$$
$$u_x(0, t) = g(t), \qquad 0 < t \leq T,$$

where F satisfies Assumption A of Sec. 20.1, φ is continuously differentiable such that φ and φ' are bounded, and g is continuous, there exists a unique bounded solution $u = u(x, t)$ that is the unique solution with bounded continu-

ous derivative, with respect to x, of

$$u(x,t) = -2\int_0^t K(x,t-\tau)g(\tau)\,d\tau + \int_0^\infty N(x,\xi,t)\varphi(\xi)\,d\xi$$
$$+ \int_0^t\int_0^\infty N(x,\xi,t-\tau)F\big(\xi,\tau,u(\xi,\tau),u_\xi(\xi,\tau)\big)\,d\xi\,d\tau,$$

$$(20.3.5)$$

where $K(x,t)$ is defined by (3.2.1) and $N(x,\xi,t)$ is defined by Eq. (F) of the Exercise section of Chapter 3. Moreover, there exists a constant $C = C(T,C_F)$, where C_F is the Lipschitz constant of F, such that

$$\|u_1 - u_2\|_T \le C\Bigg\{ \sup_{0<x<\infty} |\varphi_1(x)-\varphi_2(x)| + \sup_{0<x<\infty} |\varphi_1'(x)-\varphi_2'(x)|$$

$$+ \sup_{0\le t\le T} |g_1(t)-g_2(t)|$$

$$+ \sup_{\substack{0<x<\infty \\ 0\le t\le T \\ -\infty<u<\infty \\ -\infty<p<\infty}} |F_1(x,t,u,p)-F_2(x,t,u,p)| \Bigg\}, \quad (20.3.6)$$

where u_i, $i=1,2$, are the solutions of (20.3.4) corresponding, respectively, to the data (φ_i, g_i, F_i), $i=1,2$, and $\|u_1 - u_2\|_T$ defined by (20.2.5), with η replaced by T and $-\infty$ by 0.

    Proof. The proof follows the pattern of that given in Sec. 20.2, and is left for the reader. □

A Finite Conductor with Temperature Boundary Conditions

Theorem 20.3.3. *For the problem*

$$u_t = u_{xx} + F(x,t,u,u_x), \quad 0<x<1, \quad 0<t\le T,$$
$$u(x,0) = \varphi(x), \quad 0<x\le 1,$$
$$u(0,t) = g(t), \quad 0<t\le T,$$
$$u(1,t) = h(t), \quad 0<t\le T,$$

$$(20.3.7)$$

where F satisfies Assumption A of Sec. 20.1, φ is continuously differentiable such that φ and φ' are bounded, g is continuously differentiable, h is continuously differentiable, $\varphi(0) = g(0)$, and $\varphi(1) = h(0)$, there exists a unique bounded solution $u = u(x,t)$ that is the unique solution with bounded continuous *derivative, with respect to x, of*

$$u(x,t) = \int_0^1 \{\theta(x-\xi,t) - \theta(x+\xi,t)\} \varphi(\xi)\, d\xi$$

$$-2\int_0^t \frac{\partial\theta}{\partial x}(x,t-\tau)g(\tau)\, d\tau + 2\int_0^t \frac{\partial\theta}{\partial x}(x-1,t-\tau)h(\tau)\, d\tau$$

$$+\int_0^t \int_0^1 \{\theta(x-\xi,t-\tau) - \theta(x+\xi,t-\tau)\}$$

$$\cdot F(\xi,\tau,u(\xi,\tau),u_\xi(\xi,\tau))\, d\xi\, d\tau, \qquad (20.3.8)$$

where $\theta(x,t)$ is defined by (6.1.6). Moreover, there exists a constant $C = C(T,C_F)$, where C_F is the Lipschitz constant of F, such that

$$\|u_1 - u_2\|_T \le C \left\{ \sup_{0<x<1} |\varphi_1(x) - \varphi_2(x)| + \sup_{0<x<1} |\varphi_1'(x) - \varphi_2'(x)| \right.$$

$$+ \sup_{0\le t\le T} |g_1(t) - g_2(t)| + \sup_{0\le t\le T} |g_1'(t) - g_2'(t)|$$

$$+ \sup_{0\le t\le T} |h_1(t) - h_2(t)| + \sup_{0\le t\le T} |h_1'(t) - h_2'(t)|$$

$$\left. + \sup_{\substack{0\le x\le 1 \\ 0\le t\le T \\ -\infty < u < \infty \\ -\infty < p < \infty}} |F_1(x,t,u,p) - F_2(x,t,u,p)| \right\}, \qquad (20.3.9)$$

where u_i, $i=1,2$, are the solutions of (20.3.7) corresponding, respectively, to the data $(\varphi_i, g_i, h_i, F_i)$, $i=1,2$, and $\|u_1 - u_2\|_T$ is defined by (20.2.5), with η replaced by T, $-\infty$ by 0, and $+\infty$ by 1.

   Proof.   See Sec. 20.2.   □

A Finite Conductor with Flux-Boundary Conditions

Theorem 20.3.4.   *For the problem*

$$u_t = u_{xx} + F(x,t,u,u_x), \qquad 0<x<1, \qquad 0<t\le T,$$
$$u(x,0) = \varphi(x), \qquad 0<x<1,$$
$$u_x(0,t) = g(t), \qquad 0<t\le T,$$
$$u_x(1,t) = h(t), \qquad 0<t\le T, \qquad (20.3.10)$$

where F satisfies Assumption A of Sec. 20.1, φ is continuously differentiable such that φ and φ' are bounded, g is continuous, and h is continuous, there exists a unique, bounded solution $u = u(x, t)$ that is the unique solution with bounded, continuous derivative, with respect to x, of

$$u(x,t) = \int_0^1 \{\theta(x-\xi,t) + \theta(x+\xi,t)\} \varphi(\xi)\, d\xi - 2\int_0^t \theta(x,t-\tau) g(\tau)\, d\tau$$

$$+ 2\int_0^t \theta(x-1,t-\tau) h(\tau)\, d\tau$$

$$+ \int_0^t \int_0^1 \{\theta(x-\xi,t-\tau) + \theta(x+\xi,t-\tau)\}$$

$$\cdot F\big(\xi, \tau, u(\xi,\tau), u_\xi(\xi,\tau)\big)\, d\xi\, d\tau, \qquad (20.3.11)$$

where $\theta(x,t)$ is defined by (6.1.6). Moreover, there exists a constant $C = C(T, C_F)$, where C_F is the Lipschitz constant of F, such that

$$\|u_1 - u_2\|_T \le C \left\{ \sup_{0 < x < 1} |\varphi_1(x) - \varphi_2(x)| + \sup_{0 < x < 1} |\varphi_1'(x) - \varphi_2'(x)| \right.$$

$$+ \sup_{0 \le t \le T} |g_1(t) - g_2(t)| + \sup_{0 \le t \le T} |h_1(t) - h_2(t)|$$

$$\left. + \sup_{\substack{0 \le x \le 1 \\ 0 \le t \le T \\ -\infty < u < \infty \\ -\infty < p < \infty}} |F_1(x,t,u,p) - F_2(x,t,u,p)| \right\}, \qquad (20.3.12)$$

where u_i, $i = 1, 2$, are the solutions of (20.3.10) corresponding, respectively, to the data $(\varphi_i, g_i, h_i, F_i)$, $i = 1, 2$, and $\|u_1 - u_2\|_T$ is defined by (20.2.5), with η replaced by T, $-\infty$ by 0, and $+\infty$ by 1.

   Proof.   See Sec. 20.2.   □

EXERCISES

20.1. Demonstrate Lemma 20.1.1.

20.2. Modify the proof of Lemma 19.2.1 and Lemma 19.2.2 in order to demonstrate Lemma 20.2.1.

20.3. Prove Theorem 20.3.1.

20.4. Prove Theorem 20.3.2.

20.5. Prove Theorem 20.3.3.

20.6. Prove Theorem 20.3.4.

20.7. For

$$u_t = u_{xx} + F(x, t, u, u_x), \qquad 0 < x < \infty, \qquad 0 < t \le T,$$
$$u(x, 0) = \varphi(x), \qquad 0 < x < \infty, \qquad \qquad \text{(A)}$$
$$u_x(0, t) = G(t, u(0, t)), \qquad 0 < t \le T,$$

show that there exists a unique solution that depends continuously
upon the data (φ, F, G), provided that F satisfies Assumption A of
Sec. 20.1, φ is continuously differentiable such that φ and φ' are
bounded, and $G = G(t, u)$ is continuous in its arguments and uni-
formly Lipschitz-continuous with respect to u.

20.8. For

$$u_t = u_{xx} + F(x, t, u, u_x), \qquad s(t) < x < \infty, \qquad 0 \le t \le T,$$
$$u(x, 0) = \varphi(x), \qquad s(0) < x < \infty, \qquad \qquad \text{(B)}$$
$$u(s(t), t) = g(t), \qquad 0 < t \le T,$$

where $s = s(t)$ is a continuously differentiable function on $0 \le t \le T$,
show that, by the change of variable $x = \xi + s(t)$ and

$$U(\xi, t) = u(\xi + s(t), t), \qquad \qquad \text{(C)}$$

Problem (B) becomes

$$U_t = U_{\xi\xi} + G(\xi, t, U, U_\xi), \qquad 0 < \xi < \infty, \qquad 0 < t \le T,$$
$$U(\xi, 0) = \varphi(\xi + s(0)), \qquad 0 < \xi < \infty, \qquad \qquad \text{(D)}$$
$$U(0, t) = g(t), \qquad 0 < t \le T,$$

where

$$G(\xi, t, U, U_\xi) = F(\xi + s(t), U, U_\xi) + \dot{s}(t) U_\xi. \qquad \text{(E)}$$

20.9. For the problem

$$u_t = u_{xx} + F(x, t, u, u_x), \qquad s_1(t) < x < s_2(t), \qquad 0 < t \le T,$$
$$u(x, 0) = \varphi(x), \qquad s_1(0) < x < s_2(0),$$
$$u(s_1(t), t) = g(t), \qquad 0 < t \le T, \qquad \qquad \text{(F)}$$
$$u(s_2(t), t) = h(t), \qquad 0 < t \le T,$$

where $s_i = s_i(t)$, $i = 1, 2$, are continuously differentiable on $0 \le t \le T$,
show that under the transformation

$$\xi = \frac{x - s_1(t)}{s_2(t) - s_1(t)}, \qquad \qquad \text{(G)}$$

and the transformation

$$\eta = \int_0^t (s_2(\tau) - s_1(\tau))^{-2} \, d\tau, \qquad \qquad \text{(H)}$$

we obtain a problem of the form

$$v_\eta = v_{\xi\xi} + G_1(\xi, \eta, v, v_\xi), \qquad 0 < \xi < 1, \qquad 0 < \eta \le T_1,$$
$$v(\xi, 0) = \varphi_1(\xi), \qquad 0 < \xi < 1,$$
$$v(0, \eta) = g_1(\eta), \qquad 0 < \eta \le T_1, \tag{I}$$
$$v(1, \eta) = h_1(\eta), \qquad 0 < \eta \le T_1.$$

20.10. For the problem

$$a(u)u_t = b(t)(a(u)u_x)_x + F(x, t, u, u_x), \qquad s_1(t) < x < s_2(t),$$
$$0 < t \le T, \qquad u(x, 0) = \varphi(x), \qquad s_1(0) < x < s_2(0),$$
$$u(s_1(t), t) = g(t), \qquad 0 < t \le T, \tag{J}$$
$$u(s_2(t), t) = h(t), \qquad 0 < t \le T.$$

Set

$$v(x, t) = \int_0^{u(x,t)} a(\xi)\, d\xi. \tag{K}$$

Then,

$$v_t = a(u)u_t \tag{L}$$

and

$$b(t)(a(u)u_x)_x = b(t)v_{xx}. \tag{M}$$

Since $a(\xi) > 0$, the function

$$v = A(u) = \int_0^u a(\xi)\, d\xi \tag{N}$$

possesses an inverse function

$$u = B(v), \tag{O}$$

whence

$$u_x(x, t) = B'(v)v_x. \tag{P}$$

Thus, Problem (J) becomes

$$v_t = b(t)v_{xx} + F(x, t, B(v), B'(v)v_x), \qquad s_1(t) < x < s_2(t),$$
$$0 < t \le T, \qquad v(x, 0) = A(\varphi(x)), \qquad s_1(0) < x < s_2(0),$$
$$v(s_1(t), t) = A(g(t)), \qquad 0 < t \le T, \tag{Q}$$
$$v(s_2(t), t) = A(h(t)), \qquad 0 < t \le T.$$

Show that the transformation

$$x = (s_2(t) - s_1(t))\xi + s_1(t) \tag{R}$$

reduces (Q) to a problem of the form

$$w_t = c(t)w_{\xi\xi} + G(\xi, t, w, w_\xi), \qquad 0 < \xi < 1, \qquad 0 < t \le T,$$
$$w(\xi, 0) = \varphi_1(\xi), \qquad 0 < \xi < 1,$$
$$w(0, t) = g_1(t), \qquad 0 < t \le T,$$
$$w(1, t) = h_1(t), \qquad 0 < t \le T,$$

(S)

where $c(t) > 0$. Show that another application of a transformation of the form

$$\eta = \int_0^t c(\tau)\, d\tau \tag{T}$$

reduces Problem (J) to a problem of the form

$$z_\eta = z_{\xi\xi} + G_1(\xi, \eta, z, z_\xi), \qquad 0 < \xi < 1, \qquad 0 < \eta \le T_1,$$
$$z(\xi, 0) = \varphi_1(\xi), \qquad 0 < \xi < 1,$$
$$z(0, \eta) = g_2(\eta), \qquad 0 < \eta \le T_1,$$
$$z(1, \eta) = h_2(\eta), \qquad 0 < \eta \le T_1.$$

Translate conditions of the data $(G_1, \varphi_1, g_2, h_2)$ given in Theorem 20.3.3 into conditions on the data $(a, b, F, \varphi, g, h, s_1, s_2)$ so that Problem (J) possesses a unique bounded solution.

NOTES

Modifications and extensions of the material presented here can be found in Gevrey [2], Friedman [1], and Ladyzenskaja, Solonnikov, and Ural'ceva [3].

REFERENCES

1. Friedman, A., *Partial Differential Equations of Parabolic Type*. Prentice-Hall, Inc., Englewood Cliffs, N.J., 1964.
2. Gevrey, M., Sur les équations aux dérivées partielles du type parabolique, *J. Math. Pures Appl.*, **9** (1913), 305–471.
3. Ladyzenskaja, O. A., Solonnikov, V. A., and Ural'ceva, N. N., *Linear and Quasilinear Equations of Parabolic Type*, Vol. 23. Translations and Math. Monographs. American Math. Soc., Providence, Rhode Island, 1968.

Some References to the Literature on $\mathscr{L}(u) \equiv u_{xx} - u_t$

1. CLASSICAL: 1800–1950

1. Amerio, L., Sull'equazione del calore. I. *Atti Accad. Naz. Lincei. Rend. Cl. Sci. Fis. Mat. Nat.* (8) **1**, 346–352 (1946), MR 8, 274.
2. Amerio, L., Sull'equazione del calore. II. *Atti Accad. Naz. Lincei. Rend. Cl. Sci. Fis. Mat. Nat.* (8) **1**, 544–548 (1946), MR 8, 274.
3. Amerio, L., Sull'equazione de propagazione del calore. *Univ. Roma. Ist. Naz. Alta Mat. Rend. Mat. e App.* (5) **5**, 84–120 (1946), MR 9, 37.
4. Appell, P., Sur l'équation $(\partial^2 z/\partial x^2)-(\partial z/\partial y)=0$ et la théorie de la chaleur, *J. Math. Pures Appl.* **8** (1892), 187–216.
5. Archibald, W. J., The integration of the differential equation of the ultra-centrifuge, *Ann. New York Acad. Sci.*, **43** (1942), 211–227, MR 4, 145.
6. Avrami, M., and Little, J. B., Diffusion of heat through a rectangular bar and the cooling and insulating effect of fins. I. The steady state, *J. Appl. Phys.* **13** (1942), 255–264, MR 3, 248.
7. Awberry, J. H., The periodic flow of heat in a hollow cylinder, *Philos. Mag.* **28** (1939), 447–451, MR 1, 120.
8. Babbitt, J. D., On the differential equations of diffusion, *Canadian J. Research*, Sect. A., **28**, 449–474 (1950), MR 12, 710.
9. Barrer, R. M., Gas flow in solids, *Philos. Mag.* **28**, 148–162 (1939), MR 1, 120.
10. Barrer, R. M., Diffusion in spherical shells, and a new method of measuring the thermal-diffusivity constant, *Philos. Mag.* (7) **35** (1944), 802–811, MR 7, 162.

11. Bell, R. P., A problem of heat conduction with spherical symmetry, *Proc. Phys. Soc.* **57**, 45–48 (1945), MR 6, 178.

12. Bellman, R., On the existence and boundedness of solutions of nonlinear partial-differential equations of parabolic type, *Trans. Amer. Math. Soc.* **64** (1948), 21–44, MR 10, 43.

13. Benfield, A. E., A problem of the temperature distribution in a moving medium, *Quart. Appl. Math.* **6**, 439–443 (1949), MR 10, 301.

14. Benfield, A. E., The temperature in an accreting medium with heat generation, *Quart. Appl. Math.* **7** (1950), 436–439, MR 11, 362.

15. Bernstein, S. N., Bounds for the moduli of successive derivatives of parabolic equations, *Dokl. Akad. Nauk.* **18** (1938), 385–388 (Russian).

16. Bickley, W. G., Experiments in approximating to solutions of a partial-differential equation, *Philos. Mag.* (7) **32** (1941), 50–55, MR 3, 155.

17. Biegelmeier, G., Ein Beitrag zur klassischen Diffusions theorie, *Acta Physica Austriaca* **4** (1950), 278–289, MR 12, 831.

18. Biegelmeier, G., Ein Beitrag zur klassischen Diffusions theorie, *Osterreich Akad. Wiss. Math.-Nat. Kl. S.-B., IIa*, **158** (1950), 161–172, MR 13, 464.

19. Bononcini, V. E., Un problema della propagazione del calore. *Atti Semi. Mat. Fis. Univ. Modena* **3** (1949), 142–161, MR 11, 362.

20. Brandt, W. H., Solution of the diffusion equation applicable to the edgewise growth of pearlite, *J. Appl. Phys.* **16** (1945), 139–146, MR 6, 178.

21. Brillouin, M., Sur quelques problèmes nonresolues de la physique mathématique classique. *Propagation de la Fusion*, Ann. Inst. H. Poincare, Vol. 1 (1931), 285–308.

22. Brinkley, S. R., Jr., Heat transfer between a fluid and a porous solid generating heat, *J. Appl. Phys.* **18** (1947), 582–585, MR 9, 38.

23. Brown, H. K., Resolution of temperature problems by the use of finite Fourier transformations, *Bull. Amer. Math. Soc.* **50** (1944), 376–385, MR 5, 240.

24. Brown, R. L., A problem in nonsteady heat conduction, *Philos. Mag.* (7) **37** (1946), 318–322, MR 8, 585.

25. Carslaw, H. S., A simple application of the Laplace transformation, *Philos. Mag.* (7) **30** (1940), 414–417, MR 2, 204.

26. Carslaw, H. S., *Introduction to the Mathematical Theory of Conduction of Heat in Solids*. Dover Publications, New York, New York, 1945; xxi + 268 pp, MR 7, 450.

27. Carslaw, H. S., and Jaeger, J. C., A problem in conduction of heat, *Proc. Cambridge Philos. Soc.* **35** (1939), 394–404, MR 1, 77.

28. Carslaw, H. S., and Jaeger, J. C., Some two-dimensional problems in conduction of heat with circular symmetry, *Proc. London Math. Soc.* (2) **46** (1940), 361–388, MR 2, 56.

29. Carslaw, H. S., and Jaeger, T. C., The determination of Green's function for the equation of conduction of heat in cylindrical coordinates by the Laplace transformation, *J. London Math. Soc.* **15** (1940), 273–281, MR 2, 292.

30. Carslaw, H. S., and Jaeger, J. C., The determinatoin of Green's function for line sources for the equation of conduction of heat in cylindrical coordinates by the Laplace transformation, *Philos. Mag.* (7) **31** (1941), 204–208, MR 2, 292.

31. Carslaw, H. S., and Jaeger, J. C., *Conduction of Heat in Solids*. Oxford, at the Clarendon Press, 1947; viii + 386 pp., MR 9, 188.

32. Carstoiu, I., Sur le calcul symbolique à deux variables et ses applications, *C. R. Acad. Sci. Paris*, **226** (1948), 45–47, MR 9, 287.

33. Casci, C., Sulla distribuzione della temperatura in un anello rotante in ambienti a temperatura diversa, *Atti Accad. Naz. Lincei. Rend. Cl. Sci. Fis. Mat. Nat.* (8) **7** (1949) (1950), 297–303, MR 11, 522.

34. Cattaneo, C., Sulla conduzione del calore, *Atti. Sem. Mat. Fis. Univ. Modena* **3** (1949), 83–101, MR 11, 362.

35. Christor, C., On a problem of gas diffusion, *Annuaire [Godisnik] Univ. Sofia. Fac. Phys.-Math.*, Livre 1 (1945) **41**, 143–163 (Bulgarian, English Summary), MR 12, 264.

36. Churchill, R. V., On the problem of temperatures in a nonhomogeneous bar with discontinuous initial temperatures, *Amer. J. Math.* **61** (*1939*), 651–664, MR 1, 57.

37. Churchill, R. V., A heat-conduction problem introduced by C. J. Tranter, *Philos. Mag.* (7) **31** (1941), 81–87, MR 2, 204.

38. Commenetz, G., Continuous heating of a hollow cylinder, *Quart. Appl. Math.* **5** (1948), 503–510, MR 9, 287.

39. Cooper, J. L. B., The uniqueness of the solution of the equation of heat conduction, *J. London Math. Soc.* **25** (1950), 173–180, MR 12, 104.

40. Craggs, J. W., Heat conduction in semi-infinite cylinders, *Philos. Mag.* (7) **36** (1945), 220–222, MR 7, 162.

41. Crank. J., A diffusion problem in which the amount of diffusing substance is finite. IV. Solutions for small values of the time. *Philos. Mag.* (7) **39** (1948), 362–376, MR 9, 591.

42. Crank, J., A diffusion problem in which the amount of diffusing substance is finite. II. Diffusion with nonlinear absorption. *Philos. Mag.* (7) **39** (1948), 140–149, MR 9, 439.

43. Crank, J., and Godson, S. M., A diffusion problem in which the amount of diffusing substance is finite. III. Diffusion with nonlinear absorption into a composite circular cylinder. *Philos. Mag.* (7) **38** (1947), 794–801, MR 9, 591.

44. Crank, J., and Nicolson, P., A practical method for numerical evaluation of solutions of partial-differential equations of the heat-conduction type, *Proc. Cambridge Philos. Soc.*, **43** (1947), 50–67, MR 8, 409.

45. Dacev, A., On the cooling of a rod composed of two homogeneous rods of finite length, *Dokl. Adad. Nauk. SSSR (N.S.)* **56** (1947), 255–258 (Russian), MR 9, 147.

46. Dacev, A., On the cooling of bars composed of a finite number of homogeneous parts, *Dokl. Akad. Nauk. SSSR (N.S.)* **56** (1947), 355–358 (Russian), MR 9, 146.

47. Datzeff, A., Sur le refroidissement d'un corps nonhomogene, *C. R. (Dokl.) Acad. Sci. URSS (N.S.)* **55** (1947), 111–114, MR 8, 585.

48. Dacev, A. B., On the linear problem of Stephan, *Dokl. Akad. Nauk. SSSR. (N.S.)* **58** (1947), 563–566 (Russian), MR 9, 513.

49. Datzeff, A., Sur certaines analogies mécaniques de la théorie de la chaleur, *C. R. Acad. Bulgare Sci. Math. Nat.* **2**, No. 2–3 (1949), 25–28, MR 12, 104.

50. Datzeff, A., Sur la propagation de la chaleur dans un milieu à plusieurs couches, *Annuaire [Godisnik] Univ. Sofia. Fac. Sci.*, Livre 1, (1949) **45**, 63–91,

MR 12, 504.

51. Datzeff, A., Sur le problème linéaire géneral de propagation de la chaleur dans un milieu à plusieurs couches, *C. R. Acad. Bulgare Sci. Math. Nat.* **2**, No. 2–3 (1949), 21–24, MR 12, 104.

52. Datzeff, A., Sur le problème linéaire de Stéfan, *Annuaire* [*Godisnik*] *Univ. Sofia. Fac. Sci.*, Livre 1, (1949) **45**, 321–352 (French, Bulgarian Summary), MR 12, 504.

53. Datzeff, A., Sur le problème linéaire de Stéfan. II., *Annuaire* [*Godisnik*] *Univ. Sofia. Fac. Sci.*, Livre 1, (1950) **46**, 271–325 (French, Bulgarian Summary), MR 13, 947.

54. Dacev, A. B., On the linear problem of Stefan. The case of two phases of infinite thickness, *Dokl. Akad. Nauk. SSSR* (*N.S.*) **74** (1950), 445–448 (Russian), MR 12, 263.

55. Dacev, A. B., On the linear problem of Stefan. The case of alternating phases, *Dokl. Akad. Nauk. SSSR* (*N.S.*) **75** (1950), 631–634 (Russian), MR 12, 710.

56. Dacev, A. B., On a general linear problem of heat coduction in a stratified medium, *Izvestiya Akad. Nauk. SSSR. Ser. Geograf. Geofiz.* **14**, (1950) 113–127 (Russian), MR 12, 104.

57. Danckwerts, P. V., Unsteady-state diffusion or heat conduction with moving boundary, *Trans. Faraday Soc.* **46** (1950), 701–712, MR 12, 264.

58. Datta M., S., A note on the differential equation of the Clusius column for separation of isotopes, *Science and Culture* **15** (1950), 329, MR 13, 465.

59. Davies, C. N., The sedimentation and diffusion of small particles, *Proc. Roy. Soc. London*, *Ser. A.*, **200** (1949), 100–113, MR 11, 362.

60. Del Pasqua, D., Risoluzione, con sole integrazioni, dell'equazione differenziale di tipo parabolico, con i dati di Cauchy su una curva assegnata, *Ann. Scuola Norm. Super. Pisa* (3) **2** (1948), 55–61 (1950), MR 11, 668.

61. Ditkin, V. A., Solution of a problem of heat conduction by the method of operational calculus, *Trav. Inst. Math. Stekloff* **20** (1947), 77–86 (Russian), MR 9, 472.

62. Doetsch, G., Les équations aux dérivées partielles du type parabolique, *Enseignement Math.*, **35** (1936), 43–87.

63. Doetsch, G., Ein Zusammenkang zwischen Randwert problemen verschiedenen Typs, *Math. Z.* **46** (1940) 315–328, MR 1, 314.

64. Doetsch, G., *Theorie und Anwendung der Laplace-Transformation.* Dover, New York, 1943.

65. Dressel, F. G., and Elliott, E. R., A class of solutions for the heat equation and associated boundary-value problems, *Amer. J. Math.* **65** (1943), 408–422, MR 5, 69.

66. Erugin, N. P., A closed solution of a parabolic inhomogeneous boundary problem, *Akad. Nauk. SSSR, Prikl. Mat. Meh.* **14** (1950), 215–217 (Russian), MR 11, 667.

67. Evans, G. W., II, Isaacson, E., and MacDonald, J. K. L., Stefan-like problems, *Quart. Appl. Math.* **8** (1950), 312–319, MR 12, 263.

68. Feller, W., Zur Theorie der stochastischen Prozesse, *Math. Ann.*, **113** (1936), 113–160.

69. Fjeldstad, J. E., A generalization of Abel's integral equation, *Norsk. Mat. Tidsskr.* **22** (1940), 41–51, MR 2, 204.

70. Fourier, J., *Analytical Theory of Heat*, Great Books of the Western World, Vol. 45. Encyclopedia Britannica, Inc., Chicago-London-Toronto, 1952.

71. Fowler, C. M., Analysis of numerical solutions of transient heat-flow problems, *Quart. Appl. Math.* **3** (1946), 361–376, MR 7, 383.

72. Gaskell, R. E., A problem in heat conduction and an expansion theorem, *Amer. J. Math.* **64** (1942), 447–455, MR 3, 247.

73. George, J.-C., Résolution de l'équation de la diffusion par une méthode symbolique à deux opérateurs, *C. R. Acad. Sci. Paris* **219** (1944), 405–407, MR 7, 383.

74. Gevrey, M., Sur les équations aux dérivées partielles du type parabolique, *J. Math. Pures Appl.*, **9** (1913), 305–471.

75. Goursat, E., *Cours d'Analyse Mathématique*, Vol. 3, Chapter 29. Gauthier-Villars, Paris, 1923.

76. Griffith, M. V., and Horton, G. K., The transient flow of heat through a two-layer wall, *Proc. Phys. Soc.* **58** (1946), 481–487, MR 8, 585.

77. de Groot, S. R., Sur l'intégration de quelques problèmes aux limites régis par l'équation de Fourier dite "de la chaleur" au moyen de la méthode des transformations fonctionnelles simultanées, *Nederl. Akad. Wetensch. Proc.* **45** (1942), 643–649, 820–825, MR 5, 240.

78. Hadamard, J., Remarques sur le cas parabolique des équations aux dérivées partielles, *Publ. Inst. Mat. Univ. Nac. Litoral*, **5** (1945), 3–11, MR 7, 162.

79. Hadamard, J., *Lectures on Cauchy's Problem in Linear Partial-Differential Equations*. Dover Publications, New York, 1952.

80. Harkeevic, Y. F., The graphical solution of partial-differential equations of parabolic type, *Akad. Nauk. SSSR Prikl. Mat. Mech.* **14** (1950), 303–310 (Russian), MR 12, 134.

81. Hartman, P., and Wintner, A., On the solutions of the equation of heat conduction, *Amer. J. Math.* **72** (1950), 367–395, MR 12, 104.

82. Holmgren, E., Sur l'équation de la propagation de la chaleur, *Arkiv. Mat. Fysik*, **14** (1908), No. 4, 1–11.

83. Holmgren, E., Sur l'équation de la propagation de la chaleur, II, *Arkiv. Mat. Fysik*, **4** (1908), No. 18, 1–27.

84. Holmgren, E., Sur les solutions quasi-analytiques de l'équation de la chaleur, *Arkiv. Mat.* **18** (9) (1924), 1–9.

85. Hopf, E., The partial-differential equation $u_t + uu_x = \mu u_{xx}$, *Comm. Pure Appl. Math.* **3** (1950), 201–230, MR 13, 846.

86. Howell, W. T., A note on the solution of some partial-differential equations in the finite domain, *Philos. Mag.*, **28** (1939), 396–402, MR 1, 120.

87. Huber, A., Hauptaufsatze über das Fortschreiten der Schmelzgrenze in einem linearen Leiter, *Z. Angew. Math. Mech.*, **19** (1939), 1–21.

88. Ivancov., G. P., The temperature field around a spherical, cylindrical, or pointed crystal growing in a cooling solution, *Dokl. Akad. Nauk. SSSR* (*N.S.*) **58** (1947), 567–569 (Russian), MR 9, 439.

89. Jaeger, J. C., Radial heat flow in circular cylinders with a general boundary condition, *J. Proc. Roy. Soc. New South Wales*, **74** (1940), 342–352, MR 2, 204.

90. Jaeger, J. C., Heat conduction in composite circular cylinders, *Philos. Mag.* (7) **32** (1941), 324–335, MR 3, 247.

91. Jaeger, J. C., Conduction of heat in regions bounded by planes and cylinders, *Bull. Amer. Math. Soc.*, **47** (1941), 734–741, MR 3, 247.

92. Jaeger, J. C., Heat flow in the region bounded internally by a circular cylinder, *Proc. Roy. Soc. Edinburgh*, *Sect.* A, **61** (1942), 223–228, MR 4, 144.

93. Jaeger, J. C., Conduction of heat in a slab in contact with well-stirred fluid, *Proc. Cambridge Philos. Soc.* **41** (1945), 43–49, MR 6, 230.

94. Jaeger, J. C., Diffusion in turbulent flow between parallel planes, *Quart. Appl. Math.* **3** (1945), 210–217, MR 7, 162.

95. Jaeger, J. C., On thermal stresses in circular cylinders, *Philos. Mag.* (7) **36** (1945), 418–428, MR 7, 384.

96. Jaeger, J. C., Some applications of the repeated integrals of the error function, *Quart. Appl. Math.* **4** (1946), 100–103, MR 8, 81.

97. Jaeger, J. C., Conduction of heat in composite slabs, *Quart. Appl. Math.* **8** (1950), 187–198, MR 12, 105.

98. Jaeger, J. C., Conduction of heat in a solid with a power law of heat transfer at its surface, *Proc. Cambridge Philos. Soc.* **46** (1950), 634–641, MR 12, 264.

99. Karimov, D. H., Sur les solutions périodiques des équations différentiells nonlinéaires du type parabolique, *C. R. (Dokl.) Acad. Sci. URSS (N.S.)* **25** (1939), 3–6, MR 1, 315.

100. Karimov. D. H., Sur les solutions périodiques des équations différentielles nonlinéaires du type parabolique, *C. R. (Dokl.) Acad. Sci. URSS (N.S.)* **54** (1946), 293–295, MR 8, 465.

101. Karimov. D. H., On periodic solutions of nonlinear differential equations of parabolic type, *Dokl. Akad. Nauk. SSSR (N.S.)*, **58** (1947), 969–972 (Russian), MR 9, 287.

102. Kataoka, H., Linear conduction of heat in a compound rod. II. *Trans. Soc. Mech. Engrs. Japan* **14**, No. 46, (1948), 168–170 (Japanese), MR 11, 522.

103. Kataoka, H., Linear conduction of heat in a compound rod. III. *Trans. Soc. Mech. Engrs. Japan* **15**, No. 47, Part 2 (1948), 29–35 (Japanese), MR 11, 522.

104. Kendell, D. G., A form of wave propagation associated with the equation of heat conduction. *Proc. Cambridge Philos. Soc.* **44** (1948), 591–594, MR 10, 196.

105. Kienast, A., Über einige Falle der Green'schen Funktion der Wärmeleitung. *Viertelschr. Naturforsch. Ges. Zurich* **85** (1940), 29–34, MR 2, 55.

106. Kostitzin, V. A., Sur l'équation de la chaleur dans le cas d'une sphère soumise aux conditions spéciales, *C. R. Acad. Sci. Paris* **213** (1941), 972–974, MR 5, 98.

107. Kostitzin, V. A., Sur l'équation de la chaleur dans le cas d'une sphère stratifiée avec des sources distribuées sur les surfaces de discontinuité, *C. R. Acad. Sci. Paris* **214** (1942), 461–464, MR 4, 144.

108. Krzyzanski, M., Sur la solution élémentaire de l'équation de la chaleur, *Atti. Accad. Naz. Lincei Rend. Cl. Sci. Fis. Mat. Nat.* (8) **8** (1950), 193–199, MR 12, 105.

109. Kuznetsov, E. S., Conditions for heat flows on the boundary surface of two media, radiating heat transfer being taken into account, *Bull. Acad. Sci. SSSR, Ser. Geograph. Geophys.* [*Izvestia Akad. Nauk. SSSR* 1942, (1942) 243–248, MR 4, 199.

110. Laasonen, P., Über eine Methode zur Lösung der Wärmeleitungsgleichung, *Acta Math.* **81** (1949), 309–317, MR 11, 252.

111. Landau, H. G., Heat conduction in a melting solid, *Quart. Appl. Math.* **8** (1950), 81–94, MR 11, 441.

112. Landau, H. G., A problem in radiobiology: Diffusion and recombination of ions, *Bull. Math. Biophys.* **12** (1950), 27–34, MR 11, 598.

113. Levi, E. E., Sull'equazione del calore, *Annali de Matematica, Ser. 3*, **14** (1908), 187–264.

114. Levy, P., Sur un problème de calcul des probabilités lié à celui du refroidissement d'une barre homogène, *Ann. Scuola Norm. Sup. Pisa*, **1** (1932), 283–296.

115. Lidjaev, S., Über die Darstellbarkeit der Lösung der Wärmeleitungsgleichung durch das Poissonsche Integral, *Bull. Acad. Sci. URSS. Ser. Math.* [*Izvestia Akad. Nauk. SSSR* **5** (1941), 263–268 (Russian, German summary), MR 3, 45.

116. Lidyaev, S. F., On the representation of a solution of the equation of heat conduction in the form of a Poisson integral, *Ucenye Zapiski Moskov. Gos. Univ.* **135** (1948), *Matematika, Tom.* II, 86–109 (Russian), MR 11, 440.

117. Lowan, A. N., On Green's functions in the theory of heat conduction in spherical coordinates, *Bull. Amer. Math. Soc.*, **45** (1939), 407–413, MR 1, 76.

118. Lowan, A. N., On the problem of heat conduction in thin plates, *J. Math. Phys. Mass. Inst. Tech.* **24** (1945), 22–29, MR 6, 230.

119. Lowan, A. N., Note on the problem of heat conduction in a semi-infinite hollow cylinder, *Quart. Appl. Math.* **2** (1945), 348–350, MR 6, 156.

120. Lowan, A. N., On the problem of heat conduction in a semi-infinite radiating wire, *Quart. Appl. Math.* **3** (1945), 84–87, MR 6, 230.

121. Luchak, G., and Langstroth, G. O., Applications of diffusion theory to evaporation from droplets and flat surfaces, *Canadian J. Research., Sect. A.* **28** (1950), 574–579, MR 12, 415.

122. Ludwig, K., Wärmeausgleichsvorgange in bestrahlten Platten, *Z. Angew. Math. Mech.* **23** (1943), 259–269, MR 6, 178.

123. Ludwig, K., Das Aufheizen einer Wand durch konstante Wärmestromdichten, *Z. Angew. Math. Mech.* **23** (1943), 358–360, MR 6, 178.

124. Ludwig, K., Das Aufheizen einer Wand durch eine anlaufende Heizanlage, *Ing.-Arch.* **16** (1947), 45–50, MR 10, 458.

125. Lyubov, B. M., and Finkelstein, B. N., Designing of unstationary temperature fields in bodies of the simplest shape. I. *Akad. Nauk. SSSR. Zhurnal Eksper. Teoret. Fiz.* **13** (1943) (Russian), 35–41, MR 5, 69.

126. Lyubov, G. J., Designing of unstationary temperature fields in bodies of the simplest shape. II. *Akad. Nauk. SSSR. Zhurnal Eksper. Teoret. Fiz.* **13** (1943), 42–49, MR 5, 122.

127. Lyubov, B. Y., Solution of a nonstationary distribution problem of heat conduction for a region with uniform transmission at the boundary, *Dokl. Akad. Nauk. SSSR (N.S.)* **57** (1947), 551–554 (Russian), MR 9, 239.

128. Manarini, M., Sull'equazione del calore, *Boll. Un. Mat. Ital. (3)* **4** (1949), 117–121, MR 11, 181.

129. Matsumoto, T., Sur le principe de Duhamel–Nomitsu, *Mem. Coll. Sci. Kyoto Imp. Univ., Ser. A*, **22** (1939), 381–391, MR 7, 303.

130. McLachlan, N. W., Heat conduction in elliptical cylinder and an analogous

electromagnetic problem, *Philos. Mag.* (7) **36** (1945), 600–609, MR 8, 81.

131. Mersman, W. A., Heat conduction in an infinite composite solid, *Bull. Amer. Math. Soc.* **47** (1941), 956–964, MR 3, 128.

132. Mersman, W. A., Heat conduction in a semi-infinite slab, *Philos. Mag.* (7) **33** (1942), 303–309, MR 4, 46.

133. Mersman, W. A., Heat conduction in an infinite composite solid with an interface resistance, *Trans. Amer. Math. Soc.* **53** (1943), 14–24, MR 4, 160.

134. Miles, J. W., A note on Riemann's method applied to the diffusion equation, *Quart. Appl. Math.* **8** (1950), 95–101, MR 11, 521.

135. Minakshisundaram, S., Studies in Fourier Ansatz and parabolic equations, *J. Indian Math. Soc.* (*N.S.*) **6** (1942), 41–50, MR 6, 68.

136. Minakshisundaram, S., Studies in Fourier Ansatz and parabolic equations, *J. Madras Univ.*, **14** (1942), 73–142, MR 6, 68.

137. Minakshisundaram, S., Fourier ansatz and nonlinear parabolic equations, *J. Indian Math. Soc.* (*N.S.*) **7** (1943), 129–142, MR 6, 4.

138. Naef, R. A., Wärmeleitung im Zylinder, *Schweiz. Arch. Angew. Wiss. Tech.* **14** (1948), 156–157, MR 9, 591.

139. Nordon, J., Sur une solution nouvelle de l'équation de Fourier, *G. R. Acad. Sci. Paris* **228** (1949), 167–168, MR 10, 378.

140. Nordon, J., Une solution nouvelle de l'équation de la chaleur à $(n+1)$ variables, *C. R. Acad. Sci. Paris* **228** (1949), 459–460, MR 10, 458.

141. Obrechkoff, N., Sur un problème limité rélativement de l'équation de la chaleur, *Annuaire* [*Godisnik*] *Univ. Sofia. Fac. Phys.-Math.*, *Livre 1*, **38** (1942), 303–318 (Bulgarian, French summary), MR 12, 264.

142. Offner, F., Weinberg, A. and Young, G., Nerve conduction theory: some mathematical consequences of Bernstein's model, *Bull. Math. Biophys.* **2** (1940), 89–103, MR 1, 351.

143. Olevsky, M., Solution du problème de Cauchy et de certains problèmes limités pour l'équation des ondes, l'équation de la chaleur, et l'équation de Laplace dans les espaces à courbure constante, *C. R.* (*Dokl.*) *Acad. Sci. URSS* (*N.S.*) **33** (1941), 282–287, MR 5, 98.

144. Ornstein, L. S., and Milatz, J. M. W., Accidental deviations in the conduction of heat, *Physica* **6** (1939), 1139–1145, MR 1, 315.

145. Osida, I., On the application of graphical methods to the problems of heat conduction (A study on the graphical integration of the partial-differential equations of diffusion type), *J. Phys. Soc. Japan* **3** (1948), 223–232, MR 13, 693.

146. Panov, D. Y., *Spravocnik po cislennomu reseniyu differencial'nyh uravonenii v castnyh proizvodnyh* [*Handbook on the numerical solution of partial-differential equations*], 4th ed. Gosudarstv. Izdat. Tehn.-Teor. Lit. Moscow-Leningrad, 1950, 183 pp., MR 14, 93.

147. Patterson, S., The conduction of heat in a medium generating heat, *Philos. Mag.* (7), **32** (1941), 384–392, MR 3, 248.

148. Patterson, S., On certain integrals in the theory of heat conduction, *Quart. Appl. Math.* **4** (1946), 305–306, MR 8, 209.

149. Patterson, S., The heating or cooling of a solid sphere in a well-stirred fluid, *Proc. Phys. Soc.* **59** (1947), 50–58, MR 8, 585.

150. Parodi, H., Nouvelle solution du problème du mur plan indéfini, soumis, sur

ses deux faces, à des variations périodiques de température, *C. R. Acad. Sci. Paris* **223** (1946), 472–474, MR 8, 274.

151. Parodi, H., Sur le problème du refroidissement de la sphère, *C. R. Acad. Sci. Paris* **223** (1946), 540–542, MR 8, 274.

152. Perron, O., Eine neue Behändlung der Randwertaufgabe für $\Delta u = 0$, *Math. Zeit.*, **18** (1923), 42–54.

153. Petrovsky, I. G., Zur ersten Randwertaufgabe der Wärmeleitungsgleichung, *Composito Math.*, **1** (1935), 383–419.

154. Picone, M., Sul problema della propagazione del calore in un mezzo privo di frontiera, conduttore, isotropo e omogeneo, *Math. Ann.*, **101** (1929).

155. Pollard, H., One-sided boundedness as a condition for the unique solution of certain heat equations, *Duke Math. J.* **11** (1944), 651–653, MR 6, 87.

156. Potocek, J., La diffusion et la notion de reversibilité de M. Kolmogoroff, *Bull. Int. Acad. Sci. Bohème*, 1939 (1939), 1–10, MR 1, 247.

157. Prinetti, T., Sulle sorgenti mobili di calore, *Atti Accad. Sci. Torino. Cl. Sci. Fis. Mat. Nat.* **81–82** (1948), 182–188, MR 10, 123.

158. Pugh, H. L. D., and Harris, A. J., The temperature distribution around a spherical hole in an infinite conducting medium, *Philos. Mag.* (7) **33** (1942), 661–666, MR 4, 46.

159. Rocard, Y., and Veron, M., Sur la cousection vise d'un fluide s'écoulant en régime laminaire le long d'une plaque, *C. R. Acad. Sci. Paris*, **214** (1942), 301–304, MR 4, 176.

160. da Rocha, Miguel M., On the integration of Fourier's equation, *Anais. Acad. Brasil. Ci.* (1944), 53–56 (Portuguese), MR 6, 3.

161. Rothe, E., Zweidimensionale parabolische Randwertaufgaben als Grenzfall eindimensionaler Randwertaufgaben, *Math. Ann.*, **102** (1930), 650–670.

162. Rothe, E., Über die Grundlosung bei parabolischen Gleichungen, *Math. Zeit.*, **33** (1931), 488–504.

163. Rothe, E., Wärmeleitungsgleichung mit nichtkonstanten Koeffizienten, *Math. Ann.*, **104** (1931), 340–362.

164. Rott, N., Die Aufwarmung unbegrenzter Korper durch eine stetigwirkende Warmequelle, *Schweiz. Arch. Angew. Wiss. Tech.* **11** (1945), 164–174, 212–216, MR 7, 206.

165. Rubinstein, L. I., On the question of the process of propagation of freezing in frozen soil, *Izvestiya Akad. Nauk. SSSR. Ser. Geograf. Geofiz.* **11** (1947), 489–496 (Russian), MR 9, 287.

166. Rubinstein, L. I., On the determination of the position of the boundary which separates two phases in the one-dimensional problem of Stephan. *Dokl. Akad. Nauk. SSSR* (*N.S.*) **58** (1947), 217–220 (Russian), MR 9, 287.

167. Rubinstein, L. I., On the solution of Stefan's problem, *Bull. Acad. Sci. URSS Ser. Geograph. Geophs.* [*Izvestia Akad. Nauk. SSSR*], **11** (1947), 37–54 (Russian, English summary), MR 8, 516.

168. Rubinstein, L. I., Concerning the existence of a solution of Stefan's problem, *Dokl. Akad. Nauk. SSSR* (*N.S.*) **63** (1948), 195–198 (Russian), MR 10, 196.

169. Rubinstein, L. I., On the initial velocity of the front of crystallization in the one-dimensional problem of Stefan, *Dokl. Akad. Nauk. SSSR* (*N.S.*) **62** (1948), 753–756 (Russian), MR 10, 245.

170. Rubinstein, L. I., On the stability of the boundary of the phases in a

two-phase heat-conducting medium, *Izvestiya Akad. Nauk. SSSR. Ser. Geograf. Geofiz.* **12** (1948), 557–560 (Russian), MR 10, 458.

171. Sakadi, Z., On the thermal stress in an elastic solid body, *Proc. Phys.-Math. Soc. Japan* (3) **25** (1943), 673–685, *MR* 7, 383.

172. Samarskii, A. A., Concerning a problem of the transfer of heat. II. *Vestnik Moskov. Univ.* 1947, No. 6 (1947), 119–129 (Russian), MR 10, 301.

173. Samarsky, A., Concerning a problem of the transfer of heat, *Vestnik Moskov. Univ.* 1947, No. 3 (1947), 85–101 (Russian, English summary), MR 10, 301.

174. Satoh, T., On the mathematical analysis of the problem of the conduction of heat when emissivity is variable, *J. Phys. Soc. Japan* **5** (1950), 253–254, MR 12, 264.

175. Segal, B. I., On a problem of heat conduction, *Trav. Inst. Math. Stekloff* **20** (1947), 65–76 (Russian), MR 9, 472.

176. Sestini, G., Sopra un problema di propagazione del calore, *Ist. Lombardo Sci. Lett. Cl. Sci. Mat. Nat. Rend.* (3) **6** (75) (1942), 47–65, MR 8, 274.

177. Sestini, G., Sopra un problema si limiti in un caso non stazionario di propagazione del calore, *Univ. Roma. Ist. Naz. Alta Mat. Rend. Mat. e Appl.* (5) **6** (1947), 464–477, MR 9, 439.

178. Sestini, G., Sopra la conduzione del calore in una piastra sottile limitata da due circonferenze concentriche, *Atti. Sem. Mat. Fis. Univ. Modena* **3** (1949), 125–137, MR 11, 362.

179. Sestini, G., Su due problemi di propagazione del calore in un solido eterogeneo con simmetria cilindrica, *Rivista Mat. Univ. Parma* **1** (1950), 405–417, MR 12, 710.

180. Shvez, M., On the problem of unstationary distribution of air temperature near the underlying surface, *C. R. (Dokl.) Acad. Sci. URSS (N.S.)* **40** (1943), 144–147, MR 6, 68.

181. Soloviev, P. V., Fonctions de Green des équations paraboliques, *C. R. (Dokl.) Acad. Sci. URSS (N.S.)* **24** (1939), 107–109, MR 2, 55.

182. Sommerfeld, A., *Partielle Differentialgleichungen der Physik.* (Vorlesungen über theoretische Physik, Band VI.). Akademische Verlagsgesellschaft Geest & Portig K. G., Leipzig, 1947; xiii + 332 pp. Chapter III: Heat conduction, MR 10, 195.

183. Stefan, J., Über einige Probleme der Theorie der Wärmeleitung, *S.-B. Wien. Akad. Mat. Natur.*, **98** (1889), 173–484.

184. Stefan, J., Über die Diffusion von Sauren und Basen qeqen einander, *S.-B. Wien. Akad. Mat. Natur.* **98** (1889), 616–634.

185. Stefan, J., Über die Theorie der Eisbildung insbesondere über die Eisbildung im Polarmeere, *S.-B. Wien. Akad. Mat. Natur.*, **98** (1889), 965–983.

186. Stefan, J., Über die Verdampfung und die Auflosung als Vorgange der Diffusion, *S.-B. Wien. Akad. Mat. Natur.*, **98** (1889), 1418–1442.

187. Sternberg, W., Über die Gleichung der Wärmeleitung, *Math. Annalen*, **101** (1929), 394–398.

188. Tacklind, S., Sur les classes quasianalytiques des solutions des équations aux derivées partielles du type parabolique, *Nova Acta Regiae Soc. Sci. Upsal.*, **10** (1936), 1–57.

189. Thomas, L. H., "Stability of Solution of Partial-Differential Equations," *Symp. on Theor. Comp. Flow*, 28 June 1949. Naval Ord. Lab. White Oak,

Md., Rep. NOLR-1132, (1950) 83–94, MR 12, 361.

190. Thomas, L. H., Numerical solution of partial differential equations of parabolic type, *Proc. Semin. of Sci. Comp.*, Nov. 1949, 71–78, IBM Corp., New York, N.Y. 1950

191. Tihonov, A. N., Théorèmes d'unicité pour l'équation de la chaleur, *Mate. Sb.*, **42** (1935), 199–216.

192. Tihonov, A. N., On the heat equation in several variables, *Vestnik Moskov. Gos. Univ., Ser. A*, **1** (1937), No. 9, 1–49 (Russian).

193. Tihonov, A., On boundary conditions containing derivatives of order higher than the order of the equation, *Mat. Sbornik N.S.* **26** (68) (1950), 35–56 (Russian), MR 11, 440.

194. Tihonov, A. N., On the third boundary problem for an equation of parabolic type, *Izvestiya Akad. Nauk. SSSR Ser. Geograf. Geofiz.* **14** (1950), 193–198 (Russian), MR 12, 29.

195. de Toledo Piza, A. P., Determination of solutions of Fourier's partial-differential equation by a method called refinement of solutions, *Anais. Acad. Brasil. Ci.*, **15** (1943), 325–342 (Portuguese), MR 5, 145.

196. de Toledo Piza, A. P., Solution of Fourier's partial-differential equation, *Anais. Acad. Brasil. Ci* **16** (1944), 47–51 (Portuguese), MR 6, 3.

197. Tranter, C. J., Note on a problem in the conduction of heat, *Philos. Mag.* **28** (1939), 579–583, MR 1, 181.

198. Tranter, C. J., Note on a problem in the conduction of heat, *Philos. Mag.* (7) **31** (1941), 432, MR 2, 365.

199. Tranter, C. J., On a problem in heat conduction, *Philos. Mag* (7) **35** (1944), 102–105, MR 6, 3.

200. Tranter, C. J., Heat flow in an infinite medium heated by a cylinder, *Philos. Mag.* (7) **38** (1947), 131–134, MR 9, 147.

201. Tranter, C. J., Note on a problem in heat conduction, *Philos. Mag.* (7) **38** (1947), 530–531, MR 9, 239.

202. Traupel, W., Unsteady heat conduction in plates, cylinders and spheres, *Sulzer. Tech. Rev.* (1949) No. 3, 12–23, MR 11, 362.

203. Vecoua, N., Über den Greuzubergang von den stationaren in Randproblemen für Wärmeleitung, *Mitt. Georg. Abt. Akad. Wiss. USSR* [*Soobscenia Gruzinsbogo Filiala Akad. Nauk. SSSR*] **1** (1940), 651–657 (Russian), MR 3, 45.

204. Vernotte, P., Thermocinétique, *Publ. Sci. Tech. Ministère de l'Air*, Paris, No. 22 (1949); xxi + 459 pp., MR 11, 521.

205. Vodicka, V. Le Mouvement de la chaleur dans une matière liquide in fusée dans un vase sphérique, *Vestnik Kralovske Ceske Spolecnosti Nauk. Trida Matemat. Prirodoved* 1946 (1947) No. 11, 8 pp., MR 9, 94.

206. Vodicka, V. Der Kreiszylinder in einen zeitlich veränderlichen Temperaturfelde, *Schweiz. Arch. Angew. Wiss. Tech.*, **14** (1948), 177–180, MR 10, 123.

207. Volterra, V., Sur les équations différentielles du type parabolique, *C. R. Acad. Sci. Paris*, **139** (1904), 956–959.

208. Volterra, V., Leçons sur les équations intégrales et des équations integrodifférentielles, Chapter II. Gauthier Villars, Paris, 1913.

209. Wagner, C., Wärmeleitungs probleme für Systeme mit beheizten Rohren und Hohlkugeln in einer unendlich ausgedehnten Umgebung, *Ing.-Arch.* **14** (1944),

398–409, MR 7, 162.

210. Wallace, P. R., and LeCaine, J., "Elementary Approximations in the Theory of Neutron Diffusion." National Research Council of Canada, Division of Atomic Energy, Document No. 1480; $i + 172$ pp. (47 plates) (1946), MR 9, 590.

211. Walters, A. G., A problem on the conduction of heat, *Philos. Mag.* (7) **38** (1947), 70–78, MR 9, 147.

212. Ward, A. G., The diffusion of decomposition products through plastic materials, *Philos. Mag.* (7) **39** (1948), 621–632, MR 10, 123.

213. Weber, B. R.-H., Die partiellen Differentialgleichungen der mathematischen Physik, Vol. 2 (1901), 111–118. Vieweg, Braunschweig, Germany, 1912.

214. Westphal, H., Zur Abschatzung der Losungen nichtlinearer parabolischen Differentialgleichungen, *Math. Z.* **51** (1949), 690–695, MR 11, 252.

215. Whitehead, S., An approximate method for calculating heat flow in an infinite medium heated by a cylinder, *Proc. Phys. Soc.* **56** (1944), 357–366, MR 6, 177.

216. Widder, D. V., Singular points of functions which satisfy the partial-differential equation of the flow of heat, *Bull. Amer. Math. Soc.*, **36** (1930), 687–694.

217. Widder, D. V., Positive temperatures on an infinite rod, *Trans. Amer. Math. Soc.* **55** (1944), 85–95, MR 5, 203.

218. Wilson, A. H., A diffusion problem in which the amount of diffusing substance is finite, *I.*, *Philos. Mag.* (7) **39** (1948), 48–58, MR 9, 439.

219. Yuskov, P. P., On the application of triangular nets to the numerical integration of the equation of heat conduction, *Akad. Nauk. SSSR. Prikl, Mat. Meh.* **12** (1948), 223–226 (Russian), MR 9, 624.

220. Zanobetti, D., Sulla distribuzione transitoria disuniforme di una corrente continua in una lamiera e in un cilindro, e sulla resistenza di rotaia, *Atti Accad. Naz. Lincei. Rend. Cl. Sci. Fis. Mat. Nat.* (8) **8** (1950), 129–134, MR 12, 105.

2.   SEMI-CLASSICAL: 1950–1973

2.1.   Properties of Solutions

A.   *The Maximum Principle, a priori estimates, parabolic inequalities, and nonexistence of solutions*

1. Adler, G., Sulla caratterizzabilità dell'equazione del calore dal punto di vista del calcolo delle variazioni, *Magyar Tud. Akad. Mat. Kutato Int. Kozl.*, **2** (1957), 3/4, 153–157 (Hungarian and Russian summaries), MR 21, 1445.

2. Adler, G., Maggiorazione del gradiente delle funzioni del calore, *Atti Accad. Naz. Lincei Rend. Cl. Sci. Fis. Mat. Nat.* (8) **30** (1961), 357–361, MR 26, 485.

3. Adler, G., Maggiorazione del gradiente delle soluzioni delle equazioni $\Delta u = f$ e $\Delta u - au_t = g$, *Atti. Accad. Naz. Lincei Rend. Cl. Sci. Fis. Mat. Nat.* (8) **30** (1961), 673–676, MR 24, A2740.

4. Arnese, G., Maggiorazioni in L^p dei potenziali relativi all' equazione del calore, *Ricerche Mat.* **13** (1964), 147–191, MR 33, 6112.

5. Bobisud, L., Speed of energy propagation for parabolic equations, *Illinois J. Math.* **11** (1967), 111–113, MR 34, 7971.

6. Boley, B., Upper bounds and Saint-Vernant's principle in transient heat conduction, *Quart. Appl. Math.* **18** (1960/61), 205–207, MR 22, 3442.

7. Brandt, A., Interior Schauder estimates for parabolic differential (or difference) equations via the maximum principle, *Israel J. Math.* **7** (1969), 254–262, MR 40, 3044.

8. Cattabriga, L., Un confronto fra operatori generalizzanti l'operatore $(\partial^2/\partial x^2)$ $+(\partial/\partial y)$, *Richerche Mat.* **7** (1958), 64–70, MR 21, 767.

9. Ciesielski, Z., Heat conduction and the principle of not feeling the boundary, *Bull. Acad. Polon. Sci. Ser. Sci. Math. Astronom. Phys.* **14** (1966), 435–440 (Russian summary), MR 34, 6348.

10. Ciliberto, C., Formule di maggiorazione e teoremi di esistenzà per le soluzioni delle equazioni paraboliche in due variabili, *Ricerche Mat.* **3** (1954), 40–75, MR 16, 139.

11. Edelstein, W. S., A spatial-decay estimate for the heat equation, *Z. Angew. Math. Phys.* **20** (1969), 900–905 (German summary), MR 41, 4024.

12. Franklin, J., and Keller, H. B., *A priori* bounds for temperature in circulating-fuel reactors, *Quart. Appl. Math.* **14** (1956), 57–62, MR 17, 975.

13. Friedman, A., Remarks on maximum principle for parabolic equations and its applications, *Pacific J. Math.* **8** (1958), 201–211, MR 21, 1444.

14. Friedman, A., A strong maximum principle for weakly subparabolic functions, *Pacific J. Math.* **11** (1961), 175–184, MR 23, A427.

15. Friedman, A., Liouvilles theorem for parabolic equations of the second order with constant coefficients, *Proc. Amer. Math. Soc.* **9** (1958), 272–277, MR 20, 1843.

16. Fulks, W., A mean-value theorem for the heat equation, *Proc. Amer. Math. Soc.* **17** (1966), 6–11, MR 33, 427.

17. Gagliardo, E., Formule di maggiorazione integrale per le soluzioni dell'equazione del calore non omogenea, *Ricerche Mat.* **3** (1954), 202–219, MR 16, 1028.

18. Godunova, E. K., and Levin, V. I., Certain qualitative questions of heat conduction, *Z. Vycisl. Mat. i Mat. Fiz.* **6** (1966), 1097–1103 (Russian), MR 34, 3134.

19. Gross, W., Formula di Taylor e linee di livello delle soluzioni dell'equazione del calore, *Atti. Accad. Naz. Lincei Rend Cl. Sci. Fis. Mat. Nat.* (8) **32** (1962), 477–483.

20. Guscin, A. K., Certain estimates of the solutions of boundary-value problems for the heat equation in an unbounded domain, *Trudy Mat. Inst., Steklov.* **91** (1967), 5–18 (Russian), MR 36, 5491.

21. Hadamard, J., Équations du type parabolique dépourvues de solutions, *J. Rational Mech. Anal.* **3** (1954), 3–12, MR 15, 627.

22. Hadamard, J., Extension à l'équation de la chaleur d'un théorem de A. Harnack, *Rend. Circ. Mat. Palermo* (2) **3** (1954), 337–346 (1955), MR 16, 930.

23. Hill, C. D., A sharp maximum principle for degenerate elliptic–parabolic equations, *Indiana Univ. Math. J.* **20** (1970–71), 213–229, MR 44, 4382.

24. Mlak, W., On a linear differential inequality of parabolic type, *Bull. Acad.*

Polon. Sci. Ser. Sci. Math. Astr. Phys. **7** (1959), 653–656, MR 22, 2800.

25. Moreau, J. J., Justification statistique de la loi de la diffusion. Actes du colloque sur la diffusion, Montpellier, 1955, pp. 9–15. *Publ. Sci. Tech. Ministere de l'Air*, Notes Tech. No. 59, Paris 1956, MR 17, 1166.

26. Nicolescu, M., Sur l'équation de la chaleur, *Com. Acad. R. P. Romane* **1** (1951), 747–751 (Romanian, Russian, and French summaries), MR 17, 43.

27. Nicolescu, M., Une proprieté caracteristique de moyennes des solutions régulières de l'équation de la chaleur, *Com. Acad. R. P. Romane* **2** (1952), 677–679 (Romanian, Russian, and French summaries), MR 17, 43.

28. Nirenberg, L., A strong maximum principle for parabolic equations, *Comm. Pure Appl. Math.* **6** (1953), 167–177, MR 14, 1089.

29. Osipenko, P. A., "The continuous dependence of the solution of the first boundary-value problem for the heat equation on the initial data of the problem" (Russian), in *Certain Problems of Applied Mathematics*, No. 3 (Russian), pp. 145–152. Izdat. "Naukova Dumka," Kiev, 1967, MR 39, 3166.

30. Primicerio, M., Alcune osservazioni sull'equazione della conduzione del calore, *Rend. Ist. Mat. Univ. Trieste* **2** (1970), 34–41, MR 42, 4091.

31. Protter, M. H., and Weinberger, H. F., *Maximum Principles in Differential Equations*. Prentice-Hall, Inc., Englewood Cliffs, N.J., 1967; x + 261 pp., MR 36, 2935.

32. Pucci, C., and Weinstein, A., Sull'equazione del calore con dati subarmonici e sue generalizzazioni, *Atti. Accad. Naz. Lincei. Rend. Cl. Sci. Fis. Mat. Nat.* (8) **24** (1958), 493–496, MR 20, 5956.

33. Resch, D., Temperature bounds on the infinite rod, *Proc. Amer. Math. Soc.* **3** (1952), 632–634, MR 14, 173.

34. Trosciev, V. E., On the width of the zone of influence for the heat equation, *Z. Vycisl. Mat. i Mat. Fiz.* **5** (1965), 1135–1138 (Russian), MR 33, 1981.

35. Turan, P., A remark on the heat equation, *J. Analyse Math.* **14** (1965), 443–448, MR 31, 1476.

B.   Uniqueness and representation theorems

1. Ahiezer, N. I., The uniqueness theorem for the heat equation, *Har'kov. Politehn. Inst. Trudy. Ser. Ins.-Fiz.* **5** (1955), 51–55 (Russian), MR 20, 5340.

2. Birjukova, F. G., A uniqueness theorem for the first boundary-value problem of the heat equation with discontinuous coefficient, *Isv. Akad. Nauk. Kazah. SSR Ser. Fiz.-Mat.*, 1970, No. 3, 38–44 (Russian), MR 44, 5627.

3. Blackman, J., The inversion of solutions of the heat equation for the infinite rod, *Duke Math. J.* **19** (1952), 671–682, MR 14, 475.

4. Czipszer, J., Sur la propagation de la chaleur dans une barre infinie. I., *Magyar Tud. Akad. Alkalm. Mat. Int. Kozl.* **3** (1954), 395–408 (1955) (Hungarian, Russian, and French Summaries), MR 17, 858.

5. Ficken, F. A., Uniqueness theorems for certain parabolic problems, *J. Rational Mech. Anal.* **1** (1952), 573–578, MR 14, 282.

6. Fulks, W., On the unique determination of solutions of the heat equation, *Pacific J. Math.* **3** (1953), 387–391, MR 14, 1090.

7. Gehring, F. W., On solutions of the equation of heat conduction, *Michigan Math. J.* **5** (1958), 191–202, MR 21, 5077.

8. Gehring, F. W., The boundary behavior and uniqueness of solutions of the heat equation, *Trans. Amer. Math. Soc.* **94** (1960), 337–364, MR 22, 2790.

9. Hille, E., Quelques remarques sur l'équation de la chaleur, *Rend. Mat. e Appl.* (5) **15** (1956), 102–118, MR 21, 4297.

10. Hirschman, I. I., Jr., A note on the heat equation, *Duke Math. J.* **19** (1952), 487–492, MR 14, 172.

11. Kampe de Feriet, J., Sur une classe de solutions de l'équation de la chaleur, *C. R. Acad. Sci. Paris*, **234** (1952), 2139–2140, MR 14, 173.

12. Kampe de Feriet, J., Un théorème d'unicité pour les intégrales de l'équation de la chaleur appartenant à la classe *L*., *C. R. Acad. Sci. Paris*, **236** (1953), 1527–1529, MR 14, 878.

13. Kamynin, L. I., The difference in uniqueness theorems for the heat-conduction equation and for systems of difference–differential equations, *Dokl. Akad. Nauk. SSSR* (*N.S.*) **82** (1952), 13–16 (Russian), MR 14, 172.

14. Kato, M., On positive solutions of the heat equation, *Nagoya Math. J.* **30** (1967), 203–207, MR 35, 3293.

15. Krzyzanski, M., Sur la solution élémentaire de l'équation de la chaleur—Note complementaire, *Atti. Accad. Naz. Lincei. Rend. Cl. Sci. Fis. Mat. Nat.* (8) **13** (1952), 24–25, MR 14, 651.

16. Krzyzanski, M., Sur le problème de Fourier dans une région indéfinite, *Arch. Mech. Stos.* **5** (1953), 584–588 (1954) (Polish, Russian, and French Summaries), MR 16, 370.

17. Mustata, P., Sur la réprésentation de Poisson des solutions du problème de Cauchy pour l'équation de la chaleur, *Rev. Romaine Math. Pures Appl.* **11** (1966), 673–691, MR 34, 1725.

18. Nicolescu, M., and Foias, C., Réprésentation de Poisson et problème de Cauchy pour l'équation de la chaleur, *Atti. Accad. Naz. Lincei. Rend. Cl. Sci. Fis. Mat. Nat.* (8) **38** (1965), 466–476, MR 34, 4715.

19. Nicolescu, M., and Foias, C., Réprésentation de Poisson et problème de Cauchy pour l'équation de la chaleur. II., *Atti. Accad. Naz. Lincei Rend. Cl. Sci. Fis. Mat. Nat.* (8) **38** (1965), 621–626, MR 34, 4716.

20. Nicolescu, M., and Foias, C., Sur l'unicité du problème de Cauchy pour l'équation de la chaleur, *Atti. Accad. Naz. Lincei Rend. Cl. Sci. Fis. Mat. Nat.* (8) **40** (1966), 785–791 (Italian summary), MR 34, 4717.

21. Pagni, M., Un teorema di unicità relativo alla equazione della propagazione de calore, *Atti. Sem. Mat. Fis. Univ. Modena* **9** (1959/60), 24–31, MR 23, B239.

22. Rosenbloom, P. C., and Widder, D. V., A temperature function which vanishes initially, *Amer. Math. Monthly* **65** (1958), 607–609, MR 20, 7146.

23. Shapiro, V. L., The uniqueness of solutions of the heat equation in an infinite strip, *Trans. Amer. Math. Soc.* **125** (1966), 326–361, MR 34, 1727.

24. Spagnolo, S., Sul limite delle soluzioni di problemi di Cauchy relativ all'equazione del calore, *Ann. Scuola Norm. Sup. Pisa* (3) **21** (1967), 657–699, MR 37, 612.

25. Widder, D. V., Positive temperatures on a semi-infinite rod, *Trans. Amer. Math. Soc.* **75** (1953), 510–525, MR 15, 332.

26. Widder, D. V., Positive solutions of the heat equation, *Bull. Amer. Math. Soc.* **69** (1963), 111–112, MR 26, 1030.

27. Zolotarev, G. N., On uniqueness of solution of Cauchy's problem for the heat-conduction equation, *Dokl. Akad. Nauk. SSSR (N.S.)* **104** (1955), 349–351 (Russian), MR 17, 858.

C. Transforms, polynomials, and analyticity of solutions

1. Anandani, P., Use of generalized Legendre-associated functions and the H-function in heat production in a cylinder, *Kyungpook Math. J.* **10** (1970), 107–113, MR 42, 7954.
2. Bajpai, S. D., Associated Legendre functions and heat production in a cylinder, *Proc. Nat. Inst. Sci. India, Part* A, **35** (1969), 366–374, MR 41, 655.
3. Bajpai, S. D., The use of Bessel function and Jacobi polynomial in the cooling of a heated cylinder, *Proc. Nat. Acad. Sci. India, Sect.* A, **39** (1969), 320–322, MR 43, 7792.
4. Bajpai, S. D., Gauss' hypergeometric series and heat production in a cylinder, *Portugal. Math.* **29** (1970), 71–80, MR 43, 6606.
5. Bhonsle, B. R., Jacobi polynomials and heat production in a cylinder, *Math. Japan* **11** (1966), 83–90, MR 34, 6347.
6. Bhonsle, B. R., Heat conduction and Hermite polynomials, *Proc. Nat. Acad. Sci. India, Sect.* A, **36** (1966), 359–360, MR 40, 3080.
7. Bragg, L. R., The radial heat polynomials and related functions, *Trans. Amer. Math. Soc.* **119** (1965), 270–290, MR 31, 5996.
8. Cholewinski, F. M., and Haimo, D. T., Classical analysis and the generalized heat equation, *SIAM Rev.* **10** (1968), 67–80, MR 37, 4416.
9. Conte, S. D., On some nonlinear partial-differential equations, *Indust. Math.* **4** (1953), 17–21, MR 17, 495.
10. Dettman, J. W., The wave, Laplace, and heat equations and related transforms, *Glasgow Math. J.* **11** (1970), 117–125, MR 42, 8188.
11. Eskin, L. D., The heat equation and the Weierstrass transformation in certain symmetric Riemannian spaces, *Izv. Vyss. Ucebn. Zaved. Matematika*, No. 5 (48) (1965), 151–166 (Russian), MR 34, 3139.
12. Greiner, P., An asymptotic expansion for the heat equation, in *Global Analysis* (*Proc. Sympos. Pure Math.*, **XVI**, Berkeley, CA (1968), pp. 133–135). Amer. Math. Soc., Providence, R.I., 1970, MR 42, 693.
13. Haimo, D. T., Expansions in terms of generalized heat polynomials and of their Appell transforms, *J. Math. Mech.* **15** (1966), 735–758, MR 33, 4340.
14. Haimo, D. T., and Cholewinski, F. M., Integral representations of solutions of the generalized heat equation, *Illinois J. Math* **10** (1966), 623–638, MR 33, 7796.
15. Haimo, D. T., Series representation of generalized temperature function, *SIAM J. Appl. Math.* **15** (1967), 359–367, MR 35, 3204.
16. Haimo, D. T., Series expansions and integral representations of generalized temperatures, *Illinois J. Math.* **14** (1970), 621–629, MR 42, 3441.
17. Haimo, D. T., and Cholewinski, F. M., Series representations for dual Laguerre temperatures, *Bull. Inst. Politehn., Iasi (N.S.)* **17** (21) (1971), Faci 1–2, Sect. I, 103–108, MR 43, 5137.
18. Haimo, D. T., An integral representation for generalized temperatures in two space variables, *Proc. Amer. Math. Soc.* **30** (1971), 533–538, MR 44, 652.
19. Kempe de Feriet, J., Équation de la chaleur et polynomes d'Hermite, *C. R. Acad. Sci. Paris* **248** (1959), 883–887, MR 20, 7145.

20. Kempe de Feriet, J., Heat equation and Hermite polynomials, *Calcutta Math. Soc. Golden Jubilee Commemoration Vol.* (1958-59), Part I, pp. 193-204. Calcutta Math. Soc., Calcutta, 1963, MR 27, 5051.

21. Paterson, S., On certain types of solution of the equation of heat conduction, *Proc. Glasgow Math. Assoc.* **1** (1952), 48-52, MR 14, 381.

22. Pollard, H., and Widder, D. V., Gaussian representations related to heat conduction, *Arch. Rational Mech. Anal.* **35** (1969), 253-258, MR 39, 7356.

23. Pollard, H., and Widder, D. V., Inversion of a convolution transform related to heat conduction, *SIAM J. Math. Anal.* **1** (1970), 527-532, MR 43, 824.

24. Rosculet, M. N., and Cocarlan, P., Contributions to the study of partial-differential equations of parabolic type. VIII. Heat polynomials and the transform of Paul Appell. *Stud. Cerc. Mat.* **19** (1967), 739-746 (Romanian), MR 40, 3075.

25. Rosenbloom, P. C., and Widder, D. V., Expansions in terms of heat polynomials and associated functions, *Trans. Amer. Math. Soc.* **92** (1959), 220-226, MR 21, 5845.

26. Shirota, T., A remark on the abstract analyticity in time for solutions of a parabolic equation, *Proc. Japan Acad.* **35** (1959), 367-369, MR 22, 3887.

27. Widder, D. V., The heat equation and the Weierstrass transform, *Proc. Conf. on Differential Equations* (dedicated to A. Weinstein), pp. 227-234. Univ. of Maryland Book Store, College Park, MD, 1956, MR 18, 304.

28. Widder, D. V., Una transformazione integrale connessa con la propagazione del calore, *Atti. Accad. Naz. Lincei. Rend. Cl. Sci. Fis. Mat. Nat.* (8) **20** (1956), 750-752, MR 19, 413.

29. Widder, D. V., Integral transforms related to heat conduction, *Ann. Mat. Pura Appl.* (4) **42** (1956), 279-305, MR 19, 413.

30. Widder, D. V., Series expansions of solutions of the heat equation in n dimensions, *Ann. Mat. Pura Appl.* (4) **55** (1961), 389-409, MR 25, 331.

31. Widder, D. V., Series expansions in terms of the temperature functions of Poritsky and Powell, *Quart. Appl. Math.* **20** (1962/63), 41-47, MR 25, 942.

32. Widder, D. V., Analytic solutions of the heat equation, *Duke Math. J.* **29** (1962), 497-503, MR 28, 364.

33. Widder, D. V., Sur une classe de fonctions $u(x, y, t)$ harmoniques en (x, y) et satisfaisant l'équation de la chaleur en (x, t), *C. R. Acad. Sci. Paris* **256** (1963), 2751-2753, MR 28, 1319.

34. Widder, D. V., Some analogies from classical analysis in the theory of heat conduction, *Arch. Rational Mech. Anal.* **21** (1965), 108-119, MR 32, 2031.

35. Widder, D. V., Expansions in series of homogeneous temperature functions of the first and second kinds, *Duke Math. J.* **36** (1969), 495-509, MR 40, 4704.

36. Widder, D. V., Expansions in terms of the homogeneous solutions of the heat equation, in *Orthogonal Expansions and Their Continuous Analogues* (*Proc. Conf.*, Edwardsville, Ill., 1967), pp. 171-196. Southern Illinois Univ. Press, Carbondale, Ill., 1968, MR 41, 2233.

D. Potentials and Green's functions

1. Arnese, G., Sui potenziali relativi all'equazione del calore, *Atti. Accad. Naz. Lincei Rend. Cl. Sci. Fis. Mat. Nat.* (8) **36** (1964), 604-608, MR 31, 1445.

2. Bokov, A., The membership of the solution of the equation of heat conduction in a certain class of functions, *Godisnik Viss. Tehn. Ucebn Zaved. Mat.* **2**

(1965), Kn. 3, 81–88 (1966) (Bulgarian, Russian, and German summaries), MR 38, 1411.

3.   Feldman, J., On the Schrödinger and heat equations for nonnegative potentials, *Trans. Amer. Math. Soc.* **108** (1963), 251–264, MR 28, 3478.

4.   Gendzojan, G. V., Certain estimates of the Green's function of the first boundary-value problem for the heat equation, *Izv. Akad. Nauk. Armjan. SSSR Ser. Mat.* **1** (1966), No. 4, 238–269, MR 34, 8000.

5.   Hartman, P. and Wintner, A., On the inverse of the parabolic differential operator $(\partial^2/\partial x^2) - (\partial/\partial t)$, *Amer. J. Math.* **75** (1953), 598–610, MR 15, 227.

6.   Hawlitschek, K., Green-Funktionen für Randwertaufgaben parabolischer Differentialgleichungen, *Math. Ann.* **177** (1968), 1–22, MR 37, 5540.

7.   Kamynin, L. I., and Maslennikova, V. N., Boundary estimates for the solution of the third boundary-value problem for a parabolic equation, *Dokl. Akad. Nauk. SSSR* **153** (1963), 526–529 (Russian), MR 29, 1447.

8.   Kamynin, L. I., Liapunov theorems for heat potentials, *Dokl. Akad. Nauk. SSSR* **160** (1965), 271–273 (Russian), MR 30, 3306.

9.   Kamynin, L. I., The maximum principal and boundary estimates of a solution of the first boundary-value problem for a parabolic equation in a noncylindrical region, *Z. Vycisl. Mat. i. Mat. Fiz.* **7** (1967), 551–567 (Russian), MR 36, 1843.

10.   Kamynin, L. I., On the smoothness of thermal potentials, *Differencial'nye Uravnenija* **1** (1965), 799–839 (Russian), MR 37, 4425.

11.   Kamynin, L. I., On the smoothness of thermal potentials, II. Thermal potentials on the surface of type $L^{1,\alpha,\alpha/2}_{1,1,(1+\alpha)/2}$ *Differencial'nye Uravnenija* **2** (1966), 647–687, MR 33, 4479.

12.   Kamynin, L. I., On the smoothness of thermal potentials. III. A special single-layer thermal potential $P(x, t)$ on surfaces of type $L^{0,1(1+\alpha)/2}_{1,\alpha,\alpha/2}$ and $L^{1,\alpha,\alpha/2}_{1,1,(1+\alpha)/2}$, *Differencial'nye Uravnenija* **2** (1966), 1333–1357 (Russian), MR 34, 3126.

13.   Kamynin, L. I., On the smoothness of thermal potentials. III., *Differencial'nye Uravnenija* **2** (1966), 1484–1501 (Russian), MR 34, 3127.

14.   Kamynin, L. I., On the smoothness of thermal potentials. IV., *Differencial'nye Uravenija* **3** (1967), 1303–1312 (Russian), MR 36, 2974b.

15.   Kamynin, L. I., On the smoothness of thermal potentials. V. Thermal potentials U, V, and W on surfaces of types $\Pi^{m+1,\alpha,\alpha/2}_{2m+1,1,(1+\alpha)/2}$ and $\Pi^{m+1,1,(1+\alpha)/2}_{2m+3,\alpha,\alpha/2}$ *Differencial'nye Uravnenija* **4** (1968), 881–895 (Russian), MR 37, 5537.

16.   Kamynin, L. I., The theory of thermal potentials and its applications, *Mat. Zametki* **4** (1968), 113–123 (Russian), MR 38, 1412.

17.   Kamynin, L. I., The smoothness of thermal potentials in a Dini–Holder space, *Sibirsk. Mat. Z.* **11** (1970), 1017–1045, 1196 (Russian), MR 42, 4895.

18.   Kamynin, L. I., On the Gevrey theory for parabolic potentials, I, II., *Differencial'nye Uravnenija* **7** (1971), 312–328; *ibid.* **7** (1971), 711–726, (Russian), MR 44, 1918.

19.   Kval'vasser, V. I., and Rutner, J. F., A method of finding Green's function in boundary-value problems of heat-conduction equation for a line segment with uniformly moving boundaries, *Dokl. Akad. Nauk. SSSR* **156** (1964), 1273–1276 (Russian); translated as *Soviet Mat. Dokl.* **5** (1964), 809–812, MR 32, 6071.

20.   Lions, J. L., and Magenes, E., Espaces de fonctions et distributions du type de

Gevrey et problèmes aux limites paraboliques, *Ann. Mat. Pura Appl.* (4) **68** (1965), 341–417, MR 33, 2965.

21. Rutner, J. F., The Green's functions of the first boundary-value problem for the heat conduction equation in central-symmetric regions with uniformly moving boundaries, *Volz. Mat. Sb. Vyp.* **5** (1966), 314–320 (Russian), MR 35, 4610.

22. Rutner, J. F., and Skrjabina, L. P., The Green's function for the heat-conduction equation in the half-plane with nonuniform boundary conditions, *Volz. Mat. Sb. Vyp.* **5** (1966), 321–325 (Russian), MR 35, 4611.

23. Shapiro, V. L., Characteristic planes and pointwise solutions of the heat equation, *Indiana Univ. Math. J.* **20** (1970/71), 115–133, MR 41, 8844.

24. Van Tun, Theory of the heat potential. II. Smoothness of contour heat potentials, *Z. Vycisl. Mat. i. Mat. Fiz.* **5** (1965), 474–487 (Russian), MR 33, 1600.

25. Van Tun [Wang Tang], A theory of the thermal potential, III. Smoothness of a plane thermal potential, *Z. Vycisl. Mat. i. Mat. Fiz.* **5** (1965), 658–666 (Russian), MR 34, 481.

26. Varadhan, S. R. S., On the behavior of the fundamental solution of the heat equation with variable coefficients, *Comm. Pure Appl. Math.* **20** (1967), 431–455, MR 34, 8001.

27. Wolska-Bochenek, J., Boundary value of the normal derivative of the potential of the double layer in the heat-conductivity equation, *Zeszyty Nauk. Politech. Warszawsk. Mat.*, No. 83 (1964) Zeszyt 1, 3–11 (Polish, Russian, and French summaries), MR 32, 6072.

28. Yamabe, H., Kernel functions of diffusion equations, I., *Osaka Math. J.* **9** (1957), 201–214, MR 21, 2813.

E. Behavior of solutions at the boundary, and regularity of boundary points

1. Babuska, I., and Vyborny, R., Regulare und stabile Randpunkte für das Problem der Wärmeleitungsgleichung, *Ann. Polon. Math.* **12** (1963), 91–104, MR 34, 6346.

2. Cannon, J., Regularity at the boundary for solutions of initial-value problems for linear parabolic differential equations, *Atti del Convegno su le Equazioni alle Derivate Parziali* (Nervi, 1965), p. 36. Edizioni Cremonese, Rome, 1966, MR 34, 1713.

3. Cannon, J. R., Regularity at the boundary for solutions of linear parabolic differential equations, *Ann. Scuola Norm. Sup. Pisa* (3) **19** (1965), 415–427, MR 35, 575.

4. Eidel'man, S. D., Behavior of solutions of the heat equation in the neighborhood of an isolated singular point, *Uspehi Mat. Nauk.* (*N.S.*) **11** (1956), No. 3 (69), 207–210 (Russian), MR 19, 557.

5. Fulks, W., On the boundary values of solutions of the heat equation, *Pacific J. Math.* **2** (1952), 141–145, MR 14, 51.

6. Fulks, W., Regular regions for the heat equation, *Pacific J. Math.* **7** (1957), 867–877, MR 19, 425.

7. Hattemer, J. R., Boundary behavior of temperatures. I. *Studia Math.* **25** (1964/65), 111–155, MR 31, 6064.

8. Hattemer, J. R., Boundary behavior of temperatures. II., *Illinois J. Math.* **10** (1966), 466–469, MR 36, 4153.

9. Hughes, R. B., Boundary behavior of random-valued heat polynomial expansions, *Pacific J. Math.* **31**, (1969), 61–72, MR 40, 3157.

10. Jones, B. F., Jr., Singular integrals and parabolic equations, *Bull. Amer. Math. Soc.* **69** (1963), 501–503, MR 26, 6599.

11. Jones, B. F., Jr., Singular integrals and a boundary-value problem for the heat equation, in *Singular Integrals* (*Proc. Sympos. Pure Math.* Chicago, Ill., 1966), pp. 196–207. Amer. Math. Soc., Providence, R.I., 1967, MR 38, 3741.

12. Jones, B. F., Jr., Characterization of spaces of Bessel potentials related to the heat equation, in *Pseudo-Diff. Operators* (*C.I.M.E.*, Stresa, 1968), pp. 143–155. Edizioni Cremonese, Rome, 1969, MR 41, 659.

13. Jones, B. F., Jr., and Tu, C. C., Nontangential limits for a solution of the heat equation in a two-dimensional Lip$_\alpha$ region. *Duke Math. J.* **37** (1970), 243–254, MR 41, 4026.

14. Kemper, J. T., Kernel functions and parabolic limits for the heat equation, *Bull. Amer. Math. Soc.* **76** (1970), 1319–1320, MR 41, 8842.

15. Landis, E. M., Necessary and sufficient conditions for the regularity of a boundary point for the Dirichlet problem for the heat equation, *Dokl. Akad. Nauk. SSSR* **185** (1969), 517–520 (Russian), MR 41, 7308; *Soviet Math. Dokl.* **10** (1969), 380–384.

16. Magenes, E., Sull'equazione del calore: teoremi di unicità e teoremi di completezzà connessi col metodo di integrazione di M. Picone. I., *Rend. Sem. Mat. Univ. Padova* **21** (1952), 99–123, MR 14, 282.

17. Magenes, E., Sull'equazione del calore: teoremi di unicità e teoremi di completezzà connessi col metodo di integrazione di M. Picone. II., *Rend. Sem. Mat. Univ. Padova* **21** (1952), 136–170, MR 14, 282.

18. Mogilevskii, S. I., Certain tests of stability for the first boundary problem for the equation of heat conduction at a boundary point of the region, in *Abstracts of Lectures of the Scientific Conference Dedicated to the Review of the Scientific Research of the Academic Year 1962–63*, pp. 107–109. Kalinin. Gos. Red. Inst., Kalini 1964 (Russian), MR 31, 1468.

19. Mogilevskii, S. I., Stability criteria at a boundary point for the first boundary-value problem for the heat equation, *Kalinin. Gos. Ped. Inst. Ucen. Zap.* **39** (1964), 18–24 (Russian), MR 32, 1457.

20. Montaldo, O., Sul primo problema di valori al contorno per l'equazione del calore, *Rend. Sem. Fac. Univ. Cagliari* **25** (1955), 1–14, MR 17, 625.

21. Montaldo, O., Un'osservazione sulle equazioni paraboliche elementari, *Boll. Un. Mat. Ital.* (3) **21** (1966), 155–160, MR 34, 3128.

22. Pagni, M., Teoremi di unicità di completezzà relativi ad un problema misto tipico per l'equazione del calore, *Rend. Sem. Mat. Univ. Padova* **28** (1958), 31–39, MR 24, A1523.

23. Pini, B., Sulla soluzione generalizzata di Wiener per il primo problema di valori al contorno nel caso parabolico, *Rend. Sem. Mat. Univ. Padova* **23** (1954), 422–434, MR 16, 485.

24. Pini, B., Sulla regolarità della frontiera per il primo problema di valori al contorno relativo all'equazione del calore, *Ann. Mat. Pura Appl.* (4) **40** (1955), 69–88, MR 17, 748.

25. Seda, V., A remark on the theory of the heat equation, *Ann. Mat. Pura Appl.*
     (4) **86** (1970), 357–365, MR 42, 8101.
26. Tu, C. C., Nontangential limits of a solution of a boundary-value problem for
     the heat equation, *Math. Systems Theory* **3** (1969), 130–138, MR 40, 3081.

F.   Some roots of reaction diffusion theory

1.   Berkovskii, B. M., A class of analytic solutions of the nonlinear heat equation,
     Dokl. Akad. Nauk. USSR **13** (1969), 225–226 (Russian), MR 40, 5188.
2.   Burgers, J. M., Statistical problems connected with the solution of a simple
     nonlinear partial-differential equation, I; II; III. *Nederl. Akad. Wetensch.*
     Proc., Ser. B, **57** (1954), 403–413, 414–424, 425–433, MR 19, 150.
3.   Cerpakov, P. V., On limit relations between solutions of equations of parabolic
     and elliptic type, *Kuibysev Aviacion Inst. Trudy* **2** (1954), 3–7 (Russian), MR
     20, 2534.
4.   Ciliberto, C., Sul problem di Holmgren–Levi per l'equazione del calore, *Giorn.*
     Mat. Battaglini (4) **4** (80) (1951), 1–13, MR 12, 831.
5.   Ciliberto, C., Precisazione relativa alla memoria: Sulla equazioni non lineari di
     tipo parabolico in due variabili, *Ricerche Mat.* **7** (1958), 232–234.
6.   Dawidowicz, A., Sur l'allure asymptotique des potentials de chaleur de l'arc,
     Prace. Mat. **12** (1968), 81–98, MR 38, 4800.
7.   Drozzinov, J. N., An example of the absence of stabilization for a solution of
     the heat equation, *Z. Vycisl. Mat. i Mat. Fiz.* **9** (1969), 1198 (Russian), MR 42,
     3439.
8.   Eskin, L. D., The asymptotics of the solution of the Cauchy problem for the
     heat equation, *Izv. Vyss. Ucebn. Zaved. Matematika* 1970, No. 2 (93), 100–106
     (Russian), MR 43, 2341.
9.   Ficken, F. A., An unusual parabolic problem, *Comm. Pure Appl. Math.* **14**
     (1961), 295–307, MR 25, 1367.
10. Fujita, H., On the blowing up of solutions of the Cauchy problem for
     $u_t = \Delta u + u^{1+\alpha}$, *J. Fac. Sci. Univ. Tokyo, Sect. I,* **13** (1966), 109–124, MR 35,
     5761.
11. Fujita, H., On some nonexistence and nonuniqueness theorems for nonlinear
     parabolic equations, in *Nonlinear Functional Analysis (Proc. Symp. Pure Math.,*
     Vol. XVIII, Part 1; Chicago, Ill. 1968) pp. 105–113. Amer. Math. Soc.
     Providence R.I., 1970, MR 42, 4888.
12. Fulks, W., A note on the steady-state solutions of the heat equation, *Proc.*
     Amer. Math. Soc. **7** (1956), 766–770, MR 18, 398.
13. Fulks, W., and Guenther, R. B., Damped wave equations and the heat
     equation, *Czechoslovak Math. J.* **21** (96) (1971), 683–695, MR 44, 5626.
14. Guscin, A. K., and Mihailov, V. P., The stabilization of the solution of the
     Cauchy problem for a one-dimensional parabolic equation, *Dokl. Akad. Nauk.*
     SSSR **197** (1971), 257–260 (Russian), MR 43, 3629.
15. Hocking, L. M., Diffusion from disjoint plane domains. I. The interaction
     of two domains, *J. Inst. Math. Appl.* **8** (1971), 139–152, MR 44, 4403.
16. Hocking, L. M., Diffusion from disjoint plane domains. II. Arrays of do-
     mains, with application to response of a muscle to a drug, *J. Inst. Math. Appl.*
     8 (1971), 153–163, MR 44, 5127.

17. Isakova, E. K., The asymptotic behavior of the solution of a second-order parabolic partial differential equation with a small parameter in the highest derivative term, *Dokl. Akad. Nauk. SSSR (N.S.)* **117** (1957), 935–938 (Russian), MR 21, 203.

18. Ito, S., On blowing up of solutions of semilinear partial-differential equations of parabolic type, *Sugaku* **18** (1966), 44–47 (Japanese), MR 36 2973.

19. Kanel', J. I., The behavior of solutions of the Cauchy problem when time tends to infinity, in the case of quasilinear equations arising in the theory of combustion, *Dokl. Akad. Nauk. SSSR* **132** (1960), 268–271 (Russian); translation: *Soviet Math. Dokl.* **1**, 533–536, MR 23, A3923.

20. Kanel', J. I., Stabilization of the solutions of the equations of combustion theory with finite initial functions, *Mat. Sb. (N.S.)* **65** (107) (1964), 398–413 (Russian); MR 31, 1473.

21. Knap, Z., On a certain property of solutions of the equation $u_t = u_{xx} + f(x, t, u)$, *Ann. Polon. Mat.* **23** (1970–71), 1–5, MR 41, 8808.

22. Kryzanzki, M., Sur l'allure asymptotique des potentiels de chaleur et de l'intégrale de Fourier–Poisson, *Ann. Polon. Math.* **3** (1957), 288–299, MR 19, 40.

23. Ljubarskii, G. J., On an equation of parabolic type with rapidly changing coefficients, *Teor. Funkcii Funkcional. Anal. i Prilozen. Vyp.* **3** (1966), 183–203 (Russian), MR 35, 584.

24. NcNabb, A., Notes on criteria for the stability of steady-state solutions of parabolic equations, *J. Math. Anal. Appl.* **4** (1962), 193–201.

25. Mihailov, V. P., The stabilization of the solution of the Cauchy problem for the heat equation, *Dokl. Akad. Nauk. SSSR* **190** (1970), 38–41 (Russian), MR 41, 2232.

26. Mlak, W., An example of the equation $u_t = u_{xx} + f(x, t, u)$ with distinct maximum and minimum solutions of a mixed problem, *Ann. Polon. Math.* **13** (1963), 101–103, MR 26, 6613.

27. Mogilevskii, S. I., Stability of the first boundary-value problem for the heat equation, *Kalinin. Gos. Ped. Inst. Ucen. Zap* **29** (1963), 67–84 (Russian), MR 28, 1411.

28. Petrusko, I. M., and Eidel'man, S. D., Solvability of the Cauchy problem for parabolic equations of second order in classes of arbitrary growing functions, *Ukrain. Mat. Z.* **19** (1967) No. 1, 108–113 (Russian) MR 34, 4718.

29. Pini, B., Maggioranti e minoranti delle soluzioni delle equazioni paraboliche, *Ann. Mat. Pura Appl.* (4) **37** (1954), 249–264, MR 16, 593.

30. Podolsky, B., A problem in heat conduction, *J. Appl. Phys.* **22** (1951), 581–585, MR 12, 710.

31. Poljakova, V. M., Stabilization of the solution of the heat-conduction equation, *Dokl. Akad. Nauk. SSSR* **129** (1959), 1230–1233 (Russian), MR 25, 5293.

32. Prodi, G., Questioni di stabilità per equazioni non linear alle derivate parziali di tipo parabolico, *Atti Accad. Naz. Lincei. Rend. Cl. Sci. Fis. Mat. Nat.* (8) **10** (1951), 365–370, MR 13, 351.

33. Repnikov, V. D., and Eidel'man, S. D., A new proof of the theorem on the stabilization of the solution of the Cauchy problem for the heat equation, *Mat. Sb. (N.S.)* **73** (115) (1967), 155–159 (Russian), MR 35, 3296.

34. Rudenko, E. N., On the question of asymptotic stability of solutions of

parabolic equations, *Differencial'nye Uravnenija* **6** (1970), 204–208 (Russian), MR 43, 678.

35. Rudenko, E. N., The behavior of the solutions of mixed problems for quasilinear equations of parabolic type, *Differencial'nye Uravnenija* **7** (1971), 115–120 (Russian), MR 44, 641.

36. Szybiak, A., On the asymptotic behavior of the solutions of the equation $\Delta u - (\partial u / \partial t) + c(x)u = 0$, *Bull. Acad. Polon. Sci. Ser. Sci. Math. Astr. Phys.* **7** (1959), 183–186 (Russian summary, unbound insert), MR 22, 4879.

37. Tappert, F. D., and Varma, C. M., Asymptotic theory of self-trapping of heat pulses in solids, *Phys. Rev. Lett.* **25** (1970), 1108–1111, MR 42, 5526.

38. van der Pol, B., On a nonlinear partial-differential equation satisfied by the logarithm of the Jacobian theta functions, with arithmetical applications. I, II., *Nederl. Akad. Wetensch. Proc., Ser. A*, **54** = *Indagationes Math.* **13** (1951), 261–271, 272–284, MR 13, 135.

39. Ventcel', T. D., Quasilinear parabolic systems with increasing coefficients, *Vestnik Moskov. Univ. Ser. I. Mat. Meh.*, No. 6 (1963) 34–44 (Russian, English summary), MR 28, 4252.

40. Vitasek, W., Über die quasistationare Lösung der Wärmeleitungsgleichung, *Apl. Math.* **5** (1960), 109–140, MR 22, 1752.

41. Winer, I. M., Solution to a class of complete heat-flow equations, *Quart. Appl. Math.* **23** (1965), 82, MR 31, 1025.

42. Zelenjak, T. I., Stability of stationary solutions of a mixed problem, *Dokl. Akad. Nauk. SSSR* **171** (1966), 266–268 (Russian), MR 34, 7987.

43. Zelenjak, T. I., On the question of stability of mixed problems for a quasilinear equation, *Differencial'nye Uravnenija* **3** (1967), 19–29 (Russian), MR 35, 593.

G. Periodic solutions

1. Cerpakov, P. V., Periodic solutions of the heat equation, *Izv. Vyss. Ucebn. Zaved. Matematika*, No. 2 (9) (1959), 247–251 (Russian), MR 24, A332.

2. Corduneanu, C., Solutions presque-periodiques de certaines équations paraboliques, *Mathematica (Cluj)* **9** (32) (1967), 241–244, MR 41, 623.

3. Fife, P., Solutions of parabolic boundary problems existing for all times, *Arch. Rational Mech. Anal.* **16** (1964), 155–186, MR 29, 4999.

4. Gor'kov, J. P., Periodic solutions of parabolic equations, *Differencial'nye Uravnenija* **2** (1966), 943–952 (Russian), MR 33, 7671.

5. Kallina, C., Periodicity and stability for linear and quasilinear parabolic equations, *SIAM J. Appl. Math.* **18** (1970), 601–613, MR 42, 2144.

6. Karimov, D. H., On periodic solutions of nonlinear differential equations of parabolic type, *Akad. Nauk. Uzbek. SSR Trudy Inst. Mat. Meh.* **5** (1949), 30–35 (Russian), MR 16, 709.

7. Kochina, N. N., Periodic regimes for some distributed systems, *Dokl. Akad. Nauk. SSSR* **165** (1965), 1015–1018 (Russian); translated as *Soviet Physics Dokl.* **10** (1966), 1142–1144, MR 34, 3135.

8. Kochina, N. N., Periodic solution of the diffusion equation with nonlinear boundary condition, *Dokl. Akad. Nauk. SSSR* **179** (1968), 1297–1300 (Russian); translated as *Soviet Physics Dokl.* **13** (1968), 305–307, MR 38, 4834.

9. Kolesov, J. S., Periodic solutions of second-order quasilinear parabolic equations, *Trudy Moskov. Mat. Obsc.* **21** (1970), 103–134 (Russian), MR 42, 6401.

10. Kruzkov, S. N., Periodic solutions of nonlinear second-order parabolic equations, *Differencial'nye Uravnenija* **6** (1970), 731–740 (Russian), MR 42, 6402.

11. Kulikov, N. P., Interior boundary-value problems without initial conditions for the equation $\Delta u = (\partial u / \partial t) + qu + F$, *Izv. Vyss. Ucebn. Zaved. Matematika*, No. 6 (19) (1960), 140–149 (Russian), MR 27, 6044.

12. Kusano, T., A remark on a periodic boundary problem of parabolic type, *Proc. Japan Acad.* **42** (1966), 10–12, MR 35, 1919.

13. Kusano, T., Periodic solution of parabolic partial-differential equations, *Sugaku* **18** (1966), 104–106 (Japanese), MR 36, 500.

14. Mal'cev, A. P., The convergence and stability of the Rothe method in the search for a periodic solution of a quasilinear parabolic equation with nonlinear boundary conditions, *Izv. Vyss. Ucebn. Zaved. Radiofizika* **12** (1969), 415–424 (Russian), MR 44, 4341.

15. Minasjan, R. S., A certain problem of periodic heat flow in an infinite cylinder, *Akad. Nauk. Armjan. SSSR Dokl.* **48** (1969), 3–8 (Russian, Armenian summary), MR 39, 3165.

16. Prodi, G., Problemi al contorno non lineari per equazioni di typo parabolico non lineari in due variabili soluzioni periodiche, *Rend. Sem. Mat. Univ. Padova* **23** (1954), 25–85, MR 15, 712.

17. Smulev, I. I., Periodic solutions of boundary-value problems without initial conditions for parabolic equations, *Dokl. Akad. Nauk. SSSR* **141** (1961), 1313–1316 (Russian), MR 24, A2738.

18. Smulev, I. I., Almost periodic solutions of a parabolic equation, *Sibirsk. Mat. Z.*, 7 (1966), 685–698 (Russian), MR 34, 475.

19. Smulev, I. I., Almost periodic and periodic solutions of the problem with oblique derivative for parabolic equations, *Differencial'nye Uravnenija* **5** (1969), 2225–2236 (Russian), MR 42, 648.

20. Squire, W., A problem in heat conduction, *J. Appl. Phys.* **22** (1951), 1508–1509, MR 13, 560.

21. Stastnova, V., and Vejvoda, O., Periodic solutions of the first boundary-value problem for a linear and weakly nonlinear heat equation, *Apl. Mat.* **13** (1968), 466–477; correction, *ibid.* **14** (1969), 241 (Czech and Russian summaries), MR 39, 4512.

22. Vaghi, C., Soluzioni limitate, o quasi-periodiche, di un'equazione di tipo parabolico non lineare, *Boll. Un. Mat. Ital.* (4) **I** (1968), 559–580, MR 39, 3134.

23. Vaghi, C., Su un'equazione parabolica contermine quadratico nella derivata z_x, *Ist. Lombardo Accad. Sci. Lett. Rend. A*, **104** (1970), 3–23, MR 43, 6597.

24. Vejvoda, O., Periodic solutions of nonlinear partial-differential equations of evolution, *Diff. Equations and Their Applications*, Equadiff II (Proc. Conf., Bratislava, 1966), pp. 293–300; *Slov. Ped. Nakladatel'*, Bratislava, 1967 (survey article), MR 40, 3034.

25. Vodicka, V., Circular cylinder under periodic fluctuations of temperature, *Z. Angew. Math. Phys.* **8** (1957), 53–64., MR 18, 701.

26. White, G. W. T., On the use of matrices for solving periodic heat-flow problems, *Appl. Sci. Res.* **A** (1957), 433–444, MR 19, 213.

27. Zaidman, S., Soluzioni limitate e quasi-periodiche dell'equazione del calore non omogenea. I., *Atti Accad. Naz. Lincei Rend. Cl. Sci. Fis. Mat. Nat.* (8) **31** (1961), 362–368, MR 26, 6609a.

28. Zaidman, S., Soluzioni limitate e quasi-periodiche dell'equazione del calore non omogenea, II., *Atti. Accad. Naz. Lincei Rend. Cl. Sci. Fis. Mat. Nat.* (8) **32** (1962), 30–37, MR 26, 6609b.

2.2. Initial Boundary-Value Problems

1. Abasov, A. M., The first one-dimensional, nonideal, heat-contact boundary-value problem, *Akad. Nauk. Azerbaidzan. SSR Dokl.* **23** (1967), No. 7, 3–8 (Russian, Azerbaijani summary), MR 36, 2975.

2. Abasov, A. M., The first one-dimensional, nonideal, heat-contact boundary-value problem, *Azerbaidzan. Gos. Univ. Ucen. Zap. Ser. Fiz.-Mat. Nauk.* (1967), No. 1, 24–29 (Russian, Azerbaijani summary), MR 41, 4023.

3. Abasov, A. M., The first one-dimensional, nonideal, heat-contact boundary-value problem, *Problems Comput. Math.*, pp. 113–127, *Izdat. Akad. Nauk. Azerbaidzan. SSR*, Baku 1968 (Russian), MR 43, 8300.

4. Abasov, A. M., A certain one-dimensional, nonideal, mixed heat-contact boundary-value problem, *Azerbaidzan. Gos. Univ. Ucen. Zap. Ser. Fiz.-Mat. Nauk.* (1968), No. 1, 38–45 (Russian), MR 43, 7791.

5. Adamczyk, H., and Borzymowski, A., A boundary problem for the generalized heat equation, *Zeszyty. Nauk. Politech. Warszawak. Mat.*, No. 4 (1965), 85–96 (Polish, Russian, and English summaries), MR 35, 1964.

6. Adamczyk, H., and Borzymowski, A., The tangential, discontinuous boundary-value problem for the generalized heat equation (methods of successive approximations), *Zeszyty Nauk. Politech. Warszawsk, Mat.*, No. 4 (1965), 143–156 (Polish, Russian, and English summaries), MR 35, 571.

7. Adler, G., Un type nouveau des problèmes aux limites de la conduction de la chaleur, *Magyar Tud. Akad. Mat. Kutato Int. Kozl.* **4** (1959), 109–127 (Hungarian and Russian summaries), MR 23, B1869.

8. Adler, G., Refrigeration de retour des matières granuleuses, *Magyar Tud. Akad. Mat. Kutato Int. Kozl.* **4** (1959), 327–365, MR 24, B2346.

9. Afanas'ev, E. F., Certain problems for the heat-transfer equations with mixed boundary conditions, *Differencial'nye Uravnenija* **1** (1965), 663–670 (Russian), MR 34, 7995.

10. Agaev, G. N., and Mamazov, G. K., Solution of a mixed problem for a nonlinear parabolic equation, *Dokl. Akad. Nauk. Azerbaidzen., SSR* **14** (1958), 505–510 (Russian), MR 22, 9744.

11. Akizanov, A., Cauchy–Dirichlet problems in a multiply connected region for the heat equation, *Vestnik Akad. Nauk. Kazah. SSR*, No. 5 (230) (1964), 40–47 (Russian, Kazak summary), MR 29, 4995.

12. Albertoni, S., Su un problema di propagazione con autovalori per l'equazione del calore, *Ist. Lombardo Accad. Sci. Lett. Rend. A*, **92** (1957/58), 206–216, MR 21, 3208.

13. Allan, D., The solution of a special heat and diffusion equation, *Amer. Math. Monthly* **63** (1956), 315–323, MR 17, 1091.

14. Allen, D. N. deG, and Severn, R. T., The application of relaxation methods to the solution of nonelliptic partial-differential equations. III. Heat conduction, with change of state, in two space dimensions, *Quart. J. Mech. Appl. Math.* **15** (1962), 53–62, MR 27, 477.

15. Amos, E. D., On half-space solutions of a modified heat equation, *Quart. Appl. Math* **27** (1969), 359–369, MR 41, 622.

16. Antimirov, M. J., On a certain contact problem of thermoconductivity, *Latvian Math. Yearbook 1965* (Russian), pp. 197–226. Izdat. "Zinatne," Riga, 1966 (Russian, Latvian and English summaries), MR 33, 2113.

17. Antimirov, M. J., Solution of certain two-dimensional heat problems with boundary conditions of mixed type by the method of divergent integrals and series, *Latvian Math. Yearbook,* **6** (Russian), pp. 3–22, Izdat. "Zinatne," Riga, 1969 (Russian, Latvian, and English Summaries), MR 41, 3000.

18. Aramanovic, I. G., and Levin, V. I., *The equations of mathematical physics.* Izdat. "Nauka," Moscow, 1964, 287 pp., MR 30, 5027.

19. Arsenin, V. J., *Mathematical Physics*: Basic equations and special functions. Izdat. "Nauka," Moscow, 1966, 367 pp., Ch. 5, parabolic equations, MR 34, 8678.

20. Azizov, F. A., The problem of the stationary distribution of temperature in the case of a contact of two solid bodies, *Azerbaidzan. Gos. Univ. Ucen. Zap. Ser. Fiz.-Mat. Nauk.,* No. 2 (1968), 45–48 (Russian).

21. Baimuhanov, B. B., The second and the third boundary-value problems for the heat equation with discontinuous coefficients for a half-plane, *Proc. First Kazakh Interuniv. Sci. Conf. on Math. Mech.* (October 17–22, 1963), p. 62. Izdat. "Nauka," Kazah. SSR, Alma-Ata, 1965 (Russian), MR 33, 425.

22. Baimuhanov, B. B., The second boundary-value problem in a strip for the heat equation with discontinuous coefficients, in *Certain Problems in Differential Equations,* pp. 71–76. Izdat. "Nauka," Kazah. SSR Alma-Ata, 1969 (Russian), MR 40, 1710.

23. Baiocchi, C., Sul problema misto per l'equazione parabolica del tipi del calore, *Rend. Sem. Mat. Univ. Padova* **36** (1966), 80–121, MR 35, 572.

24. Baiocchi, C., Soluzioni deboli problemi ai limiti per le equazioni paraboliche del tipo del calore, *Ist. Lombardo Accad. Sci. Lett. Rend. A,* **103** (1969), 704–726, MR 41, 7301.

25. Bajpai, S. D., Trigonometric functions and heat production in a cylinder, *Labdev-J. Sci. Tech.,* Part A, **6** (1968), 191–195, MR 40, 3079.

26. Baratta, M. A., Sopra un problema di ripartizione del calore, *Riv. Mat. Univ. Parma* **5** (1954), 197–207, MR 16, 594.

27. Baratta, M. A., Sopra un problema non lineare di ripartizione del calore, *Riv. Mat. Univ. Parma* **5** (1954), 363–371, MR 17, 271.

28. Baratta, M. A., Sopra un problema cilindrico non lineare di propagazione del calore, *Riv. Mat. Univ. Parma* **6** (1955), 389–398, MR 18, 580.

29. Baratta, M. A., Sopra un problema non lineare di propagazione del calore in un mezzo dotato di simmetria sferica, *Boll. Un. Mat. Ital.* (3) **11** (1956), 427–431, MR 18, 358.

30. Bauer, H., Zum Cauchyschen und Dirichletschen Problem bei elliptischen und parabolischen Differentialgleichungen, *Math. Ann.* **164** (1966), 142–153, MR 34, 6135.

31. Beattie, I. R., and Davis, D. R., A solution of the diffusion equation for isotropic exchange between a semi-infinite solid and a well-stirred solution, *Philos. Mag.* (8) **1** (1956), 874–879, MR 18, 400.

32. Belleni-Morante, A., Su un problema cilindrico di convezione forzata e di conduzione del calore in un reattore nucleare, *Boll. Un. Mat. Ital.* (3) **20** (1965), 106–113, MR 31, 5465.

33. Berg, P. W., and McGregor, J. L., *Elementary Partial-Differential Equations* (Preliminary Edition). Holden-Day, Inc., San Francisco, California 1964; 383 pp.

34. Bergles, A. E., and Kaye, J., Solutions of the heat-conduction equation with time-dependent boundary conditions, *J. Aerospace Sci.* **28** (1961), 251–252, MR 22, 7679.

35. Birjukova, F. G., Solution of the heat equation in a rectangle with mixed boundary conditions, in *Equations of Math., Phys., and Funct. Anal.* pp. 97–104. Izdat. "Nauka," Kazah. SSSR, Alma-Ata, 1966 (Russian), MR 34, 7996.

36. Birjukova, F. G., The first boundary-value problem for the heat equations with a discontinuous coefficient and without consistency conditions, *Vestnik Akad. Nauk. Kazah. SSR* **22** (1966), No. 11 (259), 65–69 (Russian), MR 36, 4133.

37. Birjukova, F. G., Determination of the temperature on a curve of discontinuity of the thermal conductivity coefficients, in *Certain Problems in Differential Equations*, pp. 77–80. Izdat. "Nauka," Kazah. SSR, Alma-Ata, 1969 (Russian), MR 40, 1712.

38. Birkhoff, G., and Kotik, J., "Note on the heat equation," *Proc. Amer. Math. Soc.* **5** (1954), 162–167, MR 15, 627.

39. Bochner, S., Sturm–Liouville and heat equations whose eigenfunctions are ultraspherical polynomials or associated Bessel functions, *Proc. Conf. on Differential Equations* (dedicated to A. Weinstein), pp. 23–48. Univ. of Maryland Book Store, College Park, MD 1956, MR 18, 484.

40. Bodrecova, L. B., Solution of a diffusion problem with variable temperature, *Mat. Sb.* (*N.S.*) **64** (106) (1964), 223–233 (Russian), MR 29, 2546.

41. Bragg, L. R., The radial-heat equation and Laplace transforms, *SIAM J. Appl. Math.* **14** (1966), 986–993, MR 34, 7998.

42. Bragg, L. R., The radial-heat equation with pole-type data, *Bull. Amer. Math. Soc.* **73** (1967), 133–135, MR 34, 7999.

43. Bragg, L. R., On the solution structure of radial-heat problems with singular data, *SIAM J. Appl. Math.* **15** (1967), 1258–1271, MR 36, 6802.

44. Budak, B. M., Samarskii, A. A., and Tikhonov, A. N., *A collection of problems on mathematical physics.* Translated by A. R. M. Robson, (translation edited by D. M. Brink). A Pergamon Press Book. The Macmillan Co., New York, 1964; xii + 770 pp. Ch. II, VI: Parabolic Equations; MR 29, 4969.

45. Buikis, A. A., A remark on the Lauwerier formula in the linear case, *Latvian Math. Yearbook* **7**, pp. 17–23. Izdat. "Zinatne," Riga, 1970, (Russian), MR 43, 7793.

46. Cameron, R. A., The generalized heat-flow equation and a corresponding Poisson formula, *Ann. Math.* (2) **59** (1954), 434–462, MR 15, 799.

47. Cannon, J. R., The solution of the heat equation subject to the specification of

energy, *Quart. Appl. Math.* **21** (1963), 155–160, MR 28, 3650.

48. Cess, R. S., and Shaffer, E. C., Heat transfer to laminar flow between parallel plates with a prescribed wall heat flux, *Appl. Sci. Res.* **A8** (1959), 339–344, MR 24, B246.

49. Chambre, P. L., Nonlinear heat-transfer problem, *J. Appl. Phys.* **30** (1959), 1683–1688, MR 21, 7729.

50. Chan, C. Y., A nonlinear, second initial-boundary-value problem for the heat equation, *Quart. Appl. Math.* **29** (1971), 261–268, MR 44, 650.

51. Chow, Tse-Sun, On a problem of heat conduction with time-dependent boundary conditions, *Z. Angew. Math. Phys.* **8** (1957), 478–484, MR 19, 1127.

52. Chuang, Feng-Kan, On the radiation problem of a quasilinear parabolic differential equation in connection with the mathematical model of turbulence, *Acta Math. Sinica* **3** (1953), 316–327 (Chinese and English summaries), MR 17, 162.

53. Ciliberto, C., Su di un problema al contorno per l'equazione $u_{xx} - u_y = f(x, y, u, u_x)$, *Ricerche Mat.* **1** (1952), 295–316, MR 14, 651.

54. Cobble, M. H., Quasi-analytic solution of the diffusion equation, *J. Franklin Inst.* **279** (1965), 110–123, MR 31, 5345.

55. Coleman, B. D., Duffin, R. J., and Mizel, V. J., Instability, uniqueness, and nonexistence theorems for the equation $u_t = u_{xx} - u_{xtx}$ on a strip, *Arch. Rat. Mech. Anal.* **19** (1965), 100–116, MR 31, 1479.

56. Copson, E. T., and Keast, P., On a boundary-value problem for the equation of heat, *J. Inst. Math. Appl.* **2** (1966), 358–363, MR 34, 5308.

57. Corduneanu, C., Approximation des solutions d'une équation parabolique dans un domaine non borne, *Mathematica (Cluj)* **3** (26) (1961), 217–224, MR 28, 266.

58. Dacev, A. B., On the two-dimensional multilayer problem of heat conduction, *Dokl. Akad. Nauk. SSSR (N.S.)* **101** (1955), 813–816 (Russian), MR 17, 625.

59. Dacev, A. B., On the three-dimensional multilayer problem of heat conduction, *Dokl. Akad. Nauk. SSSR (N.S.)* **101** (1955), 1019–1021 (Russian), MR 17, 625.

60. Dacev, A. B., On the heat conduction of a nonhomogeneous bar, *Dokl. Akad. Nauk. SSSR (N.S.)* **82** (1952), 861–864 (Russian), MR 13, 656.

61. Dacev, A. B., "Sur le problème de la propagation de la chaleur dans les corps solides (préface de Louis de Broglie)," in *Memor. Sci. Phys. Fasc., LXVII.* Gauthier-Villars, Editeur, Paris, 1963, 135 pp., MR 32, 763.

62. Deckert, K. L., and Maple, C. G., Solutions for diffusion equations with integral-type boundary conditions, *Proc. Iowa Acad. Sci.* **70** (1963), 354–361, MR 29, 2537.

63. Delavault, H., Sur un problème de la théorie de la chaleur, et sa solution au moyen des transformations de Hankel, *C. R. Acad. Sci. Paris* **236** (1953), 2484–2486, MR 14, 1090.

64. Dennemyer, R., *Introduction to Partial Differential Equations and Boundary-Value Problems.* McGraw-Hill Book Co., New York-Toronto-London, 1968; viii + 376 pp., MR 37, 555.

65. Descloux, J., On the heat equation, *Math. Z.* **113** (1970), 376–382, MR 41, 8840.

66. Dicker, D., and Friedman, M. B., Solutions of heat-conduction problems with nonseparable domains, *Trans. ASME Ser. E. J. Appl. Mech.* **30** (1963), 493–499, MR 29, 912.

67. Doob, J. L., A probability approach to the heat equation, *Trans. Amer. Math. Soc.* **80** (1955), 216–280, MR 18, 76.

68. Dynkin, E. B., General lateral conditions for some diffusion processes, *Proc. Fifth Berkeley Sympos. Math., Statist. and Prob.* (Berkeley, CA, 1965/66), Vol. II, Contributions to Probability Theory, Part 2, pp. 17–49. Univ. California Press, Berkeley, CA, 1967, MR 36, 965.

69. Eskin, L. D., Solution of the Cauchy problem for the heat equation, *Izv. Vyss. Ucebn. Zaved. Mathematika*, No. 3 (82) (1969), 107–117 (Russian), MR 40, 4611.

70. Evans, G. W., II, An application of the Mauro–Picone theorem for heat conduction, *Proc. Amer. Math. Soc.* **4** (1953), 961–968, MR 15, 433.

71. Fasano, A., Una dimostrazione della convergenza del metodo do Rothe per un problema di tipo parabolico, *Riv. Mat. Univ. Parma* (2) **9** (1968), 303–326 (English summary), MR 42, 2176.

72. Fasano, A., and Primicerio, M., Esistenza e unicità della soluzione par una classe di problemi di diffusione con condizioni al contorno nonlineari, *Boll. Un. Mat. Ital.* (4) **3** (1970), 660–667 (English summary), MR 44, 651.

73. Fenyo, I., Sur une méthode de solution de quelques équations différentielles de la physique mathématique, *Magyar Tud. Akad. Alkalm. Mat. Int. Kozl.* **1** (1952–1953), 355–362, MR 15, 228.

74. Fieber, H., und Selig, F., Temperaturfelder in endlichen Korpern bei bewegten Warmequellen, *Osterreich. Ing. Arch.* **10** (1956), 96–103, MR 17, 1092.

75. Fieber, H., Über das Temperaturfeld in langs einer Richtung bewegten und zeitlich veranderlichen Bereichen, *Osterreich. Ing. Arch.* **10** (1956), 155–160, MR 18, 537.

76. Fil'cakov, P. F., and Petrenko, V. G., An application of the power-series method to the solution of the first boundary-value problem for an equation of parabolic type, *Dopovidi Akad. Nauk. Ukrain. RSR, Ser. A*, 1967, 1034–1039 (Ukrainian, Russian, and English summaries), MR 36, 3519.

77. Flaiser, N. M., A new type of problem for the heat equation, *Comment. Math. Univ. Carolinae* **9** (1968), 71–77 (Russian), MR 37, 6617.

78. Flaiser, N. M. [Fleischer, N. M.], A new type of problem for the heat equation. II, *Comment. Math. Univ. Carolinae* **9** (1968), 511–514 (Russian), MR 42, 692.

79. Flaiser, N. M., *Corrigendum*: "A new type of problem for the heat equations," *Comment. Math. Univ. Carolinae* **9** (1968), 351 (Russian), MR 38, 2454.

80. Flaiser, N. M. [Fleischer, N. M.], A nonlinear problem for the heat-conduction equation, *Comment. Math. Univ. Carolinae* **10** (1969), 353–355 (Russian), MR 41, 4025.

81. Flaiser, N. M., A generalization of the Cauchy problem for parabolic equations, *Izv. Vyss. Ucebn. Zaved. Mathematika*, No. 9 (100) (1970), 87–89 (Russian), MR 43, 7779.

82. Fleishman, B. A., Dispersion of mass by molecular and turbulent diffusion: one-dimensional case, *Quart. Appl. Math.* **14** (1956), 145–152, MR 18, 88.

83. Forster, H., Über eine Randwertaufgabe der Differentialgleichung

$$r^2 \frac{\partial^2 f}{\partial r^2} + r \frac{\partial f}{\partial r} + q^2 \frac{\partial^2 f}{\partial \phi^2} = r^2 k \frac{\partial f}{\partial t},$$

Math. Z. **57** (1953), 428–455, MR 14, 1092.

84. Freud, G., Über Wärmeleitungs- und Diffusionsprobleme mit zusammenge-setzten Randbedingungen. I, *Magyar Tud. Akad. Alkalm. Mat. Int. Kozl.* **3** (1954), 369–394 (1955), (Hungarian, Russian, and German summaries), MR 17, 858.

85. Friedman, A., *Partial-Differential Equations of Parabolic Type.* Prentice-Hall, Inc., Englewood Cliffs, N.J., 1964; xiv + 347 pp. MR 31, 6062.

86. Friedman, B., *Principles and Techniques of Applied Mathematics.* John Wiley and Sons, Inc. New York; Chapman and Hall, Ltd., London, 1956; ix + 315 pp., MR 18, 43.

87. Frjazinov, I. V., Solution of the third boundary-value problem for a two-dimensional equation of heat conduction in an arbitrary region by the locally one-dimensional method, *Z. Vycisl. Mat. i Mat. Fiz.* **6** (1966), 487–502 (Russian), MR 35, 5155.

88. Fukuda, N., and Yosikawa, T., Conduction of heat in a cylinder composed of three different materials, *J. Osaka Inst. Sci. Tech., Part I,* **2** (1950), 53–65, MR 15, 132.

89. Fulks, W., and Maybee, J. S., A singular nonlinear equation, *Osaka Math. J.* **12** (1960), 1–19, MR 23, A426.

90. Fulks, W., On the solutions of the heat equation, *Proc. Amer. Math. Soc.* **2** (1951), 973–979, MR 13, 750.

91. Fulks, W., The Neumann problem for the heat equation, *Pacific J. Math.* **3** (1953), 567–583, MR 15, 433.

92. Gagliardo, E., Problem al contorno generalizzato per l'equazione del calore, *Ricerche Mat.* **4** (1955), 74–94, MR 17, 625.

93. Gal'chuk, L. I., Certain problems on optimal control of systems which are describable by a parabolic equation, *Vestnik Moskov. Univ. Ser. I Mat. Meh.* **23** (1968), No. 3, 31–33 (Russian, English summary), MR 39, 607.

94. Gal'chuk, L. I., Optimal control of systems described by parabolic equations, *SIAM J. Control* **7** (1969), 546–558, MR 41, 8799.

95. Garnir, H. G., Sur les distributions resolvantes des opérateurs de la physique mathématique, III., *Bull Soc. Roy. Sci. Liège,* **20** (1951), 271–296, MR 13, 352.

96. Gibson, R. E., A heat-conduction problem involving a specified moving boundary, *Quart. Appl. Math.* **16** (1958), 426–430, MR 21, 2479.

97. Gibson, R. E., A linear heat problem with a moving interface, *Z. Angew. Math. Phys.* **11** (1960), 198–206 (German summary), MR 22, 10594.

98. Girifalco, L. A., and Behrendt, D. R., Mathematics of diffusion-controlled precipitation in the presence of homogeneously distributed sources and sinks, *Phys. Rev.* (2) **124** (1961), 420–427, MR 27, 2741.

99. Girsanov, I. V., Minimax problems in the theory of diffusion processes, *Dokl. Akad. Nauk. SSSR* **136** (1961), 761–764 (Russian); translated as *Soviet Math. Dokl.* **2** (1961), 118–121, MR 31, 509.

100. Gordon, I. A., A certain generalized solution of a mixed problem for the heat equation, in *Collection of Candidates' Works Math., Mech., Phys.* (Edited by B. M. Gagaev.) (Russian), pp. 33–40. Izdat. Kazan. Univ., Kazan, 1964, MR 38, 3612.

101. Greco, D., Una nuova applicazione del metodo delle transformate alla risoluzione di un problema al contorno per un'equazione di tipo parabolico, *Giorn. Mat. Battaglini* (4) 4 (80) (1951), 102–128, MR 12, 709.

102. Greenspan, D., *Introduction to Partial-Differential Equations*, in International Series in Pure and Applied Mathematics. McGraw-Hill Book Co., Inc., New York-Toronto-London, 1961; viii + 195 pp., MR 24, A314.

103. Gupta, R. K., Two problems of heat flow in solids, *Proc. Nat. Inst. Sci. India, Part A*, 29 (1963), 84–89, MR 27, 4497.

104. Gusev, M. A., A mixed boundary-value problem for the heat equation in an infinite strip, *Izv. Akad. Nauk. Kazah. SSR Ser. Fiz. Mat.*, No. 1 (1968), 58–65 (Russian, Kazakh summary), MR 38, 3613.

105. Hadamard, J., La Théorie des Équations aux Dérivées Partielles. *Editions Scientifique*, Peking. Gauthier-Villars Editeur, Paris, 1964; vi + 322 pp., MR 37, 556.

106. Hairullin, E., A general boundary problem of the equation of heat conductivity in the two-dimensional case with discontinuous coefficients for strips, *Izv. Akad. Nauk. Kazah. SSR Ser. Fiz. Mat.*, No. 1 (1971), 71–78 (Russian), MR 44, 4397.

107. Hantush, M. S., and Jacobi, C. E., Nonsteady Green's functions for an infinite strip of leaky aquifier, *Trans. Amer. Geophys. Union* 36 (1955), 101–112, MR 17, 272.

108. Hantush, M. S., and Jacobi, C. E., Nonsteady radial flow in an infinite leaky aquifier, *Trans. Amer. Geophys. Union* 36 (1955), 95–100.

109. Harin, S. N., On thermal problems with a moving boundary, *Izv. Akad. Nauk. Kazah. SSSR Ser. Fiz. Mat. Nauk.*, No. 3 (1965), 52–60, (Russian), MR 33, 428.

110. Hofmann, R., Die Lösung eines speziellen Wärmeleitungsproblems der Elektrotechnik mittels der zweidimensionalen Laplace-Transformation, *Arch. Elek. Ubertr.* 11 (1957), 278–282, MR 19, 808.

111. Hvingiya, L. V., On a solution of the differential equation of heat conduction for bodies of complicated shape, *Soobsc. Akad. Nauk. Gruzin. SSR* 20 (1958), 257–264 (Russian), MR 20, 2988.

112. Il'in, A. M., Klasnikov, A. S., and Oleinik, O. A., Second-order linear equations of parabolic type, *Uspehi Mat. Nauk.* 17 No. 3 (105) (1962), 3–146 (Russian), MR 25, 2328.

113. Imanliev, M. I., Vostrov, V. K., and Dzuraev, M. D., An application of Newton's method to the solution of a certain class of inverse problems for parabolic-type equations, *Izv. Akad. Nauk. Kirgiz. SSSR*, No. 4 (1968), 3–10 (Russian), MR 38, 3598.

114. Ivanov, T. F., The representation of solutions of certain problems in the theories of filtration and heat conduction in the form of a sum of particular self-similar solutions of the heat-conduction equation, *Izv. Akad. Nauk. SSSR Ser. Meh. Zidk. Gaza*, No. 5 (1968), 179–185. (Russian), MR 38, 2997.

115. Jackel, H., Mathematische Behändlung gesteurter Abkuhl- und Anwarmvor-

gange, *Ing.-Arch.* **26** (1958), 146–156, MR 19, 1229.

116. Jaeger, J., Conduction of heat in a solid in contact with a thin layer of a good conductor, *Quart. J. Mech. Appl. Math.* **8** (1955), 101–106, MR 16, 930.

117. John, F., Hyperbolic and parabolic equations, in *Partial-Differential Equations* (Proc. Summer Seminar, Boulder, Col., 1957), pp. 1–129. Interscience, New York, 1964, MR 29, 2533.

118. John, F., *Partial Differential Equations: Mathematics applied to physics*; pp. 229–315. Springer, New York, 1970, MR 41, 5752.

119. Kadlec, J., On a generalization of the heat equation, *Comment. Math. Univ. Carolinae* **6** (1965), 13–18 (Russian), MR 31, 1469.

120. Kadlec, J., Solution of the first boundary-value problem for a generalization of the heat equation in classes of functions possessing a fractional derivative with respect to the time variable, *Czechoslavak Math. J.* **16** (91) (1966), 91–113 (Russian, English summary), MR 32, 7960.

121. Kampe de Feriet, J., Intégrales aléatoires de l'équation de la diffusion, *C. R. Acad. Sci. Paris* **243** (1956), 929–932, MR 19, 40.

122. Kamynin, L. I., La solution d'après la méthode Fourier du premier problème extrémal pour l'équation de la conductibilité de la chaleur, *Bul. Univ. Shteteror Tirane, Ser. Shk. Nat.*, No. 1 (1959), 16–33, MR 22, 9732.

123. Kamynin, L. I., The solution of boundary-value problems for a parabolic equation with discontinuous coefficients, *Dokl. Akad. Nauk. SSSR* **139** (1961), 1048–1051 (Russian), MR 24, A1519.

124. Kamynin, L. I., Dependence upon the boundary of the solution of the mixed problem for a parabolic equation, *Dokl. Akad. Nauk. SSSR* **140** (1961), 1244–1247 (Russian), MR 24, A1520.

125. Kamynin, L. I., The method of heat potentials for a parabolic equation with discontinuous coefficients, *Sibirsk. Mat. Z.* **4** (1963), 1071–1105 (Russian), MR 28, 362.

126. Kamynin, L. I., The existence of a solution of boundary-value problems for a parabolic equation with discontinuous coefficients, *Izv. Akad. Nauk. SSSR Ser. Mat.* **28** (1964), 721–744 (Russian), MR 29, 2534.

127. Kamynin, L. I., A boundary-value problem in the theory of heat conduction with nonclassical boundary conditions, *Z. Vycisl. Mat. i Mat. Fiz.* **4** (1964), 1006–1024 (Russian), MR 30, 1316.

128. Kamynin, L. I., Solution of the fifth boundary-value problem for a second-order parabolic equation in a noncylindrical region, *Sibirsk. Mat. Z.* **9** (1968), 1153–1166 (Russian),

129. Kamynin, L. I., The solution of the fourth and fifth boundary-value problems for a one-dimensional, second-order parabolic equation in a curvilinear region, *Z. Vycisl. Mat. i Mat. Fiz.* **9** (1969), 558–572 (Russian), MR 40, 7633.

130. Kanel', J. I., Stabilization of solutions of the Cauchy problem for equations encountered in combustion theory, *Mat. Sb.* (*N.S.*) **59** (101) (1962), suppl. 245–288 (Russian), MR 28, 367.

131. Kapilevic, M. B., The solution of a mixed boundary-value problem for an equation of parabolic type by the method of S. Bergman, *Volz. Mat. Sb. Vyp.* **6** (1968), 86–89 (Russian), MR 43, 3631.

132. Keilson, J., On diffusion in an external field and the adjoint source problem, *Quart. Appl. Math.* **12** (1955), 435–438, MR 16, 710.

133. Kessler, A., Die Ewärwung von Platten und endlichen Stäben mit innerer Wärmeerzeugung, bei ein-dimensionaler Lösung der Wärmestromung, *Apl. Mat.* **3** (1958), 190–222 (Czech, Russian, and German summaries), MR 21, 1841.

134. Kigai, A. K., A verification of the solution for the third boundary problem of the heat equation in a rectangular parallelepiped, *Izv. Akad. Nauk. Kazah. SSR Ser. Fiz. Mat.*, No. 1 (1971), 42–49 (Russian), MR 44, 4398.

135. Kim, E. I., The propagation of heat in two dimensions in an infinite inhomogeneous body, *Akad. Nauk. SSSR Prikl. Mat. Meh.* 17 (1953) (Russian), MR 15, 627.

136. Kim, E. I., On a plane problem of heat-conduction equation, *Rostov. Gos. Ped. Inst. Uc. Zap.*, No. 2 (1953), 31–38 (Russian), MR 19, 40.

137. Kim, E. I., and Baimuhanov, B. B., The temperature distribution in a piecewise homogeneous semi-infinite plate, *Dokl. Akad. Nauk. SSSR* **140** (1961), 333–336 (Russian). Translation: *Soviet Phys. Dokl.* **6** (1962) 756–759, MR 25, 940.

138. Kim, E. I., and Birjukova, F. G., The first boundary-value problem for the heat equation in a rectangle with a discontinuous coefficient, in *Equations of Math., Phys., and Funct. Anal.*, pp. 3–9. Izdat. "Nauka" Kazah., SSSR, Alma-Ata, 1966 (Russian), MR 34, 7997.

139. Kim, E. I., and Birjukova, F. G., Solutions of the heat-conduction equation with a discontinuous coefficient when the initial data are inconsistent, *Izv. Akad. Nauk. Kazah. SSR Ser. Fiz. Mat. Nauk.*, No. 5 (1967), 3–15 (Russian, Kazakh summary), MR 36, 1829.

140. Kiselev, K. A., and Lazarev, A. I., Temperature gradient of an infinite plate with variable heat-loss coefficients in a medium of varying temperatures, *Z. Tehn. Fiz.* **30** (1960), 616–621 (Russian); translation: *Soviet Phys. Tech. Phys.* **5**, 579–584, MR 22, 7680.

141. Kochina, N. N., The solution of a diffusion problem with a nonlinear boundary condition, *Dokl. Akad. Nauk. SSSR* **174** (1967), 305–308 (Russian); translated as *Soviet Phys. Dokl.* **12** (1967), 425–427, MR 38, 4833.

142. Kovtun, D. G., On some series of the theory of heat conduction of Fourier–Poisson, I, *Ukrain. Mat. Z.* **7** (1955), 273–290 (Russian), MR 17, 1079.

143. Kral. J., Flows of heat and the Fourier problem, *Czech. Math. J.* **29** (95) (1970), 556–598, MR 42, 6437.

144. Kruger, M., Zur Kombination thermischer und elektro-magnetischen Felder um Falle der ebenen Platte, *Ing.-Arch.* **20** (1952), 234–246, MR 14, 476.

145. Krylov, N. V., Minimax-type equations in the theory of elliptic and parabolic equations on the plane, *Mat. Sb.* (*N.S.*) **81** (123) (1970), 3–22, MR 41, 614.

146. Kushner, H. J., On the optimal control of a system governed by a linear parabolic equation with white-noise inputs, *SIAM J. Control* **6** (1968), 596–614, MR 40, 3953.

147. Kval'vasser, V. I., Rutner, J. F., and Skrjabina, L. P., Mixed boundary-value problems for the heat equation for an infinite strip, *Comment. Math. Univ. Carolinae* **9** (1968), 1–12 (Russian), MR 38, 3614.

148. Kval'vasser, V. I., Rutner, J. F., and Skrjabina, L. P., Correction to: "Mixed boundary-value problems for the heat equation for an infinite strip," *Com-*

ment. Mat. Univ. Carolinae **10** (1969), 37–40 (Russian), MR 40, 7644.

149. Laasonen, P., Das Wärmeleitungsproblem einer linearen Mannigfaltigkeit, in *Den 11te Skandinaviske Matematikerkongress*, Trondheim 1949, pp. 148–152. Johan Grundt Tanums Forlag, Oslo, 1952, MR 14, 877.

150. Ladyzenskaya, O. A., On solvability of the fundamental boundary-value problems for equations of parabolic and hyperbolic type, *Dokl. Akad. Nauk. SSSR (N.S.)* **97** (1954), 395–398 (Russian), MR 17, 495.

151. Ladyzenskaja, O. A., Solonnikov, V. A., and Ural'ceva, N. N., Linear and quasilinear equations of parabolic type. Translated from Russian by S. Smith. *Translations of Mathematical Monographs*, Vol. 23. Amer. Math. Soc., Providence, R.I., 1968; xi + 648 pp., MR 39, 3159b.

152. Lauwerier, H. A., Optimum problems in the conduction of heat in a semi-infinite solid, *Appl. Sci. Research* **A4** (1953), 142–152, MR 15, 533.

153. Lauwerier, H. A., The transport of heat in an oil layer caused by the injection of hot fluid, *Appl. Sci. Res.* **A5** (1955), MR 16, 931.

154. Lauwerier, H. A., Diffusion from a point source into a space bounded by an impenetrable plane, *Appl. Sci. Res.* **A6** (1956), 197–204, MR 18, 400.

155. Lauwerier, H. A., Randwaarde problemen; Deel 1, 2, 3, (Dutch) [Boundary-value problems; Parts 1, 2, 3.], MC Syllabus, 3.2, Mathematisch Centrum, Amsterdam, 1968; ii + 89 pp., MR 40, 2252

156. Lawruk, B., and Rolewicz, S., The minimum time-control problem for a certain class of linear parabolic partial-differential equations controlled by the boundary condition (loose Russian summary), *Bull. Acad. Polon. Sci. Ser. Sci. Math. Astronom. Phys.* **16** (1968), 489–493, MR 39, 609.

157. Legras, J., *Techniques de Résolution des Équations aux Dérivées Partielles: Équation de la Chaleur, Équation de Laplace, Équation des Ondes.* Dunod, Paris, 1956; xv + 180 pp., MR 19, 419.

158. Letcher, J. S., Jr., On the boundary layers in the theory of heat conduction in solids, *SIAM Rev.* **11** (1969), 20–29, MR 40, 1077.

159. Lions, J.-L., Problèmes aux limites, I, II, III., *C. R. Acad. Sci. Paris* **236** (1953), 2372–2375, 2470–2472; **237**, 12–14, MR 15, 317.

160. Lions, J. L., On some optimization for linear parabolic equations, in *Functional Analysis and Optimization*; pp. 115–131. Academic Press, New York, 1966, MR 37, 6583.

161. Lovass-Nagy, V., Über die Anwendung der Matrizenrechnung zur Berechung zweidimensionaler Temperaturfelder, *Z. Angew. Math. Mech.* **42** (1962), 110–119 (English summary), MR 25, 939.

162. Lozanovskaya, I. T., and Uflyand, Y. S., A class of problems in mathematical physics with a mixed eigenvalue spectrum, *Dokl. Akad. Nauk. SSSR* **164** (1965), 1005–1007 (Russian) translation: *Soviet Phys. Dokl* **10** (1966), 918–919, MR 33, 7689.

163. MacCamy, R. C., Mizel, V. J., and Seidman, T. I., Approximate boundary controllability for the heat equation, *J. Math. Anal. Appl.* **23** (1968), 699–703, MR 37, 4384.

164. MacCamy, R. C., Mizel, V. J., and Seidman, T. I., Approximate boundary controllability of the heat equation, II., *J. Math. Anal. Appl.* **28** (1969), 482–492, MR 40, 521.

165. Magenes, E., Problemi al contorno misti per l'equazione del calore, *Rend.*

Sem. Mat. Univ. Padova **24** (1955), 1–28, MR 16, 1118.

166. Mamedov, N. M., An application of M. L. Rasulov's contour-integration method to the solution of plane mixed problems for differential equations of parabolic type, *Azerbaidzen. Gos. Univ. Ucen. Zap. Ser. Fiz. Mat. i. Him. Nauk.*, No. 4 (1964), 45–55 (Russian), MR 36, 4154.

167. Manfredi, B., Sopra un problema cilindrico non lineare di propagazione del calore, *Rivista Mat. Univ. Parma* **3** (1952), 383–396, MR 14, 1090.

168. Manfredi, B., Sopra un problema non lineare di propagazione del calore per un mezzo dotato di simmetria sferica, *Rivista Mat. Univ. Parma* **4** (1953), 123–132, MR 15, 228.

169. Manfredi, B., Osservazioni su di un problema di distribuzione della temperatura in un mezzo che si muove, *Rivista Mat. Univ. Parma* **4** (1953), 327–335, MR 16, 46.

170. Mangeron, D., and Krivosein, L. E., Teoremi di esistenza, di unicità, e di valutazione delle soluzioni di alcuni problemi al contorno concernenti equazioni integro-differenziali con operatori esterni di tipo parabolico, *Atti Accad. Naz. Lincei Rend. Cl. Sci. Fis. Mat. Nat.* (8) **36** (1964), 451–456, MR 30, 2231.

171. Mangeron, D., and Krivocheine, L. E. [Krivosein, L. E.], Problèmes aux limites de type mixte concernant une classe d'équations intégro-différentielles paraboliques, *Acad. Roy. Belg. Bull. Cl. Sci.* (5) **50** (1964), 424–433, MR 31, 1534.

172. Mann, W. R., and Wolf, F., Heat transfer between solids and gases under nonlinear boundary conditions, *Quart. Appl. Math.* **9** (1951), 163–184, MR 13, 134.

173. Marchi, E., and Fasulo, A., Heat conduction in sectors of hollow cylinders with radiation, *Atti Accad. Sci. Torino Cl. Sci. Fis. Mat. Nat.* **101** (1966/67), 373–382, MR 36, 2361.

174. *Linear Equations of Mathematical Physics*, Edited by S. G. Mihlin. English translation edited by Harry Hochstadt. Holt, Rinehart and Winston, Inc., New York-Toronto-London, 1967; xiii + 318 pp., MR 34, 6278.

175. Milgram, A. N., Parabolic equations, in *Partial-Differential Equations* (Proc. Summer Seminar, Boulder, Col., 1957), pp. 327–339. Interscience New York, 1964, MR 29, 2548.

176. Miller, K. S., *Partial-Differential Equations in Engineering Problems.* Prentice-Hall, Inc., New York, 1953, viii + 254 pp., MR 16, 364.

177. Minyatov, A. V., The heating of an infinite cylinder enclosed in a sheath, *Z. Tehn. Fiz.* **30** (1960), 611–615 (Russian) translation: *Soviet Physics Tech. Phys.* **5**, 575–578, MR 22, 7682.

178. Miranda, G., Integral-equation solution of the first initial-boundary-value problem for the heat equation in domains with nonsmooth boundary, *Comm. Pure Appl. Math.* **23** (1970), 757–765, MR 42, 694.

179. Mlak, W., The first boundary-value problem for a nonlinear parabolic equation, *Ann. Polon. Math.* **5** (1958/59), 257–262, MR 21, 1447.

180. Montaldo, O. Su un problema di valori al contorno nella teoria diffusiva dei reattori nucleari, *Rend. Sem. Fac. Sci. Univ. Cagliari*, **28** (1958), 118–120, MR 22, 9728.

181. Murakami, H., On nonlinear partial-differential equations of parabolic types

I, II, III, *Proc. Japan Acad.* **33** (1957), 530–535, 616–621, 622–627, MR 20, 6589.

182. Murakami, H., On the regularity of domains for parabolic equations, *Proc. Japan Acad.* **34** (1958), 347–348, MR 20, 6590.

183. Murakami, H., Relations between solutions of parabolic and elliptic differential equations, *Proc. Japan Acad.* **34** (1958), 349–352, MR 20, 6591.

184. Napetvaridze, O. I., On a fundamental contact boundary-value problem in heat conduction, *Soobsc. Akad. Nauk. Gruzin. SSSR* **33** (1964), 271–278 (Russian, Georgian summary), MR 29, 2547.

185. Napetvaridze, O. I., On the existence of a solution of a contact boundary-value problem in the theory of heat conduction, *Soobsc. Akad. Nauk. Gruzin. SSR* **37** (1965), 259–262 (Russian, Georgian summary) MR 31, 1475.

186. Napetvaridze, O. I., On approximate solution of the Cauchy–Neumann Problem for the heat equation, *Thbilis. Sahelmc. Univ. Srom. Mekh.-Math. Mecn. Sev.* **110** (1965), 109–114 (Russian, Georgian summary), MR 34, 1712.

187. Napetvaridze, O., Problems for the heat equation with mixed boundary conditions, *Differencial'nye Uravnenija* **4** (1968), 1283–1288 (Russian), MR 38, 432.

188. Nomura, T., and Nakamura, K., Terminal control of distributed-parameter systems, *J. Japan Assoc. Automat. Control Engrs.* **14** (1970), 354–361 (Japanese, English summary), MR 42, 4867.

189. Nomura, T., and Nakamura, K., Time-optimal control of a parabolic distributed-parameter system, *J. Japan Assoc. Automat. Control Engrs.* **14** (1970), 408–416 (Japanese, English summary), MR 42, 4868.

190. Nowacki, W., Mixed boundary-value problems in heat conduction, *Arch. Mech. Stos.* **16** (1964), 865–884 (Polish and Russian summaries), MR 34, 3885.

191. Olcer, N. Y., Note on the general solution of the heat equation, *Quart. Appl. Math.* **24** (1967), 380–383, MR 35, 3294.

192. Orynbasarov, M. O., The first and the second boundary-value problems for the heat equation in a semi-infinite strip, *Proc. First Kazakh. Interuniv. Sci. Conf. on Math. Mech.* (October 17–22, 1963) (Russian) pp. 90–94. Izdat. "Nauka" Kazah. SSR, Alma-Ata, 1965, MR 33, 436.

193. Orynbasarov, M., The first boundary-value problem for the heat-transfer equation in the case of a boundary with two angular points, *Izv. Akad. Nauk. Kazah. SSSR Ser. Fiz.-Mat. Nauk.*, No. 3 (1966), 83–89 (Russian) Kazakh summary), MR 33, 7712.

194. Orynbasarov, M., The first boundary-value problem for the heat equation when the boundary of the region has singularities, *Izv. Akad. Nauk. Kazah. SSR Ser. Fiz.-Mat. Nauk.*, No. 1 (1967), 51–56 (Russian, Kazakh summary), MR 35, 3295.

195. Orynbasarov, N., The first boundary-value problem for the heat equation when the boundary of the region has singularities, II., *Izv. Akad. Nauk. Kazah. SSR Ser. Fix.-Mat. Nauk.*, No. 5 (1967), 33–39 (Russian, Kazakh summary), MR 37, 3216.

196. Ossicini, A., Il calcolo simbolico e la propagazione del calore in una ipersfera dello spazio euclideo ad *n* dimensioni, *Ann. Scuola Norm. Sup. Pisa* (3) **5** (1951), 269–278, MR 13, 751.

197. Ostapenko, V. M., and Malahivs'ka, S. M., A solution of the heat equation for
      a plate of finite thickness, *Dopovidi Akad. Nauk. Ukrain. RSR Ser. A* (1969),
      217–222, 269, MR 40, 7645.
198. O'Sullivan, D. G., Treatment of the equations of classical diffusion in homo-
      geneous isotropic media, *J. Chem Phys.* **25** (1956), 270–274, MR 18, 130.
199. Pagni, M., Su un problema al contorno tipico per l'equazione del calore, *Ann.
      Scuola Norm. Sup. Pisa* (3) **11** (1957), 73–115, MR 25, 332.
200. Pagni, M., Su un problema al contorno tipico per l'equazione del calore in
      ($n + 1$) dimensioni, *Ann. Scuola Norm. Sup. Pisa* (3) **11** (1957), 209–216, MR
      23, A1940.
201. Parodi, M., Mathématiques appliqués a l'art de l'ingenieur. In Tome V:
      Équations aux Dérivées Partielles. Société d'Édition d'Enseignement Supérieur
      (SEDES), Paris, 1966; 323 pp., MR 34, 1126.
202. Paterson, S., Conduction of heat from local sources in a medium generating
      or absorbing heat, *Proc. Glasgow Math. Assoc.* **1** (1953), 164–169, MR 15,
      627.
203. Phillips, R. S., Semigroup methods in the theory of partial-differential equa-
      tions, in *Modern Mathematics for the Engineer:* Second series, pp. 100–132.
      McGraw-Hill, New York, 1961, MR 23, B2195.
204. Pignedoli, A., Su un problema di diffusione delle fisica-matematica, *Ann. Mat.
      Pura Appl.* (4) **32** (1951), 281–293, MR 13, 750
205. Pini, B., Sul primo problema di valori al contorno per le equazioni parabo-
      liche lineari, *Riv. Mat. Univ. Parma* **6** (1955), 215–237, MR 20, 1842.
206. Pini, B., Sul primo problema di valori al contorno per l'equazione parabolica
      non lineare del secondo ordine, *Rend. Sem. Mat. Univ. Padova* **27** (1957),
      149–161, MR 19, 1060.
207. Pogorzelski, W., Sur le problème de Fourier généralisé, *Ann. Polon. Math.* **3**
      (1956), 126–141, MR 18, 665.
208. Poritsky, H., and Powell R. A., Certain solutions of the heat-conduction
      equation, *Quart. Appl. Math.* **18** (1960–61), 97–106, MR 22, 3443.
209. Prodi, G., Teoremi di esistenza per equazioni alle derivate parziali non lineari
      di tipo parabolico. I, II; *Ist. Lombardo Sci. Lett. Rend. Cl. Sci. Mat. Nat.* (3)
      17 (86) (1953), 3–26, 27–47, MR 16, 259.
210. Radok, J. R. M., Solution of a heat-flow problem, *Australian J. Sci. Res. Ser.*
      A4 (1951), 12–15, MR 13, 42.
211. Rafalski, P., Orthogonal projection method. I. Heat conduction boundary
      problem, *Bull. Acad. Polon. Sci. Ser. Sci. Tech.* **17** (1969), 117–121 (Russian
      summary), MR 39, 7483a.
212. Rafalski, P., Orthogonal projection method. II. Thermoelastic problem, *Bull.
      Acad. Polon. Sci. Ser. Sci. Tech.* **17** (1969), 123–128. (Russian summary), MR
      39, 7483b.
213. Rainville, E. D., A heat-conduction problem and the product of two error
      functions, *J. Math. Physics* **32** (1953), 43–47, MR 14, 758.
214. Rasulov, M. L., and Asadova, O. G., The solution of a certain nonlinear
      problem of mathematical physics, *Azerbaidzan. Gos. Univ. Ucen. Zap. Ser.
      Fiz.-Mat. Nauk.*, No. 5 (1968), 3–11 (Russian), MR 44, 653.
215. Redozubov, D. V., The solution of linear thermal problems with a uniformly
      moving boundary in a semi-infinite region, *Z. Tehn. Fiz.* **30** (1960), 606–610

(Russian), translated in *Soviet Physics Tech. Phys.* **5**, 570–574, MR 22, 7683.

216. Reid, W. P., A method for solving certain boundary-value problemss, *J. Soc. Indust. Appl. Math.* **3** (1955), 259–261, MR 18, 44.

217. Reid, W. P., Heat flow in a cylinder, *Quart. Appl. Math.* **16** (1958), 147–153, MR 19, 1229.

218. Reid, W. P., Linear heat flow in a composite slab, *J. Aerospace Sci.* **29** (1962), 905–908, MR 25, 4783.

219. Reismann, H., Temperature distribution in a spinning sphere during atmospheric entry, *J. Aerospace Sci.* **29** (1962), 151–159, MR 25, 941.

220. Rektorys, K., Two theorems concerning the equation $\partial u / \partial t = \partial^2 u / \partial x^2 + \partial^2 u / \partial y^2$. *Casopis Pest. Mat.* **79** (1954), 333–366, (Czech, Russian, and English summaries), MR 18, 47.

221. Ritchie, R. H., and Skakura, A. Y., Asymptotic expansions of solutions of the heat-conduction equation in internally bounded cylindrical geometry, *J. Appl. Phys.* **27** (1956), 1453–1459, MR 18, 780.

222. Rjaben'kii, V. S., Computation of heat conductivity on a system of rods, *Z. Vycisl. Mat. i Mat. Fiz.* **10** (1970), 236–239 (Russian), MR 42, 4092.

223. Rodin, E. Y., On some approximate and exact solutions of boundary-value problem for Burgers' equation, *J. Math. Anal. Appl.* **30** (1970), 401–414, MR 41, 2236.

224. Rubinstein, L. I., Forced convection in a plane layer with axial symmetry, *Dokl. Akad. Nauk. SSSR* **135** (1960), 553–556 (Russian); translation *Soviet Physics Dokl.* **5** (1961), 1194–1197, MR 23, B806.

225. Rubinstein, L. I., On the conduction of heat in a thermo-anisotropic medium with infinitely small thermoconductivity in one direction, in *Latvian Mat. Yearbook 1965*, pp. 227–234. Izdat. "Zinatne," Riga, 1966 (Russian, Latvian, and English summaries), MR 34, 1028.

226. Rutner, J. F., and Skrjabina, L. P., Application of the Wiener–Hopf method to the solution of a boundary-value problem of the heat-transfer equation, *Differencial'nye Uravnenija* **2** (1966), 1101–1106 (Russian), MR 34, 1726.

227. Salmanova, D. G., Solution by the generalized Fourier method of a mixed problem for a second-order parabolic equation in an unbounded domain, *Izv. Akad. Nauk. Azerbaidzan. SSR Ser. Fiz.-Tehn. Mat. Nauk.*, No. 3, (1968), 118–123 (Russian, Azerbaijani summary), MR 41, 2226.

228. Samarskii, A. A., Parabolic equations with discontinuous coefficients, *Dokl. Akad. Nauk. SSSR* **121** (1958), 225–228 (Russian), MR 21, 2815.

229. Samosjuk, G. P., Solution of a one-dimensional diffusion problem with integro-differential boundary conditions, *Vestnik Leningrad. Univ.* **16** (1961), No. 1, 5–12 (Russian), MR 24, A2146.

230. Satoh, T., On the mathematical analysis of the problem of the heat conduction with the variable transfer coefficient, *J. Phys. Soc. Japan* **7** (1952), 245–249, MR 14, 283.

231. Schechter, E., Convergence considerations on a second-order method of approximate solution of parabolic equations, *Mathematica (Cluj)* **8** (31) (1966), 357–364, MR 35, 1960.

232. Seda, V., On the existence of a generalized solution to the first initial-boundary value problem for a nonlinear parabolic equation, *Ann. Mat. Pura Appl.* (4) **87** (1970), 375–388, MR 43, 3633.

233. Selig, F., and Fieber, H., Wärmeleitprobleme mit zeitlich varabler *Übergangszahl, Osterreich. Ing.-Arch.* **11** (1957), 37–40, MR 19, 213.

234. Sestini, G., Sulla risoluzione di un notevole gruppo di problemi rett i da equazioni di tipo parabolico, *Riv. Mat. Univ. Parma* **6** (1955), 349–352, MR 18, 656.

235. Sestini, G., Problemi di propagazione del calore con convezione forzata, *Riv. Mat. Univ. Parma* **8** (1957), 5–14, MR 20, 7513.

236. Sharma, O. P., *H*-function and heat production in a cylinder, *Proc. Nat. Acad. Sci. India, Sect. A,* **39** (1969), 355–360, MR 43, 7794.

237. Simko, N. G., A finite-integral Hankel transform for a hollow cylinder, *Inz.-Fiz. Z.* **3** (1960), 39–46 (Russian), MR 23, B1872.

238. Simoes-Pereora, J. M. S., The heat equation on closed surfaces, *Arquivo Inst. Gulberkian Ci. A Estud. Mat. Fis.-Mat.* **3** (1965), 1–68, MR 33, 4485.

239. Simoes, J. M. S., On the theory of the bi-dimensional diffusion equation, *Univ. Lisboa Revista Fac. Ci. A* (2) **11** (1964/65), 5–120 (Portuguese), MR 35, 587.

240. Sirao, T., A probabilistic treatment of semilinear parabolic equations, *Proc. Japan Acad.* **42** (1966), 885–890, MR 36, 983.

241. Smirnov, M. M., Some nonhomogeneous boundary problems of the equation of heat conduction, *Akad. Nauk. SSSR Prikl. Mat. Meh.* **15** (1951), 367–370 (Russian), MR 13, 134.

242. Sneddon, I. N., Solutions of the diffusion equation for a medium generating heat, *Proc. Glasgow Math. Assoc.* **1** (1952), 21–27, MR 14, 476.

243. Sobolev, S. L., *Partial-Differential Equations of Mathematical Physics*, Translated from the third Russian edition by E. R. Dawson; English translation edited by T. A. A. Broadbent. Pergamon Press, Oxford-Edinburgh-New York-Paris-Frankfurt; Addison-Wesley Publishing Co., Inc., Reading Mass.- London, 1964; x+427 pp., MR 31, 2478.

244. Sokolov, J. D., Sur l'application de la méthode des corrections fonctionnelles moyennes aux équations du type parabolique linéaires par rapport aux dérivées, *Ukrain. Mat. Z.* **12** (1960), 181–195 (Russian, French summary), MR 28, 4315.

245. Sparrow, E. M., and Eckert, E. R. G., Nonlinear problems of combined conductive–radiative heat transfer., in *Nonlinear Problems of Engineering*, pp. 104–122. Academic Press, New York, 1964, MR 30, 1741.

246. Stakgold, I., *Boundary-Value Problems of Mathematical Physics*. Vol. II. The Macmillan Co., New York; Collier-Macmillan Ltd., London, 1968; viii+408 pp., Chapters 6 and 7: Heat equation, MR 39, 4507.

247. Susea, A. A., A new boundary problem for the heat equation in two independent variables, *Stud. Cerc. Mat.* **18** (1966), 731–739 (Romanian), MR 35, 589.

248. Sutton, G. W., On one-dimensional heat conduction with an arbitrary heating rate, *J. Aero. Sci.* **24** (1957), 854–855, MR 19, 808.

249. v. Szalay, L., Anwendung des Picone-Verfahrens auf die Wärmeleitungs-differentialgleichung, *Elektrotech. Z.* **74** (1953), 141–143, MR 14, 758.

250. Szilvay, F. G., and Zergenyi, E., Über ein Wärmeleitungsproblem, *Magyar Tud. Akad. Alkalm. Mat. Int. Kozl.* **3** (1954), 253–263 (1955), (Hungarian, Russian and German summaries), MR 17, 845.

251. Taam, C. T., On nonlinear diffusion equations, *J. Differential Equations* **3** (1967), 482–499, MR 36, 501.

252. Thiruvenkatachar, V. R., and Ramakrishna, B. S., A case of combined radial and axial heat flow in composite cylinders, *Quart. Appl. Math.* **10** (1952), 255–262, MR 14, 283.

253. Tingley, A. L., On a generalization of the Poisson formula for the solution of the heat-flow equation, *Proc. Amer. Math. Soc.* **7** (1956), 846–851, MR 19, 149.

254. Tricomi, F. G., *Equazioni a Derivate Parziali*. Edizioni Cremonese, Rome 1958; xii + 392 pp., Chapter IV: Parabolic and heat equation, MR 20, 5928.

255. Tricomi, F. G., *Repertorium der Theorie der Differentialgleichungen*. Springer-Verlag, Berlin-New York, 1968; viii + 167 pp. Chapter 3: Parabolic equations and the heat equation, MR 38, 1301.

256. Trofimov, E. P., The nonstationary temperature field of an unlimited hollow cylinder, *Inz.-Fiz. Z.* **3** (1960), 47–53 (Russian), MR 23, B1873.

257. Usolcev, S. A., Solution of the heat-conduction equation for a semi-infinite rod with discontinuous coefficient, *Izv. Akad. Nauk. Kazah. SSSR Ser. Mat. Meh.*, No. 6 (10) (1957), 82–86 (Russian, Kazah summary), MR 21, 2476.

258. Vacca, M. T., Conduzione del calore in una piastra anulare, sottile, limitata da due circonferenze concentriche, *Atti Sem. Mat. Fis. Univ. Modena* **5** (1951), 190–212, MR 14, 758.

259. Ventcel', T. D., Certain quasilinear systems, *Dokl. Akad. Nauk. SSSR (N.S.)* **117** (1957), 21–24 (Russian), MR 21, 205.

260. Vladimirov, V. S., *The Equations of Mathematical Physics*. Izdat. "Nauka," Moscow, 1967, 436 pp. 0.98 r.; Chapter 3: Heat equation. Chapter 6: Parabolic problems, MR 39, 599.

261. Vodicka, V., Conduction de la chaleur dans une barre formée de plusieurs parties en matériaux différents, *Prace Mat.-Fiz.* **48** (1952), 45–52, MR 14, 983.

262. Vodicka, V., Eindimensionale Wärmeleitung in geschichteten Korpen, *Math. Nachr.* **14** (1955), 47–55, MR 17, 374.

263. Walsh, R. A., Initial-value problems associated with $u_t(x, t) = \delta u_{xx}(x, t) - u(x, t)u_x(x, t)$, *J. Math. Anal. Appl.* **26** (1969), 235–247, MR 38, 6251.

264. Wassermann, G. D., Heat conduction in solids as an eigenvalue problem, *Quart. J. Mech. Appl. Math.* **5** (1952), 466–471, MR 14, 560.

265. Yeh, J., Nonlinear Volterra functional equations and linear parabolic differential systems, *Trans. Amer. Math. Soc.* **95** (1960), 408–432, MR 23, A3917.

266. Yusupov, K. Y., Solution of a problem of heat conduction, *Kazan. Aviac. Inst. Trudy* **29** (1955), 39–46 (Russian), MR 19, 609.

267. Zeragiya, P. K., Solution of fundamental boundary problems for nonlinear differential equations of parabolic type by the method of academician S. A. Caplygin, *Soobsc. Akad. Nauk. Gruzin. SSR* **17** (1956), 103–109 (Russian), MR 17, 1091.

268. Zeragiya, P. K., Boundary problems for certain nonlinear equations of parabolic type, *Akad. Nauk. Gruzin. SSSR Trudy Tbiliss. Mat. Inst. Razmadze* **24** (1957), 195–221 (Russian), MR 20, 4082.

269. Zeuli, T., Su alcuni problemi di propagazione del calore in una sfera, *Atti Accad. Sci. Torino, Cl. Sci. Fis.-Mat. Nat.* **87** (1953), 127–145, MR 16, 45.

270. Zlateff, I., Sur la propagation de la chaleur dans une barre se composant de

deux parties du longueur variable dans des conditions linéaires aux limites, *C. R. Acad. Bulgare Sci.* **4** (1951) (1953), 5–8, MR 14, 1090.

2.3.   Stefan and Free-Boundary-Value Problems

1. Bacelis, R. D., and Melamed, V. G., Solution of a limiting boundary-value problem to which the generalized Stefan problem can be reduced, *Sibirsk. Mat. Z.* **5** (1964), 738–745 (Russian), MR 29, 4942.

2. Bacelis, R. D., Melamed, V. G., and Sljaifer, D. B., A solution of a Stefan-type problem by the method of straight lines, *Z. Vycisl. Mat. i Mat. Fiz.* **9** (1969), 585–594 (Russian), MR 40, 8290.

3. Berman, N. R., On a Stefan-type problem, *Bul. Akad. Stiince RSS Moldoven,* No. 4 (1966), 55–58 (Russian), MR 35, 5188.

4. Bocarova, I. V., The asymptotic behavior of solutions of a problem with free boundary for the heat equation, *Dokl. Akad. Nauk. SSSR* **143** (1962), 259–261 (Russian), MR 26, 5301.

5. Boley, B. A., A method of heat-conduction analysis of melting and solidification problems, *J. Math. and Phys.* **40** (1961), 300–313.

6. Boley, B. A., Upper and lower bounds for the solution of a melting problem, *Quart. Appl. Math.* **21** (1963), 1–11, MR 26, 4642.

7. Boley, B. A., Uniqueness in a melting slab with space- and time-dependent heating, *Quart. Appl. Math.* **27** (1969/70), 481–487, MR 40, 7643.

8. Budak, B. M., Vasil'ev, F. P., and Uspenskii, A. B., Difference methods for solving certain boundary-value problems of Stefan type, in *Numer. Methods in Gas Dynamics*, pp. 139–183. Izdat. Moskov. Univ., Moscow, 1965 (Russian), MR 34, 7126.

9. Budak, B. M., Gold'man, N. L., and Uspenskii, A. B., Difference schemes with linearization of fronts for solution of multifront problems of Stefan type, *Dokl. Akad. Nauk. SSSR* **167** (1966), 735–738 (Russian), MR 34, 7127.

10. Budak, B. M., Gold'man, N. L., and Uspenskii, A. B., Difference schemes with straightening of fronts for the solution of multifront problems of Stefan type, in *Comput. Methods Programming*, **VI**, pp. 206–216. Izdat. Moskov. Univ., Moscow, 1967 (Russian), MR 36, 4834.

11. Budak, B. M., and Moskal, M. Z., The classical solution of the multidimensional multifrontal Stefan problem, *Dokl. Akad. Nauk. SSSR* **188** (1969), 9–12 (Russian), MR 40, 6080.

12. Budak, B. M., and Moskal, M. Z., The classical solution of the multidimensional Stefan problem, *Dokl. Akad. Nauk. SSSR* **184** (1969), 1263–1266 (Russian), MR 41, 8845.

13. Budak, B. M., and Moskal, M. Z., The classical solution of a multidimensional, multiphase, Stefan-type problem in a region with piecewise-smooth boundary, *Dokl. Akad. Nauk. SSSR* **191** (1970), 751–754 (Russian), MR 41, 7299.

14. Cannon, J. R., and Hill, C. D., Existence, uniqueness, stability, and monotone dependence in a Stefan problem for the heat equation, *J. Math. Mech.* **17** (1967), 1–19, MR 42, 4893.

15. Cannon, J. R., Douglas, J., Jr., and Hill, C. D., A multiboundary Stefan

problem and the disappearance of phases, *J. Math. Mech.* **17** (1967), 21–33, MR 42, 4892.

16. Cannon, J. R., and Hill, C. D., Remarks on a Stefan problem, *J. Math. Mech.* **17** (1967), 433–441, MR 36, 1854.

17. Cannon, J. R., and Douglas, J., Jr., The stability of the boundary in a Stefan problem, *Ann. Scuola Norm. Sup. Pisa* (3) **21** (1967), 83–91, MR 42, 4891.

18. Cannon, J. R., and Hill, C. D., On the infinite differentiability of the free boundary in a Stefan problem, *J. Math. Anal. Appl.* **22** (1968), 385–397, MR 37, 610.

19. Cannon, J. R., Hill, C. D., and Primicerio, M., The one-phase Stefan problem for the heat equation with boundary temperature specification, *Arch. Rational Mech. Anal.* **39** (1970), 270–274, MR 42, 4894.

20. Cannon, J. R., and Hill, C. D., On the movement of a chemical reaction interface, *Indiana Univ. Math. J.* **20** (1970/71), 429–454, MR 43, 5170.

21. Cannon, J. R., and Primicerio, M., Remarks on the one-phase Stefan problem for the heat equation with the flux prescribed on the fixed boundary, *J. Math. Anal. Appl.* **35** (1971), 361–373, MR 43, 4357.

22. Chambre, P. L., On the dynamics of phase growth, *Quart. J. Mech. Appl. Math.* **9** (1956), 224–233, MR 18, 358.

23. Chan, C. Y., Continuous dependence on the data for a Stefan problem, *SIAM J. Math. Anal.* **1** (1970), 282–287, MR 41, 7307.

24. Citron, S. J., Heat conduction in a melting slab, *J. Aero/Space Sci.* **27** (1960), 219–228, MR 22, 2331.

25. Dacev, A. B., On the appearance of a phase in the linear problem of Stefan, *Dokl. Akad. Nauk. SSSR (N.S.)* **87** (1952), 353–356 (Russian), MR 14, 560.

26. Dacev, A., Sur le problème de Stéfan (probléme de congélation) au cas de deux ou trois dimensions, *Ann. Univ. Sofia Fac. Sci. Phys. Math.*, Livre 2, **48** (1953/54), 33–76 (1954), (Bulgarian summary), MR 20, 4081.

27. Dacev, A. B., On the two-dimensional Stefan problem, *Dokl. Akad. Nauk. SSSR (N.S.)* **101** (1955), 441–444 (Russian), MR 17, 624.

28. Dacev, A. B., On the two-dimensional Stefan problem, Translated by Morris D. Friedman, 572 California St., Newtonville 60, Mass., 1956; 7 pp., translated from the Russian of the article in *Dokl. Akad. Nauk. SSSR (N.S.)* **101** (1955), No. 3, 441–444, MR 19, 360.

29. Dacev, A. B., On the three-dimensional problem of Stefan, *Dokl. Akad. Nauk. SSSR (N.S.)* **101** (1955), 629–632 (Russian), MR 17, 624.

30. Douglas, J., Jr., and Gallie, T. M., Jr., On the numerical integration of a parabolic differential equation subject to a moving boundary condition, *Duke Math. J.* **22** (1955), 557–571, MR 17, 1241.

31. Douglas, J., Jr., A uniqueness theorem for the solution of a Stefan problem, *Proc. Amer. Math. Soc.* **8** (1957), 402–408, MR 19, 1060.

32. Evans, G. W., II, A note on the existence of a solution to a problem of Stefan, *Quart. Appl. Math* **9** (1951), 185–193, MR 13, 243.

33. Friedman, A., Free boundary problems for parabolic equations. I. Melting of solids, *J. Math. Mech.* **8** (1959), 499–517, MR 26, 1626.

34. Friedman, A., Free boundary problems for parabolic equations. II. Evaporation or condensation of a liquid drop, *J. Math. Mech.* **9** (1960), 19–66, MR 26, 1627.

35. Friedman, A., Free boundary problems for parabolic equations. III. Dissolution of a gas bubble in a liquid, *J. Math. Mech.* **9** (1960), 327–345, MR 26, 1628.

36. Friedman, A., Remarks on Stefan-type free-boundary problems for parabolic equations, *J. Math. Mech.* **9** (1960), 885–903, MR 26, 1629.

37. Friedman, A., The Stefan problem in several space variables, *Trans. Amer. Math. Soc.* **133** (1968), 51–87, MR 37, 3209.

38. Friedman, A., Correction to: "The Stefan problem in several space variables," *Trans. Amer. Math. Soc.* **142** (1969), 557, MR 39, 7289.

39. Friedman, A., One-dimensional Stefan problems with nonmonotone free boundary, *Trans. Amer. Math. Soc.* **133** (1968), 89–114, MR 37, 3210.

40. Friedman, A., Free-boundary problems for parabolic equations, *Bull. Amer. Math. Soc.* **76** (1970), 934–941, MR 43, 6600.

41. Frjazinov, I. V., The Stefan problem for inhomogeneous media, *Z. Vycisl. Mat. i. Mat. Fiz* **1** (1961), 927–932 (Russian), MR 24, B2349.

42. Greenberg, J. M., A free-boundary problem for the linear heat equation, *J. Differential Equations* **7** (1970), 287–306, MR 41, 8841.

43. Grigorian, S. S., On heating and melting of a solid body owing to friction, *J. Appl. Math. Mech.* **22** (1958), 815–825 (577–585 *Prikl. Mat. Meh*), MR 21, 7730.

44. Gu, L.-K., The behavior of the solution of Stefan's problem as time goes to infinity, *Dokl. Akad. Nauk. SSSR* **138** (1961), 263–266, (Russian), MR 29, 380.

45. Jackson, F., The solution of problems involving the melting and freezing of finite slabs by a method due to Portnov, *Proc. Edinburgh Math. Soc.* (2) **14** (1964/65), 109–128, MR 31, 1901.

46. Jiang, L.-S., The proper posing of free boundary problems for nonlinear parabolic differential equation, *Acta Math. Sinica* **12** (1962), 369–388 (Chinese); translated as *Chinese Math.* **3** (1963), 399–418, MR 29, 5000.

47. Jiang, L.-S., The two-phase Stefan problem. I., *Acta. Math. Sinica* **13** (1963), 631–646 (Chinese); translated as *Chinese Math.* **4** (1964), 686–702, MR 29, 2539.

48. Jiang, L.-S., The two-phase Stefan problem. II., *Acta Math. Sinica* **14** (1964), 33–49 (Chinese); translated as *Chinese Math.* **5** (1964), 36–53, MR 29, 2540.

49. Jiang, L.-S., The correctness of the problem with free boundary for nonlinear equations of parabolic type, *Sci. Sinica* **13** (1964), 193–212 (Russian), MR 30, 3305.

50. Jiang, L.-S., Existence and differentiability of the solution of a two-phase Stefan problem for quasilinear parabolic equations, *Acta Math. Sinica* **15** (1965), 749–764 (Chinese); translated as *Chinese Math.-Acta* **7** (1965), 481–496, MR 33, 7708.

51. Kamenomostskaja, S. L., On Stefan's problem, *Mat. Sb.* (*N.S.*) **53** (95) (1961), 489–514 (Russian), MR 25, 5292.

52. Kamynin, L. I., A hydraulics problem, *Dokl. Akad. Nauk. SSSR* **143** (1962), 779–781 (Russian), MR 24, A2758.

53. Kamynin, L. I., On the existence of the solution of Verigin's problem, *Z. Vycisl. Mat. i Mat. Fiz* **2** (1962), 833–858 (Russian), MR 28, 2364.

54. Kinoshita, N., and Mura, T., On Stefan's problem, *Sugaku* **8** (1956/57), 216–218 (Japanese), MR 21, 2830.

55. Kolodner, I. I., Free-boundary problem for the heat equation with applications to problems of change of phase. I. General method of solution, *Comm. Pure Appl. Math.* **9** (1956), 1–31, MR 19, 285.

56. Kruzhkov, S. N., On some problems with unknown boundaries for the heat-conduction equation, *Prikl. Mat. Meh.* **31** (1967), 1009–1020 (Russian), translated as *J. Appl. Math. Mech.* **31** (1967), 1014–1024 (1968), MR 37, 6618.

57. Kruzhkov, S. N., A class of problems with an unknown boundary for the heat-conduction equation, *Dokl. Akad. Nauk. SSSR* **178** (1968), 1036–1038 (Russian); translated as *Soviet Physics Dokl.* **13** (1968), 101–103, MR 39, 1817.

58. Kyner, W. T., On a free boundary-value problem for the heat equation, *Quart. Appl. Math.* **17** (1959), 305–310, MR 23, A1165.

59. Kyner, W. T., An existence and uniqueness theorem for a nonlinear Stefan problem, *J. Math. Mech.* **8** (1959), 483–498, MR 26, 1630.

60. Langford, D., New analytic solutions of the one-dimensional heat equation for temperature and heat-flow rate, both prescribed at the same fixed boundary (with applications to the phase-change problem), *Quart. Appl. Math.* **24** (1967), 315–322, MR 35, 1976.

61. Langford, D., Pseudosimilarity solutions of the one-dimensional diffusion equation with applications to the phase-change problem, *Quart. Appl. Math.* **25** (1967), 45–52, MR 35, 538.

62. Lefur, B., Bataille, J., and Aguirre-Puente, J., Étude de la congélation d'une lame plate dont une face est maintenue a température constante, l'autre face entant soumise a une température variable en fonction du temps (problème de Stefan unidimensional), *C. R. Acad. Sci. Paris* **259** (1964), 1483–1485, MR 29, 5529.

63. Li, V., On closed solutions of the many-dimensional Stefan problems, *Akad. Nauk. Kazah. SSR Trudy Sekt. Mat. Meh.* **2** (1963), 71–72 (Russian), MR 32, 2747.

64. Martynov, G. A., On propagation of heat in a two-phase medium for given law of motion of the phase boundary, *Z. Tehn. Fiz.* **25** (1955), 1754–1767 (Russian), MR 19, 710.

65. Martynov, G. A., On the solution of the inverse problem of Stephan for the semispace when the phase boundary moves according to a linear law, *Dokl. Akad. Nauk. SSSR (N.S.)* **109** (1956), 279–282 (Russian), MR 18, 314.

66. Martynov, G. A., Solution of the inverse Stefan problem in the case of spherical symmetry, *Z. Tehn. Fiz.* **30** (1960), 239–241 (Russian); translated as *Soviet Physics Tech. Phys.* **5**, 215–218, MR 22, 4430.

67. Melamed, V. G., Reduction of the Stefan problem to a system of ordinary differential equations, *Izv. Akad. Nauk. SSSR Ser. Geofiz.* (1958), 848–869 (Russian), MR 23, A1163.

68. Melamed, V. G., Solution of the Stefan problem in the case of the second boundary problem, *Vestnik Moskov. Univ. Ser. Mat. Meh. Astronom. Fiz. Him.*, No. 1(1959), 17–22 (Russian), MR 23, A1164.

69. Melamed, B. G., Solution of a Stefan-type problem for a certain quasilinear

parabolic system, *Z. Vycisl. Mat. i Mat. Viz.* **9** (1969), 1327–1335 (Russian), MR 42, 4034.

70. Miranker, W. L., A free-boundary-value problem for the heat equation, *Quart. Appl. Math.* **16** (1958), 121–130, MR 20, 656.

71. Miranker, W. L., and Keller, J. B., The Stefan problem for a nonlinear equation, *J. Math. Mech.* **9** (1960), 67–70, MR 22, 1331.

72. Mirzadzanzade, A. H., and Dzalilov, K. N., On approximate solution of the one-dimensional problem of Stefan, *Z. Tehn. Fiz.* **25** (1955), 1800–1803 (Russian), MR 18, 539.

73. Mura, T., and Kimosita, N., Stefan-like problem of a cylinder, *Proc. Fourth Japan National Congress Appl. Mech., 1954*, pp. 345–348. Science Council of Japan, Tokyo, 1955, MR 17, 374.

74. Nguen, D. C., On a problem with free boundary for a parabolic equation, *Vestnik Moskov, Univ. Ser. I. Mat. Meh.* **21** (1966), No. 2, 40–54 (Russian, English summary), MR 33, 2967.

75. Nguen, D. C., On a problem with a free boundary for a parabolic equation, *Vestnik Moskov. Univ. Ser. I. Mat. Meh.* **21** (1966), No. 5, 51–62 (Russian, English summary), MR 33, 7706.

76. Oleinik, O. A., A method of solution of the general Stefan problem, *Dokl. Akad. Nauk. SSSR* **135** (1960), 1054–1057 (Russian), translated as *Soviet. Math. Dokl.* **1** (1961), 1350–1354, MR 23, A2644.

77. Oleinik, O., On Stefan-type free-boundary problems for parabolic equations, *Seminari 1962* **63**, *Anal. Alg. Geom. e Topol.*, Vol. 1, *Ist. Naz. Alta. Mat.*, pp. 388–403. Ediz. Cremonese, Rome, 1965, MR 33, 434.

78. Paterson, S., Propagation of a boundary of fusion, *Proc. Glasgow Math. Assoc.* **1** (1952), 42–47, MR 14, 476.

79. Pirmamedov, V. G., On a certain problem of Stefan type, *Izv. Akad. Nauk. Azerbaidzan SSR Ser. Fiz.-Tehn. Mat. Nauk.* (1965), No. 4, 36–41 (Russian, Azerbaijani summary), MR 33, 7685.

80. Pirmamedov, V. G., Solution by the generalized method of integral relations of a certain Stefan-type problem, *Problems Comput. Math.* pp. 14–20. *Izdat. Akad. Nauk. Azerbaidzan. SSR*, Baku, 1968 (Russian), MR 39, 7835.

81. Potters, M. L., Some calculations on a parabolic differential equation with free boundary, *Math. Centrum Amsterdam, Rekenafdeling* R359b (1957), 12 pp., MR. 23, B2175.

82. Primicerio, M., Sur une classe de problème du type de Stéfan, *Annuaire Univ. Sofia Fac. Phys.* **63** (1968/69), 83–90 (1971), MR 44, 4404.

83. Quilghini, D., Un teorema di unicità per un problema del tipo di Stefan, *Boll. Un. Mat. Ital.* (3) **18** (1963), 270–278 (French summary), MR 28, 1414.

84. Quilghini, D., Su di un nuovo problema del tipo di Stefan, *Ann. Mat. Pura Appl.* (4) **62** (1963), 59–97 (English summary), MR 28, 3240.

85. Quilghini, D., Un teorema di unicità per un problema unidimensionale non lineare del tipo di Stefan, *Matematiche (Catania)* **20** (1965), 142–155 (English summary), MR 34, 4684.

86. Quilghini, D., Una analisi fisico-matematica del processo del combiamento de fase, *Ann. Mat. Pura Appl.* (4) **67** (1965), 33–74, MR 31, 4981.

87. Redozubov, D. V., On linear thermal problems with a shifting boundary and the problem of freezing for a semi-infinite medium, *Z. Tehn. Fiz.* **27** (1957),

2147–2157 (Russian), MR 28, 1412.

88. Rose, M. E., A method for calculating solutions of parabolic equations with a free boundary, *Math. Comput.* **14** (1960), 249–256, MR 22, 6085.

89. Rubinstein, L. I., On the asymptotic behavior of the phase-separation boundary in the one-dimensional problem of Stefan, *Dokl. Akad. Nauk. SSSR (N.S.)* **77** (1951), 37–40 (Russian), MR 12, 710.

90. Rubinstein, L. I., On the propagation of heat in a stratified medium with varying phase state, *Dokl. Akad. Nauk. SSSR (N.S.)* **79** (1951), 221–224 (Russian), MR 13, 134.

91. Rubinstein, L. I., On the uniqueness of solution of the homogeneous problem of Stefan in the case of a single-phase initial condition of the heat-conducting medium, *Dokl. Akad. Nauk. SSSR (N.S.)* **79** (1951), 45–47 (Russian), MR 13, 243.

92. Rubinstein, L. I., On the propagation of heat in a two-phase system having cylindrical symmetry, *Dokl. Akad. Nauk. SSSR (N.S.)* **79** (1951), 945–948 (Russian), MR 13, 243.

93. Rubinstein, L. I., The numerical solution of the integral equations of the Stefan problem, *Izv. Vyss. Ucebn. Zaved. Matematika*, No. 4 (5) (1958), 202–214, MR 25, 768.

94. Rubinstein, L. I., On a case of Stefan's problem, *Dokl. Akad. Nauk. SSSR* **142** (1962), 576–577 (Russian), translated as *Soviet Physics Dokl.* **7** (1962), 21–22, MR 25, 4258.

95. Rubinstein, L. I., On a variant of a one-dimensional Stefan problem with a strengthened nonlinearity, *Latvijas Valsts Univ. Zinatn. Raksti* **47** (1963), Laid. 1, 163–218 (Russian, English summary), MR 35, 5775.

96. Rubinstein, L. I., The two-phase Stefan problem on an interval with one-phase initial state of a heat-conducting medium, *Latvijas Valsts Univ. Zinatn. Raksti* **58** (1964), Vyp. 2, 111–148 (Russian, English summary), MR 31, 6066.

97. Rubinstein, L. I., On the uniqueness of the solution of a double-layer, single-phase problem of Stefan type, *Dokl. Akad. Nauk. SSSR* **160** (1965), 1019–1022 (Russian), MR 30, 3311.

98. Rubinstein, L. I., On the uniqueness of solution of a certain double-layer, single-phase problem of the Stefan type, *Latvian Mat. Yearbook 1965* (Russian), pp. 131–154. Izdat. "Zinatne," Riga, 1966, MR 34, 512.

99. Rubinstein, L. I., Solution of a single-phase, Stefan-type problem, *Dokl. Akad. Nauk. SSSR* **168** (1966), 770–773 (Russian), translated as *Soviet Physics Dokl.* **11** (1966), 485–487, MR 35, 585.

100. Rubinstein, L. I., The Stefan problem, *Latvian State Univ. Computing Center.* Izdat. "Zvaigzne," Riga, 1967, 457 pp. (loose errata), MR 36, 5488.

101. Ruoff, A. L., An alternate solution of Stefan's problem, *Quart. Appl. Math.* **16** (1958), 197–201, MR 19, 1230.

102. Sackett, G. G., An implicit free-boundary problem for the heat equation, *SIAM J. Numer. Anal.* **8** (1971), 80–96, MR 44, 654.

103. Schatz, A., Free-boundary problems of Stefan type with prescribed flux, *J. Math. Anal. Appl.* **28** (1969), 569–580, MR 42, 2187.

104. Selig, F., Bermerkungen zum Stefanschen Problem, *Osterreich. Ing.-Arch.* **10** (1956), 277–280, MR 18, 537.

105. Sestini, G., Esistenza di una soluzione in Problemi analoghi a quello di Stefan,

Rivista Mat. Univ. Parma **3** (1952), 3–23, MR 14, 381.

106. Sestini, G., Esistenza ed unicità nel problema di Stefan relativo a campi dotati di simmetria, *Rivista Mat. Univ. Parma* **3** (1952), 103–113, MR 14, 476.

107. Sestini, G., Sul problema unidimensional non lineare di Stefan in uno strato indefinito, *Ann. Mat. Pura Appl.* (4) **51** (1960), 203–224, MR 23, B804.

108. Sestini, G., Sul problema non lineare di Stefan in strati cilindrici o sferici, *Ann. Mat. Pura Appl.* (4) **56** (1961), 193–207, MR 25, 4259.

109. Sestini, G., Problemi analoghi a quello di Stefan e loro attualità, *Rend. Sem. Mat. Fis. Milano* **37** (1967), 39–50 (English summary), MR 37, 3206.

110. Sherman B., A free-boundary problem for the heat equation with heat input at a melting interface, *Quart. Appl. Math.* **23** (1965/66), 337–348, MR 32, 7966.

111. Sherman, B., A free-boundary problem for the heat equation with prescribed flux at both fixed face and melting interface, *Quart. Appl. Math.* **25** (1967), 53–63, MR 35, 3969.

112. Sherman, B., A general one-phase Stefan problem, *Quart. Appl. Math.* **28** (1970), 377–382, MR 43, 7795.

113. Sherman, B., Free-boundary problems for the heat equation in which the moving interface coincides initially with the fixed face, *J. Math. Anal. Appl.* **33** (1971), 449–466, MR 43, 733.

114. Sherman, B., General one-phase Stefan problems and free-boundary problems for the heat equation with Cauchy data prescribed on the free boundary, *SIAM J. Appl. Math.* **20** (1971), 555–570, MR 44, 4400.

115. Siskin, G. I., A certain heat problem with a free boundary, *Dokl. Akad. Nauk. SSSR* **197** (1971), 1276–1279 (Russian), MR 43, 3649.

116. Solomon, A., Some remarks on the Stefan problem, *Math. Comp.* **20** (1966), 347–360, MR 34, 2262.

117. Solomon, A., A steady-state, phase-change problem, *Math. Comp.* **21** (1967), 355–359, MR 36, 6166.

118. Solomon, A., The equilibrium configuration of a two-phase medium with a density change, *J. Differential Equations* **9** (1971), 549–554, MR 43, 695.

119. Tirskii, G. A., Two exact solutions of Stefan's nonlinear problem, *Dokl. Akad. Nauk. SSSR* **125** (1959), 293–296 (Russian), translated as *Soviet Physics, Dokl.* **4**, 288–292, MR 21, 7728.

120. Trench, W. F., On an explicit method for the solution of a Stefan problem, *J. Soc. Indust. Appl. Math.* **7** (1959), 184–204, MR 22, 1087.

121. Uspenskii, A. B., Method of rectifying the fronts for many-front, one-dimensional, Stefan-type problems, *Dokl. Akad. Nauk. SSSR* **172** (1967), 61–64 (Russian), translated as *Soviet Physics Dokl.* **12** (1967) 26–29, MR 35, 4550.

122. Vasil'ev, F. P., On the method of finite differences for solving a one-phase Stefan problem, *Z. Vycisl. Mat. i Mat. Fiz.* **3** (1963), 861–873 (Russian), MR 31, 897.

123. Vasil'ev, F. P., and Uspenskii, A. B., A difference method for solving a two-phase Stefan problem, *Z. Vycisl. Mat. i Mat. Fiz.* **3** (1963), 874–886 (Russian), MR 31, 1784.

124. Vasil'ev, F. P., The method of finite differences for solving a one-phase Stefan problem for a quasilinear equation, *Dokl. Akad. Nauk. SSSR* **152** (1963), 783–786 (Russian), MR 30, 5501.

125. Vasil'ev, F. P., Uspenskii, A. B., The finite-difference method for solving a two-phase Stefan problem for a quasilinear equation, *Dokl. Akad. Nauk. SSSR* **152** (1963), 1034–1037 (Russian), MR 31, 898.

126. Vasil'ev, F. P., A difference method of solving problems of Stefan type for a quasilinear parabolic equations with discontinuous coefficients, *Dokl. Akad. Nauk. SSSR* **157** (1964), 1280–1283 (Russian), MR 29, 5393.

127. Vasil'ev, F. P., The method of straight lines for the solution of a single-phase problem of Stefan type, *Z. Vycisl. Mat. i Mat. Fiz* **8** (1968), 64–78, (Russian), MR 37, 3211.

128. Vasil'ev, F. P., The existence of a solution of a certain optimal Stefan problem, in *Computing Methods and Programming*, XII, 110–114. *Izdat. Moskov. Univ. Moscow*, 1969 (Russian), MR 44, 3015.

129. Ventcel, T. D., A free boundary problem for the heat equation, *Dokl. Akad. Nauk. SSSR* **131** (1960), 1000–1003 (Russian), translated as *Soviet Math. Dokl.* **1**, 358–361, MR 22, 8225.

2.4. Not-Well-Posed and Inverse Problems

1. Arcangeli, R., Un problème de résolution rétrograde de l'équation de la chaleur, *Rev. Francaise Informat. Recherche Opérationnelle* **2** (1968), No. 13, 61–78.

2. Artole, M., Équations paraboliques à retardement, *C. R. Acad. Sci. Paris, Ser. A – B,* **264** (1967), A668–A671, MR 35, 3297.

3. Bragg, L. R., and Dettman, J. W., An operator calculus for related partial-differential equations, *J. Math. Anal. Appl.* **22** (1968), 261–271, MR 37, 561.

4. Cannon, J. R., Determination of an unknown coefficient in a parabolic differential equation, *Duke Math. J.* **30** (1963), 313–323, MR 28, 358.

5. Cannon, J. R., Douglas, J., Jr., and Jones, F. B., Jr., Determination of the diffusivity of an isotropic medium, *Internat. J. Engrg. Sci.* **1** (1963), 453–455 (French, German, Italian, and Russian summaries), MR 28, 3259.

6. Cannon, J. R., and Jones, B. F., Jr., Determination of the diffusivity of an anisotropic medium, *Internat. J. Engrg. Sci.* **1** (1963), 457–460 (French, German, Italian, and Russian summaries), MR 28, 3260.

7. Cannon, J. R., Determination of certain parameters in heat-conduction problems, *J. Math. Anal. Appl.* **8** (1964), 188–201, MR 28, 3261.

8. Cannon, J. R., A Cauchy problem for the heat equation, *Ann. Mat. Pura Appl.* (4) **66** (1964), 155–165, MR 30, 3304.

9. Cannon, J. R., A priori estimate for continuation of the solution of the heat equation in the space variable, *Ann. Mat. Pura Appl.* (4) **65** (1964), 377–387, MR 29, 6192.

10. Cannon, J. R., Some numerical results for the solution of the heat equation backwards in time, in *Numerical Solutions of Nonlinear Differential Equations* (*Proc. Adv. Sympos. Madison Wis.*, 1966), pp. 21–54. John Wiley and Sons, Inc., New York, 1966, MR 34, 7037.

11. Cannon, J. R., and Douglas, J., Jr., The Cauchy problem for the heat equations, *SIAM J. Numer. Anal.* **4** (1967), 317–336, MR 35, 7614.

12. Cannon, J. R., Determination of an unknown heat source from overspecified boundary data, *SIAM J. Numer. Anal.* **5** (1968), 275–286, MR 37, 7105.

13. Cannon, J. R., and Hill, C. D., Continuous dependence of bounded solutions of a linear, parabolic partial-differential equation upon interior Cauchy data, *Duke Math. J.* **35** (1968), 217–230, MR 36, 6757.

14. Cannon, J. R., and Douglas, J., Jr., The approximation of harmonic and parabolic functions on half-spaces from interior data, in *Numerical Analysis of Partial Differential Equations* (C.I.M.E. 2° Ciclo, Ispra, 1967), pp. 193–230. Edizioni Cremonese, Rome, 1968, MR 39, 5076.

15. Cannon, J. R., and Knightly, G. H., The approximation of the solution of the heat equation in a half-strip from data specified on the bounding characteristics, *SIAM J. Numer. Anal.* **6** (1969), 149–159, MR 40, 1048.

16. Chavent, G., Sur une méthode de résolution du problème inverse dans les équations aux dérivées partielles paraboliques, *C. R. Acad. Sci. Paris, Ser. A – B*, **269** (1969), A1135–A1138, MR 42, 684.

17. Chavent, G., Deux résultats sur la problème inverse dans les équations aux dérivées partielles du deuxième ordre en *t* et sur l'unicité de la solution du problème inverse de la diffusion, *C. R. Acad. Sci. Paris, Ser. A – B*, **270** (1970), A25–A28, MR 41, 2245.

18. Douglas, J., Jr., and Jones, B. F., Jr., The determination of a coefficient in a parabolic differential equation, II. Numerical approximation, *J. Math. Mech.* **11** (1962), 919–926, MR 27, 3949.

19. Douglas, J., Jr., Approximate continuation of harmonic and parabolic functions, in *Numerical Solution of Partial Differential Equations* (Proc. Sympos. Univ. Maryland, 1965), pp. 353–364. Academic Press, New York, 1966, MR 34, 2206.

20. Effros, E. G., and Kazdan, J. L., On the Dirichlet problem for the heat equation, *Indiana Univ. Math. J.* **20** (1970,71), 683–693, MR 42, 3440.

21. Ginsberg, F., On the Cauchy problem for the one-dimensional heat equation, *Math. Comp.* **17** (1963), 257–269, MR 28, 5266.

22. Glagoleva, R. J., The continuous dependence on the initial data of the solution of the first boundary-value problem for a parabolic equation with negative time, *Dokl. Akad. Nauk. SSSR* **148** (1963), 20–23 (Russian), MR 26, 1631.

23. Glagoleva, R. J., The three-cylinder theorem and its applications, *Dokl. Akad. Nauk. SSSR* **163** (1965), 801–804 (Russian), MR 32, 4394.

24. Glagoleva, R. J., Some properties of the solutions of a second-order, linear parabolic equation, *Mat. Sb. (N.S.)* **74** (116) (1967), 47–74 (Russian), MR 35, 7003.

25. Glück, V., Bestimmung der Diffusionskonstante in Kenntnis der Materieverteilung des mehrschichtigen Systems, *Magyar Tud. Akad. Alkalm. Mat. Int. Kozl.* **2** (1953), 361–366 (1954) (Hungarian, Russian, and German summaries), MR 16, 287.

26. Hayashida, K., Unique continuation property of nonpositive weak subsolutions for parabolic equations of higher order, *Nagoya Math. J.* **24** (1964), 241–248, MR 33, 4476.

27. Hill, C. D., A method for the construction of reflection laws for a parabolic equation, *Trans. Amer. Math. Soc.* **133** (1968), 357–372, MR 38, 3597.

28. Hurd, A. E., Backward continuous dependence for mixed parabolic problems, *Duke Math. J.* **34** (1967), 493–500, MR 35, 6964.

29. Hurd, A. E., Backward lower bounds for solutions of mixed parabolic prob-

lems, *Michigan Math. J.* **17** (1970), 97–102, MR 42, 4906.

30. Iskenderov, A. D., Inverse boundary-value problems with unknown coefficients for certain quasilinear equations, *Dokl. Akad. Nauk. SSSR* **178** (1968), 999–1002 (Russian), MR 36, 6800.

31. Iskenderov, A. D., Certain multidimensional, inverse boundary-value problems, *Aserbaidzan Gos. Univ. Univ. Ucen. Zap. Ser. Fiz-Mat. Nauk.* (1968), No. 2, 76–80 (Russian), MR 43, 7761.

32. John, F., Numerical solution of the equation of heat conduction for preceding times, *Ann. Mat. Pura Appl.* (4) **40** (1955), 129–142, MR 19, 323.

33. Johnson, R., *A priori* estimates and unique continuation theorems for second-order parabolic equations, *Trans. Amer. Math. Soc.* **158** (1971), 167–177, MR 43, 3630.

34. Jones, B. F., Jr., The determination of a coefficient in a parabolic differential equation, I. Existence and uniqueness, *J. Math. Mech.* **11** (1962), 907–918, MR 27, 3948.

35. Jones, B. F., Jr., Various methods for finding unknown coefficients in parabolic differential equations, *Comm. Pure Appl. Math.*, **16** (1963), 33–44, MR 27, 2735.

36. Komatsu, H., Abstract analyticity in time and unique continuation property of solutions of a parabolic equation, *J. Fac. Sci. Univ. Tokyo, Sect. I*, **9** (1961), 1–11, MR 29, 6193.

37. Lavrent'ev, M. M., Romanov, V. G., and Vasil'ev, V. G., *Multidimensional Inverse Problems for Differential Equations.* Izdat. "Nauka" Sibirsk. Otdel., Novosibirsk, 1969; 67 pp. Chapter IV is devoted to heat equation, MR 43, 3599.

38. Lattes, R., and Lions, J.-L., *Méthode de Quasi-Reversibilité et Applications*, *Travaux et Recherches Mathématiques*, No. 15. Dunod, Paris, 1967, xii + 368 pp., Chapter 3: Heat equation, MR 38, 874.

39. Lees, M., and Protter, M. H., Unique continuation for parabolic differential equations and inequalities, *Duke Math. J.* **28** (1961), 369–382, MR 25, 4254.

40. Lions, J. L., Sur la stabilisation de vertains problèmes mal posés, *Rend. Sem. Mat. Fis. Milano* **36** (1966), 80–87 (English summary), (Applications to Cauchy problem for parabolic equations.) MR 35, 6965.

41. Masuda, K., A note on the analyticity in time and the unique continuation property for solutions of diffusion equations, *Proc. Japan Acad.* **43** (1967), 420–422, MR 36, 5529.

42. Miller, K., Three circle theorems in partial-differential equations and applications to improperly posed problems, *Arch. Rational Mech. Anal.* **16** (1964), 126–154, MR 29, 1435.

43. Miranker, W. L., A well-posed problem for the backward heat equation, *Proc. Amer. Math. Soc.* **12** (1961), 243–247, MR 22, 11216.

44. Mizel, V. J., and Seidman, T. I., Observation and prediction for the heat equation, *J. Math. Anal. Appl.* **28** (1969), 303–312, MR 40, 569.

45. Mizohata, S., Unicité du prolongement des solutions pour quelques opérateurs différentiels paraboliques, *Mem. Coll. Sci. Univ. Kyoto. Ser. A, Math.*, **31** (1958), 219–239, MR 21, 5081.

46. Ohya, Y., Sur l'unicité du prolongement des solutions pour quelques équations différentielles paraboliques, *Proc. Japan Acad.* **37** (1961), 358–362, MR 27,

1708.

47. Philip, J., Estimates of the age of a heat distribution, *Ark. Mat.* **7** (1968), 351–358, MR 37, 6619.

48. Pistoia, A., Sul problema inverso di propagazione, *Ist. Lombardo Sci. Lett. Rend. Cl. Sci. Mat. Nat.* (3) **17** (86) (1953), 760–768, MR 16, 260.

49. Pucci, C., Teoremi di esistenza e di unicità per il problema di Cauchy nella teoria delle equazioni lineari a derivate parziali. II., *Atti Accad. Naz. Lincei. Rend. Cl. Sci. Fis. Mat. Nat.* (8) **13** (1952), 111–116, MR 15, 35.

50. Pucci, C., Studio col metodo delle differenze di un problema di Cauchy relativo ad equazioni a derivate parziali del secondo ordine di tipo parabolico, *Ann. Scuola Norm. Super. Pisa* (3) **7** (1954), 205–215, MR 16, 140.

51. Pucci, C., On the improperly posed Cauchy problems for parabolic equations, in *Symp. Numerical Treatment of Partial Differential Equations with Real Characteristics*: Proc. Rome Symp. (28,29,30 January 1959) organized by the Provisional International Computation Centre; pp. 140–144. Libreria Eredi Virgilio Veschi, Rome 1959; xii + 158 pp., MR 21, 6100.

52. Pucci, C., Alcuni limitazioni per le soluzioni di equazioni paraboliche, *Ann. Mat. Pura Appl.* (4) **48** (1959), 161–172 (English summary), MR 22, 5807.

53. Quilghini, D., Un esempio di problema inverso di quello tipico dela conduzione, *Matematiche* (*Catania*) **20** (1965), 176–184, MR 34, 4719.

54. Reznickaja, K. G., A uniqueness theorem for a certain inverse problem in the theory of heat conductivity, linearized case, *Sibirsk. Mat. Z.* **9** (1968), 898–902 (Russian), translation, *Siberian Math. J.* **9** (1968), 666–670, MR 38, 3615.

55. Shirota, T., A unique continuation theorem of a parabolic differential equation, *Proc. Japan Acad.* **35** (1959), 455–460, MR 22, 11217a.

56. Shirota, T., A remark on my paper "A unique continuation theorem of a parabolic differential equation," *Proc. Japan Acad.* **36** (1960), 133–135, MR 22, 11217b.

57. Shirota, T., A theorem with respect to the unique continuation for a parabolic differential equation, *Osaka Math. J.* **12** (1960), 377–386, MR 23, A3922.

58. Smulev, I. I., Bounded solutions of boundary-value problems without initial conditions for parabolic equations and inverse boundary-value problems, *Dokl. Akad. Nauk. SSSR* **142** (1962), 46–49 (Russian), MR 24, A2763.

59. Sparrow, E. M., Haji-Sheikh, A., and Lundgren, T. S., The inverse problem in transient heat conduction, *Trans. ASME Ser. E., J. Appl. Mech.* **31** (1964), 369–375, MR 29, 4417.

60. Tsutsumi, A., A remark on the uniqueness of the noncharacteristic Cauchy problem for equations of parabolic type, *Proc. Japan Acad.* **41** (1965), 65–70, MR 32, 296.

61. Van, T., Theory of the heat potential. I. Level curves of heat potentials and the inverse problem in the theory of the heat potential, *Z. Vycisl. Mat. i Mat. Fiz.* **4** (1964), 660–670 (Russian), MR 29, 3766.

62. Vasilach, S., Sur l'équation du type parabolique $(\lambda^2 \partial^2 u / \partial x^2) - (\mu^2 \partial u / \partial y) = f(x, y)$, *C. R. Acad. Sci. Paris. Ser. A–B*, **272** (1971), A533–A536, MR 43, 7783.

63. Yamabe, H., A unique continuation theorem of a diffusion equation, *Ann. Mat.* (2) **69** (1959), 462–466, MR 21, 206.

64. Yosida, K., Integrability of the backward diffusion equation in a compact

Riemannian space, *Nagoya Math. J.* **3** (1951), 1–4, MR 13, 560.

65. Zaidman, S., Un teorema di esistenza per un problema non bene posto, *Atti Accad. Naz. Lincei Rend. Cl. Sci. Fis. Mat. Nat.* (8) **35** (1963), 17–22, MR 30, 375.

3. RECENT: 1973–1981

3.1. Properties of Solutions

A. *A priori estimates and maximum principles*

1. Aleksandru, I. N., A certain estimate of the solution of the first boundary-value problem for the heat equation, *Mat. Issled.* **9**, No. 3 (33) (1974), 190–194, 217 (Russian), MR 51, 13453.

2. Collins, W. D., Dual extremum principles for the heat equation, *Proc. Roy. Soc. Edinburgh, Sect. A*, **77**, No. 3–4 (1977), 273–292, MR 80g, 49025.

3. Deslauriers, G., and Dubuc, S., Log concavity of the cooling of a convex body, *Proc. Amer. Math. Soc.* **74**, No. 2 (1979), 291–294, MR 80d, 35068.

4. Edelstein, W. S., Further study of spatial-decay estimates for semilinear parabolic equations, *J. Math. Anal. Appl.* **35** (1971), 577–590, MR 46, 518.

5. Ellis, R. S., and Newman, C. M., Extensions of the maximum principle: exponential preservation by the heat equation, *J. Differential Equations* **30**, No. 3 (1978), 365–379, MR 80, 35069.

6. Fasano, A., and Primicerio, M., A stability theorem for diffusion problems with sharply changing temperature-dependent coefficients, *Quart. Appl. Math.* **33** (1975), 131–141, MR 56, 9072.

7. Knowles, J. K., On the spatial decay of solutions of the heat equation, *Z. Angew. Math. Phys.* **22** (1971), 1050–1056, MR 45, 7307.

8. Kurdjumov, S. P., Mihailov, A. P., and Plohotnikov, K. E., Heat localization in multidimensional problems of nonlinear heat conduction, *Inst. Prikl. Mat. Akad. Nauk. SSSR, Moscow* **22** (1977), 50 pp. (Russian), MR 57, 18505.

9. Mazilu, P., The equation of heat conduction, *Rev. Roumaine Math. Pures Appl.* **23**, No. 3 (1978), 419–435, MR 58, 14429.

10. Rolandi, F., A maximum principle for the solution to the nonlinear heat equation that is discontinuous with respect to the unknown, *Istit. Lombardo Accad. Sci. Lett. Rend. A*, **112**, No. 1 (1978), 12–18 (Italian), MR 81d, 35040.

11. Watson, N. A., Convexity of mean values of nonnegative subtemperatures, *Arch. Rational Mech. Anal.* **73**, No. 1 (1980), 53–62, MR 81c, 35059.

12. Wilkins, J. E., Jr., Maximum internal temperatures in one-dimensional geometries with nonuniform heat generation and equal surface temperatures, *SIAM J. Appl. Math.* **21** (1971), 306–320, MR 46, 2225.

B. *Initial-value problem and uniqueness*

1. Aronson, D. G., Widder's inversion theorem and the initial distribution problems, *SIAM J. Math. Anal.* **12**, No. 4 (1981), 639–651, MR 82i, 35081.

2. Denisov, V. N., On a necessary condition for uniform stabilization of the

solution of the Cauchy problem for the heat equation, *Dokl. Akad. Nauk. SSSR* **255**, No. 6 (1980), 1310–1312 (Russian), MR 82c, 35032.

3. Dressel, F. G., Uniqueness theorems for the heat equation. U.S. Army Research Office (Durham) 73–3 (1973), 1181–1187, MR 50, 7814.

4. Fabes, E. B., and Neri, U., Characterization of temperatures with initial data in BMO, *Duke Math. J.* **42**, No. 4 (1975), 725–734, MR 53, 1023.

5. Franchi, F., A uniqueness theorem for the propagation of heat through forced convection with finite velocity in an unbounded domain, *Atti Accad. Sci. Torino Cl. Sci. Fis. Mat. Natur.* **113**, No. 1–2 (1979), 33–37 (Italian), MR 81b, 80003.

6. Goebel, J., Eine Anfangswertaufgabe für die homogene Wärmeleitungsgleichung, *Wiss. Z. Techn. Hochsch. Ilmenau* **23**, No. 5 (1977), 157–161 (German), MR 58, 23060.

7. Herrmann, R. P., and Nachlinger, R. R., A theorem on the uniqueness of solutions in nonlinear heat conduction, *Quart. Appl. Math.* **32** (1974), 329–332, MR 55, 8551.

8. Kuznecov, J. V., The uniqueness of the solution of the fundamental problem of heat conduction in a boundary layer, *Perm. Politehn. Inst. Sb. Nauchn. Trudov* **152** (1974), 37–42 (Russian), MR 53, 11228.

9. Vauthier, J., Unicité du problème de Cauchy pour l'équation de la chaleur sur le complexe de de Rham, *C. R. Acad. Sci. Paris Ser. A–B*, **281**, No. 1 (1975), Aii, A41–A43 (French, English summary), MR 52, 6798.

10. Watson, N. A., Uniqueness and representation theorems for the inhomogeneous heat equation, *J. Math. Anal. Appl.* **67**, No. 2 (1979), 513–524, MR 81c, 35017.

11. Wilcox, C. H., Positive temperatures with prescribed initial heat distributions, *Amer. Math. Monthly* **87**, No. 3 (1980), 183–186, MR 81i, 35077.

C. Representations, transformations, special functions, Runge approximations

1. Bakievich, N. I., The connection between the solutions of a fourth-order equation from the theory of the filtration of a fluid with a free surface and the solutions of the heat equation, *Ukrain. Mat. Zh.* **33**, No. 2 (1981), 241–243 (Russian), MR 82i, 35026.

2. Bluman, G. W., and Cole, J. D., The general similarity solution of the heat equation, *J. Math. Mech.* **18** (1968), 1025–1042, MR 45, 2334.

3. Bondarenko, B. A., and Hodzhanijazov, D., Polynomial and associated solutions of the heat equation and its iterates, *Izv. Akad. Nauk. UzSSR Ser. Fiz.-Mat. Nauk.* **18**, No. 5 (1974), 18–24 (Russian, Uzbek summary), MR 52, 1005.

4. Bouillet, J. E., de Saravia, D. A., and Villa, L. T., Similarity solutions of the equation of one-dimensional heat conduction, *J. Differential Equations* **35**, No. 1 (1980), 55–65, MR 81c, 35069.

5. Colton, D., Walsh's theorem for the heat equation, *Lecture Notes in Math.* **564** (1976), 54–60. Springer, Berlin, MR 58, 29110.

6. Colton, D., *Integral-Operator Methods in the Theory of Wave Propagation and Heat Conduction*. Institute for Mathematical Sciences, University of Delaware,

Newark, Del.; i + 54 pp. 1977, MR 56, 16121.

7. Colton, D., and Watzlawek, W., Complete families of solutions to the heat equation and generalized heat equation in R^n, *J. Differential Equations* **25**, No. 1 (1977), 96–107, MR 56, 854.

8. Colton, D., and Wimp, J., Asymptotic behavior of the fundamental solution to the equation of heat conduction in two temperatures, *J. Math. Anal. Appl.* **69**, No. 2 (1979), 411–418, MR 80h, 35079.

9. Diaz. R., A Runge theorem for solutions of the heat equation, *Proc. Amer. Math. Soc.* **80**, No. 4 (1980), 643–646, MR 82a, 35050.

10. Gal'perin, S. M., Asymptotic representation of the solution of a parabolic equation with a discontinuous coefficient and a discontinuous initial condition, *Izv. Vyssh. Uchebn. Zaved. Matematika 1977*, No. 7 (182) (1977), 46–53 (Russian), MR 58, 6647.

11. Greiner, P., An asymptotic expansion for the heat equation, *Arch. Rational Mech. Anal.* **41** (1971), 163–218, MR 48, 9774.

12. Gupta, P. M., and Kulshreshtha, S. K., Functions of four variables which satisfy both the heat equation and the Laplace equation in three variables. II., *Portugal. Math.* **29** (1970), 213–219, MR 49, 3324.

13. Haimo, D. T., Equivalence of integral transform and series-expansion representations of generalized tempratures, in *Analytic Methods in Mathematical Physics* (Sympos., Indiana Univ., Bloomington, Ind., 1968). Gordon and Breach, New York, 453–459, 1970, MR 50, 14084.

14. Haimo, D. T., Series expansions with convergence in the mean for dual Laguerre temperatures, *Indiana Univ. Math. J.* **22** (1972), 207–215, MR 46, 2117.

15. Haimo, D. T., Series expansions for dual Laguerre temperatures, *Canad. J. Math.* **24** (1972), 1145–1153, MR 47, 613.

16. Haimo, D. T., Widder temperature representations, *J. Math. Anal. Appl.* **41** (1973), 170–178, MR 47, 3836.

17. Haimo, D. T., Homogeneous generalized temperatures, *SIAM J. Math. Anal.* **11**, No. 3 (1980), 473–487, MR 81h, 80013.

18. Jones, B. F., Jr., An approximation theorem of Runge type for the heat equation, *Proc. Amer. Math. Soc.* **52** (1975), 289–292, MR 52, 8654.

19. Kahlig, P., Über einige Lösungen der eindimensionalen Diffusionsgleichung, *Osterreich. Akad. Wiss. Math.-Naturwiss. Kl. S.-B. II,* **186**, No. 4–7 (1977), 193–215 (German), MR 57, 10030.

20. Koprinski, S. C., Operators from the Mikusinski field related to the generalized equation of heat conductivity with constant coeffcients and polynomial boundary conditions, *Godishnik Vissh. Uchebn. Zaved. Prilozhna Mat.* **13**, No. 3 (1977), 139–147 (Bulgarian, Russian, and English summaries), MR 82c, 35009.

21. Koprinsky, S. T., An operational method for the Bessel heat equation, in *Generalized Functions and Operational Calculus* (Proc. Conf., Varna, 1975), Bulgar. Acad. Sci., Sofia, 138–143, 1979, MR 80m, 35002.

22. Llubowicz, J., Solution of the first Fourier problem for the generalized heat equation and its relation with the Laplace equation, *Demonstratio Math.* **9**, No. 4 (1976), 535–543, MR 55, 3537.

23. Meyer, H. D., A representation for a distributional solution of the heat

equation, *SIAM J. Math. Anal.* **5** (1974), 708–722, MR 50, 10519.

24. Pathak, R. S., Expansion of distributions in terms of heat polynomials, *Aligarh Bull. Math.* **7** (1977), 81–90, MR 82i, 33019.

25. Polk, J. F., Asymptotic approximations to the solution of the heat equation, *Rocky Mountain J. Math.* **6**, No. 4 (1976), 697–708, MR 57, 13141.

26. Prasad, Y. N., and Siddiqui, A., Application of generalized function in the production of heat in a cylinder, *Defence Sci. J.* **25**, No. 3 (1975), 107–114, MR 57, 12933.

27. Rutner, J. F., An inversion transformation and its application to the solution of generalized boundary-value problems of heat conduction, in *Differential Equations and Their Applications.* Kuibyshev. Gos. Ped. Inst., Kuybyshev, 88–96, 177, 1975, MR 58, 32450.

28. Scraton, R. E., Polynomial approximations to the solution of the heat equation, *Proc. Cambridge Philos. Soc.* **73** (1973), 157–165, MR 47, 1310.

29. Shah, M., A note involving partial-differential equations related to heat conduction, *Indian. J. Pure Appl. Math.* **4**, No. 3 (1973), 333–340, MR 48, 2572.

30. Singh, F., Use of special functions in the production of heat in a cylinder, *Defence Sci. J.* **22** (1972), 215–220, MR 47, 3837.

31. Tait, R. J., Additional pseudosimilarity solutions of the heat equation in the presence of moving boundaries, *Quart. Appl. Math.* **37**, No. 3 (1979), 313–324, MR 80j, 35045.

32. Watson, N. A., Positive thermic majorization of temperatures on infinite strips, *J. Math. Anal. Appl.* **68**, No. 2 (1979), 477–487, MR 80i, 35031.

33. Widder, D. V., Homogeneous solutions of the heat equation, in *Analytic Methods in Mathematical Physics* (Sympos., Indiana Univ., Bloomington, Ind., 1968). Gordon and Breach, New York, 379–398, 1970, MR 49, 5568.

34. Widder, D. V., Time-variable singularities for solutions of the heat equation, *Proc. Amer. Math. Soc.* **32** (1972), 209–214, MR 45, 3974.

35. Widder, D. V., The Huygens property for the heat equation, *Trans. Amer. Math. Soc.* **232** (1977), 239–244, MR 56, 12597.

36. Young, E. C., Basic sets of polynomials for a generalized heat equation and its iterates, *Riv. Mat. Univ. Parma* (2) **11** (1970), 97–102, MR 46, 9531.

37. Zaki, M., Almost automorphic solutions of nonhomogeneous heat equation, *Ann. Univ. Ferrara Sez. VII* (*N.S.*) **16** (1971), 149–156 (English, Italian summary), MR 46, 520.

D. Potentials, Green's functions, capacity, boundary behavior, and some applications

1. Antimirov, M. J., The Green's function of a one-dimensional equation of parabolic type when the boundary moves according to the $t^{\beta/2}$ law, *Latvian Math. Yearbook*, **17** (1976), 70–99, 277 (Russian), MR 54, 13315.

2. Arena, O., On a singular parabolic equation related to axially symmetric heat potentials, *Ann. Mat. Pura Appl.* (4) **105** (1975), 347–393, MR 52, 1011.

3. Aronszajn, N., Preliminary notes for the talk "Traces of analytic solutions of the heat equation," *Colloque International CNRS sur les Équations aux Dérivées Partielles Linéaires* (Univ. Paris-Sud, Orsay, 1972), Soc. Math. France, Paris, 5–34, 1973, MR 58, 29239a.

4. Aronszajn, N., Traces of analytic solutions of the heat equation, *Colloque International CNRS sur les Équations aux Dérivées Partielles Linéaires* (Univ. Paris-Sud, Orsay, 1972), Soc. Math. France, Paris, 35–68, 1973, MR 58, 29239b.

5. Atiyah, M., Bott, R., and Patodi, V. K., On the heat equation and the index theorem, *Invent. Math.* **19** (1973), 279–330, MR 58, 31287.

6. Baderko, E. A., The solution by the method of parabolic potentials of a certain heat-conduction problem with concentrated thermal capacities, *Differencial'nye Uravnenija* **8** (1972), 1225–1234 (Russian), MR 47, 7219.

7. Baouendi, M. S., Solvability of partial-differential equations in the traces of analytic solutions of the heat equation, *Colloque International CNRS sur les Équations aux Dérivées Partielles Linéaires* (Univ. Paris-Sud, Orsay, 1972), Soc. Math. France, Paris, 86–97, 1973, MR 58, 1564.

8. Baouendi, M. S., Solvability of partial-differential equations in the traces of analytic solutions of the heat equation, *Amer. J. Math.* **97**, No. 4 (1975), 983–1005, MR 52, 11290.

9. Bauer, K. W., Zur Lösungsdarstellung bei gewissen parabolischen Differentialgleichungen, *Rend. Ist. Mat. Univ. Trieste*, **7**, No. 2 (1975), 116–127 (Italian and English summaries), MR 53, 13804.

10. Bochner, S., Positivity of the heat kernel for ultraspherical polynomials and similar functions, *Arch. Rational Mech. Anal.* **70**, No. 3 (1979), 211–217, MR 81b, 35040.

11. Bokov, A. G., The membership of the solution of the inhomogeneous heat equation in a certain class of functions, *Godishnik Vissh. Tehn. Uchebn. Zaved. Mat.* **9**, No. 2 (1973), 109–120 (Bulgarian, Russian, and German summaries), MR 56, 6116.

12. Borzymowski, A., Continuity of tangential derivatives of a thermal potential, *Demonstratio Math.* **3** (1971), 205–221, MR 48, 2568.

13. Burykin, A. J., Construction of the Green functions of boundary-value problems for the heat equation in a half-space with a slit, *Dokl. Akad. Nauk. Ukrain. SSR Ser. A*, *1977*, No. 8 (1977), 675–678 (Russian, English summary), MR 58, 1635.

14. Cakala, S., Solution fondamentale modifiée de l'équation de la chaleur et son application à la construction de la solution fondamentale modifiée d'un système parabolique, *Demonstratio Math.* **8** (1975), 33–47 (French), MR 51, 1119.

15. Cheeger, J., and Yau, Shing Tung, A lower bound for the heat kernel, *Comm. Pure Appl. Math.* **34**, No. 4 (1981), 465–480, MR 82i, 58065.

16. Dont, M., On a heat potential, *Comment. Math. Univ. Carolinae* **14** (1973), 559–564, MR 48, 9107.

17. Dont, M., On a heat potential, *Czechoslovak Math. J.* **25** (100) (1975), 84–109, MR 51, 6147.

18. Dont, M., A note on a heat potential and the parabolic variation, *Chasopis Pest. Mat.* **101**, No. 1 (1976), 28–44 (English, Loose Russian summary), MR 57, 13202.

19. Dont, M., The heat and adjoint heat potentials, *Chasopis Pest. Mat.* **105**, No. 2 (1980), 199–203, 209 (English, Czech summary), MR 81i, 31018.

20. Dont, M., On the continuity of heat potentials, *Chasopis Pest. Mat.* **106**, No. 2 (1981), 156–167, 209 (English, Czech summary), MR 82h, 35043.

21. Dorodnicyn, V. A., Group properties and invariant solutions of an equation of nonlinear heat transport with a source or a sink, *Akad. Nauk. SSR Inst. Prikl. Mat., Preprint* 57 (1979), 31 pp. (Russian, English summary), MR 80e, 80004.

22. Evan, L. C., Regularity properties for the heat equation subject to nonlinear boundary constraints, *Nonlinear Anal.* 1, No. 6 (1976), 593–602, MR 58, 29276.

23. Hansen, W., Fegen und Dunnheit mit Anwendungen auf die Laplace- und Wärmeleitungsgleichung, *Ann. Inst. Fourier* (Grenoble) 21, No. 1 (1971), 79–121 (German, English, and French summaries), MR 54, 3002.

24. Hawlitschek, K., Green-Funktionen für Wärmeleiter mit beweglichen Randern, *Z. Angew. Math. Phys.* 29, No. 5 (1978), 777–794 (German, English summary), MR 80a, 35057.

25. Johnson, R., Representation theorems for the heat equation, *Proc. London Math. Soc.* (3) 24 (1972), 367–384, MR 45, 739.

26. Jones, B. F., Jr., Lipschitz spaces and the heat equation, *J. Math. Mech.* 18 (1968), 379–409, MR 58, 23543.

27. Jones, B. F., Jr., Tu, C. C., On the existence of kernel functions for the heat equation, *Indiana Univ. Mat. J.* 21 (1971), 857–876, MR 45, 3970.

28. Jones, B. F., Jr., A fundamental solution for the heat equation which is supported in a strip, *J. Math. Anal. Appl.* 60, No. 2 (1977), 314–324, MR 56, 9070.

29. Kaizer, V., and Mjuller, B., Removable sets for the heat equation, *Vestnik Moskov. Univ. Ser. I Mat. Meh.* 28, No. 5 (1973), 26–32 (Russian, German summary), MR 48, 11772.

30. Kannai, Y., Off diagonal short-time asymptotics for fundamental solutions of diffusion equations, *Commun. Partial Differ. Equations* 2, No. 8 (1977), 781–830, MR 58, 29247.

31. Kartashov, E. M., and Nechaev, V. M., The method of Green's functions in the solution of boundary-value problems for the heat equation in noncylindrical domains, *Z. Angew. Math. Mech* 58, No. 4 (1978), 199–208, MR 58, 14425.

32. Kemper, J. T., Temperatures in several variables: Kernel functions, representations, and parabolic boundary values, *Trans. Amer. Math. Soc.* 167 (1972), 243–262, MR 45, 3971.

33. Knerr, B. F., The behavior of the support of solutions of the equation of nonlinear heat conduction with absorption in one dimension, *Trans. Amer. Math. Soc.* 249, No. 2 (1979), 409–424, MR 81j, 35065a.

34. Knerr, B. F., Erratum to: The behavior of the support of solutions of the equation of nonlinear heat conduction with absorption in one dimension, *Trans. Amer. Math. Soc.* 258, No. 2 (1980), 539, MR 81j, 35065b.

35. Koosis, P., Green's function for the heat equation as a limit of product integrals, *Ark. Mat.* 16, No. 2 (1978), 153–159, MR 80a, 35020.

36. Krai, J., Hölder-continuous heat potentials, *Atti Accad. Naz. Lincei Rend. Cl. Sci. Fis. Mat. Natur.* (8) 51 (1971), 17–19 (English, Italian summary) MR 46, 7707.

37. Kral', I., The removable singularities of the solutions of the heat equation, in *Application of Functional Methods to the Boundary-Value Problems of Mathematical Physics* (Proc. Third Soviet-Czechoslovak Conf., Novosibirsk, 103–106, 1972 (Russian), MR 52, 3562.

38. Kral, J., and Mrzena, S., Heat sources and heat potentials, *Chasopis Pest. Mat.* **105**, No. 2 (1980), 184–191, 209, MR 81i, 31019.

39. Munier, A., Burgan, J. R., Gutierrez, J., Fijalkow, E., and Felix, M. R., Group transformations and the nonlinear heat-diffusion equation, *SIAM J. Appl. Math.* **40**, No. 2 (1981), 191–207, MR 82g, 35054.

40. Netuka, I., Thinness and the heat equation, *Chasopis Pest. Mat.* **99** (1974), 293–299, MR 50, 4991.

41. Netuka, I., and Zajichek, L., Functions continuous in the fine topology for the heat equation, *Chasopis Pest. Mat.* **99** (1974), 300–306, MR 50, 4992.

42. Ozawa, S., Perturbation of domains and Green kernels of heat equations, *Proc. Japan Acad. Ser. A Math. Sci.* **54**, No. 10 (1978), 322–325, MR 81g, 49036a.

43. Ozawa, S., Perturbation of domains and Green kernels of heat equations. II., *Proc. Japan Acad. Ser. A Math. Sci.* **55**, No. 5 (1979), 172–175, MR 81g, 49036b.

44. Ozawa, S., Perturbation of domains and Green kernels of heat equations. III., *Proc. Japan Acad. Ser. A Math. Sci.* **55**, No. 7 (1979), 227–230, MR 81g, 49036c.

45. Petrushko, I. M., The behavior, near the boundary of a ball, of a solution of the heat equation, *Trudy Moskov. Orden. Lenin. Energet. Inst. Vyp.* **290** (1976), 56–63 (Russian), MR 58, 23062.

46. Pierre, M., Capacité parabolique et équation de la chaleur avec obstacle irrégulier, *C. R. Acad. Sci. Paris Ser. A–B*, **287**, No. 3 (1978), A117–A119 (French, English summary), MR 58, 23064.

47. Sadallah, B. K., Régularité de la solution de l'équation de la chaleur dans un domaine plan non rectangulaire, *C. R. Acad. Sci. Paris Ser. A–B*, **280**, No. 22 (1975), Aii, A1523–A1526 (French, English summary), MR 52, 14653.

48. Sadallah, B. K., Régularité de la solution de l'équation de la chaleur dans un domaine plan non rectangulaire, *Boll. Un. Mat. Ital. B* (4) **13**, No. 1 (1976), 32–54 (Italian summary), MR 53, 13836.

49. Sadallah, B. K., Régularité de la solution de l'équation de la chaleur dans un domaine plan sans condition de corne, *C. R. Acad. Sci. Paris Ser. A–B*, **284**, No. 11 (1977), A599–A602 (English summary), MR 55, 6013.

50. Sokolovskii, V. B., The second boundary-value problem without initial conditions for the heat equation in the Hilbert ball, *Izv. Vyssh. Uchebn. Zaved. Matematika 1976*, No. 5 (168) (1976), 119–123 (Russian), MR 58, 6776.

51. Stanton, N. K., The fundamental solution of the heat equation associated with the ∂-Neumann problem, *J. Analyse Math.* **34** (1978), 265–274 (1979), MR 80j, 58062.

52. Suzuki, N., On the essential boundary and supports of harmonic measures for the heat equation, *Proc. Japan Acad. Ser. A Math. Sci.* **56**, No. 8 (1980), 381–385, MR 81m, 31015.

53. Taylor, P. D., Some open sets for which the heat equation is simplicial, *Canad. J. Math.* **26** (1974), 455–472, MR 49, 620.

54. Vesely, J., On the heat potential of the double distribution, *Chasopis Pest. Mat.* **98** (1973), 181–198, 213 (English, Czech summary), MR 48, 2410.

55. Vesely, J., Some properties of a generalized heat potential, *Comment. Math. Univ. Carolinae* **15** (1974), 357–360, MR 50, 7815.

56. Vesely, J., On a generalized heat potential, *Czechoslovak Math. J.* **25** (100), No.

3 (1975), 404–423, MR 52, 11086.

57. Watson, N. A., On the definition of a subtemperature, *J. London Math. Soc.* (2) **7** (1973), 195–198, MR 49, 3325.

58. Watson, N. A., A theory of subtemperatures in several variables, *Proc. London Math. Soc.* (3) **26** (1973), 385–417, MR 47, 3838.

59. Watson, N. A., Corrigendum: "Green functions, potentials, and the Dirichlet problem for the heat equation" (*Proc. London Math. Soc.* (3) **33** (1976), 251–298), *Proc. London Math. Soc.* (3) **37**, No. 1 (1978), 32–34, MR 58, 17149.

E. Asymptotic behavior of solutions

1. Abdullaeva, G. Z., and Namazov, G. K., The stability of the solution of the Stefan problem, *Izv. Akad. Nauk. Azerbaidzhan, SSR Ser. Fiz.-Tehn. Mat. Nauk.* 1974, No. 2 (1974), 132–137 (Russian, Azerbaijani, and English summaries), MR 51, 6145.

2. Ball, J. M., and Peletier, L. A., Global attraction for the one-dimensional heat equation with nonlinear, time-dependent boundary conditions, *Arch. Rational Mech. Anal.* **65**, No. 3 (1977), 193–201, MR 56, 6115.

3. Baras, P., and Veron, L., Comportement asymptotique de la solution d'une équation d'évolution sémilinéaire de la chaleur, *Comm. Partial Differential Equation* **4**, No. 7 (1979), 795–807 (French), MR 80j, 35010.

4. Benzinger, H. E., Perturbation of the heat equation, *J. Differential Equations* **32**, No. 3 (1979), 398–419, MR 80d, 35067.

5. Deuel, J., and Hess, P., Nonlinear parabolic boundary-value problems with upper and lower solutions, *Israel J. Math.* **29**, No. 1 (1978), 92–104, MR 80a, 35010.

6. Doroshenko, A. P., An asymptotic solution of the first boundary-value problem with movable boundary, *Izv. Akad. Nauk. Kazah. SSR Ser. Fiz-Mat. 1970*, No. 5 (1970), 76–78 (Russian), MR 55, 6012.

7. Drozhzhinov, J. N., Stabilization of a solution of the heat equation, *Dokl. Akad. Nauk. SSSR* **185** (1969), 13–15 (Russian), MR 58, 23059.

8. Ewer, J. P. G., The conduction of heat in bodies with varying specific-heat capacity, *Luso J. Sci. Tech.* **1**, No. 2 (1980), 31–36, MR 82a, 35051.

9. Galaktionov, V. A., Approximate self-similar solutions of equations of heat-conduction type, *Differencial'nye Uravnenija* **16**, No. 9 (1980), 1660–1676, 1726 (Russian), MR 82f, 35022.

10. Gavalas, G. R., and Yortsos, Y. C., Short-time asymptotic solutions of the heat-conduction equation with spatially varying coefficients, *J. Inst. Math. Appl.* **26**, No. 3 (1980), 209–219, MR 82k, 35053.

11. Holland, C. J., and Berryman, J. G., A nonlinear generalization of the heat equation arising in plasma physics, in *Applied Nonlinear Analysis* (Proc. Third Internat. Conf., Univ. Texas, Arlington, Tex., 1978). Academic Press, New York, 61–66, 1979, MR 80g, 82032.

12. Inoue, A., Miyakawa, T., and Yoshida, K., Some properties of solutions for semilinear heat equations with time lag, *J. Differential Equations* **24**, No. 3 (1977), 383–396, MR 56, 855.

13. Kalinichenko, V. I., and Nesenenko, G. A., The asymptotic behavior of the solution of the first boundary-value problem for the heat equation in the case

of a moving boundary, *Ukrain. Mat. Zh.* **27** (1975), 89–94, 143 (Russian), MR 57, 16928.

14. Kaljakin, L. A., Approximation of nonlinear, nonstationary problems with a small parameter, in *Mathematical Models and Numerical Methods* (Papers, Fifth Semester, Stefan Banach Internat. Math. Center, Warsaw, 1975), PWN, Warsaw, 83–88, Banach Center Publ., 3. 1978 (Russian), MR 80d, 65130.

15. Kobayashi, K., Sirao, T., and Tanaka, H., On the growing up problem for semilinear heat equations, *J. Math. Soc. Japan* **29**, No. 3 (1977), 407–424, MR 56, 9076.

16. Kruzhkov, S. N., and Jakubov, S., Solvability of a certain class of problems with an unknown boundary for the heat equation; the asymptotic behavior of solutions as $t \to \infty$, *Izv. Akad. Nauk. UzSSR Ser. Fiz.-Mat. Nauk. 1978*, No. 6 (1978), 26–28 (Russian, Uzbek summary), MR 80c, 35104.

17. Nesenenko, G. A., The asymptotic behavior of the Green's function of the heat equation with a small parameter, *Mat. Sb.* (*N.S.*) **87** (129) (1972), 204–215 (Russian), MR 46, 2224.

18. Nesenenko, G. A., The asymptotic behavior of the Green's function of the heat equation with a small parameter, *Dopovidi Akad. Nauk. Ukrain. RSR Ser. A* 1973 (1973), 181–184, **193** (Ukrainian, English and Russian summaries), MR 47, 9065.

19. Nesenenko, G. A., The symptotic behavior as $\alpha \to 0$ of the Green's function of the first boundary-value problem for the equation $\alpha \gamma_t = \alpha^2 \gamma_{xx} + V(x, t)\gamma$ in the case of a moving boundary, *Mathematical Physics*, No. 16 (Russian). Izdat. "Naukova dumka," Kiev; 152–159, 199, 1974 (Russian), MR 50, 7845.

20. Oleinik, O. A., Heat propagation in multidimensional dispersive media, *Problems in Mechanics and Mathematical Physics* (Russian). Izdat. "Nauka," Moscow; 224–236, 298, 1976 (Russian), MR 58, 1575.

21. Oulton, D. B., Wollkind, D. J., and Maurer, R. N., A stability analysis of a prototype moving-boundary problem in heat flow and diffusion, *Amer. Math. Monthly* **86**, No. 3 (1979), 175–186, MR 80i, 80006.

22. Polk, J. F., Singular perturbations in heat-conduction and diffusion problems, *Transactions of the Twenty-First Conference of Army Mathematicians* (White Sands Missile Range, White Sands, N. M., 1975), U.S. Army Res. Office, Research Triangle Park, N.C., 543–557, ARO Rep., 76-1, 1976, MR 58, 32448.

23. Redheffer, R., and Walter, W., Invariant sets for systems of partial differential equations. I. Parabolic equations, *Arch. Rational Mech. Anal.* **67**, No. 1 (1978), 41–52, MR 82h, 35050.

24. Sirao, T., On the growing up problem for semilinear heat equations, *Surikaisekikenkyusho Kokyuroku*, No. 258 (1975), 106–112, MR 58, 23065.

25. Smith, L., The asymptotics of the heat equation for a boundary-value problem, *Invent. Math.* **63**, No. 3 (1981), 467–493, MR 82m, 35117.

26. Valickii, J. N., and Eidel'man, S. D., A necessary and sufficient condition for the stabilization of positive solutions of the heat equation, *Sibirsk. Mat. Zh.* **17**, No. 4 (1976), 744–756 (Russian), MR 54, 10821.

F. Periodic solutions

1. Adler, J., Barry, P. A., and Bernal, M. J. M., Thermal-explosion theory for a slab with time-periodic surface-temperature variation, *Proc. Roy. Soc. London Ser. A*, **370**, No. 1740 (1980), 73–88, MR 81d, 80009.

2. Amann, H., Periodic solutions of semilinear parabolic equations, in *Nonlinear Analysis* (collection of papers in honor of Erich H. Rothe). Academic Press, New York, 1–29, 1978, MR 80a, 35009.

3. Ascari, A., On the determination of the periodic solution of the diffusion equation with periodic impulse, *Riv. Mat. Univ. Parma* (4) **5**, Part 2 (1979), 879–884 (1980) (Italian), MR 81j, 35052.

4. Bange, D. W., Periodic solutions of a quasilinear parabolic differential equation, *J. Differential Equations* **17** (1975), 61–72, MR 51, 3657.

5. Bange, D. W., An existence theorem for periodic solutions of a nonlinear, parabolic boundary-value problem, *J. Differential Equations* **24**, No. 3 (1977), 426–436, MR 56, 864.

6. Borisovich, J. G., and Čan Zuĭ Bond, The periodic solutions of a parabolic equation with a small parameter, *Voronezh, Gos. Univ. Trudy Mat. Fak. 1972*, No. 7 (1972), 24–29 (Russian), MR 55, 3456.

7. Brusin, V. A., Global stability of a forced periodic solution of an equation that describes the dynamics of the temperature field of the moon, *Izv. Vyssh. Uchebn. Zaved. Radiofizika* **12** (1969), 381–384 (Russian), MR 45, 8976.

8. Dvorjaninov, S. V., Periodic solution of an autonomous, singularly perturbed parabolic system, *Differencial'nye Uravnenija* **16**, No. 9 (1980), 1617–1622, 1725 (Russian), MR 82d, 35018.

9. Gaines, R., and Walter, W., Periodic solutions to nonlinear parabolic differential equations, *Rocky Mountain J. Math.* **7**, No. 2 (1977), 297–312, MR 56, 12528.

10. Gould, P. H., Oscillations in nonlinear parabolic systems, *Bull. Inst. Math. Acad. Sinica* **6**, No. 1 (1978), 107–132, MR 58, 6632.

11. Havlova, J., On periodic solutions of nonlinear parabolic equations, in *Proceedings of the Fourth Conference on Nonlinear Oscillations* (Prague, 1967). Academia, Prague; 169–172; 1968, MR 50, 774.

12. Klimov, V. S., Continuous branches of periodic solutions of quasilinear parabolic equations, *Sibirsk. Mat. Zh.* **15** (1974), 434–438, 462 (Russian), MR 49, 793.

13. Klimov, V. S., Periodic and stationary solutions of quasilinear parabolic equations, *Sibirsk. Mat. Zh.* **17**, No. 3 (1976), 692–696, 718 (Russian), MR 54, 690.

14. Lauerova, D., A note to the theory of periodic solutions of a parabolic equation, *Apl. Mat.* **25**, No. 6 (1980), 457–460, MR 82b, 35074.

15. Mal'cev, A. P., The convergence of the Rothe method in the construction of a bounded, almost periodic, and periodic solution of a boundary-value problem of parabolic type, *Izv. Vyssh. Uchebn. Zaved. Radiofizika* **15** (1972), 332–339, (Russian, English summary), MR 46, 2198.

16. Mal'cev, A. P., The construction of a bounded, almost periodic, and periodic solution of a problem of a parabolic type by the method of lines, *Izv. Vyssh. Uchebn. Zaved. Radiofizika* **15** (1972), 340–345 (Russian, English summary), MR 46, 2199.

17. Palmieri, G., Soluzioni limitate, o quasi-periodiche, di un'equazione non lineare del calore con discontinuità rispetto all'incognità, *Ist. Lombardo Accad. Sci. Lett. Rend. A*, **104** (1970), 746–757 (Italian, English summary), MR 45, 5528.

18. Petrovanu, D., Periodic and almost periodic solutions of parabolic equations and E. Rothe's method, in *Proceedings of the Conference on Differential*

Equations and Their Applications (Univ. Iasi, Iasi, 1973). Editura Acad. R. S. R., Bucharest; 81–84; 1977, MR 58, 11926.

19. Petrovanu, D., Periodic and almost periodic solutions of parabolic equations and E. Rothe's method, *Bul. Inst. Politehn. Iasi (N.S.)* **23** (27), No. 1–2, Sect. 1 (1977), 17–22, MR 56, 9078.

20. Seidman, T. I., Periodic solutions of a nonlinear, parabolic equation, *J. Differential Equations* **19**, No. 2 (1975), 242–257, MR 57, 10184.

21. Soltanov, K. N., Periodic solutions of certain nonlinear, parabolic equations with implicit degeneracy, *Dokl. Akad. Nauk. SSSR* **222**, No. 1–2 (1975), 291–294 (Russian), MR 51, 13414.

22. Steuerwalt, M., The existence, computation, and number of solutions of periodic parabolic problems, *SIAM J. Numer. Anal.* **16**, No. 3 (1979), 402–420, MR 80d, 35011.

23. Ts'ai, Long Yi, Periodic solutions of nonlinear, parabolic differential equation, *Bull. Inst. Math. Acad. Sinica* **5**, No. 2 (1977), 219–247, MR 57, 6772.

24. Ts'ai, Long Yi, Periodic solutions of parabolic equations with functional arguments, *Chinese J. Math* **6**, No. 1 (1978), 77–86, MR 80b, 35010.

25. Temkin, L., On the construction of periodic solutions of an equation of parabolic type, *Eesti NSV Tead. Akad. Toimetised. Fuus.-Mat.* **29**, No. 3 (1980), 261–271 (Russian, Estonian, and English summaries), MR 81m, 35016.

26. Vaghi, C., Soluzionilimitate, o quasi-periodiche, dell'equazione quasi'lineare del calore, *Rend. Sem. Mat. Fis. Milano* **42** (1972), 25–46 (1973) (Italian, English summary), MR 48, 11773.

27. Vasil'eva, A. B., and Dvorjaninov, S. V., Periodic solutions of singularly perturbed equations of parabolic type, *Kuibyskev Gos. Ped. Inst. Nauchn. Trudy*, 1979, **232**; *Differencial Uravnenija i Mat. Fiz.*, No. 1, 145–154, MR 82b, 35085.

28. Yamada, Y., Periodic solutions of certain nonlinear, parabolic differential equations in domains with periodically moving boundaries, *Nagoya Math. J.* **70** (1978), 111–123, MR 58, 11933.

29. Zagorskii, T. J., and Kobzarev, V. I., Periodic solutions of a parabolic system, in *Differential Equations and Their Applications* (Russian). Dnepropetrovsk, Gos. Univ., Dnepropetrovsk, 172–181, 218–219, (1976) (Russian), MR 58, 29069.

G. Applications to statistics

1. Becus, G. A., Random generalized solutions to the heat equation, *J. Math. Anal. Appl.* **60**, No. 1 (1977), 93–102, MR 56, 853.

2. Becus, G. A., Variational formulation of some problems for the random heat equation, in *Applied Stochastic Processes* (Proc. Conf., Univ. Georgia, Athens, Ga., 1978). Academic Press, New York; 19–36, 1980, MR 82c, 35082.

3. Cahlon, B., The heat equation with a stochastic, solution-dependent boundary condition, *J. Differential Equations* **15** (1974), 418–428, MR 49, 847.

4. Cahlon, B., The heat equation with stochastic nonlinear boundary conditions, *J. Differential Equations* **39**, No. 3 (1981), 426–444, MR 82e, 35082.

5. Capinski, M., A statistical approach to the heat equation, *Zeszyty Nauk. Uniw. Jagiellon. Prace. Mat.*, No. 20 (1979), 111–133, MR 80i, 35086.

6. Capinski, M., Szafirski, B., and Wozniak, M., Equivalence of two definitions of statistical solution of the heat equation, *Zeszyty Nauk. Uniw. Jagiellon. Prace Mat.*, No. 21 (1979), 71–79, MR 80j, 35043.
7. Capinski, M., Statistical solutions of the heat equation, *Zeszyty Nauk. Uniw. Jagiellon. Prace. Mat.*, No. 22 (1981), 111–118, MR 82i, 35171.
8. Divnich, N. T., The limit behavior of the solution of the Cauchy problem for a heat equation that is perturbed by a random process of "white-noise" type, *Ukrain. Mat. Zh.* **29**, No. 5 (1977), 646–650, 709 (Russian), MR 58, 17456.

3.2. Initial-Boundary Value Problems

1. Abasov, A. M., Multidimensional problems of heat propagation in a composite solid when the thermal contact is nonideal on a boundary where the composite parts of the solid are joined without aggregate transition, *Azerbaidzhan. Gos. Univ. Uchen. Zap. Ser. Fiz.-Mat. Nauk.* 1972, No. 1 (1972), 52–58 (Russian, Azerbaijani summary), MR 51, 13452.
2. Antimirov, M. J., Solution of general boundary-value problems for a one-dimensional equation of parabolic type in the case of a moving boundary subject to the law $^P\sqrt{t}$, *Latvian Mathematical Yearbook*, **12** (Russian). Izdat. "Zinatne," Riga, 3–24, 1973 (Russian, Latvian, and English summaries), MR 50, 13885.
3. Azamatov, S. Z., On a boundary-value problem for the heat equation with shifted argument, *Izv. Akad. Nauk. Kazah. SSR Ser. Fiz.-Mat.* 1981, No. 5 (1981), 61–63, 82 (Russian Kazakh summary), MR 82m, 35138.
4. Baderko, E. A., An application of the method of parabolic potentials to the solution of certain heat-contact, boundary-value problems, *Differencial'nye Uravnenija* **6** (1970), 2200–2213 (Russian), MR 46, 523.
5. Baiocchi, C., Équations différentielles abstraites: Application au problème de Cauchy mêlé pour l'équation de la chaleur, in *Publications des Séminaires de Mathématiques* (Univ. Rennes, Rennes, annee 1969-1970), Fasc. 1: Séminaires d'Analyse Fonctionnelle, Exp. No. 6. Dep. Math. et Informat., Univ. Rennes, Rennes, 14 pp., 1970 (French), MR 51, 10916.
6. Baiocchi, C., Problemi misti per l'equazione del calore, *Rend. Sem. Mat. Fis. Milano* **41** (1971), 19–54 (Italian, French summary), MR 48, 739.
7. Baiocchi, C., Sur une équation différentielle abstraite à domaine variable; applications aux problèmes mixtes, du type Cauchy-mêlé, pour l'équation de la chaleur, in *Actes du Colloque d'Analyse Fonctionnelle* (Univ. Bordeaux, Bordeaux, 1971); *Bull. Soc. Math. France Mem.*, No. 31–32 (1972), 31–33 (French), MR 50, 7860.
8. Bakievich, N. I., The relation between the solutions of the filtration equation for a fluid in cracked rocks and the heat equation, *Izv. Vyssh. Uchebn. Zaved. Matematika* 1980, No. 11 (1980), 70–72, (Russian), MR 82f, 76034.
9. Baranski, F., On a certain limit problem for the heat-conduction equation and for triangular plate, *Fasc. Math.*, No. 11 (1979), 169–172, MR 82i, 35034.
10. Baranski, F., On a certain limit problem for the heat-conduction equation and any angular time-spatial domain, *Fasc. Math.*, No. 11 (1979), 173–177, MR 82i, 35035.

11. Belikov, K. V., Gleibman, B. J., and Samarin, J. P., The formulation of boundary-value problems of heat conduction associated with vibrational contraction of materials, in *Differential Equations and Their Applications*, No. 2 (Russian). Kuibyshev. Gos. Ped. Inst., Kuybyshev; 21–27, 175–176, 1975 (Russian), MR 58, 32437.

12. Belonosov, S. M., and Karachun, V. J., Boundary-value problems for the heat equation inside a trihedral angle, *Vychisl. Prikl. Mat. (Kiev) Vyp.* **30** (1976), 113–116, 157 (Russian, English summary), MR 57, 10234.

13. Bizhanova, G. I., A problem of conjugation for the heat equation in a spherical region, in *Theoretical and Applied Problems of Mathematics and Mechanics* (Russian). "Nauka" Kazah. SSR, Alma-Ata; 63–69, 239, 1979 (Russian), MR 81a, 35041.

14. Bluman, G. W., Applications of the general similarity solution of the heat equation to boundary-value problems, *Quart. Appl. Math.* **31** (1973), 403–415, MR 55, 859.

15. Bove, A., Franchi, B., and Obrecht, E., An initial-boundary-value problem with mixed lateral conditions for heat equation, *Ann. Mat. Pura Appl.* (4) **121** (1979), 277–307 (English, Italian summary), MR 81c, 35058.

16. Burch, B. C., and Goldstein, J. A., Nonlinear semigroups and a problem in heat conduction, *Houston J. Math.* **4**, No. 3 (1978), 311–328, MR 80d, 35079.

17. Burykin, A. J., Boundary-value problems for the heat equation in domains with cuts, *Ukrain Mat. Zh.* **27**, No. 5 (1975), 646–650, 717 (Russian), MR 52, 11324.

18. Carrassi, M., Linear heat equations from kinetic theory, *Nuovo Cimento B* (11) **46**, No. 2 (1978), 363–379 (English, Italian, and Russian summaries), MR 80a, 82029.

19. Chan, C. Y., A first initial-boundary value problem for a semilinear heat equation, *SIAM J. Appl. Math.* **22** (1972), 529–537, MR 46, 2222.

20. Diaz, J. B., and Means, C. S., An initial-value problem for a class of higher-order, partial-differential equations related to the heat equation, *Ann. Mat. Pura Appl.* (4) **97** (1973), 115–187, MR 49, 5567.

21. Doman'skii, Z., Piskorek, A., and Roek, Z., Application of the Fischer–Riesz–Kupradze method to the solution of the first Fourier problem, *Comment. Math. Prace. Mat.* **16** (1972), 137–147 (Russian), MR 48, 11770.

22. Dont, M., On a boundary-value problem for the heat equation, *Czechoslovak Math. J.* **25** (100) (1975), 110–133, MR 51, 6148.

23. Dorfman, I. J., The heat equation on a Hilbert space, *Vestnik Moskov. Univ. Ser. I Mat. Meh.* **26**, No. 4 (1971), 46–51 (Russian, English summary), MR 58, 30202.

24. Ducourtioux, J.-L., Temps de vie des solutions de l'équation de la chaleur de Eells–Sampson, *C. R. Acad. Sci. Paris, Ser. A–B*, **286**, No. 7 (1978), A333–A336 (French, English summary), MR 80f, 58048.

25. Fabes, E. B., and Riviere, N. M., Dirichlet and Neumann problems for the heat equation in C^1-cylinders, in *Harmonic Analysis in Euclidean Spaces (Proc. Sympos. Pure Math.*, Williams Coll., Williamstown, Mass., 1978), Part 2. Amer. Math. Soc., Providence, R.I.; 179–196; Proc. Sympos. Pure Math., XXXV, Part 2, 1979, MR 81b, 35044.

26. Fasano, A., and Primicerio, M., Su un problema unidimensionale di diffu-

sione in un mezzo a contorno mobile con condizioni ai limiti non lineari, *Ann. Mat. Pura Appl.* (4) **93** (1972), 333–357 (English summary), MR 53, 13834.

27. Filippov, V. M., and Skorohodov, A. N., A quadratic functional for the heat equation, *Differencial'nye Uravnenija* **13**, No. 6 (1977), 1113–1123, 1158 (Russian), MR 58, 6696.

28. Filippov, V. M., and Skorohodov, A. N., The principle of a minimum of a quadratic functional for a boundary-value problem of heat conduction, *Differencial'nye Uravnenija* **13**, No. 8 (1977), 1434–1445, 1540 (Russian), MR 57, 6839.

29. Galaktionov, V. A., and Mihailov, A. P., Ob odnoi avtomodel'noi zadache dlja uravnenija nelineinoi teploprovodnosti (A self-similar problem for the equation on nonlinear heat conduction), *Inst. Prikl. Mat. Akad. Nauk. SSSR*, Moscow, 37 pp., 1977 (Russian), MR 57, 13192.

30. Galicyn, A. S., and Zhukovskii, A. N., Explicit form of the solutions of some boundary-value problems for the heat equation in elliptic coordinates, in *Analytic, Numerical, and Analogue Methods in Heat-Conduction Problems* (Russian). Izdat. "Naukova Dumke," Kiev; 18–28, 236, 1977 (Russian), MR 58, 22978.

31. Galicyn, A. S., and Zhukovskii, A. N., The solutions of the first and second boundary-value problems of heat conduction for the exterior of an elliptic cylinder, in *Analytic, Numerical, and Analogue Methods in Heat-Conduction Problems* (Russian). Izdat. "Naukova Dumka," Kiev, 62–70, 237, 1977 (Russian), MR 58, 11913.

32. Ghelardoni, G., Alcune considerazioni sull'approssimazione della soluzione dell'equazione del calore, *Boll. Un. Mat. Ital.* (4) **11**, No. 3, Suppl. (1975), 403–412 (Italian, English summary), MR 53, 6104.

33. Gopala Rao, V. R., and Ting, T. W., Solutions of pseudo-heat equations in the whole space, *Arch. Rational Mech. Anal.* **49** (1972), 57–78, MR 48, 9111.

34. Graffi, S., and Levoni, S., Sull'integrazionne dell'equazione del calore a coefficienti variabili in R^n, *Atti. Sem. Mat. Fis. Univ. Modena* **23**, No. 1, (1974), 161–177 (1975) (Italian), MR 52, 3734.

35. Grevceva, V. M., A boundary-value problem of heat conduction for a semibounded rod with a variable boundary condition of the third kind, in *Differential Equations and Their Applications*, No. 2 (Russian). Kuibyshev. Gos. Ped. Inst., Kuybyshev, 50–54, 176, 1975 (Russian), MR 58, 32655.

36. Grevceva, V. N., and Igonin, V. I., The nonstationary heat-exchange problem for a plane wall, in *Differential Equations and Their Applications*, No. 2 (Russian). Kuibyshev. Gos. Ped. Inst., Kuybyshev, 142–145, 179, 1975 (Russian), MR 58, 32441.

37. Hagenmeyer, B., *Randwertproblem und Grenzubergang in der Theorie der Wärmsleitung.* Universitat Stuttgart, Stuttgart, 41 pp., 1971, MR 55, 10849.

38. Hairullin, E., The general boundary-value problem of the heat equation for the one-dimensional case with a discontinuous coefficient, *Akad. Nauk. Kazah. SSSR Trudy Inst. Mat. i Meh.* **1** (1970), 233–239, 331–332 (Russian), MR 56, 12594.

39. Hairullin, E., The general boundary-value problem of the heat equation in a multidimensional space for a strip, *Akad. Nauk. Kazah. SSSR Trudy Inst. Mat. i Meh.* **2** (1971), 103–107, 343–344 (Russian), MR 56, 12595.

40. Hairullin, E., The general boundary-value problem of the heat equation in a multidimensional space with a discontinuous coefficient for a strip, *Akad. Nauk. Kazah. SSR Trudy Inst. Mat. i Met.* **2** (1971), 107–112, 344 (Russian), MR 56, 12596.

41. Hartka, J. E., Temperature of a semi-infinite rod which radiates both linearly and nonlinearly, *Quart. Appl. Math.* **32** (1974), 101–111, MR 55, 3544.

42. Ionkin, N. I., The solution of a certain boundary-value problem of the theory of heat conduction with a nonclassical boundary condition, *Differencial'nye Uravnenija*, **13**, No. 2 (1977), 294–304, 361 (Russian), MR 58, 29240a.

43. Ionkin, N. I., Letter to the editors: "The solution of a certain boundary-value problem of the theory of heat conduction with a nonclassical boundary condition" (*Differencial'nye Uravnenija* **13** (1977), No. 2, 294–304), *Differencial'nye Uravnenija* **13**, No. 10 (1977), 1903 (Russian), MR 58, 29240b.

44. Ionkin, N. I., The stability of a problem in the theory of heat conduction with nonclassical boundary conditions, *Differencial'nye Uravnenija* **15**, No. 7 (1979), 1279–1283, 1343 (Russian), MR 80i, 35092.

45. Ionkin, N. I., and Moiseev, E. I., A problem for a heat equation with two-point boundary conditions, *Differencial'nye Uravnenija* **15**, No. 7 (1979), 1284–1295, 1343 (Russian), MR 80i, 35087.

46. Kalashnikov, A. S., The heat equation with long-range action, *Differencial'nye Uravnenija* **15**, No. 9 (1979), 1653–1660, 1726 (Russian), MR 80j, 35044.

47. Kalla, S. L., and Kushwaha, R. S., Production of heat in an infinite cylinder, *Acta Mexicana Ci. Tech.* **4** (1970), 89–93, MR 48, 1595.

48. Karachun, V. J., A boundary-value problem for an equation of parabolic type, *Visnik Kiiv. Univ. Ser. Mat. Meh.*, No. 18 (1976), 24–25, 141 (Ukrainian, English, and Russian summaries), MR 58, 6698.

49. Katekov, E. A., Boundary-value problems for the generalized heat equation, in *Differential Equations and Their Application* (Russian). Izdat. "Nauka" Kazah. SSR, Alma-Ata, 163–168, 216, 1970 (Russian), MR 51, 13454.

50. Kim, E. I., and Kavokin, A. A., Solution in the small of a problem with a nonlinear boundary condition for the heat equation in an expanding domain, *Izv. Akad. Nauk. Kazah. SSR Ser. Fiz.-Mat.* 1977, No. 1 (1977), 41–46, 91 (Russian, Kazakh summary), MR 58, 11929.

51. Kim, E. I., Krasnov, J. A., and Harin, S. N., A class of nonlinear problems of heat conduction in the theory of electrical contacts, in *Partial-Differential Equations* (Proc. Conf., Novosibirsk, 1978) (Russian). "Nauka" Sibirsk. Otdel., Novosibirsk, 83–86, 250, 1980 (Russian), MR 82i, 35083.

52. Kiro, S. N., An equation of M. V. Ostrogradskii in the mathematical theory of heat conduction, *Voprosy Istor. Estestvoznan. i Tehn.* 1972, No. 1(38) (1972), 31–32, 125, 134 (Russian, English summary), MR 58, 27017.

53. Kobayashi, K., On the semilinear heat equations with time lag, *Hiroshima Math. J.* **7**, No. 2 (1977), 459–472, MR 56, 12680.

54. Kochina, N. N., Certain nonlinear problems for the equation of heat conduction, *Zh. Prikl. Meh. i Tehn. Fiz.* 1972, No. 3 (1972), 123–128 (Russian), MR 46, 9529.

55. Kotel'nikov, J. V., The heat equation for a moving medium, *Izv. Akad. Nauk. Turkmen. SSSR Ser. Fiz.-Tehn. Him. Geol. Nauk.* 1969, No. 6 (1969), 104–105 (Russian, English summary), MR 46, 519.

56. Kulahmetova, S. A., The first boundary-value problem of the heat equation with discontinuous coefficients for the disc, *Izv. Akad. Nauk. Kazah. SSSR Ser. Fiz.-Mat.* 1975, No. 3 (1975), 59–66, 92 (Russian, Kazakh summary), MR 57, 16982.

57. Kulahmetova, S. A., Solution of a boundary-value problem for a heat equation with discontinuous coefficients in a domain bounded by a curve of Ljapunov type when the curve of discontinuity of the coefficient passes the boundary, *Izv. Akad. Nauk. Kazah. SSSR Ser. Fiz.-Mat.* 1977, No. 1 (1977), 76–79, 93 (Russian, Kazakh summary), MR 58, 23061.

58. Kul'chickii, V. L., Integration of a nonlinear, mixed partial-differential boundary-value problem, *Ukrain. Mat. Zh.* 25 (1973), 382–386, 431 (Russian), MR 49, 7603.

59. Lenjuk, M. P., and Fedoruk, V. V., The generalized heat equation, *Dopovidi Akad. Nauk. Ukrain. SSR Ser. A,* 1972 (1972), 1075–1078, 1149 (Ukranian, English, and Russian summaries), MR 48, 6689.

60. Lenjuk, M. P., and Seredjuk, Z. L., Generalized temperature fields in a spherical cone, *Dokl. Akad. Nauk. Ukrain. SSR Ser. A* 1977, No. 5 (1977), 405–409, 481 (Russian, English summary), MR 56, 12671.

61. Lenjuk, M. P., Fedoruk, V. V., and Fedoruk, L. O., Dynamic temperature fields in a space with a solid cylinder insertion, *Dokl. Akad. Nauk. Ukrain. SSR Ser. A,* No. 2 (1980), 12–16, 94 (Russian, English summary), MR 81d, 80007.

62. Ljubich, L. V., A heat-conduction problem that contains the time derivative in the boundary condition (Russian), *Ukrain. Mat. Zh.* 28, No. 6 (1976), 819–822, 863, MR 58, 29241.

63. Ljubich, L. V., Solution of the heat-conduction problem in a right dihedral angle with time derivatives in the boundary condition, *Dokl. Akad. Nauk. Ukrain, SSR Ser. A,* 1976, No. 8 (1976), 691–693, 766 (Russian, English summary), MR 55, 860.

64. Lukesh, J., A new type of generalized solution of the Dirichlet problem for the heat equation, in *Nonlinear Evolution Equations and Potential Theory* (Proc. Summer School, Podhradi, 1973). Academia, Prague, 117–123, 1975, MR 58, 6699.

65. MacCamy, R. C., An integro-differential equation with application in heat flow, *Quart. Appl. Math.* 35, No. 1 (1977), 1–19, MR 56, 10465.

66. Malkin, L. M., Application of the Wiener–Hopf method to the solution of a mixed boundary-value problem for the heat equation in a half-plane, *Moskov. Gos. Ped. Inst. Uchen. Zap.,* No. 375 (1971), 195–200 (Russian), MR 49, 7604.

67. Malyshev, I. G., Boundary-value problems for a parabolic equation with a variable coefficient in a plane domain with moving boundary, in *Linear Boundary-Value Problems of Mathematical Physics* (Russian). Izdanie Inst. Mat. Akad. Nauk. Ukrain. SSR, Kiev, 88–104, 1973 (Russian, English summary), MR 50, 13888.

68. Markov, V. G., and Oleinik, O. A., On propagation of heat in one-dimensional disperse media, *J. Appl. Math. Mech.* 39, No. 6 (1975), 1028–1037 (1976), MR 57, 11382.

69. Meyer, J. P., and Kostin, M. D., Diffusion through a membrane: approach to

equilibrium, *Bull. Math. Biology* **38**, No. 5 (1976), 527–534, MR 58, 3992.

70. Mikhlin, S. G., On the approximate solution of the heat-conduction integral equations, *Rend. Mat.* (6) 8, No. 2 (1975), 355–362 (English, Italian summary), MR 54, 10823.

71. Minasjan, R. S., A mixed-value problem of heat conduction for a rectangle, *Akad. Nauk. Armjan. SSR Dokl.* **63**, No. 3 (1976), 146–151 (Russian, Armenian Summary), MR 58, 29242.

72. Minasjan, R. S., A mixed boundary-value problem of heat conduction for a sphere, *Akad. Nauk. Armjan. SSR Dokl.* **67**, No. 3 (1978), 132–137 (Russian, Armenian summary), MR 80g, 35025.

73. Nakao, M., Initial-boundary-value problem for a nonlinear heat equation, *Mem. Fac. Sci. Kyushu Univ. Ser. A*, **26**, No. 2 (1972), 219–239, MR 53, 13850.

74. Njashin, J. I., Variational formulation of the nonstationary problem of heat conduction, *Perm. Politehn. Inst. Sb. Nauchn. Trudov*, No. 152 (1974), 3–9 (Russian), MR 53, 12234.

75. Pashkovskii, V. I., A problem of temperature distribution in ribbed bodies, *Differencial'nye Uravnenija*, **15**, No. 1 (1979), 106–111, 189 (Russian), MR 80f, 35019.

76. Pol'skii, B. S., and Rubinstein, L. I., A certain mixed problem that concerns the heat-conduction equation, in *Latvian Mathematical Yearbook*, **11** (Russian). Izdat. "Zinatne," Riga, 143–155, 1972 (Russian, Latvian and English summaries), MR 48, 4514.

77. Pol'skii, B. S., The alternating Schwarz method for the heat-conduction equation, in *Latvian Mathematical Yearbook*, **11** (Russian). Izdat. "Zinatne," Riga, 133–142, 1972 (Russian, Latvian, and English summaries), MR 48, 4513.

78. Pol'skii, B. S., A nonlinear, boundary-value problem for the heat equation, *Latviisk. Mat. Ezhegodnik*, *Vyp.* **19** (1976), 194–197, 247, (Russian), MR 57, 10237.

79. Powazka, Z., Lösung eines Fourier–Neumann Problems für die Differentialgleichung $\Delta u - u_t = 0$ für einen unbegrenzten Raum, *Wyz. Szkol. Ped. Krakow. Rocznik Nauk.-Dydakt.*, No. 51, *Prace. Mat.*, No. 7 (1974), 127–143 (German), MR 51, 1120.

80. Prudnikov, A. P., Watson operators and their application to the solution of certain problems of the theory of heat conduction, in *Seminar on Numerical Methods for Solving Balance Equations* (Berlin, 1980). Akad. Wiss. DDR, Berlin; 79–82, Rep. 1980, 5, 1980 (Russian), MR 82a, 35052.

81. Pyhteev, G. N., and Shokamolov, I., The exact and approximate solution of plane problems of the theory of heat conduction with boundary conditions of the first and second kind, *Izv. Akad. Nauk. Tadzhik. SSR Otdel. Fiz.-Mat. i Geolog.-Him. Nauk.* 1974, No. 2(52) (1974), 10–16, 106 (Russian, Tajiki summary), MR 51, 13455.

82. Pytel-Kudela, M., On Fourier's problem for the heat equation and for the exterior of the sphere, *Fasc. Math.*, No. 11 (1979), 179–192, MR 82g, 35022.

83. Rafalski, P., On the application of the orthogonal projection method to a heat-conduction boundary problem, *Indiana Univ. Math. J.* **21** (1971), 715–727, MR 45, 2335.

84. Raynal, M.-L., On some nonlinear problems of diffusion, in *Volterra Equations* (Proc. Helsinki Sympos. Integral Equations, Otaniemi, 1978). Springer, Berlin; 251–266, Lecture Notes in Math., 737, 1979, MR 81b, 45024.

85. Sabodash, P. F., and Cheban, V. G., Solution of the heat equation with a finite velocity of heat diffusion for domains with variable boundaries, *Bul. Akad. Shtiince RSS Moldoven* 1971, No. 2 (1971), 11–16 (Russian), MR 45, 741.

86. Shopolov, N. N., A boundary-value problem for the heat equation, *C. R. Acad. Bulgare Sci.* **32**, No. 8 (1979), 1027–1028 (Russian), MR 81d, 35038.

87. Shopolov, N. N., Some boundary-value problems for the heat equation, *C. R. Acad. Bulgare Sci.* **33**, No. 1 (1980), 47–50 (Russian), MR 82f, 35088.

88. Sierpinski, K., The mixed problem for the heat-conductivity equation with a nonlinear boundary condition. I, II, *Biul. Wojsk. Akad. Tech.* **19**, No. 12(220) (1970), 51–62, 63–69 (Polish, Russian, and English summaries), MR 45, 3972.

89. Skrjabina, L. P., Uniqueness of the solution of certain boundary-value problems for the heat equation with discontinuous boundary conditions, *Volzh. Mat. Sb.*, No. 15 (1973), 137–143 (Russian), MR 49, 9415.

90. Slesarenko, A. P., The solution of linear and nonlinear heat-conduction problems for nonclassical domains, in *Analytic, Numerical and Analog Methods in Heat-Conduction Problems* (Russian). Izdat. "Naukova Dumka," Kiev, 28–38, 236, 1977 (Russian), MR 58, 17535.

91. Slesarenko, A. P., Multidimensional problems of heat conduction with a heat-exchange coefficient that varies with time, *Dokl. Akad. Nauk. Ukrain SSR, Ser. A*, 1980, No. 9 (1980), 68–71 (Russian, English summary), MR 82b, 80014.

92. Strnad, J., A hundred years of Stefan's law, *Obzornik Mat. Fiz.* **26**, No. 3 (1979), 65–73 (Slovenian, English summary), MR 80e, 01016.

93. Stroock, D. W., On a conjecture of M. Kac., *Bull. Amer. Math. Soc.* **79** (1973), 770–775, MR 48, 707.

94. Swan, G. W., A mathematical model for the density of malignant cells in the spread of cancer in the uterus, *Math. Biosci.* **25**, No. 3/4 (1975), 319–329, MR 54, 2252.

95. Tao, L. N., Heat conduction with nonlinear boundary condition, *Z. Angew. Math. Phys.* **32**, No. 2 (1981), 144–155 (English, Germany summary), MR 82d, 80019.

96. Ting, T. W., A cooling process according to two-temperature theory of heat conduction, *J. Math. Anal. Appl.* **45** (1974), 23–31, MR 48, 9108.

97. Tkachenko, N. M., Solution of the heat equation when the boundary has an angular point, *Vestnik Har'kov. Politehn. Inst.*, No. 2 (50) (1965), 30–37 (Russian), MR 45, 9013.

98. Vigak, V. M., Construction of the solution of the heat equation for a piecewise-homogeneous body, *Dokl. Akad. Nauk. Ukrain. SSR Ser. A*, No. 1 (1980), 30–32, 93 (Russian, English summary), MR 81k, 65107.

99. Volkov, I. A., The connection between the solutions of a class of boundary-value problems for the heat equation and for the wave equation, in *Analytic, Numerical and Analogue Methods in Heat-Conduction Problems* (Russian). Izdat. "Naukova Dumka," Kiev, 39–42, 236, 1977 (Russian), MR 58, 1637.

100. Wake, G. C., Nonlinear heat generation with reactant consumption, *Quart. J. Math. Oxford Ser.* (2) **22** (1971), 583–595, MR 45, 3973.

101. Watzlawek, W., Zum Cauchy–Problem bei der verallgemeinerten Wärmeleitungsgleichung, *Monatsch. Math.* **81**, No. 3 (1976), 225–233 (German), MR 54, 3157.

102. Weissler, F. B., Existence and nonexistence of global solutions for a semilinear heat equation, *Israel J. Math.* **38**, No. 1–2 (1981), 29–40, MR 82g, 35059.

103. Widder, D. V., *The Heat Equation.* Academic Press (Harcourt Brace Jovanovich, Publishers), New York-London; xiv + 267 pp., 1975, MR 57, 6840.

104. Zischka, K. A., and Chow, P. S., On nonlinear initial-boundary-value problems of heat conduction and diffusion, *SIAM Rev.* **16** (1974), 17–35, MR 48, 9109.

3.3. Stefan and Free-Boundary-Value Problems

1. Abdullaeva, G. Z., The continuous dependence of the solution of the Stefan problem, *Izv. Akad. Nauk. Azerbaidzhan. SSR Ser. Fiz.-Tehn. Mat. Nauk.* 1977, No. 1 (1977), 92–97 (Russian, Azerbaijani, and English summaries), MR 58, 11912.

2. Abramov, A. A., and Gaipova, A. N., The numerical solution of certain systems for Stefan-type problems, *Zh. Vychisl. Mat. i Mat. Fiz.* **11** (1971), 121–128 (Russian), MR 45, 1411.

3. Alexiades, V., A three-phase free-boundary problem arising in alloy solidification, in *Nonlinear Partial-Differential Equations in Engineering and Applied Science* (Proc. Conf., Univ. Rhode Island, Kingston, R. I., 1979). Dekker, New York, 101–109, 1980, MR 81m, 35137.

4. Alexiades, V., and Cannon, J. R., Free-boundary problems in solidification of alloys, *SIAM J. Math. Anal.* **11**, No. 2 (1980), 254–264, MR 81b, 35096.

5. Baiocchi, C., Variational inequalities and free-boundary problems, in *Variational Inequalities and Complementarity Problems* (Proc. Internat. School, Erice, 1978). Wiley, Chichester, 25–33, 1980, MR 81i, 35161.

6. Bazalii, B. V., and Shelepov, V. J., A certain stationary Stefan problem, *Dopovidi Akad. Nauk. Ukrain. RSR Ser. A* 1974 (1974), 5–8, 91 (Ukrainian, English, and Russian summaries), MR 49, 7602.

7. Bazalii, B. V., and Shelepov, V. J., A generalization of the stationary Stefan problem, *Mat. Fiz. Vyp.* **17** (1975), 65–81, 189 (Russian), MR 57, 10232.

8. Bazalii, B. V., A certain quasistationary Stefan problem, *Dokl. Akad. Nauk. Ukrain. SSR Ser. A*, 1976, No. 1 (1976), 3–5, 92 (Russian, English summary), MR 54, 8026.

9. Bazalii, B. V., A proof for the existence of the solution of a two-phase Stefan problem, in *Mathematical Analysis and Probability Theory* (Russian). "Naukova Dumka," Kiev, 7–11, 204, 1978 (Russian), MR 80k, 35042.

10. Bazalii, B. V., and Shelepov, V. J., The asymptotic behavior of the solution of a Stefan problem, *Dokl. Akad. Nauk. Ukrain. SSR Ser. A*, 1978, No. 12 (1978), 1059–1061, 1150 (Russian, English summary), MR 80b, 35015.

11. Bensoussan, A., Brezis, H., and Friedman, A., Estimates on the free boundary for quasivariational inequalities, *Comm. Partial-Differential Equations* **2**, No.

3 (1977), 297–321, MR 57, 13175.

12. Bizhanova, G. I., Stabilization of the solution of Stefan's second boundary-value problem, *Izv. Akad. Nauk. Kazah. SSR Ser. Fiz.-Mat.*, 1980, No. 5 (1980), 12–17, 85 (Russian), MR 82b, 35151.

13. Bojarchuk, A. K., and Le Chong Vin', Numerical solution of the Stefan problem by a difference method without smoothing of the coefficients, *Vychisl. Prikl. Mat.* (Kiev) No. 25 (1975), 14–25, 142 (Russian, English summary), MR 52, 7165.

14. Boley, B. A., An applied overview of moving-boundary problems, in *Moving-Boundary Problems* (Proc. Sympos. and Workshop, Gatlinburg, Tenn., 1977). Academic Press, New York; 205–231, 1978, MR 58, 1513.

15. Bonnerot, R., and Jamet, P., Numerical computation of the free boundary for the two-dimensional Stefan problem by space–time finite elements, *J. Computational Phys.* **25**, No. 2 (1977), 163–181, MR 57, 14506.

16. Bonnerot, R., and Jamet, P., A third-order, accurate, discontinuous, finite-element method for the one-dimensional Stefan problem, *J. Comput. Phys.* **32**, No. 2 (1979), 145–167, MR 80d, 65114.

17. Bonnerot, R., and Jamet, P., A conservative, finite-element method for one-dimensional Stefan problems with appearing and disappearing phases, *J. Comput. Phys.* **41**, No. 2 (1981), 357–388, MR 82h, 65081.

18. Borgioli, G., Di Benedetto, E., and Ughi, M., Stefan problems with nonlinear boundary conditions: the polygonal method, *Z. Angew. Math. Mech.* **58**, No. 12 (1978), 539–546 (English, German, and Russian summaries), MR 80a, 65208.

19. Borodin, M. A., An existence theorem for the solution of a single-phase quasistationary Stefan problem, *Dokl. Akad. Nauk. Ukrain. SSR Ser. A* 1976, No. 7 (1976), 582–585 (Russian, English summary), MR 55, 847.

20. Borodin, M. A., A one-phase quasistationary Stefan problem, *Dokl. Akad. Nauk. Ukrain. SSR Ser. A* 1977, No. 9 (1977), 775–777, 863 (Russian, English summary), MR 58, 11851.

21. Borodin, M. A., A one-phase, quasistationary Stefan problem, in *Boundary-Value Problems for Partial-Differential Equations* (Russian). "Naukova Dumka," Kiev, 13–21, 162, 1978 (Russian), MR 80j, 35095.

22. Borodin, M. A., and Fel'genhauer, U., A one-phase, quasilinear Stefan problem, *Dokl. Akad. Nauk. Ukrain. SSR Ser. A* 1978, No. 2 (1978), 99–101, 191 (Russian, English summary), MR 57, 16978.

23. Borodin, M. A., and Fel'genhauer, U., An axially symmetric, single-phase Stefan problem, *Mat. Fiz.*, No. 24 (1978), 74–76, 122 (Russian), MR 80h, 35069.

24. Brauner, C. M., and Nicolaenko, B., Singular perturbations and free-boundary-value problems, in *Computing Methods in Applied Sciences and Engineering* (Proc. Fourth Internat. Sympos., Versailles, 1979). North-Holland, Amsterdam, 699–724, 1980, MR 82c, 35005.

25. Budak, B. M., and Uspenskii, A. B., Difference-iteration schemes for the solutions of one-dimensional boundary-value problems and problems with unknown boundary for a quasilinear parabolic equation, in *Solutions of Stefan Problems* (Russian). Moskov. Gos. Univ., Moscow, 3–64, 1970 (Russian), MR 57, 3637.

26. Budak, B. M., and Gaponenko, J. L., The solution of problems of Stefan type for a differential or an integro-differential equation, in *Difference Methods of Solution of Boundary-Value Problems for Integro-differential Equations* (Russian). Moskov. Gos. Univ., Moscow, 83–232, 1970 (Russian), MR 54, 3165.

27. Budak, B. M., and Moskal, M. Z., The classical solution of the first boundary-value problem of Stefan for the multidimensional heat equation in a coordinate parallelepiped, in *Solutions of Stefan Problems* (Russian). Moskov. Gos. Univ., Moscow, 87–133, 1970 (Russian), MR 58, 1634.

28. Budak, B. M., and Gaponenko, J. L., The solution of the Stefan problem for a quasilinear parabolic equation with quasilinear boundary conditions, in *Solutions of Stefan Problems* (Russian). Moskov. Gos. Univ., Moscow, 235–312, 1970 (Russian), MR 57, 6844.

29. Caffarelli, L. A., The regularity of elliptic and parabolic free boundaries, *Bull. Amer. Math. Soc.* **82**, No. 4 (1976), 616–618, MR 53, 8662.

30. Caffarelli, L. A., and Rivière, N. M., Asymptotic behavior of free boundaries at their singular points, *Ann. Math.* (2) **106**, No. 2 (1977), 309–317, MR 57, 3634.

31. Caffarelli, L. A., The regularity of free boundaries in higher dimensions, *Acta Math.* **139**, No. 3–4 (1977), 155–184, MR 56, 12601.

32. Caffarelli, L. A., Some aspects of the one-phase Stefan problem, *Indiana Univ. Math. J.* **27**, No. 1 (1978), 73–77, MR 57, 6838.

33. Caffarelli, L. A., and Friedman, A., The one-phase Stefan problem and the porous-medium diffusion equation: continuity of the solution in n space dimensions, *Proc. Nat. Acad. Sci.* (U.S.A.) **75**, No. 5 (1978), 2084, MR 81h, 35021.

34. Caffarelli, L. A., and Friedman, A., Continuity of the temperature in the Stefan problem, *Indiana Univ. Math. J.* **28**, No. 1 (1979), 53–70, MR 80i, 35104.

35. Caffarelli, L. A., and Friedman, A., Regularity of the free boundary of a gas flow in an n-dimensional porous medium, *Indiana Univ. Math. J.* **29**, No. 3 (1980), 361–391, MR 82a, 35096.

36. Caffarelli, L. A., Compactness methods in free-boundary problems, *Comm. Partial-Differential Equations* **5**, No. 4 (1980), 427–448, MR 81e, 35121.

37. Caffarelli, L. A., and Friedman, A., A free-boundary problem associated with a semilinear parabolic equation, *Comm. Partial-Differential Equations* **5**, No. 9 (1980), 969–981, MR 81m, 35138.

38. Caffarelli, L. A., Friedman, A., and Visintin, A., A free-boundary problem describing transition in a superconductor, *SIAM J. Math. Anal.* **12**, No. 5 (1981), 679–690, MR 82k, 81100.

39. Caffarelli, L. A., A remark on the Hausdorff measure of a free boundary, and the convergence of coincidence sets, *Boll. Un. Mat. Ital. A* (5) **18**, No. 1 (1981), 109–113 (English, Italian summary), MR 82i, 35078.

40. Cannon, J. R., and Primicerio, M., A two-phase Stefan problem with temperature-boundary conditions, *Ann. Mat. Pura Appl.* (4) **88** (1971), 177–191 (English, Italian summary), MR 46, 9525.

41. Cannon, J. R., and Primicerio, M., A two-phase Stefan problem with flux-boundary conditions, *Ann. Mat. Pura Appl.* (4) **88** (1971), 193–205 (English, Italian summary), MR 46, 9526.

42. Cannon, J. R., and Primicerio, M., A two-phase Stefan problem: regularity of the free boundary, *Ann. Mat. Pura Appl.* (4) **88** (1971), 217–228 (English, Italian summary), MR 46, 9527.

43. Cannon, J. R., and Primicerio, M., A Stefan problem involving the appearance of a phase, *SIAM J. Math. Anal.* **4** (1973), 141–148, MR 48, 2570.

44. Cannon, J. R., Henry, D. B., and Kotlow, D. B., Continuous differentiability of the free boundary for weak solutions of the Stefan problem, *Bull. Amer. Math. Soc.* **80** (1974), 45–48, MR 48, 11768.

45. Cannon, J. R., Henry, D. B., and Kotlow, D. B., Classical solutions of the one-dimensional, two-phase Stefan problem, *Ann. Mat. Pura Appl.* (4) **107** (1975), 311–341 (1976), MR 53, 11231.

46. Cannon, J. R., and Fasano, A. A nonlinear parabolic free-boundary problem, *Ann. Mat. Pura Appl.* (4) **112** (1977), 119–149, MR 57, 886.

47. Cannon, J. R., Multiphase, parabolic, free-boundary value problems, in *Moving-Boundary Problems* (Proc. Sympos. and Workshop, Gatlingburg, Tenn., 1977). Academic Press, New York, 3–24, 1978, MR 57, 16986.

48. Cannon, J. R., and Di Benedetto, E., On the existence of weak solutions to an n-dimensional Stefan problem with nonlinear boundary conditions, *SIAM J. Math. Anal.* **11**, No. 4 (1980), 632–645, MR 81j, 35058.

49. Cannon, J. R., Di Benedetto, E., and Knightly, G. H., The steady-state Stefan problem with convection, *Arch. Rational Mech. Anal.* **73**, No. 1 (1980), 79–97, MR 81a, 35085.

50. Chan, C. Y., Uniqueness of a nonmonotone free-boundary problem, *SIAM J. Appl. Math.* **20** (1971), 189–194, MR 45, 5584.

51. Chang, Kung Ching, and Jiang, Li Shang, Free-boundary problems of the steady water cone in the well of a petroleum reservoir, *Kexue Tongbao* **23**, No. 11 (1978), 647–650, 669 (Chinese), MR 82g, 35044.

52. Chernyshov, A. D., Solution of a plane, axisymmetric, and three-dimensional single-phase Stefan problem, *Inzh.-Fiz. Zh.* **27**, No. 2 (1974), 341–350 (Russian), translated as *J. Engrg. Phys.* **27**, No. 2 (1974), 1022–1029 (1976), MR 54, 4367.

53. Ciavaldini, J. F., Analyse numérique d'un problème de Stéfan à deux phases par une méthode d'éléments finis, *SIAM J. Numer. Anal.* **12** (1975), 464–487 (French, English summary), MR 52, 12561.

54. Comincioli, V., A theoretical and numerical approach to some free-boundary problems, *Ann. Mat. Pura Appl.* (4) **100** (1974), 211–238, MR 50, 9164.

55. Crank, J., A nostalgic look at the mechanical solution of Stefan problems, *Bull. Inst. Math. Appl.* **11**, No. 1–2 (1975), 32–33, MR 56, 7516.

56. Crowley, A. B., and Ockendon, J. R., A Stefan problem with a nonmonotone boundary, *J. Inst. Math. Appl.* **20**, No. 3 (1977), 269–281, MR 56, 14189.

57. Curko, V. A., Difference schemes with advancement for the solution of multifront problems of Stefan type, *Vesci Akad. Navuk BSSR Ser. Fiz.-Mat. Navuk* 1977, No. 2 (1977), 36–40, 140 (Russian), MR 58, 13764.

58. Curko, V. A., An economical method for the forward calculation for the multidimensional Stefan problem, *Differencial'nye Uravnenija i Primenen.— Trudy Sem. Processy Optimal. Upravlenija I Sekcija*, No. 21 (1978), 91–101, 107 (Russian, Lithuanian, and English summaries), MR 80c, 65228.

59. Curko, V. A., A difference method with rectification of the fronts for the

solution of nonlinear problems of Stefan type, *Differencial'nye Uravnenija* **14**, No. 8 (1978), 1487–1489, 1532 (Russian), MR 80d, 65132.

60. Damlamian, A., Le problème de Stéfan avec contraintes au bord et les intégrandes convexes, in *Séminaire sur les Sémi-groupes et les Équations d'Évolution* (années 1972–1973, 1973–1974), Exp. No. 3, Univ. Paris XL, Orsay, 11 pp. Publ. Math. Orsay, No. 166-75.47, 1975 (French), MR 58, 2494.

61. Damlamian, A., Some results on the multi-phase Stefan problem, *Comm. Partial-Differential Equations* **2**, No. 10 (1977), 1017–1044, MR 58, 6694.

62. Damlamian, A., A variational formulation for the Stefan problem; existence of weak solutions, in *Free- and Mixed-Boundary-Value Problems* (Proc. Conf. Oberwolfach, 1978). Lang, Frankfurt, 81–89. *Methoden Verfahren Math. Phys.*, **18**, 1979, MR 82c, 35081.

63. Damlamian, A., Homogéneisation de problème de Stéfan, *C. R. Acad. Sci. Paris Ser. A–B*, **289**, No. 1 (1979), A9–A11 (French, English summary), MR 80f, 35067.

64. Damlamian, A, Une généralisation concernant le problème de Stéfan, in *Proceedings of the International Meeting on Recent Methods in Nonlinear Analysis* (Rome, 1978). Pitagora, Bologna, 25–28, 1979 (French), MR 81j, 80002.

65. Damlamian, A., and Kenmochi, N., Le problème de Stéfan avec conditions latérales variables, *Hiroshima Math. J.* **10**, No. 2 (1980), 271–293 (French), MR 81j, 35053.

66. Damlamian, A., How to homogenize a nonlinear diffusion equation: Stefan's problem, *SIAM J. Math. Anal.* **12**, No. 3 (1981), 306–313, MR 82f, 35187.

67. Daniljuk, I. I., A certain variant of the two-phase Stefan problem, *Dopovidi Akad. Nauk. Ukrain. RSR Ser. A*, 1973 (1973), 783–787, 860 (Ukrainian, English and Russian summaries), MR 50, 9187.

68. Daniljuk, I. I., and Salei, S. V., A certain version of the two-phase Stefan problem in the presence of heat sources, *Dopovidi Akad. Nauk. Ukrain. RSR Ser. A*, 1975, No. 11 (1975), 972–975, 1050 (Ukrainian, English, and Russian summaries), MR 53, 7239.

69. Daniljuk, I. I., On a quasistationary problem of Stefan type, *Zap. Nauchn. Sem. Leningrad, Otdel. Mat. Inst. Steklov.* (LOMI) **84** (1979), 24–36, 310, 317 (Russian, English summary), MR 81a, 80003.

70. Daniljuk, I. I., Variational approach to a quasistationary Stefan problem: Differential and integral equations, in *Boundary-Value Problems* (Russian). Tbilis. Gos. Univ., Tbilisi, 75–88, 1979 (Russian), MR 82g, 80010.

71. Daniljuk, I. I., On a multidimensional, nonstationary Stefan problem in domains with a nonsmooth boundary, *Dokl. Akad. Nauk. Ukrain. SSR Ser. A*, 1980, No. 11 (1980), 9–15 (Russian, English summary), MR 82i, 35082.

72. Duvaut, G., Résolution d'un problème de Stéfan (fusion d'un bloc de glace à zéro dégré), *C. R. Acad. Sci. Paris Ser. A–B*, **276** (1973), A1461–A1463 (French), MR 48, 6688.

73. Duvaut, G., Two-phase Stefan problem with varying specific-heat coefficients, *An. Acad. Brasil. Ci.* **47**, No. 3–4 (1975), 377–380, MR 58, 19876.

74. Elliott, C. M., On the finite-element approximation of an elliptic variational inequality arising from an implicit time descretization of the Stefan problem, *IMA J. Numer. Anal.* **1**, No. 1 (1981), 115–125, MR 82d, 65072.

75. Fasano, A., and Primicerio, M., Il problema di Stefan con condizioni al contorno non lineari, *Ann. Scuola Norm. Sup. Pisa* (3) **26** (1972), 711–737 (Italian), MR 51, 2488.

76. Fasano, A., and Primicerio, M., Convergence of Huber's method for heat-conduction problems with change of phase, *Z. Angew. Math. Mech.* **53**, No. 6 (1973), 341–348 (German and Russian summaries), MR 53, 4760.

77. Fasano, A., Alcune osservazioni su una classe di problemi a contorno libero per l'equazione del calore, *Matematiche (Catania)* **29**, No. 2 (1974), 397–411 (1975); (English summary), MR 53, 13835.

78. Fasano, A., and Primicerio, M., One-phase and two-phase free-boundary problems of general type for the heat equation, *Atti Accad. Naz. Lincei Rend. Cl. Sci. Fis. Mat. Natur.* (8) **57**, No. 5 (1974), 387–390 (1975) (Italian summary), MR 53, 1024.

79. Fasano, A., and Primicerio, M., General free-boundary problems for the heat equation. I., *J. Math. Anal. Appl.* **57**, No. 3 (1977), 694–723, MR 58, 6695a.

80. Fasano, A., and Primicerio, M., General free-boundary problems for the heat equation. II., *J. Math. Anal. Appl.* **58**, No. 1 (1977), 202–231, MR 58, 6695b.

81. Fasano, A., and Primicerio, M., General free-boundary problems for the heat equation. III., *J. Math. Anal. Appl.* **59**, No. 1 (1977), 1–14, MR 58, 6695c.

82. Fasano, A., Primicerio, M., and Kamin, S., Regularity of weak solutions of one-dimensional, two-phase Stefan problems, *Ann. Mat. Pura App.* (4) **115** (1977), 341–348 (1978) (English, Italian summary), MR 57, 16987.

83. Fasano, A., and Primicerio, M., Convexity of the free boundary in some classical, parabolic free-boundary problems, *Riv. Mat. Univ. Parma* (4) **5**, Part 2 (1979), 635–645 (1980), MR 82g, 35060.

84. Fasano, A., and Primicerio, M., Free-boundary problems for nonlinear parabolic equations with nonlinear free-boundary conditions, *J. Math. Anal. Appl.* **72**, No. 1 (1979), 247–273, MR 82j, 35091.

85. Fasano, A., and Primicerio, M., Cauchy-type free-boundary problems for nonlinear parabolic equations, *Riv. Mat. Univ. Parma* (4) **5**, Part 2 (1979), 615–634 (1980), MR 81h, 35026.

86. Fasano, A., and Primicerio, M., New results on some classical parabolic free-boundary problems, *Quart. Appl. Math.* **38**, No. 4 (1980), 439–460, MR 82g, 35061.

87. Fasano, A., Primicerio, M., and Rubinstein, L., A model problem for heat conduction with a free boundary in a concentrated capacity, *J. Inst. Math. Appl.* **26**, No. 4 (1980), 327–347, MR 82j, 35139.

88. Fernandez-Diaz, J., and Williams, W. O., A generalized Stefan condition, *Z. Angew. Math. Phys.* **30**, No. 5 (1979), 749–755 (English, German summary), MR 82c, 80006.

89. Finn, W. D. L., and Varoglu, E., Finite-element solution of the Stefan problem, in *Mathematics of Finite Elements and Applications, III* (Proc. Third MAFELAP Conf., Brunel Univ., Uxbridge, 1978). Academic Press, London, 117–122, 1979, MR 81h, 65111.

90. Fisher, I., and Medland, I. C., The multidimensional Stefan problem: a finite-element approach, in *Finite-Element Methods in Engineering* (Proc. Internat. Conf., Univ. New South Wales, Kensington, 1974). Unisearch,

Kensington, 767–783, 1974, MR 57, 8071.

91. Fremond, M., Variational formulation of the Stefan problem—coupled Stefan—frost propagation in porous media, in *Computational Methods in Nonlinear Mechanics* (Proc. Internat. Conf., Austin, Tex., 1974). Texas Inst. Comput. Mech., Austin, Tex., 341–349, 1974, MR 53, 2152.

92. Friedman, A., and Kinderlehrer, D., A one-phase Stefan problem, *Indiana Univ. Math. J.*, **24**, No. 11 (1974), 1005–1035, MR 52, 6190.

93. Friedman, A., Parabolic variational inequalities in one space dimension and smoothness of the free boundary, *J. Functional Analysis* **18** (1975), 151–176, MR 57, 16988.

94. Friedman, A., and Torelli, A., A free-boundary problem connected with nonsteady filtration in porous media, *Nonlinear Anal.* **1**, No. 5 (1976), 503–545, MR 58, 29277.

95. Friedman, A., Analyticity of the free boundary for the Stefan problem *Arch. Rational Mech. Anal.* **61**, No. 2 (1976), 97–125, MR 53, 11227.

96. Friedman, A., One-phase, moving-boundary problems, in *Moving-Boundary Problems* (Proc. Sympos. and Workshop, Gatlinburg, Tenn., 1977). Academic Press, New York, 25–40, 1978, MR 58, 6697.

97. Friedman, A., and Jensen, R., Convexity of the free boundary in the Stefan problem and in the dam problem, *Arch. Rational Mech. Anal.* **67**, No. 1 (1978), 1–24, MR 82i, 35100.

98. Friedman, A., Time-dependent free-boundary problems, *SIAM Rev.* **21**, No. 2 (1979), 213–221, MR 81g, 76097.

99. Friedman, A., and Kamin, S., The asymptotic behavior of gas in an n-dimensional porous medium, *Trans. Amer. Math. Soc.* **262**, No. 2 (1980), 551–563, MR 81j, 35054.

100. Friedman, A., Correction to: "On the free boundary of a quasivariational inequality arising in a problem of quality control" (*Trans. Amer. Math. Soc.* **246** (1978), 95–110; MR 80f: 93086c), *Trans. Amer. Math. Soc.* **257**, No. 2 (1980), 535–537, MR 81h, 93119.

101. Gastaldi, F., About the possibility of setting Stefan-like problems in variational form, *Boll. Un. Mat. Ital. A* (5) **16**, No. 1 (1979), 148–156 (English, Italian summary), MR 80c, 35103.

102. Gilardi, G., A new approach to evolution-free-boundary problems, *Comm. Partial Differential Equations* **4**, No. 10 (1979), 1099–1122, MR 80i, 76041.

103. Gilardi, G., Correction to the paper: "A new approach to evolution-free-boundary problems" (*Comm. Partial Differential Equations* **4**, No. 10 (1979), 1099–1122; MR 80i: 76041), *Comm. Partial-Differential Equations* **5**, No. 9 (1980), 983–984, MR 82c, 76097.

104. Gourgeon, H., and Mossino, J., Sur un problème à frontière libre de la physique des plasmas, *Ann. Inst. Fourier* (Grenoble) **29**, No. 4 (1979), ix, 127–141 (French, English summary), MR 81f, 82023.

105. Golubeva, A. A., A standard program for the numerical solution of the Stefan boundary-value problem for one-dimensional parabolic equations by a difference method with "fractional steps in x and t," in *Methods for the Solution of Boundary-Value and Inverse Problems of Heat Conduction* (Russian). Izdat. Moskov. Univ., Moscow, 38–56, 1975 (Russian), MR 55, 4716.

106. Gorodnichev, S. P., Improving the accuracy of the integral method for the

solution of one-phase Stefan problems, *Izv. Akad. Nauk. Kazah. SSR Ser. Fiz.-Mat.*, 1977, No. 5 (1977), 16–21, 91 (Russian, Kazakh summary), MR 58, 14421.

107. Greenspan, D., A particle model of the Stefan problem, *Comput. Methods Appl. Mech. Engrg.* **13**, No. 1 (1978), 95–104, MR 57, 14531.

108. Grigor'ev, S. G., Kosolapov, V. N., Pudovkin, M. A., and Chugunov, V. A., Scheme of the generalized method of integral relations for multidimensional, single-phase Stefan problems and its applications, in *Applied Problems of Theoretical and Mathematical Physics* (Russian). Latv. Gos. Univ., Riga, 43–52, 126, 1980 (Russian), MR 82f, 65136.

109. Gu, Lian Kun, The asymptotic behavior of the solution of the multiphase Stefan problem, *Acta Math. Sinica* **23**, No. 2 (1980) 203–214 (Chinese), MR 82f, 35085.

110. Hanzawa, Ei-ichi, Classical solutions of the Stefan problem, *Tohoku Math. J.* (2) **33**, No. 3 (1981), 297–335, MR 82k, 35065.

111. Hill, C. D., and Kotlow, D. B., Classical solutions in the large, of a two-phase free-boundary problem. I., *Arch. Rational Mech. Anal.* **45** (1972), 63–78, MR 51, 6149.

112. Hill, C. D., and Kotlow, D. B., Classical solutions in the large, of a two-phase free-boundary problem. II., *Arch. Rational Mech. Anal.* **47** (1972), 369–379, MR 48, 11771.

113. Hoffman, K.-H., Monotonei bei nichtlinearen Stefan-Problemen, in *Numerische Behändlung von Differentialgleichungen mit besonderer Berucksichtigung freier Randwertaufgaben* (Tagung Math. Forschungsinst., Oberwolfach, 1977). Birkhauser, Basel, 162–190. *Internat. Ser. Numer. Math.*, **39**, 1978 (German, English summary), MR 81h, 65106.

114. Hoffmann, K.-H., Monotonei bei Zweiphasen-Stefan-Problemen, *Numer. Funct. Anal. Optim.* **1**, No. 1 (1979), 79–112 (German, English summary), MR 80b, 65125.

115. Hohn, W., Konvergenzordnung bei einem expliziten Differenzenverfahren zur numerischen Lösung des Stefan-Problems, in *Numerische Behändlung von Differentialgleichungen mit besonderer Berucksichtigung freier Randwertaufgaben* (Tagung Math. Forschungsinst., Oberwolfach, 1977). Birkhauser, Basel, 191–213. *Internat. Ser. Numer. Math.*, **39**, 1978 (German, English summary), MR 80m, 65069.

116. Ichikawa, Y., and Kikuchi, N., A one-phase multidimensional Stefan problem by the method of variational inequalities, *Internat. J. Numer. Methods Engrg.* **14**, No. 8 (1979), 1197–1220, MR 80e, 65113.

117. Ishii, H., Asymptotic stability of almost periodic solutions of a free-boundary problem arising in hydraulics, *Bull. Fac. Sci. Engrg. Chuo Univ.* **22** (1979), 73–95, MR 81k, 35016.

118. Ishii, H., Erratum: "Asymptotic stability of almost periodic solutions of a free-boundary problem arising in hydraulics" (*Bull. Fac. Sci. Engrg. Chuo Univ.* 22 (1979), 73–95; MR 81k: 35016), *Bull. Fac. Sci. Engrg. Chuo Univ.* **23** (1980), 83, MR 82g, 35008.

119. Ishii, H., Asymptotic stability and existence of almost-periodic solutions for the one-dimensional, two-phase Stefan problem, *Math. Japon.* **25**, No. 4 (1980), 379–393, MR 81m, 35139.

120. Ishii, H., On a certain estimate of the free boundary in the Stefan problem, *J. Differential Equations* **42**, No. 1 (1981), 106–115, MR 82k, 35102.

121. Jakobov, S., Solvability of a class of problems with free boundary for linear equations of parabolic type, *Dinamika Sploshn. Sredy*, No. 43, *Mat. Problemy Meh.* (1979), 178–183, 188 (Russian), MR 82h, 35099.

122. Jensen, R., Smoothness of the free boundary in the Stefan problem with supercooled water, *Illinois J. Math.* **22**, No. 4 (1978), 623–629, MR 81g, 35057.

123. Jerome, J. W., Existence and approximation of weak solutions of the Stefan problem with nonmonotone nonlinearities, in *Numerical Analysis* (Proc. 6th Biennial Dundee Conf., Univ. Dundee, Dundee, 1975). Springer, Berlin, 143–165. Lecture Notes in Math., Vol. 506, 1976, MR 57, 8077.

124. Jerome, J. W., Nonlinear equations of evolution and a generalized Stefan problem, *J. Differential Equations* **26**, No. 2 (1977), 240–261, MR 58, 1659a.

125. Jerome, J. W., Corrigendum: "Nonlinear equations of evolution and a generalized Stefan problem" (*J. Differential Equations* **26** (1977), No. 2, 240–261), *J. Differential Equations* **28**, No. 3 (1978), 452, MR 58, 1659b.

126. Jochum, P., Differentiable dependence upon the data in a one-phase Stefan problem, *Math. Methods Appl. Sci.* **2**, No. 1 (1980), 73–90, MR 82k, 35054.

127. Junusov, M., The variational approach to the solution of a Stefan-type problem, *Dokl. Akad. Nauk. Tadzhik. SSR* **19**, No. 8 (1976), 8–10 (Russian, Tajiki summary), MR 55, 3539.

128. Kashkaha, V. E., The Ritz method for the study of a two-phase quasistationary problem of Stefan type, *Mat. Fiz., Vyp.* 17 (1975), 128–137, 190 (Russian), MR 56, 7257.

129. Katz, H., A large-time expansion for the Stefan problem, *SIAM J. Appl. Math.* **32**, No. 1 (1977), 1–20, MR 55, 7144.

130. Kavokin, A. A., The asymptotic continuity of the solution of the Stefan problem with respect to boundary conditions for small time, *Izv. Akad. Nauk. Kazah. SSR Ser. Fiz.-Mat.* 1976, No. 1 (1976), 62–66, 93 (Russian, Kazakh summary), MR 57, 13186.

131. Kawarada, H., Stefan-type free-boundary problems for heat equations, *Publ. Res. Inst. Math. Sci.* 9 (1973), 517–533, MR 52, 14652.

132. Kawarada, H., and Natori, M., On numerical solutions of Stefan problems. I, *Mem. Numer. Math.*, No. 1 (1974), 43–54, MR 52, 2238.

133. Kawarada, H. and Natori, M., On numerical solutions of Stefan problems. II. Unique existence of numerical solution, *Mem. Numer. Math.*, No. 2 (1975), 1–20, MR 52, 2239.

134. Kinderlehrer, D., Variational inequalities and free-boundary problems, *Bull. Amer. Math. Soc.* **84**, No. 1 (1978), 7–26, MR 57, 6759.

135. Kinderlehrer, D., and Nirenberg, L., Hodograph methods and the smoothness of the free boundary in the one-phase Stefan problem, in *Moving-Boundary Porblems* (Proc. Sympos. and Workshop, Gatlinburg, Tenn., 1977). Academic Press, New York, 57–69, 1978, MR 58, 1636.

136. Kinderlehrer, D., and Nirenberg, L., The smoothness of the free boundary in the one-phase Stefan problem, *Comm. Pure Appl. Math.* **31**, No. 3 (1978), 257–282, MR 82b, 35152.

137. Klykov, V. E., Kulagin, V. L., and Morozov, V. A., The Stefan-type problem

occurring in the investigation of salt dissolution and transport process in soil, *Prikl. Mat. Meh.* **44** (1981), No. 1, 104–112 (Russian); translated as *J. Appl. Math. Mech.* **44**, No. 1 (1980), 70–75 (1981), MR 82f, 35095.

138. Knerr, B. F., A singular free-boundary problem, *Illinois J. Math.* **23**, No. 3 (1979), 438–458, MR 82c, 62111.

139. Kotlow, D. B., A free-boundary problem connected with the optimal stopping problem for diffusion processes, *Trans. Amer. Math. Soc.* **184** (1973), 457–478 (1974), MR 51, 1981.

140. Kovaljov, O. B., Lar'kin, N. A., Fomin, W. M., and Yanenko, N. N., The solution of nonhomogeneous thermal problems and the Stefan single-phase problem in arbitrary domains, *Comput. Methods Appl. Mech. Engrg.* **22**, No. 2 (1980), 259–271, MR 81d, 80006.

141. Kruzhkov, S. N., and Jakubov, S., Solvability of a class of problems with unknown boundary for the heat equation and behavior of the solutions as the time increases to infinity, *Dinamika Sploshn. Sredy*, No. 36, *Dinamika Zhidkosti so Svobodnymi Granicami* (1978), 46–70, 160 (Russian), MR 80d, 35070.

142. Kyoto University: Evolution systems and free-boundary-value problems, *Res. Inst. Math. Sci.*, 1976, pp. i–ii and 1–218 (Japanese), MR 58, 29035.

143. Lièvre, C., and Rothen, F., New approach to the morphological stability of the Stefan problem, *Helv. Phys. Acta* **53**, No. 1 (1980), 119–133, MR 82c, 80008.

144. Macovei, D., Some typical free-boundary problems, in *Variational Inequalities and Optimization Problems* (Proc. Summer School, Constanta, 1979). Minist. Ed. Invatamiintului, Constanta, 191–217, 1979, MR 81m, 35140.

145. Magenes, E., Topics in parabolic equations: some typical free-boundary problems, in *Boundary-Value Problems for Linear Evolution: Partial-Differential Equations* (Proc. NATO Advanced Study Inst., Liege, 1976). Reidel, Dordrecht, 239–312. NATO Advanced Study Inst. Ser., Ser. C: *Math. and Phys. Sci.*, Vol. 29, 1977, MR 57, 10250.

146. McNabb, A., Asymptotic behavior of solutions of a Stefan problem, *J. Math. Anal. Appl.* **51**, No. 3 (1975), 633–642, MR 51, 10833.

147. Meirmanov, A. M., A generalized solution of the Stefan problem with lumped capacity, *Dinamika Sploshn. Sredy*, Vyp. 10 (1972), 85–101, 247 (Russian), MR 56, 17529.

148. Meirmanov, A. M., The Stefan problem that arises in the theory of combustion, *Dinamika Sploshn. Sredy*, Vyp. 11 (1972), 74–81, 124 (Russian), MR 56, 10462.

149. Meirmanov, A. M., Multiphase Stefan problem for quasilinear parabolic equations, *Dinamika Sploshn. Sredy*, Vyp. 13, *Dinamika Zhidk. so Svobod. Granicami* (1973), 74–86, 187 (Russian), MR 56, 12602.

150. Meirmanov, A. M., Classical solvability of the multidimennsional Stefan problem, *Dokl. Akad. Nauk. SSSR* **249**, No. 6 (1979), 1309–1312 (Russian), MR 81a, 35042.

151. Meirmanov, A. M., The classical solution of a multidimensional Stefan problem for quasilinear parabolic equations, *Mat. Sb.* (*N.S.*) **112** (154), No. 2(6) (1980), 170–192 (Russian), MR 81j, 35055.

152. Meirmanov, A. M., and Puhnachev, V. V., Lagrangian coordinates in the Stefan problem, *Dinamika Sploshn. Sredy*, No. 47, *Mat. Problemy Meh.*

Sploshn. Sred (1980), 90–111, 165–166 (Russian), MR 82m, 35089.

153. Meyer, G. H., A numerical method for two-phase Stefan problems, *SIAM J. Numer. Anal.* **8** (1971), 555–568, MR 45, 9518.

154. Meyer, G. H., The numerical solution of multidimensional Stefan problems—a survey, in *Moving-Boundary Problems* (Proc. Sympos. and Workshop, Gatlinburg, Tenn., 1977). Academic Press, New York, 73–89, 1978, MR 57, 14502.

155. Meyer, G. H., The numerical solution of Stefan problems with front-tracking and smoothing methods, *Appl. Math. Comput.* **4**, No. 4 (1978), 283–306, MR 80a, 65244.

156. Mikulec, A. T., The classical solution of the Stefan problem for systems of equations of parabolic type with exterior fronts, *Bul. Akad. Shtiince RSS Moldoven 1976*, No. 1 (1976), 14–24, 94 (Russian), MR 54, 8029.

157. Mikulec, A. T., Generalized solution of the Stefan problem for a system of equations of parabolic type, in *Studies in Algebra, Mathematical Analysis, and Their Applications* (Russian). Izdat. "Shtiinca," Kishinev, 53–61, 122, 1977 (Russian), MR 58, 23074.

158. Milinazzo, F., and Bluman, G. W., Numerical-similarity solutions to Stefan problems, *Z. Angew. Math. Mech.* **55**, No. 7–8 (1975), 423–429 (English, German, and Russian summaries), MR 52, 4832.

159. Mori, M., Numerical solution of the Stefan problem by the finite-element method, *Mem. Numer. Math.*, No. 2 (1975), 35–44, MR 54, 6522.

160. Mori, M., Stability and convergence of a finite-element method for solving the Stefan problem, *Publ. Res. Inst. Math. Sci.* **12**, No. 2 (1976), 539–563, MR 55, 6889.

161. Mori, M., A finite-element for solving the two-phase Stefan problem in one space dimension, *Publ. Res. Inst. Math. Sci.* **13**, No. 3 (1977), 723–753, MR 57, 14479.

162. Moskal, M. Z., and Mikulec, A. T., The existence of the classical solution of the Stefan problem for a system of parabolic equations, *Mat. Issled.* **10**, No. 2 (36) (1975), 192–211, 286 (Russian), MR 53, 13841.

163. Moskal, M. Z., and Mikulec, A. T., The uniqueness and stability of the classical solution of the Stefan problem for systems of parabolic equations, *Mat. Issled.* **10**, No. 3 (37) (1975), 115–133, 242 (Russian), MR 53, 13842.

164. Moskov Gosuderstv Univ. Reshenija zadach Stefana (Solutions of Stefan problems), 1970; 313pp (Russian), MR 52, 16052.

165. Neustadter, S. F., The time-dependent Poisson queue and moving-boundary problems for the heat and wave equation, *Indian J. Pure Appl. Math.* **10**, No. 5 (1979), 567–580, MR 82k, 35055.

166. Nitsche, J. A., Finite-element approximations to the one-dimensional Stefan problem, in *Recent Advances in Numerical Analysis* (Proc. Sympos., Math. Res. Center, Univ. Wisconsin, Madison, Wis., 1978). Academic Press, New York, 119–142, Publ. Math. Res. Center Univ. Wisconsin, 41, 1978, MR 81h, 65114.

167. Nitsche, J. A., Approximation des eindimensionalen Stefan-Problems durch finite Elemente, in *Proceedings of the International Congress of Mathemeticians* (Helsinki, 1978). Acad. Sci. Fennica, Helsinki, 923–928, 1980 (German), MR 81f, 65085.

168. Nitsche, J. A., Finite-element approximations for free-boundary problems, in

Computational Methods in Nonlinear Mechanics (Proc. Second Internat. Conf., Univ. Texas, Austin, Tex., 1979). North-Holland, Amsterdam, 341-360, MR 81i, 65095.

169. Nogi, T., A difference scheme for solving the Stefan problem, *Publ. Res. Inst. Math. Sci.* **9** (1973), 543-575, MR 52, 4657.

170. Nogi, T., The Stefan problem, *Sugaku* **30**, No. 1 (1978), 1-11 (Japanese), MR 80c, 35040.

171. Nogi, T., A difference scheme for solving two-phase Stefan problems of heat equation, *Publ. Res. Inst. Math. Sci.* **16**, No. 2 (1980), 313-341, MR 82d, 65082.

172. Ohlopkov, N. M., and Evlev, V. P., Numerical solutions of Stefan-type problem, in *Differential and Integral Equations*, No. 1 (Russian). Irkutsk. Gos. Univ., Irkutsk, 59-72, 1972 (Russian), MR 52, 9837.

173. Pavlov, A. R., A difference method of solution of a single-phase Stefan-type problem for a certain integrodifferential equation of parabolic type, in *Differential and Integral Equations*, No. 1 (Russian). Irkutsk. Gos. Univ., Irkutsk, 179-187, 1972 (Russian), MR 52, 9838.

174. Pavlov, A. R., A difference method for the solution of a problem of Stefan type for a certain system of heat-mass transfer equations, in *Differential and Integral Equations*, No. 2 (Russian). Irkutsk. Gos. Univ., Irkutsk, 18-26, 305-306, 1973 (Russian), MR 56, 17116.

175. Petrenko, V. G., Petrenko, R. P., and Efimov, V. O., An application of the method of straight lines to the solution of the modified Stefan problem, *Dopovidi Akad. Nauk. Ukrain. RSR Ser. A* 1973 (1973), 414-418, 477 (Ukrainian, English and Russian summaries), MR 49, 6654.

176. Petrenko, V. G., and Petrenko, R. P., Solution of the two-front modified Stefan problem by the differential-difference method, *Dopovidi Akad. Nauk. Ukrain. RSR Ser. A* 1973 (1973), 340-244, 382 (Ukrainian, English, and Russian summaries), MR 50, 6183.

177. Pierre, M., Un résultat d'existence pour l'équation de la chaleur avec obstacle s.c.s., *C. R. Acad. Sci. Paris Ser.* A–B **287**, No. 2 (1978), A59–A61 (French, English summary), MR 58, 23063.

178. Poznjak, A. A., An application of Biot's variational method to the solution of problems of Stefan type with variable boundary conditions, *Latvijas PSR Zinatnu Akad. Vestis. Fiz. Tehn., Zinatnu Ser.* 1976, No. 1 (1976), 45-53, 126-127 (Russian, English summary), MR 53, 12237.

179. Primicerio, M., Stefan-like problems with space-dependent latent heat, *Meccanica* **5** (1970), 187-190 (English, Italian summary), MR 51, 8633.

180. Primicerio, M., Diffusion problems with a free boundary, *Boll. Un. Mat. Ital.* A (5) **18**, No. 1 (1981), 11-68 (Italian), MR 82i, 35166.

181. Puhnachev, V. V., Generation of a singularity in the solution of a Stefan-type problem, *Differencial'nye Uravnenija* **16**, No. 3 (1980), 492-500, 574 (Russian), MR 81f, 35049.

182. Reemtsen, R., and Lozano, C. J., An approximation technique for the numerical solution of a Stefan problem, *Numer. Math.* **38**, No. 1 (1981), 141-154, MR 82k, 65056.

183. Rogers, J. C., Berger, A. E., and Ciment, M., The alternating-phase truncation method for numerical solution of a Stefan problem, *SIAM J. Numer. Anal.*

16, No. 4 (1979), 563–587, MR 80f, 65136.

184. Rubinstein, L., The Stefan problem: comments on its present state, *J. Inst. Math. Appl.* **24**, No. 3 (1979), 259–277, MR 80j, 35096.

185. Rubinstein, L., Fasano, A., and Primicerio, M., Remarks on the analyticity of the free boundary for the one-dimensional Stefan problem, *Ann. Mat. Pura Appl.* (4) **125** (1980), 295–311, MR 82f, 35087.

186. Rundell, W., The Stefan problem for a pseudo-heat equation, *Indiana Univ. Math. J.* **27**, No. 5 (1978), 739–750, MR 80b, 35086.

187. Salei, S. V., Global solvability of a Stefan problem, *Dokl. Akad. Nauk. Ukrain. SSR Ser. A*, No. 6 (1979), 424–428, 493 (Russian, English summary), MR 80i, 35088.

188. Salei, S. V., On the rate of convergence of a free boundary to the limit value in a Stefan problem, *Dokl. Akad. Nauk. Ukrain. SSR Ser. A*, 1981, No. 6 (1981), 19–23, 108 (Russian, English summary), MR 82m, 35067.

189. Santos, V. R. B., A direct method for solving two-dimensional, one-phase Stefan problems, *Mem. Numer. Math.*, No. 6 (1979), 39–63, MR 81a, 65105.

190. Santos, V. R. B., A direct method for solving two-dimensional, one-phase Stefan problems, *Comput. Methods Appl. Mech. Engrg.* **25**, No. 1 (1981), 51–64, MR 81k, 80001.

191. Schaeffer, D. G., A new proof of the infinite differentiability of the free boundary in the Stefan problem, *J. Differential Equations* **20**, No. 1 (1976), 266–269, MR 52, 11325.

192. Sero-Guillaume, O., Problème à frontière libre de type Stéfan, *C. R. Acad. Sci. Paris Ser. A–B*, **287**, No. 10 (1978), A759–A761 (French, English summary), MR 80b, 35072.

193. Sherman, B., Continuous dependence and differentiability properties of the solution of a free-boundary problem for the heat equation, *Quart. Appl. Math.* **27** (1969), 427–439, MR 58, 22968.

194. Sherman, B., Limiting behavior in some Stefan problems as the latent heat goes to zero, *SIAM J. Appl. Math.* **20** (1971), 319–327, MR 45, 2336.

195. Shishkin, G. I., A certain problem with a free boundary for a system of parabolic equations, *Chisl. Metody Meh. Sploshnoi Sredy* **3**, No. 2 (1972), 105–121 (Russian), MR 51, 13460.

196. Shishkin, G. I., A certain problem of Stefan type with a discontinuous movable boundary, *Dokl. Akad. Nauk. SSSR* **224**, No. 6 (1975), 1276–1278 (Russian), MR 54, 8030.

197. Slepjan, L. I., and Fishkov, A. L., A two-dimensional mixed problem with an irregularly moving, boundary-condition separation point, in *Studies in Elasticity and Plasticity*, No. 13 (Russian). Leningrad. Univ., Leningrad, 172–181, 314, 1980 (Russian), MR 82c, 73075.

198. Socolescu, D., On one free-boundary problem for the stationary Navier–Stokes equations, in *Free and Mixed Boundary-Value Problems* (Proc. Conf., Oberwolfach, 1978). Lang, Frankfurt, 41–60; *Methoden Verfahren Math. Phys.*, **18**, 1979, MR 82b, 35130.

199. Solonnikov, V. A., A problem with a free boundary for a system of Navier–Stokes equations, in *Partial-Differential Equations* (Russian). Akad. Nauk. SSSR Sibirsk. Otdel., Inst. Math., Novosibirsk, 127–140, 143, Trudy Sem. S. L. Soboleva, No. 2, 1978, 1978 (Russian), MR 81e, 35106.

200. Solov'eva, E. N., and Uspenskii, A. B., Direct calculation schemes for the numerical solution of the Stefan problem for the two-dimensional heat equation, in *Methods for the Solution of Boundary-Value and Inverse Problems of Heat Conduction* (Russian). Izdat. Moskov. Univ., Moscow, 24-37, 1975 (Russian), MR 55, 6881.

201. Soward, A. M., A unified approach to Stefan's problem for spheres and cylinders, *Proc. Roy. Soc. London*, Ser. A, **373**, No. 1752 (1980), 131-147, MR 82c, 80012.

202. Stewartson, K., and Waechter, R. T., On Stefan's problem for spheres, *Proc. Roy. Soc. London*, Ser. A, **348**, No. 1655 (1976), 415-426, MR 53, 4756.

203. Streit, U., Zur iterativen Lösung nichtlinearer Differenzenschemata für eindimensionale Stefan-Aufgaben, *Wiss. Z. Tech. Hochsch. Karl-Marx-Stadt* **22**, No. 4 (1980), 347-353 (German), MR 82b, 80015.

204. Suslov, A. I., On Stefan's problem occurring in the theory of powder burning, *Prikl. Mat. Meh.* **41**, No. 1 (1977), 95-101 (Russian); translated as *J. Appl. Math. Mech.* **41**, No. 1 (1977), 87-92, MR 57, 16994.

205. Tabisz, K., On a kind of the Stefan problem, *Bull. Acad. Polon. Sci. Ser. Sci. Math.* **27**, No. 11-12 (1979), 847-852 (1981) (English, Russian summary), MR 82i, 35170.

206. Tabisz, K., Local and global solutions of the Stefan-type problem, *J. Math. Anal. Appl.* **82**, No. 2 (1981), 306-316, MR 82k, 35103.

207. Tao, L. N., The Stefan problem with arbitrary initial and boundary conditions, *Quart. Appl. Math.* **36**, No. 3 (1978), 223-233, MR 80b, 80003.

208. Tao, L. N., The analyticity of solutions of the Stefan problem, *Arch. Rational Mech. Anal.* **72**, No. 3 (1979), 285-301, MR 80i, 35184.

209. Tao, L. N., On free-boundary problems with arbitrary initial and flux conditions, *Z. Angew. Math. Phys.* **30**, No. 3 (1979), 416-426, MR 81f, 35051.

210. Tao, L. N., On solidification problems including the density jump at the moving boundary, *Quart. J. Mech. Appl. Math.* **32**, No. 2 (1979) 175-185, MR 80h, 80006.

211. Tao, L. N., Free-boundary problems with radiation boundary conditions, *Quart. Appl. Math.* **37**, No. 1 (1979), 1-10, MR 80f, 35021.

212. Tao, L. N., Solidification of a binary mixture with arbitrary heat flux and initial conditions, *Arch. Rational Mech. Anal.* **76**, No. 2 (1981), 167-181, MR 82j, 80003.

213. Tarzia, D. A., Sur le probléme de Stéfan á deux phases, *C. R. Acad. Sci. Paris* Ser. A-B, **288**, No. 20 (1979), A941-A944 (French, English summary), MR 80c, 35045.

214. Tarzia, D. A., Application of variational methods in the steady-state case of the two-phase Stefan problem, *Math. Notae* **27** (1979), 145-156 (Spanish), MR 82d, 80020.

215. Tarzia, D. A., A family of problems which converges towards the steady-state case of the two-phase Stefan problem, *Math. Notae* **27** (1979), 157-165 (Spanish), MR 82d, 80021.

216. Torelli, A., Existence and uniqueness of the solution for a nonsteady free-boundary problem, *Boll. Un. Math. Ital. B* (5) **14**, No. 2 (1977), 423-466 (English, Italian summary), MR 58, 29243.

217. Turland, B. D., and Peckover, R. S., The stability of planar melting fronts in

two-phase, thermal Stefan problems, *J. Inst. Math. Appl.* **25**, No. 1 (1980), 1–15, MR 80m, 80012.

218. van Moerbeke, P., On optimal stopping and free-boundary problems, *Arch. Rational Mech. Anal.* **60**, No. 2 (1975), 101–148, MR 54, 1367.

219. Wilson, D. G., Existence and uniqueness for similarity solutions of one-dimensional, multiphase Stefan problems, *SIAM J. Appl. Math.* **35**, No. 1 (1978), 135–147, MR 57, 13187.

220. Wilson, D. G., Solomon, A. D., and Boggs, P. T., *Moving-Boundary Problems.* Academic Press (Harcourt Brace Jovanovich, Publishers), New York-London, x + 329 pp., 0-12-757350-X, 1978, MR 57, 6761.

221. Zaharov, V. G., and Petrenko, V. G., Solution of the Stefan problem by the collocation method, *Dopoviki Akad. Nauk. Ukrain. RSR Ser. A* 1972 (1972), 400–404, 476 (Ukrainian, English, and Russian summaries), MR 47, 7221.

222. Zhadaeva, N. G., Difference schemes for nonlinear Stefan problems, *Differencial'nye Uravnenija* **12**, No. 9 (1976), 1712–1714, 1727–1728 (Russian), MR 56, 17124.

3.4. Control of Various Heat Problems

1. Atanbaev, S. A., A necessary and sufficient condition for the optimality of a certain conditionally well-posed problem. *Izv. Akad. Nauk. Kazah. SSR Ser. Fiz.-Mat.* 1975, No. 1 (1975), 76–78, 95 (Russian), MR 51, 6530.

2. Bachoi, G. S., The synthesis of an optimal control of the heat-conduction process, *Math. Issled.*, No. 46 (1978), 20–31, 96–97 (Russian), MR 80a, 93086.

3. Baumeister, J., On the treatment of free-boundary problems with the heat equation via optimal control, *Math. Methods Appl. Sci.* **1**, No. 1 (1979), 40–61, MR 80h, 35131.

4. Berezovs'kii, A. A., and Metesova, T. M., The reducibility of one-dimensional, quasilinear, parabolic and hyperbolic equations, in *Analytic, Numerical and Analog Methods in Heat Conduction Problems* (Russian). Izdat. "Naukova Dumka," Kiev, 71–81, 237, 1977 (Russian), MR 57, 13115.

5. Budak, B. M., and Gol'dam, N. L., The optimal Stefan problem for a quasilinear parabolic equation, in *Solutions of Stefan Problems* (Russian). Moskov. Gos. Univ., Moscow, 156–179, 1970 (Russian), MR 57, 16981.

6. Chewning, W. C., and Seidman, T. I., A convergent scheme for boundary control of the heat equation, *SIAM J. Control Optimization* **15**, No. 1 (1977), 64–72, MR 55, 9557.

7. Collins, W. D., Dual extremum principles for the heat equation, *Proc. Roy. Soc. Edinburgh, Sect. A*, **77**, No. 3–4 (1977), 273–292, MR 80g, 49025.

8. Cuvelier, C., Optimal control of a system governed by the Navier–Stokes equations coupled with the heat equation, in *New Developments in Differential Equations* (Proc. 2nd Scheveningen Conf., Scheveningen, 1975). North-Holland, Amsterdam, 81–98; North-Holland Math. Studies, Vol. 21, 1976, MR 58, 17594.

9. Egorov, A. I., Conditions for optimality in a certain problem of control of a heat-transfer process, *Zh. Vychisl. Mat. i Mat. Fiz.* **12** (1972), 791–799 (Russian), MR 46, 7686.

10. Egorov, A. I., and Naval, E. S., Optimal control of a process of heat conduc-
     tion, *Prikl. Mat. i Programmirovanie*, Vyp. 13 (1975), 44–63, 147 (Russian),
     MR 56, 9355.
11. Elliott, C. M., and Janovsky, V., A variational-inequality approach to
     Hele–Shaw flow with a moving boundary, *Proc. Roy. Soc. Edinburgh, Sect. A*,
     88, No. 1–2 (1981), 93–107, MR 82d, 76031.
12. Fasano, A., Un esempio controllo ottimale in un problema del tipo di Stefan,
     Boll. Un. Mat. Ital. (4) **4** (1971), 846–858 (Italian, English summary), MR 45,
     8977.
13. Fattorini, H. O., The time-optimal problem for boundary control of the heat
     equation, in *Calculus of Variations and Control Theory* (Proc. Sympos., Math.
     Res. Center, Univ. Wisconsin, Madison, Wis., 1975; dedicated to Laurence
     Chisholm Young on the occasion of his 70th birthday). Academic Press, New
     York, 305–320, Math. Res. Center, Univ. Wisconsin, Publ. No. 36, 1976, MR
     54, 5933.
14. Fattorini, H. O., Reachable states in boundary control of the heat equation are
     independent of time, *Proc. Roy. Soc. Edinburgh, Sect. A*, **81**, No. 1–2 (1978),
     71–77, MR 80g, 49040.
15. Follmer, H., Martingale criteria for stochastic stability, in *Probability Theory*
     (Papers, VIIth Semester, Stefan Banach Internat. Math. Center, Warsaw,
     1976). PWN Warsaw, 89–96, Banach Center Publ. 5, 1979, MR 81c, 93088.
16. Giurgiu, M., A feedback solution of a linear quadratic problem for boundary
     control of heat equation, *Rev. Roumaine Math. Pures Appl.* **20**, No. 8 (1975),
     927–954, MR 51, 13810.
17. Giurgiu, M., Linear feedback for optimal stabilization by boundary control of
     heat equation, *Rev. Roumaine Math. Pures Appl.* **22**, No. 6 (1977), 777–796,
     MR 58, 26397.
18. Jamet, P., Stability and convergence of a generalized Crank–Nicolson scheme
     on a variable mesh for the heat equation, *SIAM J. Numer. Anal.* **17**, No. 4
     (1980), 530–539, MR 82j, 65073.
19. Junusov, M., The solution of a certain optimal problem of Stefan type, *Zh.
     Vychisl. Mat. Fiz.* **15**, No. 2 (1975), 345–357, 539 (Russian), MR 51, 8632.
20. Jurii, A. B., An optimal problem of Stefan type, *Dokl. Akad. Nauk. SSSR* **251**,
     No. 6 (1980), 1317–1321 (Russian), MR 81k, 49025.
21. Kononova, A. A., On the problem of control for the heat equation in the
     uniform metric, *Ural. Gos. Univ. Mat. Zap.* **10**, No. 2, *Issled. Sovremen. Mat.
     Anal.* (1977), 74–82, 216 (Russian), MR 58, 12587.
22. Kostyleva, V. D., Existence of a solution of Stefan's optimal problem, in
     Mathematics Collections (Russian). Izdat. "Naukova Dumka," Kiev, 86–90,
     1976 (Russian), MR 56, 13019.
23. Malanowski, K., Application of mathematical programming methods to the
     determination of the optimal control for systems that are described by heat
     equations, *Arch. Automat. i Telemech.* **18** (1973), 3–18 (Polish, Russian, and
     English summaries), MR 49, 6002.
24. Mandl, P., On the adaptive control of countable Markov chains, in *Probability
     Theory* (Papers, VIIth Semester, Stefan Banach Internat. Math. Center, Warsaw,
     1976). PWN, Warsaw, 159–173, Banach Center Publ. 5, 1979, MR 81b, 93063.
25. Mizel, V. J., and Seidman, T. I., Observation and prediction for the heat

equation,   II., *J. Math. Anal. Appl.* **38** (1972), 149–166, MR 46, 2223.

26.   Nisio, M., On a nonlinear semigroup associated with stochastic optimal control and its excessive majorant, in *Probability Theory* (Papers, VIIth Semester, Stefan Banach Internat. Math. Center, Warsaw, 1976), 175–202. Banach Center Publ. 5, 1979, MR 82j, 93057.

27.   Sachs, E., A parabolic control problem with a boundary condition of the Stefan–Boltzmann type, *Z. Angew. Math. Mech.* **58**, No. 10 (1978), 443–449 (English, German, and Russian summaries), MR 80a, 49037.

28.   Saguez, C., and Larrecq, M., Controle de systèmes à frontière libre. Application à la coulée continué d'acier, in *Computing Methods in Applied Sciences and Engineering* (Proc. Fourth Internat. Sympos., Versailles, 1979), 385–404, 1980 (French), MR 81i, 80007.

29.   Seidman, T. I., Boundary observation and control for the heat equation, in *Calculus of Variations and Control Theory* (Proc. Sympos., Math. Res. Center, Univ. Wisconsin, Madison, Wis., 1975; dedicated to Laurence Chisholm Young on the occasion of his 70th birthday), 321–351. Math. Res. Center, Univ. Wisconsin, Publ. No. 36, 1976, MR 54, 5946.

30.   Seidman, T. I., Observation and prediction for the heat equation. III., *J. Differential Equations* **20**, No. 1 (1976), 18–27, MR 52, 11326.

31.   Seidman, T. I., Observation and prediction for the heat equation. IV. Patch observability and controllability, *SIAM J. Control Optimization* **15**, No. 3 (1977), 412–427, MR 55, 8941.

32.   Schmidt, E. J. P. G., The "bang-bang" principle for the time-optimal problem in boundary control of the heat equation, *SIAM J. Control Optim.* **18**, No. 2 (1980), 101–107, MR 81b, 93006b.

33.   Schmidt, E. J. P. G., Boundary control for the heat equation with steady-state targets, *SIAM J. Control Optim.* **18**, No. 2 (1980), 145–154, MR 81b, 93006a.

34.   Sivan, R., On the controllability of the heat equation, in *Automatic and Remote Control* III (Proc. Third Congr. Internat. Federation Automat. Control (IFAC), London, 1966), Vol. 1, p. 12, Paper 6D. Inst. Mech. Engrs., London, 4 pp., 1967, MR 52, 1488.

35.   van Moerbeke, P., An optimal stopping problem with linear reward, *Acta Math.* **132** (1974), 111–151, MR 51, 12405.

36.   Wilson, D. A., and Rubio, J. E., A controllability problem for a system governed by the heat equation, *J. Inst. Math. Appl.* **17**, No. 1 (1976), 43–51, MR 55, 2246.

37.   Winther, R., Error estimates for a Galerkin approximation of a parabolic control problem, *Ann. Mat. Pura Appl.* (4) **117** (1978), 173–206, MR 80a, 49067.

3.5.   Not-Well-Posed and Inverse Problems

1.   Abdullaeva, G. Z., A certain inverse problem for the heat equation, *Akad. Nauk. Azerbaidzhan. SSR Dokl.* **32**, No. 6 (1976), 8–10 (Russian, Azerbaijani, and English summaries), MR 54, 10903.

2.   Ahundov, A. J., A multidimensional inverse problem for an equation of parabolic type in an unbounded domain, *Izv. Akad. Nauk. Azerbaidzhan.*

SSR Ser. Fiz.-Tehn. Mat. Nauk. 1977, No. 1 (1977), 86–91 (Russian, Azerbaijani, and English summaries), MR 58, 17613.

3. Ahundov, A. J., An inverse problem for a quasilinear system of parabolic equations in a domain with a moving boundary, *Izv. Akad. Nauk. Azerbaidzhan. SSR Ser. Fiz.-Tehn. Mat. Nauk.* 1979, No. 5 (1979), 119–123 (Russian, Azerbaijani, and English summaries), MR 81g, 35056.

4. Aliev, G. G., A certain inverse problem for the heat equation on a semi-infinite domain, *Izv. Akad. Nauk. Azerbaidzhan. SSR Ser. Fiz.-Tehn. Mat. Nauk.* 1970, No. 6 (1970), 42–48 (Russian, Azerbaijani summary), MR 45, 3968.

5. Aliev, G. G., The problem of finding the functions that occur in the boundary and supplementary conditions in the inverse problem for the heat equation on a semi-infinite region, *Azerbaidzhan. Gos. Univ. Uchen. Zap. Ser. Fiz.-Mat. Nauk.* 1972, No. 4 (1972), 25–31 (Russian, Azerbaijani summary), MR 58, 12039.

6. Aliev, G. G., A certain nonlinear inverse problem of nonstationary heat conduction, *Izv. Akad. Nauk. Azerbaidzhan. SSR Ser. Fiz.-Tehn. Mat. Nauk.* 1973, No. 2 (1973), 121–128 (Russian, Azerbaijani summary), MR 52, 11323.

7. Anger, G., and Czerner, R., The extreme measures in the inverse problem of the heat equation, in *Elliptische Differentialgleichungen* (Meeting, Rostock, 1977), Wilhelm-Pieck-Univ., Rostock, 53–57, 1978, MR 80g, 31002.

8. Anger, G., Ein inverses Problem der Wärmeleitungsgleichung. I., in *Romanina-Finnish Seminar on Complex Analysis* (Proc., Bucharest, 1976). Springer, Berlin, 485–507, Lecture Notes in Math., 743, 1979, MR 81e, 31004.

9. Anger, G., and Czerner, R., Solution of an inverse problem for the heat equation by methods of modern potential theory, in *Inverse and Improperly Posed Problems in Differential Equations* (Proc. Conf., Math. Numer. Methods, Halle, 1979). Akademie,-Verlag, Berlin, 9–23; *Math. Research*, 1, 1979, MR 80i, 35176.

10. Atahodzhaev, M. A., The inverse problems of the theory of generalized parabolic potentials, in *Boundary-Value Problems for Differential Equations*, No. 2 (Russian). Izdat. "Fan" Uzbek. SSR, Tashkent, 24–39, 185, 1972 (Russian), MR 56, 16178.

11. Atamanov, E. R., The inverse problem for the heat equation with aftereffect, in *Studies in Integro-Differential Equations in Kirgizia* (Russian). Izdat. "Ilim," Frunze, 258–264, 434, 1977 (Russian), MR 58, 29482.

12. Atamanov, E. R., Uniqueness and estimate of the stability of the solution of a problem for a heat equation with moving transducer, in *Inverse Problems for Differential Equations of Mathematical Physics* (Russian). Akad. Nauk. SSSR Sibirsk. Otdel., Vychisl. Centr, Novosibirsk, 35–44, 155–156, 1978 (Russian), MR 82b, 35149.

13. Azarkevich, N. N., The reconstruction of the initial condition in the Cauchy problem for the heat equation in a class of analytic functions, in *Differential Equations and Their Applications*, No. 2 (Russian). Kuibyshev. Gos. Ped. Inst., Kuybyshev, 3–8, 175, 1975 (Russian), MR 58, 29487.

14. Azarkevich, N. N., A certain inverse problem of heat conduction, *Differencial'nye Uravnenija* (Rjazan'), Vyp. 8 (1976), 3–9 (Russian), MR 56, 12683.

15. Beznoshchenko, N. J., The determination of the coefficients of the lowest

terms in a parabolic equation, *Sibirsk. Mat. Zh.* **16**, No. 3 (1975), 473–482, 643 (Russian), MR 54, 8027.

16. Beznoshchenko, N. J., Certain problems on the determination of the coefficients of the lower terms in parabolic equations, *Sibirsk. Mat. Zh.* **16**, No. 6 (1975), 1135–1147, 1369 (Russian), MR 53, 1068.

17. Beznoshchenko, N. J., The determination of a coefficient in the lowest-order term of a general parabolic equation, *Differencial'nye Uravnenija* **12**, No. 1 (1976), 175–176, 191 (Russian), MR 58, 1774.

18. Beznoshchenko, N. J., and Prilepko, A. I., Inverse problems for equations of parabolic type, in *Problems in Mathematical Physics and Numerical Mathematics* (Russian). Nauka, Moscow, 51–63, 323, 1977 (Russian), MR 58, 29488.

19. Biollay, Y., On the backward heat problem: evaluation of the norm of u_t, *Internat. J. Math. Sci.* **3**, No. 3 (1980), 599–603, MR 81j, 35106.

20. Budak, B. M., Gaponenko, J. L., and Sidorovich, V. G., The direct method of solving a certain ill-posed inverse problem, in *Solutions of Stefan Problems* (Russian). Moskov. Gos. Univ., Moscow, Gos. Univ., Moscow, 226–234, 1970 (Russian), MR 58, 3401.

21. Budak, B. M., and Vasil'eva, V. N., Solutions of the inverse Stephan problem, *Dokl. Akad. Nauk. SSSR* **204** (1972), 1292–1295 (Russian), MR 46, 3989.

22. Budak, B. M., and Vasil'eva, V. N., Solutions of the inverse Stefan problem, *Zh. Vychisl. Mat. i Mat. Fiz.* **13** (1973), 103–118, 268 (Russian), MR 48, 2569.

23. Budak, B. M., and Vasil'eva, V. N., Solution of the inverse Stefan problem. II., *Zh. Vychisl. Mat. i Mat. Fiz.* **13** (1973), 897–906, 1090 (Russian), MR 51, 6146.

24. Cannon, J. R., and DuChateau, P., Determining unknown coefficients in a nonlinear heat-conduction problem, *SIAM J. Appl. Math.* **24** (1973), 298–314, MR 48, 706.

25. Cannon, J. R., and Klein, R. E., On the observability and stability of the temperature distribution in a composite heat conductor, *SIAM J. Appl. Math.* **24** (1973), 596–602, MR 51, 6545.

26. Cannon, J. R., and DuChateau, P., An inverse problem for an unknown source in a heat equation, *J. Math. Anal. Appl.* **75**, No. 2 (1980), 465–485, MR 81k, 35161.

27. Carasso, A., The backward beam equation: Two A-stable schemes for parabolic problems, *SIAM J. Numer. Anal.* **9** (1972), 406–434, MR 47, 4459.

28. Carasso, A., The backward beam equation and the numerical computation of dissipative equations backwards in time, in *Improperly Posed Boundary-Value Problems* (Conf. Univ. New Mexico, Albuquerque, N.M., 1974). Pitman, London, 124–157. Res. Notes in Math., No. 1, 1975, MR 57, 14529.

29. Carasso, A., Computing small solutions of Burgers' equation backwards in time, *J. Math. Anal. Appl.* **59**, No. 1 (1977), 169–209, MR 56, 4187.

30. Chabrowski, J., and Johnson, R., Backwards uniqueness in time for solutions of possible degenerate equations of parabolic type, *J. Differential Equations*, **22**, No. 1 (1976), 209–226, MR 55, 3545.

31. Colton, D., The inverse Stefan problem, in *Ergebnisse einer Tagung über "Funktionentheoretische Methoden bei partiellen Differentialgleichungen"* (Bonn, 1972). Gesellsch. Math. Datenverarbeitung, Bonn, 29–41, Gesellsch.

Math. Datenverarbeitung, Bonn, Ber., No. 77, 1973, MR 57, 13135.

32. Colton, D., The inverse Stefan problem for the heat equation in two space variables, *Mathematika*, **21** (1974), 282–286, MR 51, 3692.

33. Colton, D., and Wimp, J., The construction of solutions to the heat equation backward in time, *Math. Methods Appl. Sci.* **1**, No. 1 (1979), 32–39, MR 80k, 35011.

34. Colton, D., The approximation of solutions to the backwards heat equation by solutions of pseudoparabolic equations, in *Inverse and Improperly Posed Problems in Differential Equations* (Proc. Conf., Math. Numer. Methods, Halle, 1979). Akademie-Verlag, Berlin, 59–71, Math. Research, 1, 1979, MR 80i, 35007.

35. Colton, D., The approximation of solutions to the backwards heat equation in a nonhomogeneous medium, *J. Math. Anal. Appl.* **72**, No. 2 (1979), 418–429, MR 81e, 35056.

36. Dikinov, H. Z., Kerefov, A. A., and Nahushev, A. M., A certain boundary-value problem for a loaded heat equation, *Differencial'nye Uravnenija* **12**, No. 1 (1976), 177–179, 191–292 (Russian), MR 58, 11918.

37. Dolecki, S., Observability for the one-dimensional heat equation, *Studia Math.* **48** (1973), 291–305, MR 48, 11769.

38. Elden, L., Regularization of the backward solution of parabolic problems, in *Inverse and Improperly Posed Problems in Differential Equations* (Proc. Conf., Math. Numer. Methods, Halle, 1979). Akademie-Verlag, Berlin, 73–81, Math. Research, 1, 1979, MR 80e, 65109.

39. Elubaev, S., Uniqueness theorems for an inverse problem for a system of third-order parabolic equations, in *Boundary-Value Problems for Differential Equations*, **5** (Russian). Izdat. "Fan" Uzbek. SSR, Tashkent, 65–77, 184, 1975 (Russian), MR 58, 17615.

40. Elubaev, S., A uniqueness theorem for an inverse problem for the heat equations, *Izv. Akad. Nauk. Kazah. SSR Ser. Fiz.-Mat.* 1978, No. 5 (1978), 41–44, 93 (Russian, Kazakh summary), MR 80g, 35123.

41. Elubaev, S., A uniqueness theorem of an inverse problem for an equation of parabolic type, *Izv. Akad. Nauk. Kazah. SSR Ser. Fiz.-Mat.* 1979, No. 1 (1979), 77–79, 93 (Russian), MR 80c, 35098.

42. Elubaev, S., Regularization of the inverse problem for the heat equation, *Izv. Akad. Nauk. Kazah. SSR Ser. Fiz.-Mat.* 1980, No. 3 (1980), 21–26, 88 (Russian, Kazakh summary), MR 81i, 35159.

43. Ewing, R. E., The approximation of certain parabolic equations backward in time by Sobolev equations, *SIAM J. Math. Anal.* **6** (1975), 283–294, MR 50, 13892.

44. Fasano, A., Sulla inversione di un classico problema di diffusion del calore, *Matematiche (Catania)* **26** (1971), 50–61 (1972) (Italian, English summary), MR 47, 7220.

45. Fedotov, A. M., Identification of linear models, in *Mathematical Models and Numerical Methods* (Papers, Fifth Semester, Stefan Banach Internat. Math. Center, Warsaw, 1975), PWN, Warsaw, 383–387, Banach Center Publ., 3, 1978, MR 80b, 93154.

46. Filatove, V. V., Analytic continuation from a discrete set of solutions of the heat equation, *Differencial'nye Uravnenija* **16**, No. 2 (1980), 322–327, 382

(Russian), MR 81d, 35039.

47. Gol'dman, N. L., *A priori* estimates of the solutions of differential–difference, linear parabolic equations, in *Methods for the Solution of Boundary-Value and Inverse Problems of Heat Conduction* (Russian). Izdat. Moskov. Univ., Moscow, 57–78, 1975 (Russian), MR 55, 8594.

48. Gol'dman, N. L., A class of inverse problems for multidimensional, quasilinear parabolic equations, *Differencial'nye Uravnenja* **14**, No. 7 (1978), 1245–1254, 1341 (Russian), MR 80a, 35115.

49. Horii, M., An inverse problem for the heat equation, *J. Math. Anal. Appl.* **65**, No. 2 (1978), 257–277, MR 58, 23192.

50. Iskenderov, A. D., Multidimensional inverse problems for linear and quasilinear parabolic equations, *Dokl. Akad. Nauk. SSSR* **225**, No. 5 (1975), 1005–1008 (Russian), MR 53, 1069.

51. Iskenderov, A. D., An inverse problem for parabolic equations in a multidimensional domain, *Azerbaidzhan Gos. Univ. Uchen. Zap. Ser. Fiz.-Mat. Nauk.* 1977, No. 1 (1977), 81–89 (Russian, Azerbaijani summary), MR 56, 16180.

52. Ivanov, V. K., The problem of quasi-inversion for the heat equation in a uniform metric, *Differencial'nye Uravnenija* **8** (1972), 652–658 (Russian), MR 46, 9528.

53. Ivanov, V. T., Smirnov, G. P., and Lubyshev, F. V., The direct and inverse boundary-value problems for the heat equation, *Differencial'nye Uravnenija* **8** (1972), 2023–2028, 2112 (Russian), MR 47, 5446.

54. Ivanovich, M. D., The problem of quasireversibility for a higher-order parabolic equation in a uniform metric, in *Boundary-Value Problems for Partial-Differential Equations* (Russian). "Naukova Dumka," Kiev, 60–66, 164, 1978 (Russian), MR 81a, 35096.

55. Jahno, V. G., An inverse problem for a system of parabolic equations, *Differencial'nye Uravnenija* **15**, No. 3 (1979), 566–569, 576 (Russian), MR 80f, 35124.

56. Jochum, P., The numerical solution of the inverse Stefan problem, *Numer. Math.* **34**, No. 4 (1980), 411–429, MR 82j, 65089.

57. Klibanov, M. V., A certain class of inverse problems for a parabolic equation and for problems of integral geometry, *Dokl. Akad. Nauk. SSSR* **222**, No. 1 (1975), 29–31 (Russian), MR 52, 1070.

58. Klibanov, M. V., and Rutner, J. F., An inverse problem for the heat equation, in *Differential Equations and Their Applications*, No. 2 (Russian). Kuibyshev. Gos. Ped. Inst., Kuybyshev, 55–57, 176, 1975 (Russian), MR 58, 29490.

59. Klibanov, M. V., A certain problem of integral geometry, and the inverse problem for a parabolic equation, *Sibirsk. Mat. Zh.* **17**, No. 1 (1976), 75–84, 238 (Russian), MR 58, 6780.

60. Klibanov, M. V., The inverse problem for a parabolic equation, and a certain problem of integral geometry, *Sibirsk. Mat. Zh.* **17**, No. 3 (1976), 564–570, 716 (Russian), MR 58, 12045.

61. Klibanov, M. V., Inverse problems for a quasilinear, parabolic equation, *Dokl. Akad. Nauk. SSSR* **245**, No. 3 (1979), 530–532 (Russian), MR 80j, 35093.

62. Klibanov, M. V., A formulation of the inverse problem for a parabolic equation, *Differencial'nye Uravnenija* **15**, No. 6 (1979), 1132–1134, 1151–1152 (Russian), MR 80j, 35092.

63. Kononova, A. A., The Cauchy problem for the inverse heat equation in an infinite strip, *Ural. Gos. Univ. Mat. Zap.* **8**, No. 4 (1974), 58–63, 134 (Russian), MR 51, 3693.

64. Kralchev, B. N., An inverse problem for a heat equation, *Godishnik Vissh. Uchebn. Zaved. Prilozhna Mat.* **14**, No. 3 (1978), 145–152 (1979) (Russian, Bulgarian, and English summaries), MR 81m, 35063.

65. Kralchev, B. N., Estimates of control functions in an inverse problem for the heat equation, *Godishnik Vissh. Uchebn. Zaved. Prilozhna Mat.* **14**, No. 3 (1978), 153–158 (1979) (Bulgarian, Russian, and English summaries), MR 82e, 35042.

66. Lanconelli, E., Sul problema de Dirichlet per l'equazione del calore, *Ann. Mat. Pura Appl.* (4) **97** (1973), 83–114 (Italian), MR 51, 8442.

67. Langmach, H., On the determination of functional parameters in some parabolic differential equations, in *Theory of Nonlinear Operators* (Proc. Fifth Internat. Summer School, Central Inst. Math. Mech. Acad. Sci. GDR, Berlin, 1977). Akademie-Verlag, Berlin, 175–184, Abh. Akad. Wiss. DDR, Abt. Math. Naturwiss. Tech., 1978, 6, 1978, MR 80g, 35125.

68. Lavrent'ev, M. M., and Reznickaja, K. G., Uniqueness theorems for certain nonlinear, inverse problems of parabolic-type equations, *Dokl. Akad. Nauk. SSSR* **208** (1973), 531–532 (Russian), MR 47, 5474.

69. Lavrent'ev, M. M., and Klibanov, M. V., A certain integral equation of the first kind, and the inverse problem for a parabolic equation, *Dokl. Akad. Nauk. SSSR* **221**, No. 4 (1975), 782–783 (Russian), MR 52, 6340.

70. Lavrent'ev, M. M., and Klibanov, M. V., A certain inverse problem for an equation of parabolic type, *Differencial'nye Uravnenija* **11**, No. 9 (1975), 1647–1651, 1717 (Russian), MR 52, 6225.

71. Lavrent'ev, M. M., and Amonov, B. K., Determination of the solutions of the diffusion equation from its values on discrete sets, *Dokl. Akad. Nauk. SSSR* **228**, No. 6 (1976), 1284–1285 (Russian), MR 54, 731.

72. Lazuchenkov, N. M., and Shmukia, A. A., Solution of a two-dimensional inverse problem for a differential equation of parabolic type, in *Differential Equations and Their Applications* (Russian). Dnepropetrovsk. Gos. Univ., Dnepropetrovsk, 96–100, 214–215, 1976 (Russian), MR 58, 29491.

73. Levine, H. A., and Payne, L. E., Nonexistence theorems for the heat equations with nonlinear boundary conditions and for the porous-medium equation backward in time, *J. Differential Equations* **16** (1974), 319–334, MR 57, 10235.

74. Manselli, P., and Miller, K., Dimensionality-reduction methods for efficient numerical solutions, backward in time, of parabolic equations with variable coefficients, *SIAM J. Math. Anal.*, **11**, No. 1 (1980), 147–159, MR 81h, 65097.

75. Manselli, P., and Miller, K., Calculation of the surface temperature and heat flux on one side of a wall from measurements on the opposite side, *Ann. Mat. Pura Appl.* (4) **123** (1980), 161–183 (English, Italian summary), 81m, 80004.

76. Malishev, I. G., A certain inverse problem for a parabolic equation with a variable coefficient, *Visnik Kiiv. Univ. Ser. Mat. Meh.*, No. 16 (1974), 32–37, 173 (Ukrainian, English, and Russian summaries), MR 55, 13105.
77. Malishev, I. G., Reduction of the inverse problem for a parabolic equation to the spectral-inverse problem of the Schrödinger equation, *Visnik Kiiv. Univ. Ser. Mat. Meh.*, No. 18 (1976), 125–129, 145 (Ukrainian, English, and Russian summaries), MR 58, 6781.
78. Malyshev, I. G., The second inverse problem for the heat equation in a noncylindrical domain, *Vychisl. Prikl. Mat.* (Kiev), No. 25 (1975), 56–64, 144 (Russian, English summary), MR 52, 3766.
79. Malyshev, I. G., Inverse problems for the heat equation in a domain with a moving boundary, *Ukrain. Mat. Zh.* **27**, No. 5 (1975), 687–691, 718 (Russian), MR 52, 8694.
80. Malyshev, I. G., and Malysheva, G. M., The approximate determination of the coefficients in the heat equation, *Vychisl. Prikl. Mat.* (Kiev), Vyp. 31 (1977), 132–136, 159 (Russian, English summary), MR 56, 12684.
81. Mehta, B. S., Time-reversal problem of heat conduction in solids of spherical symmetry, *Defence Sci. J.* **26**, No. 2 (1976), 81–86, MR 54, 6704.
82. Miller, K., Efficient numerical methods for backward solution of parabolic problems with variable coefficients, in *Improperly Posed Boundary-Value Problems* (Conf., Univ. New Mexico, Albuquerque, N.M., 1974). Pitman, London, 54–64, Res. Notes in Math., No. 1, 1975, MR 57, 14503.
83. Muzylev, N. V., Uniqueness theorems for some inverse problems of heat transfer, *Zh. Vychisl. Mat. i Mat. Fiz.* **20**, No. 2 (1980), 388–400, 550 (Russian), MR 81i, 35160.
84. Pagani, C. D., and Talenti, G., On a forward–backward parabolic equation, *Ann. Mat. Pura Appl.* (4) **90** (1971), 1–57, MR 47, 2189.
85. Patel, S. R., Time-reversal problems in transient heat conduction with radiation as boundary conditions, *Labdev. J. Sci. Tech.*, Part A, **13**, No. 3–4 (1975), 232–240 (1978), MR 57, 14950.
86. Pierce. A., Unique identification of eigenvalues and coefficients in a parabolic problem, *SIAM J. Control Optim.* **17**, No. 4 (1979), 494–499, MR 80e, 35066.
87. Prilepko, A. I., Inverse problems of potential theory (elliptic, parabolic, hyperbolic equation and transport equations), *Mat. Zametki* **14** (1973), 755–767 (Russian), MR 50, 2527.
88. Pagani, C. D., On forward–backward parabolic equations in bounded domains, *Boll. Un. Mat. Ital. B* (5) **13**, No. 2 (1976), 336–354 (English, Italian summary), MR 58, 17514.
89. Quilghini, D., Problema inverso di quello di Stefan in uno strato materiale piano e possibilità di controllo, *Confer. Sem. Mat. Univ. Bari*, No. 131 (1975), 3–15 (1976) (Italian), MR 58, 8844.
90. Romanov, V. G., A certain inverse problem for an equation of parabolic type, *Mat. Zametki* **19**, No. 4 (1976), 595–600 (Russian), MR 57, 17046.
91. Schuss, Z., Backward and degenerate parabolic equations, *Applicable Anal.* **7**, No. 2 (1977), 111–119, MR 80a, 35060.
92. Seidman, T. I., A well-posed problem for the heat equation, *Bull. Amer. Math. Soc.* **80** (1974), 901–902, MR 54, 5621.

93. Shatalov, J. S., An example of the inapplicability of the quasireversibility method to finding boundary functions that control the solution of the heat equation, *Differencial'nye Uravnenija* **10** (1974), 1335–1337, 1344 (Russian), MR 51, 6150.

94. Shishko, N. P., Inverse problem for a parabolic equation, *Mat. Zametki* **29**, No. 1 (1981), 55–62, 155 (Russian), MR 82i, 35164.

95. Stoyan, G., Numerical experiments on the identification of heat-conduction coefficients, in *Theory of Nonlinear Operators* (Proc. Fifth Internat. Summer School, Central Inst. Math. Mech. Acad. Sci. GDR, Berlin, 1977). Akademie-Verlag, Berlin, 259–268, Abh. Akad. Wiss. DDR, Abt. Math. Naturwiss. Tech., 1978, 6, 1978, MR, 80i, 35183.

96. Stoyan, G., On the identification of diffusion coefficients, in *Mathematical Models and Numerical Methods* (Papers, Fifth Semester, Stefan Banach Internat. Math. Center, Warsaw, 1975). PWN, Warsaw, 367–377, Banach Center Publ., 3, 1978, MR 80b, 93146.

97. Straughan, B., Global nonexistence of solutions to Ladyzhenskaya's variants of the Navier–Stokes equations backward in time, *Proc. Roy. Soc. Edinburgh*, *Sect. A*, **75**, No. 2 (1975), 165–170, MR 56, 3479.

98. Suleimanov, N. M., Theorems of Wiman–Valiron type for solutions of inverse parabolic equations, *Differencial'nye Uravnenija* **16**, No. 10 (1980), 1816–1825, 1916 (Russian), MR 82c, 30037.

99. Suzuki, T., and Murayama, R., A uniqueness theorem in an identification problem for coefficients of parabolic equations, *Proc. Japan Acad.*, *Ser. A*, *Math. Sci.* **56**, No. 6 (1980), 259–263, MR 82c, 35080.

100. Tabaldyev, B. R., An inverse problem for a quasilinear heat equation, *Izv. Akad. Nauk. Kirgiz. SSR* 1974, No. 1 (1974), 12–17 (Russian), MR 49, 9456.

101. Tabaldyev, B. R., Reconstruction of the unknown function from the coefficients of a partial-differential equation of parabolic type, in *Studies in Integro-Differential Equations in Kirgizia* (Russian). Izdat. "Ilim," Frunze, 275–297, 435, 1977 (Russian), MR 58, 23195.

102. Tsutsumi, M., On nonlinear parabolic equations backwards in time, *Bull. Sci. Engrg. Res. Lab. Waseda Univ.*, No. 68 (1975), 50–55, MR 57, 10249a.

103. Tsutsumi, M., Errata: "On nonlinear parabolic equations backwards in time" (*Bull. Sci. Engrg. Res. Lab. Waseda Univ.*, No. 68 (1975), 50–55), *Bull. Sci. Engrg. Res. Lab. Waseda Univ.*, No. 69 (1975), 110, MR 57, 10249b.

104. Uzlov, A. E., Existence and uniqueness theorems for the solutions of inverse problems for second-order equations of parabolic type, *Differencial'nye Uravnenija* **15**, No. 2 (1979), 370–372, 384 (Russian), MR 80c, 35102.

105. Vahrameev, S. S., Solution of the inverse heat problem of regulating the temperature of a smelting furnace, in *Latvian Mathematical Yearbook*, **10** (Russian). Izdat. "Zinatne," Riga, 3–14, 1972 (Russian, Latvian, and English summaries), MR 47, 9066.

106. Vasil'eva, V. N., On the problem of reconstructing the boundary conditions for the heat equation, in *Optimization Methods and Operations Research*, *Applied Mathematics* (Russian). Akad. Nauk. SSSR Sibirsk. Otdel. Sibirsk. Energet. Inst., Irkutsk, 105–114, 189, 1976 (Russian) MR 58, 3560.

4. MORE RECENT: 1981-1982

4.1. Properties of Solutions

A. *A priori estimates, maximum principle, comparison results, means*

1. Ahmad, S. (with Vatsala, A. S.), Comparison results of reaction–diffusion equations with delay, in abstract cones, *Rend. Sem. Mat. Univ. Padova* **65** (1981), 19–34 (1982).

2. Alikakos, N. D. (with Rostamian, R.), Gradient estimates for degenerate diffusion equations. I., *Math. Ann.* **259** (1982), No. 1, 53–70.

3. Di Benedetto, E. (with Pierre, M.), On the maximum principle for pseudo-parabolic equations, *Indiana Univ. Math. J.* **30** (1981), No. 6, 821–854.

4. Benilan, P. (with Crandell, M. G.), The continuous dependence on ϕ of solutions of $u_t - \Delta\phi(u) = 0$, *Indiana Univ. Mat. J.* 30 (1981), No. 2, 161–177.

5. Brown, A., A comparison of mean concentrations in a diffusion problem, *Proc. Edinburgh Math. Soc.* (2) **24** (1981), No. 3, 153–170.

6. Chabrowski, J. (with Vyborny, R.), Maximum principle for nonlinear degenerate equations of the parabolic type, *Bull. Austral. Math. Soc.* **25** (1982), No. 2, 251–263.

7. Chan, C. Y. (with Schaefer, P. W.), Some extremum principles in semilinear parabolic problems, *J. Math. Anal. Appl.* **87** (1982), No. 2, 517–527.

8. Cosner, C., Pointwise *a priori* bounds for strongly coupled semilinear systems of parabolic partial-differential equations, *Indiana Univ. Math. J.* **30** (1981), No. 4, 607–620.

9. Diaz, J. I. (with Herrero, M. A.), Estimates on the support of the solutions of some nonlinear elliptic and parabolic problems, *Proc. Roy. Soc. Edinburgh, Sect. A*, **89** (1981), No. 3–4, 249–258.

10. Dubois, R. M., Problèmes mixtes abstraits et principe du maximum parabolique [Abstract mixed problems and parabolic maximum principle], in *Seminar on Potential Theory*, Paris, No. 6, pp. 103–113. Lecture Notes in Math. 906, Springer, Berlin, 1982.

11. Galaktionov, V. A., Some properties of solutions of quasilinear parabolic equations, *Akad. Nauk. SSSR Inst. Prikl. Mat. Preprint* 1981, No. **16**, 24 pp. (Russian, English summary).

12. Gardner, R. A., Comparison and stability theorems for reaction–difusion systems, *SIAM J. Math. Anal.* **12** (1981), No. 4, 603–616.

13. Kalasnikov, A. S., Conditions for the positivity and boundedness of solutions of the second boundary-value problem for parabolic equations, *Sibirsk. Mat. Z.* **22** (1981), No. 6, 72–80, 225 (Russian).

14. Knerr, B. S., Parabolic interior Schauder estimates by the maximum principle, *Arch. Rational Mech. Anal.* **75** (1980), No. 1, 51–58.

15. Korobcuk, I. V. (with Harcenko, O. P.), Estimation of solutions of second-order parabolic equations in unbounded domains, in *Studies in the Qualitative Theory of Differential Equations and Its Applications*, pp. 26–27, 72. Akad. Nauk. Ukrain. SSR, Inst. Mat., Kiev, 1978 (Ukrainian, Russian summary).

16. Kupcov, L. P., Property of the mean for the heat equation, *Mat. Zametki* **29**

(1981), No. 2, 211–223, 317 (Russian).

17. Lavrent'ev, M. M., Jr., *A priori* estimates of solutions of nonlinear parabolic equations, *Differencial'nye Uravnenija* **18** (1982), No. 5, 868–877, 917–918 (Russian).

18. McCoy, P. A., Bernstein theorems for a parabolic equation in one space variable, in *Approximation Theory, III* (Proc. Conf., Univ. Texas, Austin, Texas, 1980), pp. 643–646. Academic Press, New York, 1980.

19. Rothe, F., Uniform bounds from bounded L_p-functionals in reaction–diffusion equations, *J. Differential Equations* **45** (1982), No. 2, 207–233.

20. Schonbek, M. E., *A priori* estimates of higher-order derivatives of solutions to the FitzHugh–Nagumo equations, *J. Math. Anal. Appl.* **82** (1981), No. 2, 553–565.

21. Sperb, R. P., Growth estimates in diffusion–reaction problems, *Arch. Rational Mech. Anal.* **75** (1980–81), No. 2, 127–145.

22. Sperb, R. P., Comparison results in a class of reaction–diffusion problems, *Z. Angew. Math. Phys.* **32** (1981), No. 1, 12–21 (German summary).

23. Uchiyama, K., Spatial growth of solutions of a nonlinear equation, *Proc. Japan Acad. Ser. A. Math. Sci.* **57** (1981), No. 2, 90–94.

B.   Green's functions, potential theory, transforms, elementary solutions

1. Aronson, D. G., Widder's inversion theorem and the initial distribution problems, *SIAM J. Math. Anal.* **12** (1981), No. 4, 639–651.

2. Atkinson, C., Some elementary solutions of the heat equation and related equations, *Internat. J. Engrg. Sci.* **19** (1981), No. 5, 713–728.

3. Barański, F. (with Musiałek, J), On the Green functions for the heat equation over the *m*-dimensional cuboid, *Demonstratio Math.* **14** (1981), No. 2, 371–382.

4. Cheeger, J. (with Yau, Shing Tung), A lower bound for the heat kernel, *Comm. Pure Appl. Math.* **34** (1981), No. 4, 465–480.

5. Ebong, L. I., A note on the kernel approximation in a certain axially symmetric temperature problem, *Internat. J. Engrg. Sci.* **17** (1979), No. 4, 345–348.

6. Kaufman, R. (with Wu, Jang Mei), Parabolic potential theory, *J. Differential Equations* **43** (1982), No. 2, 204–234, MR 83 d: 31006.

7. Musiałek, J., On the Green function for the heat equation and for the plane rectangular wedge, *Comment. Math. Prace. Mat.* **22** (1980–81), No. 2, 301–303.

8. Paquet, L., Formule de Duhamel et problème de la chaleur [Duhamel's formula and the heat problem], in *Seminar on Potential Theory*, Paris, No. 6, pp. 282–307, Lecture Notes in Math., 906. Springer, Berlin, 1982.

9. Pathak, R. S., Expansion of distributions in terms of heat polynomials, *Aligarh Bull. Math.* **7** (1977), 81–90, (1980).

10. Sablovskii, O. N., Analytic solutions of parabolic and hyperbolic heat-transfer equations for nonlinear media, *Inz.-Fiz. Z.* **40** (1981), No. 3, 510–517 (Russian), translated as *J. Engrg. Phys.* **40** (1981), No. 3, 319–324.

11. Salsa, S., Some properties of nonnegative solutions of parabolic differential operators, *Ann. Mat. Pura Appl.* (4) **128** (1981), 193–206 (Italian summary).

12. Talaga, P. C., The Hukuhara–Kneser property for parabolic systems with nonlinear boundary conditions, *J. Math. Anal. Appl.* **79** (1981), No. 2, 461–488.

13. Zdankiewicz, Z., On the Green function for the diffusion equation and for rectangular parallelpiped, *Fasc. Math.*, No. 12 (1980), 65–68.

C.   *Continuity and differentiability of solutions*

1. Azzam, A. (with Kreyszig, E.), Smoothness of solutions of parabolic equations in regions with edges, *Nagoya Math. J.* **84** (1981), 159–168.
2. Azzam, A. (with Kreyszig, E.), Über regularitatseigenschaften der Lösungen von Anfangs-Randwertproblemen für parabolische Gleichungen [On regularity properties of solutions of initial-boundary-value problem for parabolic equations], *Math. Nachr.* **101** (1981), 141–151.
3. Benilan, P. (with Crandall, M. G.), Regularizing effects of homogeneous evolution equations, in *Contributions to Analysis and Geometry* (Baltimore, Md., 1980), pp. 23–39. Johns Hopkins Univ. Press, Baltimore, Md., 1981.
4. Biroli, M., Régularité Hölderienne pour la solution d'une inéquation parabolique [Hölder continuity for the solution to a parabolic variational inequality], *C. R. Acad. Sci. Paris Ser. I Math.* **293** (1981), No. 6, 323–325 (English summary).
5. Campanato, S., Partial Hölder continuity of solutions of quasilinear parabolic systems of second order with linear growth, *Rend. Sem. Mat. Univ. Padova* **64** (1981), 59–75.
6. Chiarenza, F., Semicontinuity of the subsolutions of parabolic equations of variational type, *Boll. Un. Mat. Ital. B* (6) **1** (1982), No. 2, 467–485 (Italian, English summary).
7. Crandall, M. (with Pierre, M.), Regularizing effects for $u_t + A\phi(u) = 0$ in L^1, *J. Funct. Anal.* **45** (1982), No. 2, 194–212.
8. Diaz, J. I., Qualitative properties of some nonlinear parabolic problems: a classification for the models of heat diffusion, *Mem. Real. Acad. Cienc. Exact. Fis. Natur. Madrid* **14**, i+43 pp. (1980) (Spanish, English summary).
9. Di Benedetto, E., Continuity of weak solutions to certain singular parabolic equations, *Ann. Mat. Pura Appl.* (4), **130** (1982), 131–176 (Italian summary).
10. Dobrowolski, M., $C^{\alpha, \alpha/2}$-regularity, *Boll. Un. Mat. Ital. B.* (5) **18** (1981), No. 1, 345–354 (Italian summary).
11. Giaquinta, M. (with Struwe, M.), An optimal regularity result for a class of quasilinear parabolic systems, *Manuscripta Math.* **36** (1981–82), No. 2, 223–239.
12. Giaquinta, M. (with Struwe, M.) On the partial regularity of weak solutions of nonlinear parabolic systems, *Math. Z.* **179** (1982), No. 4, 437–451.
13. Gilding, B. H. (with Peletier, L. A.), Continuity of solutions of the porous-media equation, *Ann. Scuola Norm. Sup. Pisa Cl Sci* (4) **8** (1981), 659–675.
14. Herrero, M. A., On the behavior of the solutions of certain nonlinear parabolic problems, *Rev. Real Acad. Cienc. Exact. Fis. Natur. Madrid* **74** (1981), No. 5, 1165–1183 (Spanish, English summary).
15. Ibragimov, A. I., On some qualitative properties of solutions of second-order equations of parabolic type with continuous coefficients, *Differencial'nye Uravnenija* **18** (1982), No. 2, 306–319, 366 (Russian).
16. Krasnoborov, N. A., Behavior of solutions of a quasilinear operator-differential equation of parabolic type, *Differencial'nye Uravnenija* **18** (1982), No. 9,

1575–1585, 1655 (Russian).

17. Lavrent'ev, M.M., Jr. Properties of approximate solutions of nonlinear equations of variable type, *Sibirsk. Mat. Z.* **21** (1980), No. 6. 165–175, 223 (Russian).

18. Struwe, M., On the Hölder continuity of bounded weak solutions of quasilinear parabolic systems, *Manuscripta Math.* **35** (1981), No. 1–2, 125–145.

19. Tanabe, H., On regularity properties of some nonlinear parabolic equations, *Proc. Japan. Acad. Ser. A. Math. Sci.* **57** (1981), No. 10, 473–476.

20. Tanabe, H., Differentiability of solutions of some unilateral problems of parabolic type, *J. Math. Soc. Japan* **33** (1981), No. 3, 367–403.

21. Winiarska, T., Differentiability with respect to the parameter of solutions of operator equations of parabolic type, *Zeszyty Nauk. Uniw. Jagiellon, Prace. Mat.*, No. 23 (1982), 183–188.

D. Boundary behavior of solutions

1. Baranski, F., On a certain limit problem for the heat-conduction equation and for triangular plate, *Fasc. Math.*, No. 11 (1970), 169–172.

2. Baranski, F., On a certain limit problem for the heat-conduction equation and any angular time–spatial domain, *Fasc. Math.*, No. 11, (1979), 173–177.

3. Barlow, M. T. (with Watson, N. A.), Irregularity of boundary points in the Dirichlet problem for the heat equation, *Rend. Circ. Mat. Palermo* (2) **3** (1982), No. 2, 300–304 (Italian summary).

4. Bear, H. S., Initial values for parabolic functions, *J. Math. Anal. Appl.* **86** (1982), No. 2, 425–441.

5. Denisov, V. N., On the question of necessary conditions for the stabilization of the solution of the Cauchy problem for the heat equation in the whole space E^N and on any compactum in it, *Dokl. Akad. Nauk. SSSR* **260** (1981), No. 4, 780–783 (Russian).

6. Denisov, V. N., On the question of necessary condition for the stabilization of the solution of the Cauchy problem for a heat equation that is uniform on any compact, *Differencial'nye Uravnenija* **18** (1982), No. 1, 7–16, 178 (Russian).

7. Denisov, V. N., On a necessary condition for the uniform stabilization of the solution of the Cauchy problem for the heat equation, *Differencial'nye Uravnenija* **18** (1982), No. 2, 290–305 (Russian), 290–305, 366.

8. Evans, L. C. (with Gariepy, R. F.), Wiener's criterion for the heat equation, *Arch. Rational Mech. Anal.* **78** (1982), No. 4, 293–314.

9. Gariepy, R. (with Ziemer, W. P.), Thermal capacity and boundary regularity, *J. Differential Equations* **45** (1982), No. 3, 374–388.

10. Hamatsuka, T. (with Mo'omen, A.-A., and Akashi, H.), On pole assignment and stabilization for the heat equation, *Mem. Fac. Engrg. Kyoto Univ.* **43** (1981), No. 3, 319–327.

11. Kamynin, L. I. (with Himcenko, B. N.), *A priori* estimates of the solution of a second-order parabolic equation in the neighborhood of the lower cap of the parabolic boundary, *Sibirsk. Mat. Z.* **22** (1981), No. 4, 94–113, 230 (Russian).

12. Teman, R., Behavior at time $t = 0$ of the solutions of semilinear evolution equations, *J. Differential Equations* **43** (1982), No. 1, 73–92.

E. Periodic solutions

1. Bahvalov, N. S., Averaging of the heat-transfer process in periodic media in the presence of radiation, *Differencial'nye Uravnenija* **17** (1981), No. 10, 1765–1773, 1916 (Russian).
2. Capasso, V. (with Maddalena, L.), Periodic solutions for an epidemic model, in *Evolution Equations and Their Applications* (Schloss Retzhof, 1981), pp. 16–29, Res. Notes in Math., 68. Pitman, Boston, Mass. 1982.
3. Ganapathisubramanian, N. (with Ramaswamy, R, and Kuriacose, J. C.), Limit-cycle behavior in Belousov–Zhabotinskii oscillatory reaction, *Proc. Indian Acad. Sci., Sect. A., Chem. Sci.* **89** (1980), No. 3, 235–239.
4. Howard, L. N., Time-periodic and spatially irregular patterns, in *Dynamics and Modelling of Reactive Systems* (Proc. Adv. Sem, Math. Res. Center, Univ. Wisconsin, Madison, Wis., 1979), pp. 195–209. Publ. Math. Res. Center Univ. Wisconsin, 44, Academic Press, New York, 1980.
5. Kopell, N., Time-periodic but spatially irregular solutions to a model reaction–diffusion equation, in *Nonlinear Dynamics* (Internat. Conf., New York, 1979), pp. 397–409. Ann. New York Acad. Sci., 357, New York Acad. Sci., New York, 1980.
6. Liu, Bao Ping, (with Pao, C. V.), Periodic solutions of coupled semilinear parabolic boundary-value problems, *Nonlinear Anal.* **6** (1982), No. 3, 237–252.
7. Naito, K., On the almost-periodicity of solutions of a reaction–diffusion system, *J. Differential Equations* **44** (1982), No. 1, 9–20.
8. Pankov, A. A., Boundedness and almost-periodicity with respect to time of solutions of evolutionary variational inequalities, *Izv. Akad. Nauk. SSSR Ser. Mat.* **46** (1982), No. 2, 314–346, 431 (Russian).
9. Pankov, A. A., Bounded and almost-periodic solutions of some nonlinear evolution equations, *Uspehi Mat. Nauk.* **37** (1982), No. 2, (224), 223–224 (Russian).
10. Pikulin, V. P., Periodic and almost-periodic solutions of weakly nonlinear and parabolic equations, *Trudy Moskov. Orden. Lenin. Energet. Inst.*, No. 412 (1970), 107–111, 161 (Russian).
11. Pikulin, V. P., Periodic and almost-periodic solutions for a class of quasilinear parabolic equations, *Differencial'nye Uravneniya* **18** (1982), No. 8, 1412–1417, 1470 (Russian).
12. Temkin, L. A., On the construction of periodic solutions of an equation of parabolic type, *Eesti NSV Tead. Akad. Toimetised Fuus.-Mat* **29** (1980), No. 3, 261–271 (Russian, Estonian, and English summaries).
13. Zikov, V. V. (with Kozlov, S. M., and Oleinik, O. A.), Averaging of parabolic operators with almost-periodic coefficients, *Mat. Sb. (N.S.)* **117** (159) (1982), No. 1, 69–85 (Russian).

F. Reaction diffusion problems

1. Ablowitz, M. J. (with Zeppetella, A.), Explicit solutions of Fisher's equation for a special wave speed, *Bull. Math. Biol.* **41** (1979), No. 6, 835–840.
2. Aronson, D. G., Density-dependent interaction–diffusion systems, in *Dynamics and modelling of reactive systems* (Proc. Adv. Sem. Math. Res. Center, Univ. Wisconsin, Madison, Wis., 1979), pp. 161–176. Publ. Math. Res. Center Univ.

Wisconsin, 44, Academic Press, New York, 1980.

3. Beck, K., A model of the population genetics of cystic fibrosis in the United States, *Math. Biosci.* **58** (1982), No. 2, 243–257.

4. Beretta, E. (with Solimano, F., and Lazzari, C.), Introduction to the analysis of qualitative behavior of chemical reaction systems. I. *Boll. Un. Mat. Ital. Suppl.* **1** (1981), 131–172 (Italian).

5. Beretta, E. (with Solimano, F., and Lazzari, C.), Introduction to the analysis of qualitative behavior of chemical reaction systems. II. The Horn–Frienberg–Jackson theory, *Boll. Un. Mat. Ital. Suppl.* **1** (1981), 173–236 (Italian).

6. Bobisud, L. E., Quasistationary approximation for reaction–diffusion systems, *J. Math. Anal. Appl.* **79** (1981), No. 1, 224–235.

7. Chabrowski, J., Representation theorems for parabolic systems, *J. Austral. Math. Soc. Ser A*, **32** (1982), No. 2, 246–288.

8. Conley, C. (with Smoller, J.), Topological techniques in reaction—diffusion equations, in *Biological Growth and Spread* (Proc. Conf., Heidelberg, 1979), pp. 473–483. Lecture Notes in Biomath., 38. Springer, Berlin 1980.

9. Evans, L. C., A convergence theorem for a chemical diffusion–reaction system, *Houston J. Math.* **6** (1980), No. 2, 259–267.

10. Gajewski, H. (with Zacharias, K.), On a system of diffusion–reaction equations, *Z. Angew. Math. Mech.* **60** (1980), No. 9, 357–370 (German and Russian summaries).

11. Galaktionov, V. A. (with Kurdjumov, S. P., Mihailov, A. P., and Samarskii, A. A.), Localization of heat in nonlinear media, *Differencial'nye Uravnenija* **17** (1981), No. 10, 1826–1841, 1918 (Russian).

12. Hastings, S. P., Single- and multiple-pulse waves for the FitzHugh–Nagumo equations, *SIAM J. Appl. Math.* **41** (1982), No. 2, 247–260.

13. Hill, J. M., On the solution of reaction–diffusion equations, *IMA J. Appl. Math.* **27** (1981) , No. 2, 177–194.

14. Hoppensteadt, F. C., Mathematical methods of population biology, in *Cambridge Studies in Mathematical Biology*, 4. Cambridge University Press, Cambridge-New York, 1982, viii + 149 pp. ISBN 0-521-23846-3, 0-521-28256-X.

15. Kath, W. L. (with Cohen, Donald S.), Waiting-time behavior in a nonlinear diffusion equation, *Stud. Appl. Math.* **67** (1982), No. 2, 79–105.

16. Koga, S., Rotating spiral waves in reaction–diffusion systems, Phase singularities of multi-armed waves, *Progr. Theoret. Phys.* **67** (1982), No. 1, 164–178.

17. Lakshmikantham, V., Some problems of reaction–diffusion equations, in *Nonlinear Differential Equations* (Proc. Internat. Conf., Trento, 1980), pp. 243–258. Academic Press, New York, 1981.

18. Lee, A. I. (with Hill, J. M.), A note on the solution of reaction–diffusion equations with convection, *IMA J. Appl. Math.* **29** (1982), No. 1., 39–43.

19. Leung, A., A semilinear reaction–diffusion pre-predator system with nonlinear coupled boundary conditions: equilibrium and stability, *Indiana Univ. Math. J.* **32** (1982), No. 2, 223–241.

20. Levin, S. A., Models of population dispersal, in *Differential Equations and Applications in Ecology, Epidemics, and Population Problems* (Claremont, CA, 1981), pp. 1–18. Academic Press, New York, 1981.

21. MacCamy, R. C., A population model with nonlinear diffusion, *J. Differential Equations* **39** (1981), No. 1, 52–72.

22. Maslennikova, V. N. (with Gapeeva, N. V.), A problem with a directional derivative for multicomponent diffusion and chemical kinetics systems, *Sibirsk. Mat. Z.* **21** (1980), No. 6, 61–70 (Russian).

23. Matano, H., Nonincrease of the lap-number of solution for a one-dimensional, semilinear parabolic equation, *J. Fac. Sci. Univ. Tokyo, Sect. IA, Math.* **29** (1982), No. 2, 401–441.

24. Mcedlov-Petrosjan, P. O. (with Tanatarov, L. V.), A problem of linear reactive diffusion, *Ukrain. Fiz. Z.* **27** (1982), No. 1, 117–126, 159 (Russian, English summary).

25. de Mottoni, P. (with Tesei, A.), A comparative analysis of some semilinear, parabolic systems modelling the dynamics of fast nuclear reactors, *Z. Angew. Mat. Mech.* **60** (1980), No. 11, 615–622 (German and Russian summaries).

26. Pao, C. V., Reaction–diffusion equations with nonlinear boundary conditions, *Nonlinear Anal.* **5** (1981), No. 10, 1077–1094.

27. Pao, C. V., Coexistence of stability of a competition–diffusion system in population dynamics, *J. Math. Anal. Appl.* **83** (1981), No. 1, 54–76.

28. Pao, C. V., On nonlinear reaction–diffusion systems, *J. Math. Anal. Appl.* **87** (1982), No. 1, 165–198.

29. Rosen, G., On the Fisher and the cubic-polynomial equations for the propagation of species properties, *Bull. Math. Biol.* **42** (1980), No. 1, 95–106.

30. Sarafjan, V. V. (with Safarjan, R. G.), Diffusion processes with averaging, and propagation of concentrated waves, *Akad. Nauk. Armjan. SSR Dokl.* **70** (1980), No. 5, 274–279 (Russian, Armenian summary). 82d: 35023.

31. Shiga, T., Diffusion processes in population genetics, *J. Math. Kyoto Univ.* **21** (1981), No. 1, 133–151.

32. Sleeman, B. D., Analysis of diffusion equations in biology, *Bull. Inst. Math. Appl.* **17** (1981), No. 1, 7–13.

33. Webb, G. F., An age-dependence epidemic model with spatial diffusion, *Arch. Rational Mech. Anal.* **75** (1980), No. 1, 91–102.

34. Webb, G. F., A reaction–diffusion model for a deterministic diffusive epidemic, *J. Math. Anal. Appl.* **84** (1981), No. 1, 150–161.

35. Williams, F. A., Current problems in combustion research, in *Dynamics and Modelling of Reactive Systems* (Proc. Adv. Sem., Math. Res. Center. Univ. Wisconsin, Madison, Wisconsin, 1979), pp. 293–314. Publ. Math. Res. Center Univ. Wisconsin, 44, Academic Press, New York, 1980.

36. Williams, L. R. (with Leggett, R. W.), Unique and multiple solutions of a family of differential equations modeling chemical reactions, *SIAM J. Math. Anal.* **13** (1982), No. 1, 122–133.

G. Bifurcation, nonuniqueness, and chaos in reaction–diffusion problems

1. Bissett, E. J., Interaction of steady and periodic bifurcating modes with imperfection effects in reaction–diffusion systems, *SIAM J. Appl. Math.* **40** (1981), No. 2, 224–241.

2. Galaktionov, V. A., Conditions for the absence of global solutions of a class of quasilinear, parabolic equations, *Z. Vycisl. Mat. i. Mat. Fiz.* **22** (1982), No. 2, 322–338, 492 (Russian).

3. Hale, J. K., Stability and bifurcation in a parabolic equation, in *Dynamical Systems and Turbulence*, Warwick 1980 (Coventry, 1979-80), pp. 143-153; Lecture Notes in Math. 898. Springer, Berlin, 1981.

4. Haraux, A. (with Weissler, F. B.), Non-unicité pour un problème de Cauchy sémi-linéaire [Nonuniqueness for a semilinear Cauchy problem], in *Nonlinear Partial-Differential Equations and Their Applications*, College de France Seminar, Vol. III (Paris, 1980-81), pp. 220-233, 428, Res. Notes in Math., 70. Pitman, Boston, Mass 1982.

5. Haraux, A. (with Weissler, F. B.), Nonuniqueness for a semilinear initial-value problem, *Indiana Univ. Math. J.* **31** (1982), No. 2, 167-189.

6. Keener, J. P., Infinite-period bifurcation in simple chemical reactors, in *Modelling of Chemical Reaction Systems* (Heidelberg, 1980), pp. 126-137. Springer Ser. Chem. Phys, 18, Springer, Berlin, 1981.

7. de Mottoni, P. (with Schiaffino, A.), Bifurcation results for a class of periodic, quasilinear, parabolic equations, *Math. Methods Appl. Sci.* **3** (1981), No. 1, 11-20.

8. Nishiura, Y., Global structure of bifurcating solutions of some reaction–diffusion systems, *SIAM J. Math. Anal.* **13** (1982), No. 4, 555-593.

9. Schaeffer, D. G. (with Golubitsky, M. A.), Bifurcation analysis near a double eigenvalue of a model chemical reaction, *Arch. Rational Mech. Anal.* **75** (1980-81), No. 4, 315-347.

10. Schiffmann, Y., Classification in bifurcation theory and reaction–diffusion systems, *Phys. Rep.* **64** (1980), No. 3, 87-169.

11. Tyson, J. J., On the appearance of chaos in a model of the Belousov reaction, *J. Math. Biol.* **5** (1977-78), No. 4, 351-362.

H. Stability of solutions

1. Baiocchi, C., Stabilité uniforme et correcteurs dans la discrétisation des problèmes paraboliques [Uniform stability and "correctors" in the discretization of parabolic problems], in *Nonlinear Partial Differential Equations and Their Applications*, College de France Seminar, Vol. III (Paris, 1980-81), pp. 50-67, 424. Res. Notes in Math., 70; Pitman, Boston, Mass. 1982 (English summary).

2. Bell, J., A note on the stability of traveling-wave solutions to a class of reaction–diffusion systems, *Quart. Appl. Math.* **38** (1980-81), No. 4, 489-496.

3. Berryman, J. G. (with Holland, C. J.), Stability of the separable solution for fast diffusion, *Arch. Rational Mech. Anal.* **74** (1980), No. 4, 379-388.

4. Chow, P. L., Note on the stability of stochastic reaction–diffusion equations, in *Transactions of the Twenty-sixth Conference of Army Mathematicians* (Hanover, N.H., 1980), pp. 67-74, ARO Rep. 81, 1. U. S. Army Res. Office, Research Triangle Park, N.C., 1981.

5. Ermentrout, G. B., Stable small-amplitude solutions in reaction–diffusion systems, *Quart. Appl. Math.* **39** (1981-82), No. 1, 61-86.

6. Gardner, R. A., Existence and stability of traveling-wave solutions of competition models: a degree-theoretic approach, *J. Differential Equations* **44** (1982), No. 3, 343-364.

7. Greenberg, J. M., Stability of equilibrium solutions for the Fisher equation,

Quart. Appl. Math. **39** (1981–82), No. 2, 239–247.

8. Gushchin, A. K., A criterion for uniform stabilization of solutions of the second mixed problem for a parabolic equation, *Dokl. Akad. Nauk. SSSR* **264** (1982), No. 5, 1041–1045 (Russian).

9. Hagan, P. S., The instability of nonmonotonic wave solutions of parabolic equations, *Stud. Appl. Math.* **64** (1981), No. 1, 57–88.

10. Hernandez, J., Some existence and stability results of solutions of reaction–diffusion systems with nonlinear boundary conditions, in *Nonlinear Differential Equations* (Proc. Internat. Conf., Trento, 1980) pp. 161–173. Academic Press, New York, 1981.

11. Lakshmikantham, V. (with Vatsala, A. S.), Stability results for solutions of reaction–diffusion systems by the method of quasisolutions, *Applicable Anal.* **12** (1981), No. 3, 229–235.

12. Maginu, K., Stability of periodic traveling-wave solutions of a nerve-conduction equation, *J. Math. Biol.* **6** (1978), No. 1, 49–57.

13. Maginu, K., Stability of periodic traveling-wave solutions with large spatial periods in reaction–diffusion systems, *J. Differential Equations* **39** (1981), No. 1, 73–99.

14. Triggiani, R., Boundary stabilizability for diffusion processes, in *Recent Advances in Differential Equations* (Trieste, 1978), pp. 431–447. Academic Press, New York, 1981.

I. Asymptotic behavior

1. Abdelkader, M. A., Traveling-wave solutions for a generalized Fisher equation, *J. Math. Anal. Appl.* **85** (1982), No. 2, 287–290.

2. Alikakos, N. D., A Liapunov functional for a class of reaction–diffusion systems, in *Modeling and Differential Equations in Biology* (Conf. Southern Illinois Univ. Carbondale, Ill, 1978), pp. 153–170; Lecture Notes in Pure and Appl. Math. 58. Dekker, New York, 1980.

3. Alikakos, N. D. (with Rostamian, R.), Large-time behavior of solutions of Neumann boundary-value problem for the porous-medium equation, *Indiana Univ. Mat. J.* **30** (1981) No. 5, 749–785.

4. Atkinson, C. (with Reuter, G. E. H., and Ridler-Rowe, C. J.), Traveling-wave solution for some nonlinear diffusion questions, *SIAM J. Math. Anal.* **12** (1981), No. 6, 880–892.

5. Bebernes, J. W. (with Kassoy, D. R.), A mathematical analysis of blowup for thermal reactions—the spatially nonhomogeneous case, *SIAM J. Appl. Math.* **40** (1981), No. 3, 476–484.

6. Berryman, J. G. (with Holland, C. J.), Asymptotic behavior of the nonlinear diffusion equation $n_t = (n^{-1}n_x)_x$, *J. Math. Phys.* **23** (1982), No. 6, 983–987.

7. Bertsch, M., Asymptotic behavior of solutions of a nonlinear diffusion equation, *SIAM J. Appl. Math.* **42** (1982), No. 1, 66–76.

8. Chen, Qing Yi, On the blowing-up and quenching problems for semilinear heat equations, *Acta Math. Sci.* **2** (1982), No. 1, 17–23 (Chinese, English summary).

9. Damlamian, A. (with Li, Da Qian), Comportement limité des solutions de certains problèmes mixtes pour des équations paraboliques. II. [Limiting behavior of the solutions of certain mixed problems for parabolic equations.

II], *Acta Math. Sci.* **2** (1982), No. 1, 85–104.

10. Dias, J.-P. (with Haraux, A.), Smoothing effect and asymptotic behavior for the solutions of a nonlinear time-dependent system, *Proc. Roy. Soc. Edinburgh, Sect. A*, **87**, (1980–81), No. 3–4, 289–303.

11. Ermentrout, G. B. (with Rinzel, J), Waves in a simple excitable or oscillatory, reaction–diffusion model, *J. Math. Biol.* **11** (1981), No. 3, 269–294.

12. Ewer, J. P. G., On the asymptotic properties of a class of nonlinear parabolic equations, *Applicable Anal.* **13** (1982), No. 4, 249–260.

13. Fife, P. C., Wave-fronts and target patterns, in *Applications of Nonlinear Analysis in the Physical Sciences* (Bielefeld, 1979), pp. 206–228, Surveys of reference works in Math., 6. Pitman, Boston, Mass. 1981.

14. Fife, P. C., On the question of the existence and nature of homogeneous-center target patterns in the Belousov–Zhabotinskii reagent, in *Analytical and Numerical Approaches to Asymptotic Problems in Analysis* (Proc. Conf., Univ. Nijmegen, Nijmegen, 1980), pp. 45–56. North-Holland Math. Studies, 47, North-Holland, Amsterdam, 1981.

15. Fife, P. C. (wth McLeod, J. B.), A phase-plane discussion of convergence to traveling fronts for nonlinear diffusion, *Arch. Rational Mech. Anal.* **75** (1980–81), No. 4, 281–314.

16. Filar, M. (with Musiałek, J.), On a certain limit problem for parabolic equations, *Comment. Math. Prace. Mat.* **22** (1980–81), No. 2, 205–225.

17. Gajewski, H. (with Gartner, J.), On the asymptotic behavior of some reaction–diffusion processes, *Math. Nachr.* **102** (1981), 141–155.

18. Gusarov, A. L., Asymptotic behavior of solutions of a parabolic equation in a strip, *Dokl. Akad. Nauk. SSSR* **257** (1981), No. 1, 18–21 (Russian).

19. Hagan, P. S., Target patterns in reaction–diffusion systems, *Adv. in Appl. Math.* **2** (1981), No. 4, 400–416.

20. Hagan, P. S., Traveling-wave and multiple-traveling-wave solutions of parabolic equations, *SIAM J. Math. Anal.* **13** (1982), No. 5, 717–738.

21. Herrero, M. A. (with Vazquez, J. L.), Asymptotic behavior of the solutions of a strongly nonlinear parabolic problem, *Ann. Fac. Sci. Toulouse Math.* (5) **3** (1981), No. 2, 113–127 (French summary).

22. Il'in, A. M., Asymptotic behavior of a boundary-value problem on the half-line for a parabolic equation, *Trudy Inst. Mat. i Meh. Ural. Nauch. Centr. Akad. Nauk. SSSR*, No. 28, *Primen. Metoda Soglasovan. Asimptot. Razlozenii k Kraev. Zadacam dl'ja Differencial. Urav* (1979), 81–92, 109 (Russian).

23. Kadyrov, R. R., On the asymptotic behavior of solutions of second-order parabolic equations, *Dokl. Akad. Nauk. UzSSR*, 1981, No. 3, 7–10 (Russian).

24. Kannai, Y., Short-time asymptotic behavior for parabolic equations, in *Analytical and Numerical Approaches to Asymptotic Problems in Analysis* (Proc. Conf., Univ. Nijmegen, Nijmegen, 1980), pp. 117–132. North-Holland, Math. Studies, 47, North-Holland, Amsterdam, 1981.

25. Klainerman, S., Long-time behavior of solutions to nonlinear evolution equations, *Arch. Rational Mech. Anal.* **78** (1982), No. 1, 73–98.

26. Kopell, N. (with Howard, L. N.), Target patterns and horseshoes from a perturbed central-force problem: some temporally periodic solutions to reaction–diffusion equations, *Stud. Appl. Math.* **64** (1981), No. 1, 1–56.

27. Kopell, N. (with Howard, L. N.), Target pattern and spiral solutions to

reaction–diffusion equations with more than one space dimension, *Adv. in Appl. Math.* **2** (1981), No. 4, 417–449.

28. Kopell, N., Target-pattern solutions to reaction–diffusion equations in the presence of impurities, *Adv. in Appl. Math.* **2** (1981), No. 4, 389–399.

29. Larson, D. A. (with Murray, J. D.), Finite-amplitude traveling solitary waves in a model for the Belousov–Zhabontinskii reaction mechanism; Professor P. L. Bhatnagar commemoration volume, pp. 24–42. Nat. Acad. Sci., Allahabad, 1979.

30. Levin, S. A., Nonuniform stable solutions to reaction–diffusion equations: applications to ecological-pattern formation, in *Pattern Formation by Dynamic Systems and Pattern Recognition* (Proc. Internat. Sympos. Synerget., Garmisch-Partenkirchen, 1979), pp. 210–222. Springer Ser. in Synergetics, 5; Springer, Berlin, 1979.

31. Levine, H. A. (with Montgomery, J. T.), The quenching of solutions of some nonlinear parabolic equations, *SIAM J. Math. Anal.* **11** (1980), No. 5, 842–847.

32. Mika, J. (with Stankiewicz, R.), Asymptotic relationship between telegraphic and diffusion equations, *Math. Methods Appl. Sci.* **3** (1981), No. 1, 21–37.

33. Milicer-Gruzewska, H. (with Cąkała, S.), Sur l'existence de limite, $t \to \infty$, de la solution du problème lié au système parabolique [On the existence of the limit, as $t \to \infty$, of the solution of a problem associated with the parabolic system], *Demonstratio Math.* **14** (1981), No. 3, 531–539 (1982).

34. Pao, C. V., Asymptotic stability of reaction–diffusion systems in chemical reactor and combustion theory, *J. Math. Anal. Appl.* **82** (1981), No. 2, 503–526.

35. Pao, C. V., Asymptotic stability of a coupled diffusion system arising from gas–liquid reactions, *Rocky Mountain J. Math.* **12** (1982) No. 1, 55–73.

36. Pao, C. V., Asymptotic behavior of solutions for a parabolic equation with nonlinear boundary conditions, *Proc. Amer. Math. Soc.* **80** (1980), No. 4, 587–593.

37. Peregudov, A. N., Asymptotic behavior of the solution of the second boundary-value problem for a quasilinear parabolic system of chemical kinetics, *Izv. Akad. Nauk. SSSR Ser. Mat.* **45** (1981), No. 1, 227–238, 240 (Russian).

38. Proka, D. V., On the asymptotic behavior of solutions of boundary-value problems for the heat equation, *Mat. Issled.*, No. 63 (1981), 116–120, (Russian).

39. Ramanujam, N. (with Srivastava, U. N.), On the asymptotic behavior of solutions of parabolic systems with a small parameter, *Boll. Un. Mat. Ital. B* (5) **18** (1981), No. 2, 557–574 (Italian summary).

40. Rothe, F., Asymptotic behavior of the solutions of the Fisher equation, in *Biological Growth and Spread* (Proc. Conf., Heidelberg, 1979), pp. 279–289; Lecture Notes in Biomath., 38. Springer, Berlin, 1980.

41. Samarskii, A. A., New methods for investigating the asymptotic properties of parabolic equations, in *Analytic Number Theory, Mathematical Analysis, and Their Applications.* Trudy Mat. Inst. Steklov, 158 (1981), 153–162, 230 (Russian).

42. Scott, J. F., The long-time asymptotics of solutions to the generalized Burgers' equation, *Proc. Roy. Soc. London Ser. A*, **373**, No. 1755, 443–456 (1981).

43. Shi, Xian Liang (with Li, Ming De, and Qin, Yu Chun), On the asymptotic

behavior of solutions of the Cauchy problem for nonlinear parabolic equations, *Acta Math. Sinica* **24** (1981), No. 3, 451–463 (Chinese).

44. Tarsi-Santolini, A., Asymptotic behavior of the solution of a quasilinear parabolic inequality, *Riv. Mat. Univ. Parma* (4) **7** (1981). 251–265 (1982), (Italian, English summary).

45. Uchiyama, K., On the growth of solutions of semilinear diffusion equation with drift, *Osaka J. Math.* **17** (1980), No. 2, 281–301.

46. Weinberger, H. F., Long-time behavior of a class of biological models, *SIAM J. Math. Anal.* **13** (1982), No. 3, 353–396.

47. Zitarasu, N. V., Asymptotic representation and necessary and sufficient conditions for stabilization of the solutions of model parabolic boundary-value problems, in *Partial-Differential Equations* (Proc. Conf., Novosibirsk, 1978), pp. 120–121, 251. "Nauka" Sibirsk. Otdel., Novosibirsk, 1980 (Russian).

J. Applications to probability and statistics

1. Becus, G. A., Variational formulation of some problems for the random heat equation, in *Applied Stochastic Processes* (Proc. Conf., Univ. Georgia, Athens, Ga., 1978), pp. 19–36. Academic Press, New York, 1980.

2. Capinski, M., On turbulent statistical solutions of the heat equations, *Zeszyty Nauk. Uniw. Jagiellon. Prace. Mat.*, No. 23 (1982), 111–115.

3. Capinski, M. (with Szafirski, B.), Statistical investigation of the heat equation, *Bull. Acad. Polon. Sci. Ser. Sci. Math.* **29** (1981), No. 11–12, 553–560 (1982), (Russian summary)

4. Dombrovskii, V. A. (with Lenjuk, M. P.), Generalized stochastic-temperature fields in solid spherical bodies, *Mat. Metody i Fiz.-Meh. Polja*, No. 13 (1981), 88–91, 115 (Russian).

5. Gor'kov, J. P., Behavior of the solution of a boundary-value problem for a stationary equation of Brownian motion, *Trudy Inst. Mat. i Meh. Ural. Naucn. Centr. Akad. Nauk. SSSR, No. 28, Primen. Metoda Soglasovan. Asimptot. Razlozenii k Kraev. Zadacam dl'ja Differencial. Urav.* (1979), 73–80, 109 (Russian).

6. Knight F. B., On Brownian motion and certain heat equations, *Z. Wahrsch. Verw. Gebiete* **55** (1981), No. 1, 1–10.

7. Nakazawa, H., Stochastic Burgers' equation in the inviscid limit, *Adv. in Appl. Math.* **3** (1982), No. 1, 18–42.

8. Veretennikov, A. J., Parabolic equations and stochastic equations of Ito with coefficients that are discontinuous with respect to time, *Mat. Zametki* **31** (1982), No. 4, 549–557 (Russian).

9. Wang, P. Z., Probabilistic analysis of reaction–diffusion processes, *Proc. of Symp. on Nonequilibrium Statistical Physics* (Xian, 1979), pp. 77–97. Northwest Univ, Xian, 1979 (Chinese).

4.2. Initial-Boundary-Value Problems

1. Alexiades, V. (with Chan, C. Y.), A singular Fourier problem with nonlinear radiation in a noncylindrical domain, *Nonlinear Anal.* **5** (1981), No. 8, 835–844.

2. Alexiades, V., Generalized axially symmetric heat potentials and singular

parabolic initial-boundary-value problems, *Arch. Rational Mech. Anal.* **79** (1982), No. 4, 325–350.

3. Alihanova, R. I. (with Atakisieva, R. H.), Unique solvability of a certain class of parabolic, variational inequalities, *Izv. Akad. Nauk. Azerbaidzan. SSR Ser. Fiz.-Tehn. Mat. Nauk.* **2** (1981), No. 5, 41–46 (Russian, Azerbaijani, and English summaries).

4. Alliney, S., Solution of a diffusion problem with radial symmetry, *Atti Accad. Sci. Istit. Bologna Cl. Sci. Fis. Rend.* (13) **7** (1979–90), No. 2, 41–57 (Italian).

5. Anderson, D. (with Lisak, M.), Approximate solutions of some nonlinear diffusion equations, *Phys. Rev. A* (3) **22** (1980), No. 6, 2761–2768.

6. Anderson, R. M. (with Ford, W. T.), Test solutions of porous-media problems, *J. Math. Anal. Appl.* **79** (1981), No. 1, 26–37.

7. Atamanov, E. R., Uniqueness and stability of the solution of a general mixed problem for the heat equation, in *Studies in Integrodifferential Equations*, No. 13, pp. 329–334, 385. "Ilim," Frunze, 1980 (Russian).

8. Azamatov, S. Z., On a boundary-value problem for the heat equation with shifted argument, *Izv. Akad. Nauk. Kazah. SSR Ser. Fiz.-Mat* 1981, No. 5, 61–63 (Russian, Kazakh summary).

9. Aziz, A. (with Na, T. Y.), Solution of heat-conduction equation with temperature-dependent generation and thermal conductivity by GPM, *Indust. Math.* **31** (1981), No. 2, 67–73.

10. Azzam, A. (with Kreyszig, E.), Linear parabolic equations in regions with re-entrant edges, *Hokkaido Math. J.* **11** (1982), No. 1, 29–34.

11. Barvinok, V. A. (with Bogdanovic, V. I.), Solutions of the unsteady heat-conduction problem with boundary conditions of the first, second, and third kind, *Izv. Vyss. Ucebn. Zaved. Aviacion. Tehn.* **23** (1980), No. 2, 14–19 (Russian); translated as *Soviet Aeronaut.* **23** (1980), No. 2, 11–15.

12. Bebernes, J. (with Bressan, A.), Thermal behavior for a confined reactive gas, *J. Differential Equations* **44** (1982), No. 1, 118–133.

13. Belov, J. J. (with Janenko, N. N.), The Cauchy problem for a pseudoparabolic equation in a Banach space, *Cisl. Metody Meh. Splosn. Sredy* **11** (1980), No. 7, *Mat. Modelirovanie*, 12–22.

14. Benedek, A. (with Panzone, R.), Remarks on a problem in the flow of heat for a solid in contact with a fluid, *Rev. Un. Mat. Argentina* **29** (1979–80), No. 3, 120–130.

15. Bicadze, A. V., A nonlinear equation of parabolic type, *Dokl. Akad. Nauk. SSSR* **264** (1982), No. 6, 1293–1295 (Russian).

16. Biener, K., Zur Lösung eines nichtlinearen Randwertproblems für parabolische Differentialgleichungen aus der Mikrobiologie, in *Lectures from the Problem Seminar: Hybrid Computing* (Weissig, 1979), pp. 21–30, Weiterbildungszentrum Mat. Kybernet. Rechentech. Informationsverarbeitung, 41. Tech. Univ. Dresden, Dresden, 1980 (German).

17. Bonacina, C. (with Comini, G, Fasano, A, and Primicerio, M.), Temperature effects of thermophysical-property variations in nonlinear conductive heat transfer, *Bull. Calcutta Math. Soc.* **71** (1979), No. 5, 301–308.

18. Bragg, L. R., The radial wave and Euler–Poisson–Darboux equations with singular data, *SIAM J. Math. Anal.* **12** (1981), No. 4, 489–501.

19. Brezis, H., Problèmes elliptiques et paraboliques nonlinéaires avec données

mésures [Nonlinear elliptic and parabolic problems with measure data], *Goulaouic-Meyer-Schwartz Seminar*, 1981–82, Exp. No. XX, 13 pp. Ecole Polytech. Palaiseau, 1982.

20. Bubnov, B. A., Boundary-value problems for a class of equations that contains a derivative with respect to time, *Dokl. Akad. Nauk. SSSR* **265** (1982), No. 6, 1292–1297 (Russian).

21. Bulgaru, O. E., An asymptotic method for solving problems of heat conduction for composite materials; boundary-value problems for partial differential equations, *Mat. Issled.*, No. 63 (1981), 170–174, 185–196 (Russian).

22. Cannon, J. R. (with van der Hoek, J.), The existence of, and a continuous-dependence result for, the solution of the heat equation subject to the specification of energy, *Boll. Un. Mat. Ital. Suppl.* 1 (1981), 253–282.

23. Colton, D., Schwarz reflection principles for solutions of parabolic equations, *Proc. Amer. Math. Soc.* **82** (1981), No. 1, 87–94.

24. Daniljuk, I. I., A mixed problem for a quasilinear heat-conduction equation with piecewise-discontinuous coefficients, *Dokl. Akad. Nauk. Ukrain. SSR Ser. A.*, 1981, No. 7, 3–7 (Russian, English summary).

25. Di Benedetto, E. (with Showalter, R. E.), Implicit degenerate evolution equations and applications, *SIAM J. Math. Anal.* **12** (1981), No. 5, 731–751.

26. Dodziuk, J., Eigenvalues of the Laplacian and the heat equation, *Amer. Math. Monthly* **88** (1981), No. 9, 686–695.

27. Donati, F., A penalty-method approach to strong solutions of some nonlinear, parabolic unilateral problems, *Nonlinear Anal.* **6** (1982), No. 6, 585–597.

28. Dont, M., Third boundary-value problem for the heat equation, I. *Casopis Pest. Mat.* **106** (1981), No. 4, 376–394, 436 (Czech. summary with a loose Russian summary).

29. Dont, M., Third boundary-value problem for the heat equation, II. *Casopis Pest Mat.* **107** (1982), No. 1, 7–22, 90 (Czech summary with a loose Russian summary).

30. Douglas, J., Jr. (with Russell, T. F.), Numerical methods for convection-dominated diffusion problems based on combining the method of characteristics with finite-element or finite-difference procedures, *SIAM J. Numer. Anal.* **19** (1982), No. 5, 871–885.

31. Dzavadov, M. G. (with Guseinov, G. M.), On a boundary-value problem for a parabolic equation with a small parameter, *Dokl. Akad. Nauk. SSR* **258** (1981), No. 2, 280–283 (Russian).

32. Enachescu, D. M., Methods for solving parabolic equations based on time series, *Bull. Math. Soc. Sci. Math. R. S. Roumaine* (*N.S.*) **26** (74) (1982), No. 1, 27–34.

33. Ene, H. I. (with Sanchez-Palencia, E.), Sur la propagation de la chaleur dans les milieux poreux, *C. R. Acad. Sci. Paris Ser. II Mec. Phys. Chim. Sci. Terre Sci. Univers* **292** (1981), No. 17, 1181–1184.

34. Fikiina, I. K., Investigation of the change in the temperature field of an unbounded plate in the presence of an internal heat source. *Godisnik Viss. Ucebn. Zaved. Prilozna Mat.* **16** (1980), No. 4, 143–152 (1981), (Bulgarian, Russian and German summaries).

35. Galaktionov, V. A., Approximate self-similar solutions of equations of nonlinear heat-conduction type, *Akad. Nauk. SSSR Inst. Prikl. Mat. Preprint* 1980,

No. 106, 28 pp (Russian, English summary).

36. Galaktionov, V. A., Approximate self-similar solutions of the heat equation, *Dokl. Akad. Nauk. SSSR* **265** (1982), No. 4, 784–789 (Russian).

37. Galaktionov, V. A. (with Samarskii, A. A.), Methods for constructing approximate self-similar solutions of nonlinear equations of heat conduction, I., *Mat. Sb. (N.S.)* **118** (160) (1982), No. 3, 291–322, 431 (Russian).

38. Gonzalez-Velasco, E. A., The second boundary-value problem for the heat equation, *Internat. J. Math. Ed. Sci. Tech.* **13** (1982), No. 3, 291–294.

39. Gusakov, V. M., Solvability of a boundary-value problem that describes the process of heating a shallow cylinder, in *Boundary-Value Problems for Differential Equations*, pp. 60–65, 241. "Naukova Dumka," Kiev, 1980 (Russian).

40. Jaworski, A. R., Boundary elements for heat conduction in composite media, *Appl. Math. Modelling* **5** (1981), No. 1, 45–59.

41. Kalasnikov, A. S., Quasilinear degenerate parabolic equations with a finite rate of propagation of perturbations, in *Partial-Differential Equations* (Proc. Conf. Novosibirsk, 1978), pp. 80–83, 249. "Nauka" Sibirsk. Otdel., Novosibirsk, 1980 (Russian).

42. Kalla, S. L. (with Jain, S., Goyal, S. P., and Vasistha, S. K.), Heat conduction in a truncated wedge of semi-infinite height with circular boundaries, *Rev. Tecn. Fac. Ingr. Univ. Zulia* **4** (1981), no. 1–2, 59–80 (Spanish summary).

43. Kamynin, V. L., *A priori* estimates and solvability in the large for quasilinear parabolic equations, *Vestnik Moskov. Univ. Ser. I, Mat. Meh.* 1981, No. 1, 33–38, **107** (Russian, English summary).

44. Kapur, J. N., Oxygen-concentration profiles in capillaries and living tissues for general linear kinetics when axial diffusion is considered, *Indian J. Pure Appl. Math.* **13** (1982), No. 7, 847–860.

45. Kawohl, B., On nonlinear parabolic equations with abruptly changing nonlinear boundary conditions, *Nonlinear Anal.* **5** (1981), No. 10, 1141–1153.

46. Kim, E. I. (with Krasnov, J. A., and Harin, S. N.), A class of nonlinear problems of heat conduction in the theory of electric contacts, in *Partial Differential Equations* (Proc. Conf. Novosibirsk, 1978), pp. 83–86, 250. "Nauka" Sibirsk. Otdel., Novosibirsk, 1980 (Russian).

47. Lasiecka, I., "Unified theory for abstract parabolic boundary problems—a semigroup approach, *Appl. Math. Optim.* **6** (1980), No. 4, 287–333.

48. Lei, J. G., A remark on constructing approximate solutions of initial-boundary-value problems for parabolic equations in one space variable, *J. Math.* (Wuhan) **1** (1981), No. 1, 40–45 (Chinese, English summary).

49. Lesaint, P., Solving the heat equation on noncylindrical domains, in *Numerical Methods in Thermal Problems* (Proc. First Internat. Conf. Univ. College, Swansea, 1979), pp. 12–19. Pineridge, Swansea, 1979.

50. Lobuzov, A. A. The first boundary-value problem for a parabolic equation in an abstract Wiener space, *Mat. Zametki* **30** (1981), No. 2, 211–223, 314 (Russian).

51. Mamatov, A. Z., An *a priori* estimate of the error of an approximate solution to some problems of heat conduction, *Vestnik Leningrad. Univ. Mat. Meh. Astronom. 1981*, Vyp. 1, 47–53, 120 (Russian, English summary).

52. Mamontov, E. V., *On the well-posedness of problems of mathematical physics*, Novosibirsk. Gos. Univ., Novosibirsk, 1980; 63 pp (Russian).

53. Mataev, S., A general boundary-value problem of the heat equation for the one-dimensional case, *Izv. Akad. Nauk. Kazah. SSR Ser. Fiz.-Mat.* 1982, No. 1, 31–36, 84 (Russian, Kazakh summary).

54. McNabb, A. (with Weir, G. J.), Heat losses from an insulated pipe, *J. Math. Anal. Appl.* 77 (1980), No. 1, 270–277.

55. Meike, Z., Existence of an evolution equation with a nonmonotone continuous nonlinearity, *Nonlinear Anal.* 5 (1981), No. 9, 1021–1035.

56. Meike, Z., Existence for a parabolic equation with nonlinear boundary-value conditions, *An. Stiint. Univ. "Al. I. Cuza" Iasi Sect. Ia Mat. (N.S.)* 28 (1982), No. 1, 109–122.

57. Munier, A. (with Burgan, J. R., Gutierrez, J., Fijalkow, E, and Feix, M. R.), Group transformations and the nonlinear heat-diffusion equation, *SIAM J. Appl. Math.* 40 (1981), No. 2, 191–207.

58. Orynbasarov, M. O., Solution of the third boundary-value problem for a parabolic equation with variable coefficient in a domain with a nonregular boundary, in *Differential Equations and Their Applications*, pp. 67–74. Kazah. Gos. Univ., Alma-Ata, 1980 (Russian).

59. Peregudov, A. N., Second boundary-value problem for quasilinear parabolic systems of chemical kinetics type, *Z. Vycisl. Mat. i Mat. Fiz.* 21 (1981), No. 1, 18–28, 252 (Russian).

60. Petrushko, I. M., Boundary and initial values of solutions of second-order parabolic equations, *Dokl. Akad. Nauk. SSSR* 266 (1982), No. 3, 557–560 (Russian).

61. Pytel-Kudela, M., On Fourier's problem for the heat equation and for the exterior of the sphere, *Fasc. Math.*, No. 11 (1979), 179–192.

62. Raynal, M.-L., Homogenéisation d'un problème de diffusion non linéaire du type de Volterra, *C. R. Acad. Sci. Paris Ser. I Math.* 292 (1981), No. 7, 421–424 (English summary).

63. Sen, P. K., Flow of heat in composite cylindrical regions in presence of contact resistance and its effects on the temperature distributions, *Indian J. Math.* 18 (1976), No. 2, 105–112.

64. Siskov, A. E. (with Skripnik, I. V.), Acute angle between pairs of linear parabolic operators, *Dokl. Akad. Nauk. Ukrain. SSR Ser A* 1980, No. 10, 27–30, 93 (Russian, English summary).

65. Slesarenko, A. P., Nonstationary heat-conduction problems for bodies of complex cross section with heterogeneous through-inclusions, *Dokl. Akad. Nauk. Ukrain. SSR. Ser. A* 1980, No. 10, 19–22, 93 (Russian, English summary).

66. Soltanov, K. N., Solvability of some nonlinear, parabolic problems in Orlicz–Sobolev spaces, in *Partial-Differential Equations* (Proc. Conf., Novosibirsk, 1978), pp. 99–104, 250. "Nauka" Sibirsk. Otdel., Novosibirsk, 1980 (Russian).

67. Soltanov, K. N., Nonlinear parabolic problems in spaces of Sobolev–Orlicz type, in *Investigations in the Theory of Differential Equations and Its Applications*, No. II, pp. 99–115. "Elm," Baku, 1981 (Russian).

68. Sopolov, N. N., A boundary-value problem for a mixed parabolic equation with nonlocal initial conditions. I., *Godisnik Viss. Ucebn. Zaved. Prilozna Mat.* 15 (1979), No. 2, 137–146 (1980), (Bulgarian, Russian, and French

Summaries).

69. Sopolov, N. N., A boundary value problem for a mixed parabolic equation
     with nonlocal initial conditions. II., *Godisnik Viss. Ucebn. Zaved. Prilozna
     Mat.* **15** (1979), No. 2, 147–158 (1980), Bulgarian, Russian, and French
     summaries).

70. Sopolov, N. N., A mixed problem for the heat equation with a nonlocal initial
     condition, *C. R. Acad. Bulgare Sci.* **34** (1981), No. 7, 935–935 (Russian).

71. Tamurov, N. G. (with Odinec, V. F.), Nonstationary, plane axisymmetric
     temperature field of a heat-insulated disk of variable and piecewise-variable
     thickness, *Izv. Vyss. Ucebn. Zaved. Matematika* 1981, No. 1, 79–83 (Russian).

72. Tao, L. N., Heat conduction with nonlinear boundary condition, *Z. Angew.
     Math. Phys.* **32** (1981), No. 2, 144–155 (German summary).

73. Terzijan, S. A., The mixed problems for the Berstein equation $\partial^2 u / \partial t^2 =
     k(\int_0^1 (\partial u / \partial x)^2 dx) \partial^2 u / \partial x^2$, *Godisnik Viss. Ucebn. Zaved. Prilozna Mat.* **15**
     (1979), No. 2, 181–190 (1980), Bulgarian, Russian, and English summaries).

74. Terzijan, S. A. (with Docev, D. T.), Existence, uniqueness, and boundedness of
     solutions of a mixed problem for parabolic equations with time lag, *Izv. Akad.
     Nauk. Armjan. SSR Ser. Mat.* **15** (1980), No. 6, 433–442, 507 (Russian,
     Armenian, and English summaries).

75. Timofeev, J. A., An approximate method for calculating temperature fields of
     piecewise-homogeneous bodies, *Differencial'nye Uravnenija* **16** (1980), No. 8,
     1492–1503, 1534 (Russian).

76. Watson, N. A., Solutions of parabolic equations with initial values locally in
     L^p, *J. Math. Anal. Appl.* **89** (1982), No. 1, 86–94.

4.3. Stefan-Free Boundary

1. Free-boundary problems, Vol. I, in *Proceedings of a Seminar held in Pavia,
    September–October 1979*. Istituto Nazionale de Alta Matematica Francesco
    Severi, Rome, 1980. 523 pp.

2. Free Boundary problems, Vol. II, in *Proceedings of a Seminar Held in Pavia,
    September–October 1979*. Istituto Nazionale de Alta Matematica Francesco
    Severi, Rome, 1980. 606 pp.

3. Alexander, R. (with Manselli, P. and Miller, K.), Moving finite-elements for
    the Stefan problem in two dimensions, *Atti Accad. Naz. Lincei Rend. Cl. Sci.
    Fis. Mat. Natur.* (8) **67** (1979), No. 1–2, 57–61 (Italian summary).

4. Bai, D. H. (with Sun, S. H.), A free-boundary problem for an oil reservoir
    with bottom water, *Sichuan Daxue Xuebao* 1979, No. 1, 5–31 (Chinese,
    English summary).

5. Bai, Dong Hua (with Sun, Shun Hua) A free-boundary problem for fluid flow
    through a porous medium, *Sichuan Daxue Xuebao* 1981, No. 1, 1–10 (Chinese,
    English summary).

6. Baumeister, J., Zur optimalen Steuerung von freien Randwertaufgaben, *Z.
    Angew. Math. Mech.* **60** (1980), No. 7, T333–T335.

7. Bazalii, B. V. (with Selepov, V. J.), Estimates of the rate of stabilization for
    the solution of the Stefan problem, *Dokl. Akad. Nauk. Ukrain. SSR Ser. A*,
    1981, No. 4, 3–6 (Russian, English summary).

8. Bazalii, B. V. (with Selepov, V. J.), Estimation of the rate of stabilization of the solution of the Stefan problem, *Mat. Fiz.*, No. 30 (1981), 35–41 (Russian).

9. Bazalii, B. V., Stability of smooth solutions of a two-phase Stefan problem, *Dokl. Akad. Nauk. SSSR* **262** (1982), No. 2, 265–269 (Russian).

10. Beckett, P. M., On the use of series solutions applied to solidification problems, *Mech. Res. Comm.* **8** (1981), No. 3, 169–174.

11. Boieri, P. (with Gastaldi, F.), Convexity of the free boundary in a filtration problem, *J. Differential Equations* **42** (1981), No. 1, 25–46.

12. Borodin, M. O., On the solvability of a two-phase quasistationary Stefan problem, *Dokl. Akad. Nauk. Ukrain. SSR Ser. A*, 1982, No. 2, 3–5 (Russian, English summary).

13. Borodin, M. A., Solvability of a two-phase nonstationary Stefan problem, *Dokl. Akad. Nauk. SSR* **263** (1982), No. 5, 1040–1052 (Russian).

14. Bossavit, A. (with Damlamian, A.), Homogenization of the Stefan problem and application to magnetic composite media, *IMA J. Appl. Math.* **27** (1981), No. 3, 319–334.

15. Briere, T., Application des méthodes variationnelles à la cristallisation d'un métal par passage dans une gaine de refroidissement, with an introduction and bibliography by Georges Duvaut, and Paul Rougee; *Ann. Fac. Sci. Toulouse Math.* (5) **2** (1980), No. 3–4, 219–247 (English summary).

16. Caffarelli, L. A., Some common aspects of several free-boundary problems, in *Free-Boundary Problems*, Vol. II (Pavia, 1979), pp. 101–108. Ist. Naz. Alta Mat. Francesco Severi, Rome, 1980.

17. Caffarelli, L. A., A remark on the Hausdorff measure of a free boundary, and the convergence of coincidence sets, *Boll. Un. Mat. Ital.* A. (5) **18** (1981), No. 1, 109–113 (Italian summary).

18. Caffarelli, L. A. (with Friedman, A., and Visintin, A.), A free-boundary problem describing transition in a superconductor, *SIAM J. Math. Anal.* **12** (1981), No. 5, 679–690.

19. Cannon, J. R. (with Di Benedetto, E.), The steady-state Stefan problem with convection, with mixed temperature and nonlinear heat-flux boundary conditions, in *Free-Boundary Problems*, Vol. I (Pavia, 1979) pp. 231–265. Ist. Naz. Alta Mat. Francesco Severi, Rome, 1980.

20. Cannon, J. R. (with van der Hoek, J.), The classical solution of the one-dimensional, two-phase Stefan problem with energy specification, *Ann. Mat. Pura Appl.* (4) **130** (1982), 385–398.

21. Cannon, J. R. (with van der Hoek, J.), The one-phase Stefan problem subject to the specification of energy, *J. Math. Anal. Appl.* **86** (1982), No. 1, 281–291.

22. Cimatti, G., A free-boundary problem in the theory of lubrication, *Internat. J. Engrg. Sci.* **18** (1980), No. 5, 703–711.

23. Cimatti, G., Remark on a free-boundary problem of the theory of lubrication, *Boll. Un. Mat. Ital.* A (6) **1** (1982), No. 2, 249–251. (Italian summary).

24. Damlamian, A., The homogenization of the Stefan problem and related topics, in *Free-Boundary Problems*, Vol. I (Pavia, 1979), pp. 267–275. Ist. Naz. Alta Mat. Francesco Severi, Rome, 1980.

25. Damlamian, A., How to homogenize a nonlinear diffusion equation: Stefan's problem, *SIAM J. Math. Anal.* **12** (1981), No. 3, 306–313.

26. Danieljan, J. S. (with Janickii, P. A.), Stability of the interphase surface in the

freezing of moist ground, *Inz.-Fiz. Z.* **39** (1980), No. 1, 96–101 (Russian); translated as *J. Engrg. Phys.* **39** (1980), No. 1, 781–785 (1981).

27. Daniljuk, I. I., On a multidimensional, nonstationary Stefan problem in domains with a nonsmooth boundary, *Dokl. Akad. Nauk. Ukrain. SSR Ser. A* 1980, No. 11, 9–15 (Russian, English summary).

28. Daniljuk, I. I., A variant of the multidimensional Stefan problem in domains with an irregular boundary, *Mat. Fiz.*, No. 30 (1981), 48–49, 123 (Russian).

29. Daniljuk, I. I., A two-phase, quasistationary Stefan problem, *Dokl. Akad. Nauk. Ukrain. SSR Ser A*, 1982, No. 1, 6–10 (Russian, English summary).

30. Davis, G. B (with Hill, J. M.), A moving-boundary problem for the sphere, *IMA J. Appl. Math.* **29** (1982), No. 1, 99–111.

31. Di Benedetto, E. (with Showalter, R. E.), A pseudoparabolic variational inequality and Stefan problem, *Nonlinear Anal.* **6** (1982), No. 3, 279–291.

32. Evans, L. C., A chemical diffusion–reaction free-boundary problem, *Nonlinear Anal.* **6** (1982), No. 5, 455–466.

33. Fage, D. M., Convergence of approximations of the solution and domain in a problem with a free boundary, in *Theory of Cubature Formulas and Numerical Mathematics* (Proc. Conf., Novosibirsk, 1978) pp. 45–48, 250. "Nauka" Sibirsk. Otdel., Novosibirsk, 1980 (Russian).

34. Fasano, A., Free-boundary problems, for parabolic equations with Cauchy data on the free boundary, in *Free-Boundary Problems* Vol. II (Pavia, 1979), pp. 237–247. Ist. Naz. Alta Mat. Francesco Severi, Rome, 1980.

35. Fasano, A. (with Primicerio, M.), New results on some classical parabolic free-boundary problems, *Quart. Appl. Math.* **38** (1980–81), No. 4, 439–460.

36. Fasano, A. (with Primicerio, M., and Rubinstein, L.), A model problem for the heat conduction with a free boundary in a concentrated capacity, *J. Inst. Math. Appl.* **26** (1980), No. 4, 327–347.

37. Fasano, A. (with Villa, L. T.), Some remarks on the curvature of the free boundary in a Stefan problem with appearance of a phase, *Boll. Un. Mat. Ital. B* (6) **1** (1982), No. 2, 743–752 (Italian summary).

38. Fel'genhauer, U., Monotonie und Glatte des freien Randes bei einem Stefanschen Problem, *Wiss. Z. Tech. Hochsch. Magdeburg* **34** (1980), No. 5, 29–31.

39. Fel'genhauer, U., A one-phase, nonstationary Stefan problem, *Dokl. Akad. Nauk. Ukrain. SSR Ser. A*, 1981, No. 1, 30–32, 91 (Russian, English summary).

40. Fel'genhauer, U., An existence theorem for a certain quasilinear Stefan problem, *Dokl. Akad. Nauk. Ukrain. SSR Ser. A*, 1982, No. 5, 25–27 (Russian, English summary).

41 Friedman, A. (with Kamin, S.), The asymptotic behavior of gas in an *n*-dimensional porous medium, *Trans. Amer. Math. Soc.* **262** (1980), No. 2, 551–563.

42. Gilardi, G., The evolution dam problem, in *Free-Boundary Problems*, Vol. I (Pavia, 1979), pp. 209–217. Ist. Naz. Alta Mat. Francesco Severi, Rome, 1980.

43. Gliko, A. O. (with Efimov, A. B.), The method of a small parameter in the classical Stefan problem, *Inz.-Fiz. Z.* **38** (1980), No. 2, 329–335 (Russian); translated as *J. Engrg. Phys.* **38** (1980), No. 2, 211–216.

44. Grigor'ev, S. G. (with Kosolapov, V. N., Pudovkin, M. A., and Cugunov,

V. A.), Scheme of the generalized methods of integral relations for multidimensional, single-phase Stefan problems and its applications, in *Applied Problems of Theoretical and Mathematical Physics*, pp. 43–52, 126. Latv. Gos. Univ., Riga, 1980 (Russian).

45. Gupta, S. C. (with Lahiri, A. K.), Heat conduction with a phase change in a cylindrical mould, *Internat. J. Engrg. Sci.* **17** (1979), No. 4, 401–407.

46. Hanzawa, Ei-ichi, Classical solutions of the Stefan problem, *Tohoku Math. J.* (2) **33** (1981), No. 3, 297–335.

47. Hoffmann, K.-H., Fixpunktprinzipien und freie Randwertaufgaben, [Fixed-point principles and free boundary-value problems], in *Numerical Solution of Nonlinear Equations* (Bremen, 1980), pp 162–181, Lecture Notes in Math., 878. Springer, Berlin, 1981 (English summary).

48. Ishii, H., Erratum: "Asymptotic stability of almost periodic solutions of a free-boundary problem arising in hydraulics" [*Bull. Fac. Sci. Engrg. Chuo Univ.* **22** (1979), 73–95; MR 81k:35016], *Bull. Fac. Sci. Engrg. Chuo Univ.* **23** (1980), 83.

49. Ishii, H., On a certain estimate of the free boundary in the Stefan problem, *J. Differential Equations* **42** (1981), No. 1, 106–115.

50. Jakubov, S., Solvability of a class of problems with free boundary for linear equations of parabolic type, *Dinamika Splosn. Sredy*, No. 43, *Mat. Problemy Meh.* (1979), 178–183, 188 (Russian).

51. Jakubov, S., Solvability of some classes of problems with unknown boundary for equations of parabolic type, in *Dinamika Splosn. Sredy*, No. 46, *Dinamika Zidkosti so Svobod. Granicami* (1980), 137–159, 175 (Russian).

52. Jakubov, S. J., Solvability of a boundary-value problem for a parabolic equation in a domain with an unknown boundary, in *Boundary-Value Problems for Equations of Mathematical Physics*, pp. 44–48, 168. "Fan," Tashkent, 1980 (Russian).

53. Karatzas, I. (with Benes, V. E.), A degree method for free boundaries in stochastic control, *SIAM J. Control Optim.* **19** (1981), No. 3, 283–332.

54. Kenmochi, N., Free-boundary problems for a class of nonlinear parabolic equations: an approach by the theory of subdifferential operators, *J. Math. Soc. Japan* **34** (1982), No. 1, 1–13.

55. Kim, E. I. (with Bizanova, G. I.), Investigation of the second Stefan boundary-value problem for small time values, *Vestnik Akad. Nauk. Kazah. SSR*, 1981, No. 6, 76–86 (Russian, Kazakh summary).

56. Klibanov, M. V., A class of inverse problems, *Dokl. Akad. Nauk. SSSR* **265** (1982), No. 6, 1306–1309 (Russian).

57 Lacey, A. A., Moving-boundary problems in the flow of liquid through porous media, *J. Austral. math. Soc. Ser. B*, **24** (1982–83), No. 2, 171–193.

58. Li, D. Q., On a free-boundary problem, *Chinese Ann. Math.* **1** (1980), No. 3–4, 351–358 (Chinese summary).

59. Lions, P. L., Minimization problems in $L^1(R^3)$ and applications to free-boundary problems, in *Free-Boundary Problems*, Vol. II (Pavia, 1979), pp. 385–399. Ist. Naz. Alta Mat. Francesco Severi, Rome, 1980.

60. Lynch, D. R. (with O'Neill, K.), Continuously deforming finite elements for the solutions of parabolic problems, with and without phase change, *Internat. J. Numer. Methods Engrg.* **17** (1981), No. 1, 81–96.

61. Macovei, D. (with Pascali, D.), Variational inequality techniques in free-boundary problems, in *Proceedings of the International Symposium on Applications of Mathematics in System Theory* (Univ. Brasov, Brasov, 1978), II, pp. 357–363. Univ. Brasov, Brasov, 1979.

62. Macovei, D. (with Pascali, D.), General problems of Stefan type, *Stud. Cerc. Mat.* **33** (1981), No. 6, 611–618 (Romanian, English summary).

63. Meirmanova, A. M. (with Puhnacev, V. V.), Lagrangian coordinates in the Stefan problem, *Dinamika Splosn. Sredy*, No. 47, *Mat. Problemy Meh. Splosn. Sred* (1980), 90–111, 165–166 (Russian).

64. Meirmanov, A. M., On a problem with a free boundary for parabolic equations, *Mat. Sb.* (*N.S.*) **115** (157), (1981), No. 4, 532–543 (Russian).

65. Meirmanov, A. M., An example of the nonexistence of a classical solution to the Stefan problem, *Dokl. Akad. Nauk. SSSR* **258** (1981), No. 3, 547–549 (Russian).

66. Meirmanov, A. M., Solutions of a two-dimensional, two-phase Stefan problem that are close to one-dimensional solutions, *Dinamika Splosn. Sredy*, No. 50. Kraev. Zadaci dija Uravnenii Gidrodinamiki (1981), 135–149, 177 (Russian).

67. Meirmanov, A. M., A problem on the advance of a contact-discontinuity surface in the filtration of an immiscible compressible fluid (Verigin's problem), *Sibirsk. Mat. Z.* **23** (1982), No. 1, 85–102, 221 (Russian).

68. Meyer, G. H., On the computational solution of elliptic and parabolic free-boundary problems, in *Free-boundary Problems*, Vol. I (Pavia, 1979), pp. 151–173. Ist. Naz. Alta Mat. Francesco Severi, Rome, 1980.

69. Niezgodka, M., On some properties of two-phase parabolic free boundary-value control problems, *Control Cybernet.* **8** (1979), No. 1, 23–42 (Polish and Russian summaries).

70. Niezgodka, N. (with Pawlow, I, and Visintin, A.), Remarks on the paper by A. Visintin "Sur le problème de Stéfan avec flux nonlinéaire " [*Boll. Un. Mat. Ital. C* (5) **18** (1981), No. 1, 63–86]; *Boll. Un. Mat. Ital. C* (5) **18** (1981), No. 1, 87–88.

71. Nitsche, J. A., A finite-element method for parabolic free-boundary problems, in *Free-Boundary Problems*, Vol. I (Pavia, 1979), pp. 277–318. Ist. Naz, Alta Mat. Francesco Severi, Rome, 1980.

72. Pawlow, I., On some properties of two-layer parabolic free-boundary value problems, *Control Cybernet.* **7** (1978), No. 4, 19–37 (1979), (Polish and Russian summaries).

73. Pawlow, I., Parabolic problems with free boundaries: existence properties of solutions, optimal-control problems, in *Nonlinear Analysis* (Berlin, 1979), pp. 221–231, Abh. Akad. Wiss. DDR, Abt. Math Naturwiss. Tech., 1981, 2. Akademie-Verlag, Berlin 1981.

74. Popa, V., Sur le mouvement à surface libre dans les milieux poreux non homogènes, *J. Mécanique* **19** (1980), No. 4, 663–678 (English summary).

75. Primicerio, M., Qualitative properties of some one-dimensional, parabolic free-boundary problems, in *Free-Boundary Problems*, Vol. I, (Pavia, 1979), pp. 451–460. Ist. Naz. Alta Mat. Francesco Severi, Rome, 1980.

76. Primicerio, M., Diffusion problems with a free boundary, *Boll. Un. Mat. Ital. A* (5) **18** (1981), No. 1, 11–68 (Italian).

77. Rodriques, J.-F., Free-boundary convergence in the homogenization of the one-phase Stefan problem, *Trans. Amer. Math. Soc.* **274** (1982), No. 1, 297–305.

78. Rogers, J. C. W., The formation of selected free-boundary problems as conservation laws, in *Free-Boundary Problems*, Vol. I (Pavia, 1979), pp. 319–331. Ist. Naz. Alta Mat. Francesco Severi, Rome, 1980.

79. Rogers, J. C. W., Relation of the one-phase Stefan problem to the seepage of liquids and electrochemical machining, in *Free-Boundary Problems*, Vol. I (Pavia, 1979), pp. 333–382. Ist. Naz Alta Mat. Francesco Severi, Rome, 1980.

80. Rubinstein, L., Application of the integral-equation technique to the solution of several Stefan problems, in *Free-Boundary Problems*, Vol. I, (Pavia, 1979), pp. 383–450. Ist. Naz. Alta Mat. Francesco Severi, Rome, 1980.

81. Rubinstein, Lev. (with Fasano, A., and Primicerio, M.), Remarks on the analyticity of the free boundary for the one-dimensional Stefan problem, *Ann. Mat. Pura Appl.* (4) **125** (1980), 295–311.

82. Rubinstein, L. (with Shillor, M.), Analyticity of the free boundary for the one-phase Stefan problem, *Boll. Un. Mat. Ital. Suppl.* 1 (1981), 47–68

83. Rubinstein, L. (with Geiman, H., and Shachaf, M.), Heat transfer with a free boundary moving within a concentrated thermal capacity, *IMA J. Appl. Math.* **28** (1982), No. 2, 131–147.

84. Rubinstein, L., Global stability of the Neumann solution of the two-phase Stefan problem *IMA J. Appl. Math.* **28** (1982), No. 3, 287–299.

85. Salei, S. V., Analyticity of the free boundary in a Stefan problem, in *Boundary-Value Problems for Differential Equations*, pp. 157–166, 245. "Naukova Dumka," Kiev, 1980 (Russian).

86. Salei, S. V., On the rate of convergence of a free boundary to the limit value in a Stefan problem, *Dokl. Akad. Nauk. Ukrain. SSR Ser. A*, 1981, No. 6, 19–23 (Russian, English summary).

87. Salei, S. V., A Stefan problem in a cylindrical domain with axial symmetry, *Mat. Fiz.*, No. 30 (1981), 90–96 (Russian).

88. Sestini, G., On a one-dimensional nonlinear problem of Stefan type in a given symmetric medium, *Rev. Roumaine Math. Pures Appl.* **27** (1982), No. 3, 411–418 (Italian, French summary).

89. Showalter, R. E., Mathematical formulation of the Stefan problem, *Internat. J. Engrg. Sci.* **20** (1982), No. 8, 909–912.

90. Solomon, A. D. (with Alexiades, V., and Wilson, D. G.), The Stefan problem with a convective boundary condition, *Quart. Appl. Math.* **40** (1982–83), No. 2, 203–217.

91. Soward, A. M., A unified approach to Stefan's problem for spheres and cylinders, *Proc. Roy. Soc. London Ser. A*, **373** (1980), No. 1752, 131–147.

92. Stakgold, I., Estimates for some free-boundary problems, in *Ordinary and Partial Differential Equations* (Proc. Sixth Conf., Univ. Dundee, Dundee, 1980), pp. 333–346. Lectures Notes in Math. 846; Springer, Berlin, 1981.

93. Stedry, M. (with Vejvoda, O.), Time-periodic solutions of a one-dimensional, two-phase Stefan problem, *Ann. Mat. Pura Appl.* (4) **127** (1981), 67–78.

94. Streit, U., Zur iterativen Lösung nichtlinearer Differenzenschemata für eindimensionale Stefan-Aufgaben, *Wiss. Z. Tech. Hochsch. Karl-Marx-Stadt* **22** (1980), No. 4, 347–353.

95. Tabisz, K., On a kind of the Stefan problem, *Bull. Acad. Polon. Sci. Ser. Sci. Math.* **27** (1979), No. 11–12, 847–852 (1981) (Russian).

96. Tabisz, K., Local and global solutions of the Stefan-type problem, *J. Math. Anal. Appl.* **82** (1981), No. 2, 306–316.

97. Tao, L. N., Solidification of a binary mixture with arbitrary heat flux and initial conditions, *Arch. Rational Mech. Anal.* **76** (1981), No. 2, 167–181.

98. Tao, L. N., The exact solutions of some Stefan problems with prescribed heat flux, *Trans. ASME Ser. EJ. Appl. Mech.* **48** (1981), No. 4, 732–736.

99. Tarzia, D. A., An inequality for the coefficient σ of the free boundary $s(t) = 2\sigma\sqrt{t}$ of the Neumann solutions for the two-phase Stefan problem, *Quart. Appl. Math.* **39** (1981–82), No. 4, 491–497.

100. Tokuda, N., Numerical and series solution of a Stefan problem, in *Numerical Methods in Thermal Problems* (Proc. First Internatl. Conf, Univ. College, Swansea, 1979), pp. 159–171. Pineridge, Swansea, 1979.

101. van Duyn, C. J., On the continuity of the free boundary arising in a problem of porous media, *Delft Progr. Rep.* **6** (1981), No. 2, 83–86.

102. van Duyn, C. J., Nonstationary filtration in partially saturated porous media: continuity of the free boundary, *Arch. Rational Mech. Anal.* **79** (1982), No. 3, 261–265.

103. Visintin, A., Existence results for some free-boundary filtration problems, *Ann. Mat. Pura Appl.* (4) **124** (1980), 293–320 (Italian summary).

104. Visintin, A., Sur le problème de Stéfan avec flux non linéaire. [The Stefan problem with nonlinear flux], *Boll. Un. Mat. Ital. C* (5) **18** (1981), No. 1, 63–86 (Italian summary).

105. Voller, V. R. (with Cross, M., and Walton, P. G.), Assessment of weak-solution numerical techniques for solving Stefan problems, in *Numerical Methods in Thermal Problems* (Proc. First Internat. Conf., Univ. College Swansea, 1979), pp. 172–181. Pineridge, Swansea, 1979.

106. Wolska-Bochenek, J., On some Stefan-like problem, in *Nonlinear Analysis* (Berlin, 1979), pp. 433–435; Abh. Akad. Wiss. DDR, Abt. Math. Naturwiss. Tech., 1981, 2. Akademie-Verlag, Berlin, 1981.

107. Iska-Bochenek, J., On a Stefan-like problem for the system of one-dimensional diffusion equation, *Bull. Acad. Polon. Sci. Ser. Sci. Tech.* **29** (1981), No. 7–8, 413–318 (1982).

108. Wolska-Bochenek, J. (with v. Wolfersdorf, L.), On some generalized free-boundary-value problem for a system of one-dimensional diffusion equations, *Z. Angew. Math. Mech.* **61** (1981), No. 11, 541–545 (German and Russian summaries).

109. Yotsutani, S., Stefan problems with the unilateral boundary condition on the fixed boundary, I. *Osaka J. Math.* **19** (1982), No. 2, 365–403.

4.4. Control of Various Heat Problems

1. Abbasov, A. A. (with Sapozhnikov, M. E.), Existence of a solution of an optimal problem of Stefan type for a quasilinear equation, in *Investigations in Some Questions of the Constructive Theory of Functions and Differential Equations*, pp. 131–135. Azerbaidzhan. Inst. Nefti i Khimii, Baku, 1981 (Russian).

2. Bublik, B. M, (with Danilov, V. J.), A problem of control of a heat-conduction process, *Vycisl. Prikl. Mat. (Kiev)*, No. 42 (1980), 142–152 (Russian, English summary).

3. Buzurnjuk, S. N., A time-optimality problem for a process described by a system of differential equations of parabolic type, in *Mathematical Methods in Mechanics*. Mat. Issled. No. 57 (1980), 9–12, 133. (Russian).

4. Egorov, A. I. (with Kapustjan, V. E.), Optimal control by processes described by parabolic equations with time lag, *Dokl. Akad. Nauk. Ukrain. SSR Ser. A*, 1981, No. 3, 69–73, 95 (Russian, English summary).

5. Glashoff, K. (with Sprekels, J.), The regulation of temperature by thermostats and set-valued integral equations, *J. Integral Equations* **4** (1982), No. 2, 95–112.

6. Kawohl, B., On a nonlinear heat-control problem with boundary conditions changing in time, *Z. Angew. Math. Mech.* **61** (1981), No. 5, T248-T249.

7. Li, X. J., Time-optimal boundary-control for systems governed by parabolic equations, *Chinese Ann. Math.* **1** (1980), No. 3–4, 453–458 (Chinese, English summary).

8. Mackenroth, U., Convex parabolic boundary-control problems with point wise state constraints, *J. Math. Anal. Appl.* **87** (1982), No. 1, 256–277.

9. Musienko, E. I., Control of the solution of a parabolic problem in the neighborhood of an unstable stationary solution, *Dinamika Splosn. Sredy*, No. 51, *Mekh. Bystroprotekayushch. Protesessov* (1981), 68–83, 191 (Russian).

10. Nababan, S. (with Teo, K. L.), Necessary conditions for optimality of Cauchy problems for parabolic, partial delay-differential equations, *J. Optim. Theory Appl.* **34** (1981), No. 1, 117–155.

11. Nambu, T., Feedback stabilization of diffusion equations by a functional observer, *J. Differential Equations* **43** (1982), No. 2, 257–280.

12. Nambu, T., Feedback stabilization for distributed-parameter systems of parabolic type. II, *Arch. Rational Mech. Anal.* **79** (1982), No. 3, 241–259.

13. Niezgodka, M., Control of a parabolic systems with free boundaries—application of inverse formulations, *Control Cybernet.* **8** (1979), No. 3, 213–225 (Polish and Russian summaries).

14. Noussair, E. S. (with Nababan, S., and Teo, K. L.), On the existence of optimal controls for quasilinear parabolic partial-differential equations, *J. Optim. Theory Appl.* **34** (1981), No. 1, 99–115.

15. Seidman, T. I., Regularity of optimal boundary controls for parabolic equations, I. Analyticity, *SIAM J. Control Optim.* **20** (1982), No. 3, 428–453.

16. Tohyama, T. (with Araki, M., and Miyoshi, T.), Existence, stability and convergence in a heat-control problem, *Kumamoto J. Sci. (Math.)* **15** (1982), No. 1, 11–32.

17. Zhou, Hong Xin, A note on approximate controllability for semilinear, one-dimensional heat equation, *Appl. Math. Optim.* **8** (1982), No. 3, 275–285.

4.5. Not-Well-Posed and Inverse Problems

1. Abdullaeva, G. Z., Classical solvability in the large, of a nonlinear one-dimensional inverse boundary-value problem for second-order parabolic equations, *Akad. Nauk. Azerbaidzan.SSR Dokl.* **37** (1981) No. 11, 3–6 (Russian).

2. Abdullaeva, G. Z., Investigation of the classical solution of a nonlinear one-dimensional inverse boundary-value problem for second-order parabolic equations, *Akad. Nauk. Azerbaidzan. SSR Dokl.* **38** (1982), No. 3, 3–5 (Russian, Azerbaijani, and English summaries).

3. Anger, G., Einige Betrachtungen über inverse Probleme, Identifikationsprobleme, und inkorrekt gestellte Probleme [Some reflections on inverse problems, identification problems, and ill-posed problems], in *Yearbook: Surveys of Mathematics*, 1982, pp. 55–71. Bibliographisches Inst., Mannheim, 1982.

4. Anger, G. (with Czerner, R.), An inverse problem for the heat-conduction equation. II, *Math. Nachr.* **105** (1982), 163–170.

5. Artjuhin, E. A. (with Rumjancev, S. V.), Optimal choice of descent steps in gradient methods of solution of inverse heat-conduction problems, *Inz.-Fiz. Z.* **39** (1980), No. 2, 264–269 (Russian); translated as *J. Engrg. Phys.* **39** (1980), No. 2, 865–869 (1981).

6. Atamanov, E. R., Uniqueness and estimate of the stability of the solution of a problem for a heat equation with moving transducer, in *Inverse Problems for Differential Equations of Mathematical Physics*, pp. 35–44, 155–156. Akad. Nauk. SSSR Sibirsk. Otdel., Vycisl. Centr. Novosibirsk, 1978 (Russian).

7. Atamanov, E. R., Uniqueness and estimation of the stability of the solution of an ill-posed problem for a pseudoparabolic equation, in *Studies in Integro-Differential Equations*, pp. 218–223, 261. "Illim," Frunze, 1979 (Russian).

8. Atamanov, E. R., Regularization of the Cauchy problem for the heat equation with inverse time, in *Studies in Integro-Differential Equations*, pp. 224–227, 261. "Illim," Frunze, 1979 (Russian).

9. Baranski, W. (with Furmanczyk, S.), On *a posteriori* error bounds for the heat-flow problem, *Bull. Acad. Polon. Sci. Ser. Sci. Tech.* **29** (1981), No. 7–8, 363–368 (1982) (Russian summary).

10. Bell, J. B., The noncharacteristic Cauchy problem for a class of equations with time dependence.   I. Problems in one space dimension, *SIAM J. Math. Anal.* **12** (1981), No. 5, 759–777.

11. Bell, J. B., The noncharacteristic Cauchy problem for a class of equations with time dependence.   II. Multidimensional problems, *SIAM J. Math. Anal.* **12** (1981), No. 5, 778–797.

12. Beznoscenko, N. J., Existence of a solution of problems of determining coefficients multiplying the higher derivatives of parabolic equations, *Differencial'nye Uravnenija* **18** (1982), No. 6, 996–1000, 1101 (Russian).

13. Beznoscenko, N. J., Determination of the coefficient q from the solution of the second boundary-value problem for the equation $u_t - \Delta u + qu = f$ in a half-space (existence "in the large"), *Sibirsk. Mat. Z.* **23** (1982), No. 1, 3–11, 219 (Russian).

14. Blagovescenskii, A. S., On a case of explicit solution of a nonstationary inverse problem, *Prikl. Mat. Meh.* **43** (1979), No. 1, 188–190 (Russian); translated as *J. Appl. Math. Mech.* **43** (1979), No. 1, 207–209, (1980).

15. Bolt, B. A., What can inverse theory do for applied mathematics and the sciences? *Austral. Math. Soc. Gaz.* **7** (1980), No. 3, 69–78.

16. Cannon, J. R. (with DuChateau, P. C), Weak solutions $u(x, t)$ to parabolic partial-differential equations with coefficients that depend upon $u(y_i, \psi_i(t, u(x, t)))$, $i = 1, \ldots, k$; *J. Differential Equations* **42** (1981), No. 3, 438–446.

17. Cannon, J. R. (with Zachmann, D.), Parameter determination in parabolic partial-differential equations from overspecified boundary data, *Internat. J. Engrg. Sci.* **20** (1982), No. 6, 779–788.
18. Carasso, A., Determining surface temperatures from interior observations, *SIAM J. Appl. Math.* **42** (1982), No. 3, 558–574.
19. Carroll, R. (with Santosa, F.), Stability for the one-dimensional inverse problem via the Gel'fand–Levitan equation, *Applicable Anal.* **13** (1982), No. 4, 271–277.
20. Chrzanowski, E. M., An inverse problem for a combined system of diffusion equations, *Demonstratio Math.* **14** (1981), No. 2, 427–436.
21. Colton, D., Continuation and reflection of solutions to parabolic partial-differential equations, in *Ordinary and Partial-Differential Equations* (Proc. Fifth Conf., Univ. Dundee, Dundee, 1978), pp. 54–82, Lecture Notes in Math. 827. Springer, Berlin, 1980.
22. Denisov, A. M., On the approximate solution of a Volterra equation of the first kind, related to an inverse problem for the heat equation, *Vestnik Moskov. Univ. Ser. XV Vycisl. Mat. Kibernet.* 1980, No. 3, 49–52, 70 (Russian).
23. Denisov, A. M., Uniqueness of the solutions of some inverse problems for the heat equation with a piecewise-constant coefficient, *Zh. Vychisl. Mat. i Mat. Fiz.* **22** (1982), No. 4, 858–864 (Russian).
24. DuChateau, P. Monotonicity and uniqueness results in identifying an unknown coefficient in a nonlinear diffusion equation, *SIAM J. Appl. Math.* **41** (1981), No. 2, 310–323.
25. Fujita, W. (with Kuhara, S., and Hisamura, T.), Observability of parabolic-type distributed-parameter systems with periodic scanning detection, *Bull. Sci. Engrg. Res. Lab. Waseda Univ.*, No. 92 (1980), 75–81.
26. Galanov, B. O. (with Danilenko, V. M.), Application of the method of boundary integral equations to two inverse boundary-value problems of the theory of diffusion in a solid, *Vycisl. Prikl. Mat. (Kiev)*, No. 42, (1980), 74–82 (Russian, English summary).
27. Gol'dman, N. L., Some inverse problems for determining the boundary conditions for the Stefan problem, *Vestnik Moskov. Univ. Ser. XV. Vycisl. Mat. Kibernet.* 1982, No. 1, 16–23, 79 (Russian).
28. Grigor'ev, E. A., On stability in the class of positive solutions of the time-inverse Cauchy problem for the heat equation, *Differencial'nye Uravnenija* **17** (1981), No. 7, 1250–1255, 1343.
29. Grysa, K. (with Ciałkowski, M. J.), Inverse problems of temperature fields— A survey, *Mech. Teoret. Stos.* **18** (1980), No. 4. 535–554 (1981), (Polish, Russian, and English summaries).
30. Han, H., The finite-element method in the family of improperly posed problems, *Math. Comp.* **38** (1982), No. 157, 55–56.
31. Hoffmann, K.-H. (with Kornstaedt, H.-J.), Zum inversen Stefan-Problem, in *Numerical Treatment of Integral Equations* (Workshop, Math. Res. Inst., Oberwolfach, 1979), pp. 115–143, Internat. Ser. Numer. Math., 53. Birkhauser, Basel, 1980 (English summary).
32. Isakov, V. M., A class of inverse problems for parabolic equations, *Dokl. Akad. Nauk. SSSR* **263** (1982), No. 6, 1296–1299 (Russian).
33. Jochum, P., The inverse Stefan problem as a problem of nonlinear approximation theory, *J. Approx. Theory* **30** (1980), No. 2, 81–98.

34. Klibanov, M. V., Some inverse problems for parabolic equations, *Mat. Zametki* **30** (1981), No. 2, 203–210, 314 (Russian).
35. Kolbel, K., Eine inverse Aufgage für parabolische Potentiale, *Wiss. Z. Tech. Hochsch. Karl-Marx-Stadt* **22** (1980), No. 5, 467–475.
36. Krzeminski, S.K., Identification of the thermal conductivity by the method of penalty function, in *Numerical Methods in Thermal Problems*. (Proc. First Internat. Conf., Univ. College, Swansea, 1979), pp. 1077–1087. Pineridge, Swansea, 1979.
37. Lavrent'ev, M. M. (with Romanov, V. G., and Sisatskii, S. P.), *Ill-posed Problems of Mathematical Physics and Analysis*. "Nauka," Moscow, 1980, 287pp. (Russian).
38. Lavrent'ev, M. M. (with Amonov, B. K.), Uniqueness and an estimate of the stability of the solution of an interior problem for the heat equation, *Dokl. Akad. Nauk. SSSR* **262** (1982), No. 3, 528–530 (Russian).
39. Lousberg, P., Backward parabolic equations, in *Analytic Solutions of Partial-Differential Equations* (Trento, 1981), pp. 213–221. Asterisque, 89–90, Soc. Math. France, Paris, 1981.
40. Milstein, J., The inverse problem: estimation of kinetic parameters, in *Modeling of Chemical Reaction Systems* (Heidelberg, 1980), pp. 92–101. Springer, Ser. Chem. Phys., 18 Springer, Berlin, 1981.
41. Murayama, R., The Gel'fand–Levitan theory and certain inverse problems for the parabolic equation, *J. Fac. Sci. Univ. Tokyo, Sect. IA, Math.* **28** (1981), No. 2, 317–330.
42. Murio, D. A., Numerical methods for inverse, transient heat-conduction problems, *Rev. Un. Mat. Argentina* **30** (1981), No. 1, 25–46.
43. Murio, D. A., On the estimation of the boundary temperature on a sphere from measurements at its center, *J. Comput. Appl. Math.* **8** (1982), No. 2, 111–119.
44. Reznickaja, K. G., A source problem. The linearized case when $a \neq 0$, in *Inverse Problems for Differential Equations of Mathematical Physics*, pp. 69–78, 157–158. Akad. Nauk. SSSR Sibirsk. Otdel, Vycisl. Centr. Novosibirsk, 1978 (Russian).
45. Romanov, V. G., Inverse problems and energy inequalities, in *Partial-Differential Equations* (Proc. Conf., Novosibirsk, 1978), pp. 148–153, 253. "Nauka" Sibirsk. Otdel, Novosibirsk, 1980 (Russian).
46. Sisko, N. P., Inverse problem for a parabolic equation, *Math. Zametki* **29** (1981), No. 1, 55–62, 155.
47. Spivak, S. I. (with Gorskii, V. G), The nonuniqueness of the solution to the problem of the reconstruction of kinetic constants, *Dokl. Akad. Nauk. SSSR* **257** (1981), No. 2, 412–415 (Russian).
48. Suzuki, T., Remarks on uniqueness in an inverse problem for the heat equation. I., *Proc. Japan Acad. Ser. A. Math. Sci.* **58** (1982), No. 3, 93–96.
49. Suzuki, T., Remarks on uniqueness in an inverse problem for the heat equation. II., *Proc. Japan Acad. Ser. A. Math. Sci.* **58** (1982), No. 5, 175–177.
50. Suzuki, T., On a certain inverse problem for the heat equation on the circle, *Proc. Japan Acad. Ser. A. Math. Sci.* **58** (1982), No. 6, 243–245.
51. Suzuki, T., Inverse problems for parabolic equations, *Sugaku* **34**, (1982), No. 1, 55–64 (Japanese).
52. Talenti, G. (with Vessella, S.), A note on an ill-posed problem for the heat

equation, *J. Austral. Math. Soc. Ser. A.*, **32** (1982), No. 3, 358–368.

53. Tarzia, D. A., Determination of the unknown coefficients in the Lame–Clapeyron problem (or one-phase Stefan problem), *Adv.in Appl. Math.* **3** (1982), No. 1, 74–82.

54. Temirbulatov, S. I., Ill-posed mixed problems of heat conductivity, *Dokl. Akad. Nauk. SSSR* **264** (1982), No. 1, 45–47 (Russian).

55. Zeinalov, I. S, An ill-posed problem for a parabolic system, *Akad. Nauk. Azerbaidzan. SSR Dokl.* **36** (1980), No. 3, 3–5 (Russian, Azerbaijani, and English summaries).

Symbol Index

q   Flow rate of heat energy

k   Thermal conductivity

$u_x \equiv \dfrac{\partial u}{\partial x}$ Partial derivative of u with respect to x

$u_t \equiv \dfrac{\partial u}{\partial t}$ Partial derivative of u with respect to t

$u_{xx} \equiv \dfrac{\partial^2 u}{\partial x^2}$ Second partial derivative of u with respect to x

c   Heat capacity

ρ Density of the conductor

Δx Change in x

Δt Change in t

κ Thermal diffusivity $(\rho c)^{-1} k$

$\exp\{x\} \equiv e^x, \ e = 2.71828\ldots$

$\sin x \equiv \dfrac{e^{ix} - e^{-ix}}{2i}, \ i = \sqrt{-1}$

$\cos x \equiv \dfrac{e^{ix} + e^{-ix}}{2}$

$\sinh x \equiv \dfrac{e^x - e^{-x}}{2}$

$$\cosh x \equiv \frac{e^x + e^{-x}}{2}$$

$$p_n(x,t) \equiv n! \sum_{k=0}^{[n/2]} \frac{t^k}{k!} \frac{x^{n-2k}}{(n-2k)!} \qquad \text{Heat polynomials}$$

$$\Gamma(a) = \int_0^\infty x^{a-1} e^{-x} dx, \, 0 < a < \infty \qquad \text{The gamma function}$$

$n!$ $1 \cdot 2 \cdot 3 \cdots n = \Gamma(n+1)$

$\mathbf{R}$ Real numbers

D Open domain in $\mathbf{R}^2 = \mathbf{R} \times \mathbf{R}$

D_T $= D \cap \{(x,t) \in \mathbf{R}^2 | t \le T\} \cup$ certain boundary points (x,T) to form the parabolic domain

B_T The parabolic boundary of D_T

$\in$ Belongs to

ε Positive real number

$\subset$ Inclusion of a set in a set. For example, for $A \subset B$ read: A is a subset of B.

$\cup$ Union of sets

$\cap$ Intersection of sets

$\{x \in A | P(x)\}$ Set-builder notation: $x \in A$ such that property $P(x)$ is satisfied.

$f^{(j)}(t)$ jth derivative of f with respect to its argument

$$K(x,t) \equiv (4\pi t)^{-1/2} \exp\left\{-\frac{x^2}{4t}\right\}, \, t > 0 \qquad \text{The fundamental solution of the heat equation}$$

$G(x,\xi,t) \equiv K(x-\xi,t) - K(x+\xi,t)$ The Green's function for the semi-infinite conductor

$N(x,\xi,t) \equiv K(x-\xi,t) + K(x+\xi,t)$ The Neumann function for the semi-infinite conductor

$$\theta(x,t) = \sum_{n=-\infty}^{\infty} K(x+2n,t) \qquad \text{The theta function}$$

$\|\psi\|_t = \sup\limits_{0 < \tau < t} |\psi(\tau)|$ or $\max\limits_i \{\|\psi_i\|_t\}$ for a vector function ψ

$\|f\|_{[a,b]} = \sup\limits_{a \le x \le b} |f(x)|$

$\|E\|_j = \max\limits_{0 \le i \le m} |E_{i,j}|$ Vector norm for the vector $E \equiv E_j = (E_{i,j}, \ldots, E_{m,j})$, $E_{i,j} \in \mathbf{R}$.

$\|\psi\|_2 = \left\{\int_D |\psi|^2\right\}^{1/2}$ The L^2 norm

$C, C_i, i =$ positive integer Positive constants

$C(I)$ Where I is an interval or domain, denotes the continuous functions on I

$C^1(I)$ Denotes the space of continuously differentiable functions on I

$C^\beta(I), \, 0 < \beta \le 1$ Denotes the space of Hölder continuous functions on I with Hölder exponent β. The space of Lipschitz continuous functions, $\beta = 1$, presents some confusion which is removed by the context of the usage below.

$|\psi|_\beta = \sup\limits_{\substack{t,t+\delta\in I \\ \delta>0}} \delta^{-\beta}|\psi(t+\delta)-\psi(t)|$ The Hölder seminorm or constant for ψ

$C^0_{(\nu)}(I),\, 0<\nu\le 1$ The subspace of $C(I),\, I=(0,T]$, such that $\|\psi\|_T^{(\nu)} = \sup\limits_{t\in I} t^{1-\nu}|\psi(t)| < \infty$

$C^\varepsilon_{(\nu)}(I),\, \varepsilon>0$ The subspace of $C^0_{(\nu)}(I),\, I=(0,T]$, such that
$$|\psi|_\varepsilon^{(\nu)} = \sup\limits_{\substack{t,t+\delta\in I \\ \delta>0}} t^{1-\nu+\varepsilon}\delta^{-\varepsilon}|\psi(t+\delta)-\psi(t)| < \infty$$

P Used in Chapter 16 to denote a closed polygonal region

Q Used in Chapter 1 to denote heat content and in Chapter 16 to denote a point (x,t)

$\mathscr{L}(u)\equiv u_{xx}-u_t$ Used to assist in the repetitive usage of the heat operator

$M_P\psi$ Denotes the function used in Chapter 16 formed from the function ψ defined in D as
follows:
$$M_P\psi(Q)=\begin{cases}\psi, & Q\in D-P^0 \ (P^0 \text{ is the interior of } P),\\ \mathscr{L}(\psi)=0, & Q\in P^0.\end{cases}$$

$s(h,t)=\sup\limits_{0<\tau<t} h(\tau)$ Used in Chapter 13

$i(h,t)=\inf\limits_{0<\tau<t} h(\tau)$ Also used in Chapter 13

$\dot{s}(t)$ Derivative of s with respect to t

Subject Index